Biological Magnetic Resonance
Volume 21

EPR: Instrumental Methods

A Continuation Order Plan is available for this series. A continuation order will bring delivery of each new volume immediately upon publication. Volumes are billed only upon actual shipment. For further information please contact the publisher.

Biological Magnetic Resonance
Volume 21

EPR: Instrumental Methods

Edited by

Lawrence J. Berliner
University of Denver
Denver, Colorado

and

Christopher J. Bender
Fordham University
Bronx, New York

KLUWER ACADEMIC / PLENUM PUBLISHERS
NEW YORK, BOSTON, DORDRECHT, LONDON, MOSCOW

ISBN: 0-306-47864-1

233 Spring Street, New York, New York 10013

http://www.wkap/com

10 9 8 7 6 5 4 3 2 1

A C.I.P. record for this book is available from the Library of Congress

Printed in the United States of America

Preface

Practitioners of electron magnetic resonance (EMR) spectroscopy are now enjoying something akin to a renaissance in this branch of magnetic resonance, much of which can be attributed to improved technology for microwave device fabrication. Principle among these advances are solid-state circuits, and in particular the monolithic microwave integrated circuit (MMIC). These devices on the one hand are fast, enabling one to work in the nanosecond time domain (e.g., pulsed EMR). MMICs also enable one to obtain the performance characteristics of electron tube devices (e.g., reflex klystrons) in a monolithic package (e.g., Gunn device), thus affording new capabilities to the spectroscopist in the way of multi- and high frequency techniques.

With these advances as a background, we felt that the time was appropriate for devoting one or more volumes in this successful series to EMR instrumentation. With this, our first volume devoted to instrumentation, and introductory chapter on general topics of microwave instrumentation, as applied to EMR, sets the stage. This general introduction is followed by a chapter that addresses the specific problem of low-frequency spectrometer design, which is relevant to such advanced EMR techniques as imaging. Chapter 3 treats the crucial problem of sensitivity, and how this is affected by spectrometer operating frequency. Chapter 3 is therefore important to any practitioner of multi-frequency EMR. The next two chapters treat the often-called ‘advanced EMR’ techniques, namely electron nuclear double resonance (ENDOR) and electron spin echo (envelope) modulation (ESEEM). The ENDOR chapter focuses on a topic that is absent from the EMR literature, namely the rf field-generating coil and how one renders it compatible with the microwave resonator for optimum performance. The ESEEM chapter takes a more systemic viewpoint and describes the manner in which the spin echo and spin coherence effects can be invoked and measured. These first five chapters provide the practical background for the remaining chapters, which concern several of the current applications of EMR.

The remaining three chapters in this first volume on instrumentation deal with unique applications of the modern EMR spectrometer. First among these is a chapter describing an algorithmic approach to the analysis of spectral parameters that is generally applicable. This is followed by a chapter on EPR imaging and, finally, by a pulsed electron spin echo method of determining crystallographic distances between paramagnetic centers.

Finally, during the preparation of this volume we were saddened to receive news that our colleague Jerry Babcock had passed away. Jerry is known to many in the EMR community for his work in flash kinetic and ENDOR spectroscopies, and to the scientific community at large for his investigations into the oxygen chemistry of photosynthesis and oxidative phosphorylation. Besides the published work, however, many of us remember

Jerry as someone who brought enjoyment to cutting-edge scientific research. His enthusiasm was infectious, and he brought out the best in those who worked in his laboratory. This can be best exemplified by something he used to say to his graduate students as they neared completion of their thesis; he would remind them that their work represented something totally new, and that they, as the person who did the work, knew the most about that topic. One could not help but get excited about science in Jerry's laboratory.

The editors and authors therefore dedicate these volumes on EMR instrumentation to the memory of Jerry Babcock.

Christopher J. Bender
Lawrence J. Berliner

CONTENTS

Chapter 1

Microwave Engineering Fundamentals and Spectrometer Design

Christopher J. Bender
Department of Chemistry, Fordham University, 441 E. Fordham Road, Bronx, New York 10458 USA

1. SCOPE & RATIONALE

This introductory chapter is intended as a concise overview of microwave technology and the various devices that comprise a conventional electron magnetic resonance spectrometer. It is hoped that it will serve the reader with background information that is relevant to the necessary understanding of instrument operation, as well as be a precursor to the chapters that follow and describe many of the new advances in EMR spectrometer technology. The three major sections of this chapter correspond to the functional subsystems that would exist in a typical spectrometer, namely the microwave source, transmission line devices, and the receiver.

I prefer to approach this subject from an historical perspective because doing so imposes a logical ordering of the facts, and therefore much of what I have written in this chapter is distilled from the high frequency electrical engineering and EMR literature dating back to the 1940s. The reference section is consequently written as a bibliography of many now out-of-print books and reviews because these may often be consulted for insights regarding application-specific microwave components. Among the EMR literature, there are several textbooks devoted exclusively to instrument design and technique. Poole's (1967; 1983) classic treatise is still an indispensable guide to EMR spectrometer design and technique, and its second (1983) edition has been reissued by the Dover press. Algers' (1968) text is a more elementary presentation of spin resonance and includes useful information on the construction 'arts' of spectrometer design, and it likewise has recently been reissued by University Microforms (Ann Arbor) after many years of unavailability. Harvey's (1963) microwave engineering textbook is an excellent advanced guide to systems design that includes chapters on spectroscopy, and Wilhurst's (1968) text is a good general guide to EMR instrumentation.

2. MICROWAVE SOURCES

2.1 Overview

Oscillatory signals can be obtained from static voltages by using reactive circuit elements. A common textbook example is the charged capacitor or energized inductor (*i.e.* an *RC* or *RL* circuit) that is switched at some threshold voltage (Horowitz & Hill, 1989, p284ff). *RLC* circuits combine two reactive elements in series or parallel, and the solutions to Kirchhoff's equations resemble that of the harmonic oscillator. In other words, when an *RLC* circuit is closed, the response is oscillatory and characterized by natural frequencies that are determined by the circuit elements (Reintjes & Coate, 1952, p115ff; Lancaster, 1980, Ch. 4). These *RLC* circuits are self-oscillatory, and it is a common engineering design practice to couple these resonant circuits to an amplifier-like device in order to produce a signal that is a high fidelity sine wave with a characteristic frequency. All of the high frequency oscillators that are described in this section operate on this principle.

In general, the ideal waveform will be sinusoidal because it is easy to generate and not distorted by any linear network. This means that one can pass a sinusoidal signal to a linear amplifier, and the output signal from the amplifier will differ from the input signal only with respect to the amplitude and phase. A second linear network can then be added so that it alters (or samples) the amplified sine wave output and exactly duplicates (both amplitude and phase) the original signal input to the amplifier. In this manner one can therefore use a linear network as a means to provide a feedback that maintains a constant signal input to the amplifier stage. A practical oscillator consequently consists of a resonant circuit, a linear amplifier, and a linear feedback network that compensates for damping (resistive dissipative losses) in the resonator.

Figure 1 depicts two representative oscillator circuits that illustrate the interplay between a resonant *RLC* circuit and an amplifier. These Hartley and Colpitts oscillators feature recognizable interactive resonator, amplifier

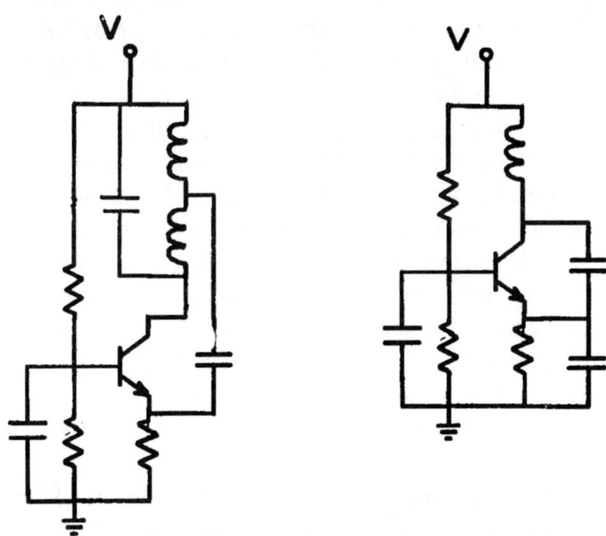

Figure 1. The Hartley (left) and Colpitts (right) oscillators, illustrating how a practical source of sine waves is obtained by coupling an amplifier to an *RLC* circuit, which determines the frequency of the oscillation.

and feedback 'stages'; in some versions the *RLC* resonator is replaced by a quartz crystal and a capacitor that permits a limited tuning range. The most important feature of these circuits (and all oscillators) is the feedback between the amplifier stage and the resonator circuit: the small (ca. 25%) amount of the amplified signal is returned to the resonator in order to overcome damping effects that would otherwise cause the signal amplitude to decay. Ordinarily the two simple circuits depicted in Figure 1 would be used to generate radiofrequency signals, but their design is easy to comprehend for illustrative purposes. As for spectroscopic applications, these oscillators find applications in nuclear quadrupole resonance spectrometers, and the pioneering experiments of EMR spectroscopy were conducted with such oscillators. These low frequency (~150 MHz) EMR experiments are described by Bruin (1961) and Kochelaev & Yablokov (1995).

High frequency electromagnetic waves are generated from electron tubes or solid-state devices (semiconductors). Experiments conducted during the nineteenth century demonstrated that electromagnetic waves were emitted from rarefied gas atmospheres subjected to an electrical discharge, and the technological application of these devices followed from the discoveries that the device could be used as transmitters or receivers of electromagnetic radiation. By definition, an electron tube is a device whose function follows from the conduction of electrons through a vacuum or low pressure gas, and it is therefore a gas tight device. Electron tubes may be housed in glass, but modern versions are often constructed of a glass ceramic such as beryllium oxide. The tube type is generally designated by the number of electrodes (diode, triode, tetrode, *etc.*), and the behavior of a given tube is derived from the breakdown of the insulating layer (*i.e.* the gap between the two electrodes) subject to a sufficiently high voltage. The basic elements of an electron tube are the cathode, the anode, and the filament. The anode collects the electrons emitted by the cathode. The filament is a heater element that is local to the cathode, and it is used to aid thermionic emission from the cathode. In a vacuum tube, such as the reflex klystron or traveling wave tube, the heater filament must make physical contact with the cathode, and either series or parallel arrangements of filament and cathode are used. More electrodes can be added to the cathode-anode pair in order to control the electron flow; the control of an electron beam by a grid electrode is described elsewhere.

Studies of semiconductors and semiconductor interfaces in the 1940s led to the discovery of the transistor and the realization that many functions of the electron tube could be duplicated in the solid state. Semiconductors have gradually come to replace electron tubes in many applications, and many low power microwave tubes, such as the reflex klystron, are considered as obsolete for many facets of the communications and navigation industries. Semiconductors are now achieving higher output levels and operating voltages, but electron tubes still retain specialized niches where their durability is a factor: applications that involve high voltage, unstable or unpredictable loads, or high speed control of the beam current (and therefore the output).

2.2 The Reflex Klystron

2.2.1 Overview and Basic Principles

EMR spectrometers have traditionally used reflex klystron oscillators,[1] as a microwave source. The reflex klystron is a type of electron tube (*i.e.* triode) that delivers stable signals of moderate power (0.1–1 W) and at one time was widely used at the stable low-power reference signal for radar receivers (Ridenour, 1964), but its excellent qualities as a moderate power microwave source made it popular for general microwave test and

[1] Now considered an obsolete technology, principles of klystron operation nonetheless provide an excellent foundation for understanding all tube and solid-state devices.

measurement applications. As the microwave source of choice for EMR, its

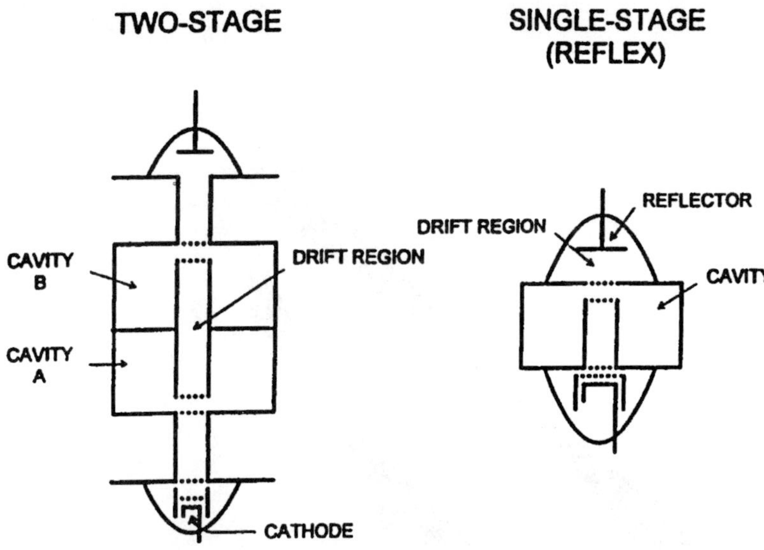

Figure 2. The functional anatomy of the klystron oscillator. An electron beam is the source of power for small signal amplification, whereas a cavity resonator serves as a lumped parameter *RLC* element that defines the frequency of operation.

stability renders it ideal as a source of coherent monochromatic radiation, and its power output is adequate for serving both as the excitation source and receiver reference.

Like the Hartley and Colpitts low-frequency oscillators, the klystron is a self-oscillatory resonant structure that is coupled to an energy source. An electron beam passes parallel to the internal electric field lines of a cavity resonator. The beam behaves analogously to a DC current passed along a solid conductor and thus generates a wall current within the klystron body (the origins of eddy currents are described at length in Golding & Widdis, 1963; and Skilling, 1952). The wall current likewise produces its own field, but that electromagnetic field is constrained to a specific resonant mode by the klystron cavity, and therefore the resultant electromagnetic waves are generated at a fixed frequency. The oscillatory electromagnetic field that is induced within the klystron cavity in turn interacts with the electron beam and tends to alter the velocity of the individual electrons. In other words, the klystron cavity provides feedback to the electron beam.

The functional anatomy of the basic two-stage and reflex klystron oscillators is illustrated in Figure 2, which depicts a cross-sectional view

(Harrison, 1944; 1947)[2] of three distinct regions through which the electron beam travels. The beam electrons emitted from the cathode encounter a

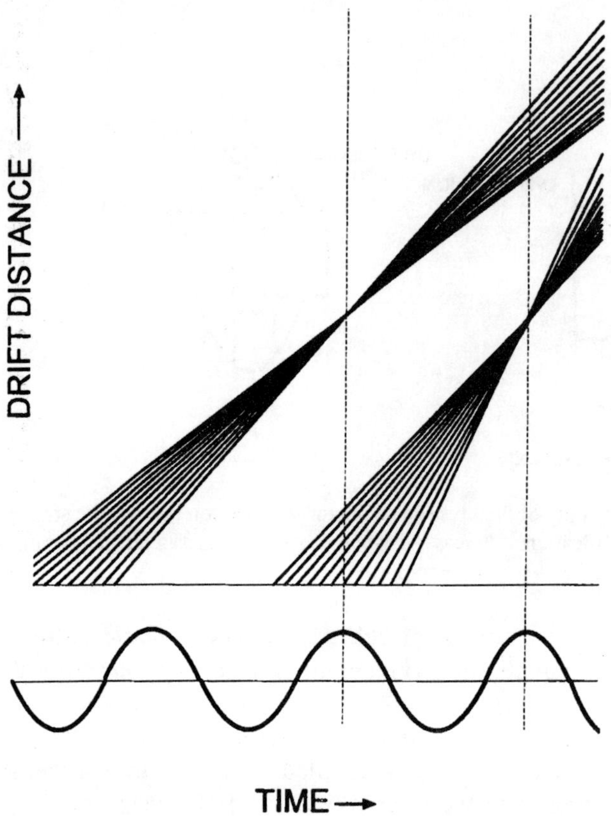

Figure 3. The Applegate Diagram, which depicts electron trajectories in a beam that is subjected to a time-varying driving force (the electromagnetic field within the cavity resonator). Electrons are accelerated or decelerated by the oscillating field and therefore form bunches as the trajectories converge (intersection on diagram).

gridded wall section of a toroidal cavity, which is well suited to this application because its internal electric (resonant) field lines are dense and axially oriented in the gap between the axial 'bulge' and the end wall. Passage of the electrons into the gap region is made possible by replacing this portion of the cavity wall by a grid, and they enter parallel to the resonator's internal electric field lines with a velocity $\mathbf{v}_0$. Mechanical tuning of the klystron is achieved by adjusting the position of the end wall or by

[2] For detailed cutaway drawings of the reflex klystron, see Pierce & Shepard (1947), or Nelson (1960).

inserting a plunger into the cavity. Oscillations are rapidly generated in the klystron cavity after the electron beam is engaged. The cavity is a resonant load (an *RLC* structure) on the electron beam, and one may regard the two grids and the cavity body 'toroid' as capacitor and inductor, respectively (Pierce & Shepard, 1947). Shot noise from the electron beam initiates the wall current (Pierce & Shepard, 1947), which is then maintained via convection as the beam traverses the cavity gap. The beam electrons are accelerated from the cavity by a static voltage V_B (the beam voltage) and enter the gap region where they sense an oscillatory voltage $V(t) = v_0 \sin\omega t$. The transit time of the electron through the gap region is short relative to the period of the oscillatory wave, and therefore electrons entering the gap region at different times are subject to different fields. The electrons are accelerated by the instantaneous field and exit the gap region with a velocity $v = v_0[1 + \frac{1}{2}\xi\sin\omega t]$. The terminology is derived from Shevchik (1963), where ξ is related to the magnitude of the oscillating voltage and a coupling parameter.[3]

The electrons possess a velocity $v = v_0[1 + \frac{1}{2}\xi\sin\omega t]$ as they exit the gap region, and bunching occurs as the faster electrons catch up with the slower electrons of the previous half cycle. This bunching occurs in the drift region and may be pictorially represented by an Applegate diagram (Figure 3), in which a plot of electron trajectory is superimposed over the waveform of the perturbing electromagnetic field. The slope of the trajectory is the electron velocity, and bunching is depicted by a convergence of the electron trajectories.

The amount of useful high frequency power that is available by tapping into the bunching resonator is low because the energy of the electron beam is not appreciably transferred to the cavity. A second 'catcher resonator' that is tuned to the same frequency as the buncher resonator and situated at the end of the drift region would couple to the modulated electron beam and accrue the beam's energy so long as the beam arrived in phase with the cavity's natural oscillations. The basic power klystron is therefore a two-resonator device that consists of a cathode, a buncher resonator, a drift region, and a catcher resonator, the latter of which is tapped for output. Additional cavity stages can be linearly added in order to further amplify the effect (*e.g.* high power klystrons).

The electrons with the highest exit velocity are those that are subject to an accelerating voltage $V_B + V_0$, and the electrons with the slowest exit velocity sense a voltage $V_B - V_0$. The electrons of the beam therefore bunch around those that are unperturbed, that is, those that arrive in the gap region when

[3] In Shevchik's terminology, the velocity amplitude v_o is determined by the beam voltage V_B, and the time-dependent component of the velocity $\frac{1}{2}v_o\xi\sin(\omega t)$ corresponds to the voltage term $V_o \sin(\omega t)$.

$V(t) = V_0 \sin\omega t$. In order to maximize the energy transfer to the catcher cavity, the arrival of the electron bunch should coincide with the oscillatory wave maximum. The bunch 'center' should therefore have a transit time (through the drift region) that corresponds to $n + \frac{3}{4}$ cycles of the oscillatory

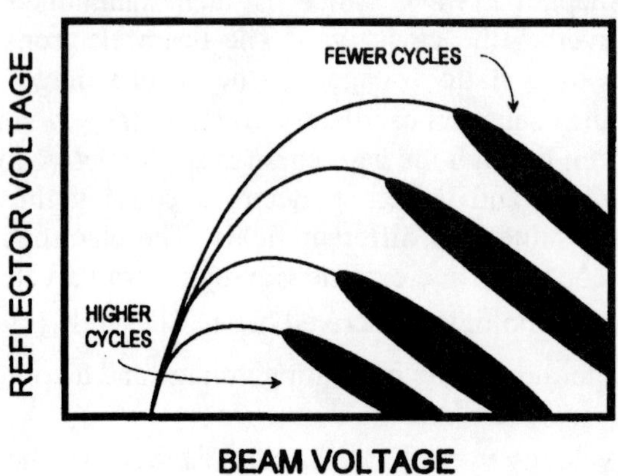

Figure 4. Operating modes of a reflex klystron oscillator as a function of beam and reflector voltages. Only specific ranges of the voltages (indicated by the shaded regions in the diagram) successfully start oscillations in the klystron. Adapted from Harrison, 1947.

wave (n = 1, 2, 3...), and the resultant drift region transit time can be computed by the relation $fT = n - \frac{1}{4}$ (n now defined as an ordinal number). Discrete transit times of klystron operation can therefore be found by adjusting the physical variables of the klystron oscillator, for example, the beam voltage or length of the drift space.

A two-stage klystron is cumbersome because it requires the coordinated tuning of two resonators and the beam velocity in order to ensure that the electron bunches arrive in phase (the length of the drift region is usually fixed and transit times are adjusted via the accelerating voltages). The reflex klystron combines buncher and catcher functions in a single resonator, and, by reflecting the bunched electron beam, requires only half the drift space of a two-stage klystron. An electrode poised at negative potential relative to the cathode is situated near the exit of the gap region, and the exiting electron beam is turned back into the cavity body. Since the exiting electron beam is oscillatory, it can be returned to the resonator in phase with the cavity's internal electromagnetic field, which is then reinforced by the energy of the beam. For applications that require frequent tuning, the reflex klystron is superior to the two-stage device.

The magnitude of the reflector voltage determines the location at which the electron bunch turns back towards the klystron cavity, and this region is generally called the reflector region, in contrast to the drift region of the two-stage klystron. Higher reflector voltages push the turning point closer to the grid, but there are only discrete reflector voltages at which the klystron will oscillate because the turning point must correspond to a distance over which the electron bunch must travel and arrive in phase with the cavity field. It therefore follows that the beam voltage, the reflector voltage, and the cavity resonant frequency must all balance in order for a reflex klystron to operate. This is typically represented by a mode diagram, as illustrated in Figure 4.[4]

2.2.2 Analysis of Electrical Parameters

The operating modes that are illustrated in Figure 4 correspond to discrete drift regions and transit times, determined by the reflector and beam voltages, respectively, that return the electron bunches in phase with the cavity's internal oscillating field. Each mode is characterized by a limited range of both V_B and V_R that coincides with the cavity parameters (represented by a quality factor Q, see below) and enable the klystron to oscillate. There is in each case (*i.e.* mode), however, a lower limit of the beam current beyond which no reflector voltage can induce the klystron to oscillate. This threshold current is also called the starting current and represents a point at which the electron beam compensates for reactive losses within the resonator and the klystron can begin to generate useful power.

The threshold current is affected by the load, which can be demonstrated by the network analysis of the klystron resonator. To begin, the resonator is represented by a parallel array of L, C, and R. The inherent resistance of the klystron is designated as a shunt resistance, R_S, and accounts for losses; the load resistance R_L is added in parallel. Electron bunching in the gap region of the cavity is described kinematically. The current is described in terms of a velocity, dx/dt that is proportional to a retarding 'friction' or resistance. In a constant field defined by the beam voltage V_B, one can kinematically describe electron bunching through the resistance term, which in the klystron is taken to be a spatial variable because of the oscillating field within the gap region. If the resistance is negative in some region between x and $x + dx$, then the electron will accelerate upon reaching the point x and leave behind a region of low electron density within $x + dx$. If the electron encounters a positive resistance at the point $x + dx$, the electron will decelerate and bunch.

The kinematic description of electron bunching requires that the current density of the electron beam be simulated along its trajectory. In order to reinforce the field within the klystron cavity, the electron beam current must be spatially periodic (*i.e.* a standing wave) with the boundary condition

[4] Adapted from Ginzton, 1957; Gewartowski & Watson, 1965; Harrison, 1947; Shevchik, 1963.

being that the spatial wave arrives in phase with the cavity oscillation.[5] This spatial analysis employs a first order Bessel function of the form $J_1(x) = \frac{1}{2}x - \frac{1}{16}x^2 + \frac{1}{364}x^4$, which resembles a damped sinusoid and reduces to $\frac{1}{2}x$ as $x \to 0$. The term x is a unitless quantity equal to $\pi N V(t)/V_B$ where N is the number of cycles completed in the bunching process, and the voltage terms are those defined in the preceding paragraphs. The rf (*i.e.* bunched electron) current is $I = I_B[1 + 2J_1(x)\sin(\omega t - \phi)]$, and I_B is the beam or DC current.[6]

The threshold or starting current is the point at which the power of the electron beam exceeds the power losses of the resonator circuit, and at this threshold DC current the power losses equal the power delivered by the returning electron bunches to the klystron cavity. The power loss is defined as the function of the voltage and shunt resistance, that is, $P_L = V / R_S$, whereas the power delivered by the returning electron bunches is $P = VI$. The current I is the sum of the DC (beam) current and the rf current, the latter of which is expressed in terms of the Bessel function $J_1(x)$. The desired starting current is obtained by treating the rf current as a small signal (*i.e.* setting $J_1(x) = \frac{1}{2}x$, where x has been defined as $\omega N V / V_B$), equating the two power expressions, and solving for the DC current I_B. One finds that $I_{B,\text{THRESHOLD}} = V / \pi R_S N$, where $V \to V_B$ in this small signal limit. This analysis demonstrates how the threshold beam current (the leftmost end of the operating modes illustrated in Figure 4) is determined by the beam voltage, the reflector voltage,[7] and the klystron cavity losses. If the cavity were loaded by an external circuit, R_L is added parallel to R_S, and the expression is modified accordingly.

A second type of mode diagram is illustrated in Figure 5 and depicts the power output and electrical tuning curves as a function of the reflector voltage (for a fixed beam voltage). This diagram can be envisaged as a slice through Figure 4 at a fixed beam voltage, and it indicates that the frequency can be varied by the electrical parameters, namely via changes in the reactance of the electron beam. This diagram and its associated network analysis can be used to establish the optimum operating conditions for a reflex klystron. For example, the tuning curves (Figure 5, top) show variation in slope and range; the analysis in the following paragraphs will demonstrate how these tuning properties vary with the klystron load and quality factor (Ginzton, 1957).

[5] This constraint to spatial variables will be relaxed in the analysis of the electrical tuning range.

[6] This equation represents both the spatial and time-dependent properties of the electron beam current as a standing wave. The sine term comes from the velocity expression $v = v_0 + \frac{1}{2}\xi\sin(\omega t)$, and the Bessel function $J_1(x)$ corresponds to the spatial variation.

[7] The reflector voltage is manifest in the term N, implicitly specifying V_R ensures that electron bunches arrive in phase with the klystron cavity field.

The Applegate diagram illustrated in Figure 3 is an idealization in the sense that the klystron operation is contingent upon the electron bunches arriving at a specific point among the electromagnetic waveform (shown as

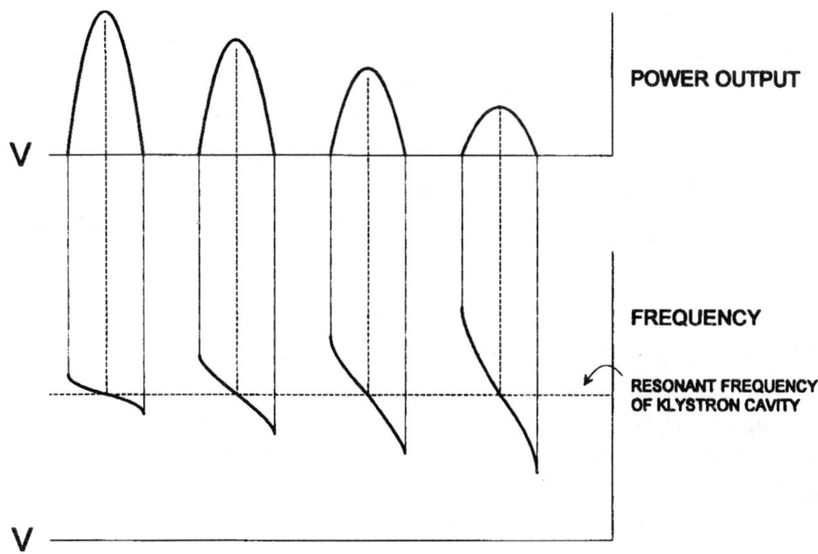

Figure 5. Electrical tuning curves of a reflex klystron, depicting the power output as a function of reflector voltage. This plot would be generated by taking a vertical section (constant beam voltage) through Figure 4 and each of the regions in which the klystron oscillates. In each of the oscillating modes, however, the electrical tuning range of the reflector voltage and the power delivered by the klystron varies; the greater the tuning range, the less power delivered. Adapted from Harrison, 1947.

the positive crest in the figure). In actuality, the klystron will operate as long as the electron bunches arrive during the retarding phase of the cycle, which is represented by the positive modes of the waveform in Figure 3. The degree of coupling and energy transfer from the beam to the cavity is affected by the relative phase of the electron bunch and klystron field, and this is expressed in terms of an impedance $Z = V / I$.[8] From elementary circuit theory and AC signals (*cf.* Nilson, 1983), the oscillatory voltage and current signals are out of phase when traveling through either a capacitor or an inductor, and this phase shift is typically expressed as a function of either C or L, respectively. In the case of electron bunches, the phase factor ϕ is

[8] Impedance is defined as $Z = R + iX$, where R is the DC resistance and X is the reactance (e.g. $X = \omega L$ or $1/\omega C$). Admittance $Y = Z^{-1} = (R - iX) / (R^2 + X^2) = G + iB$, where G is the DC conductance and B is the susceptance. If there is a phase difference ϕ between the current and the voltage ($I = I_0 \cos(\omega t)$ and $V = V_0 \cos(\omega t + \phi)$, then $\cos\phi = R / (R^2 + X^2)^{1/2}$. See Lancaster, 1980; p50–60.

determined by the transit time of the bunches through the drift region, and this is controlled by the reflector voltage. One can therefore conceptually imagine the reflector voltage and its effect on ϕ as a beam susceptance, which can then be incorporated into the mathematical expression for the klystron impedance Z.

The impedance is written as a sum of real and imaginary parts in the following manner. The bunched beam current is written in the form $I = I_0 \cos(\omega t + \phi)$, where ϕ is the phase factor defining the arrival time of the electron bunches. The current derived from the electron bunches is a complex quantity, however, and so $I = I_0 \cos(\omega t + \phi) + i \sin(\omega t + \phi)$. One then uses the $Z = V/I$ relationship and the definitions for V and I to obtain

$$Z = \frac{-V}{2IJ_1(x)[\cos\phi - i\sin\phi]}.$$

This derivation of Z is similar to the expression for the starting current with the exception that the coupling variation has been introduced by including the trigonometric terms, and ϕ is the phase difference between the electron bunch and oscillating klystron field.[9] One can separate the real and imaginary parts of the equation defining Z (see footnote 8), which leads to the realization that both the resistance and reactance are hyperbolic functions with poles[10] at $\phi = 0$ or $\pm\frac{1}{2}\pi$. When $\phi = 0$, the sine term vanishes and the impedance term is purely resistive; the converse is true when $\phi = \pm\frac{1}{2}\pi$. For intermediate values of ϕ, Z is at least partly reactive. The conditions of $\phi = 0$ and $\frac{1}{2}\pi$ constitute two special cases from which the tuning range and optimal operating conditions of a reflex klystron can be derived.

Real and imaginary components of the klystron impedance are separated, and the poles establish bounds of the electronic tuning range because of the relationship between ϕ and the reflector voltage. This concept is intuitive because as the impedance of the loaded klystron approaches infinity the current should cease. At each electrical tuning limit (*i.e.* the boundary values), the current and voltage terms in the impedance expression assume the starting current values that were derived in preceding paragraphs, and the real component, which corresponds to the klystron resistance, is therefore written as

$$\operatorname{Re} Z = R_{S,L} = \frac{V_B}{\pi I_B N \cos\phi}.$$

A similar term is derived in $\sin\phi$ that defines the pure reactance limit, and from these real and imaginary terms one can determine ϕ as a function of the

[9] The trigonometric term comes from the phasor representation (see Nilson, 1983) of the AC Ohm's Law, $V = RI \exp(i\,\theta)$, and the expansion of the complex exponent.

[10] A pole is a term to describe a point where a function approaches infinity, i.e. a functional singularity (Churchill, 1995).

loaded cavity resistance ($R_{S,L}$) and the susceptance of the klystron cavity and electron beam (Ginzton, 1957).

Reflex klystron tunability arises from the expressions for the operating range because of the effect of the beam susceptance on the loaded cavity and because the klystron cavity susceptance is assumed fixed unless the load changes. It was originally stated that the klystron is represented as a parallel *RLC* network, and this network can be assigned some susceptance that is expressed as a sum of the capacitive, inductive, and beam susceptance, that is, ωC, $1/\omega L$, and $\tan\phi$, respectively.[11] The total susceptance is the sum of the cavity and the beam susceptance,

$$B = R_{S,L}^{-1} \tan\phi - \frac{1}{\omega L} + \omega C\,,$$

and can be rewritten in terms of a quality factor Q (see Lancaster, 1992, p126; or Poole, 1983, Ch.5), which can be expressed in terms of a tuning range, Δf.

The resultant expression for the electrical tuning is $\Delta f/f = \frac{1}{2}Q\tan\phi$, where the effect of varying the reflector voltage is expressed implicitly through ϕ. This expression demonstrates how the tuning range varies with Q; a high loaded Q decreases the range of frequencies in a given mode. The slope of the tuning curve (Figure 5, top) decreases as Q increases because the increased Q represents a larger susceptance. The net result of the flattened tuning curve is a broader electrical tuning range, that is, there is more latitude in the reflector voltage needed to cover the frequency range. Increasing Q therefore increases the range of voltages available to span the frequencies in a given mode.

The network parameters of the reflex klystron have been summarized in this section in order to provide a theoretical basis for optimizing the oscillator's performance for a specific type of experiment. For example, given the relationship between the quality factor and the klystron tuning range, one would optimize a swept frequency mode of operation by increasing Q. Conversely, high stability in the output frequency is achieved by operating the reflex klystron with a high beam voltage, low current, and high reflector voltage. Barring control by the user over circuit variables, the mode diagram of Figure 5 is used to identify which of the reflector voltage modes are best suited to an application.

[11] The $\tan\phi$ term is obtained from the imaginary component of the impedance; the coefficient, written in terms of the starting current, is factored by using the real component, and a ratio of $\sin\phi$ and $\cos\phi$ is the resultant. The remnant of the coefficient is $R_{S,L}$.

2.2.3 An Alternative Electron Beam Oscillator

Reflex klystron oscillators are size-limited in the sense that the dimensions of the cavity resonator and the electron beam path determine the

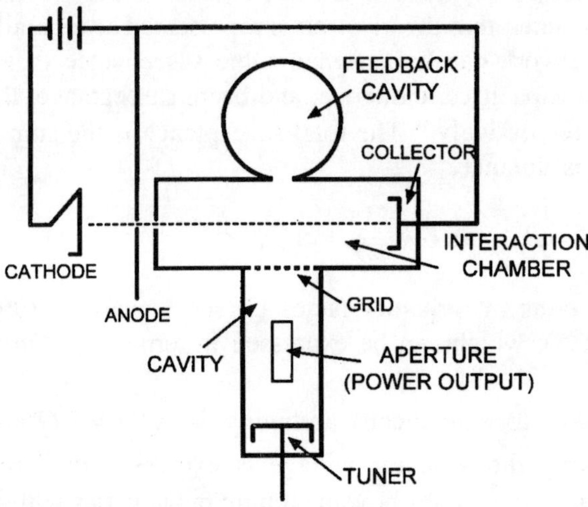

Figure 6. A variation of the electron beam oscillator in which the beam passes transverse to the E-field lines of the cavity. The output cavity is stabilized by a high-Q spherical reference cavity. Adapted from L. Wharton, *NASA Tech. Briefs*, July 1997, p44.

operating frequency of the oscillator. Reflex klystrons offer moderate power in the millimeter-wave frequency range (*e.g.* the Varian VRY-2131A yields 10mW at 170–220 GHz), and are unsuitable for spectroscopy without the use of amplifiers.

A variation of the klystron oscillator has been described by Wharton (1997) and provides improvements over the conventional reflex klystron by changing the orientation of coupling between the electron beam and the rf field. Rather than longitudinal interactions (coupling of the electron beam to the rf modes along the trajectory of the electrons), the modified design employs a transverse interaction. Two cavity resonators are situated at opposite sides of the electron gun (Figure 6); one is a spherical reference cavity, whereas the second receives and transmits the amplified signal in the same manner as the second stage of a two-resonator klystron. Both cavities resonate at a low order mode of the device's operating frequency, which allows one to generate high frequency carriers (unlike a klystron) because the electron beam path does not have to be rendered so small as to become non-functional (no bunching).

2.3 Semiconductor Devices

2.3.1 The Transferred Electron (Gunn) Device

The most important feature of an electron tube acting as an oscillator is the modulation of the electron velocities. As has been described in the preceding section, the electron beam in the klystron is modulated by its interaction with the resonant fields within the cavity. Electrons traversing the cavity are slowed in a periodic manner, or, in other words, the cavity's electromagnetic field introduces regions of negative resistance. The same phenomenon occurs in semiconductors.

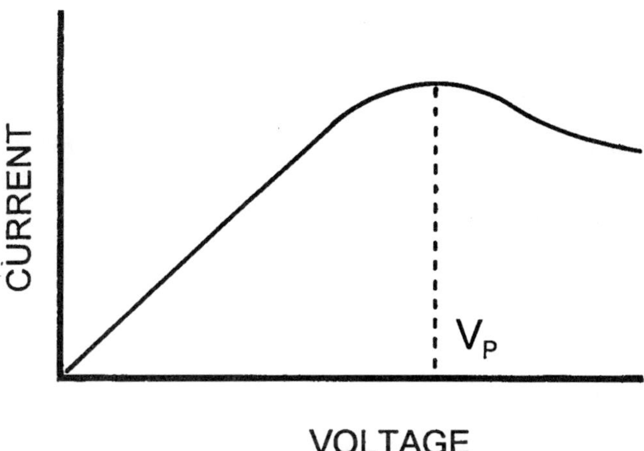

Figure 7. The concept of negative resistance. In this current-voltage curve, which corresponds to the interface between ohmic and semiconductor materials, there is a critical voltage V_p, above which the current decreases as the voltage increases.

Negative resistance simply means that the *I–V* plot has a negative slope at some point. By contrast, an ohmic device is characterized by a linear *I–V* plot with a positive slope. GaAs and InP are semiconductors whose steady-state *I–V* plots are non-linear and possess a region of negative resistance beyond some threshold value, V_P (Figure 7). This behavior is a consequence of the energetic properties of the conduction band, which in the case of GaAs and InP is not uniform and leads to a region of higher electron velocity (Swaminathan & Macrander, 1991; Thim, 1993).

The transferred-electron, or Gunn, effect is observed when a semiconductor such as GaAs of InP is sandwiched between two ohmic materials, the latter of which can be a (doped) semiconductor/metal or a semiconductor/semiconductor pair. Such a device is depicted schematically in Figure 8. If the device is biased by a voltage $V > V_P$, the electrons in the central region move with a velocity that is slower than those electrons at either end, and a depletion layer[12] is created. Assuming that the electrons move from left to right in the figure, electrons from the leftmost ohmic region are crossing the interface and slowing down, which leads to electron bunching.

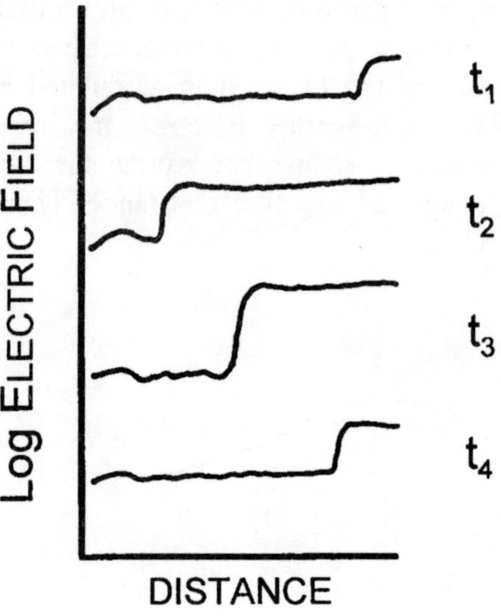

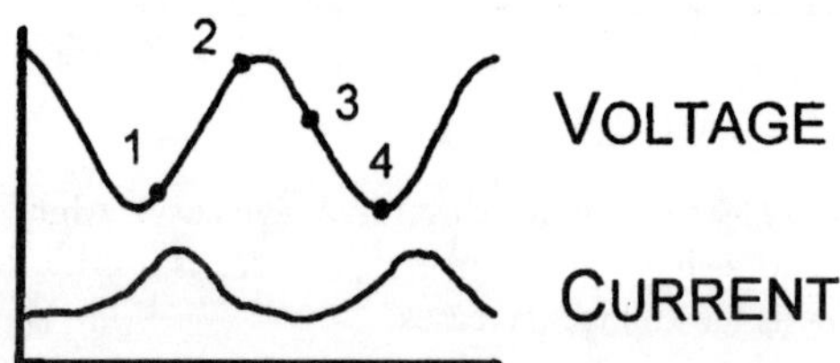

Figure 8. Waveform generation in transferred electron device. As electrons bunch, there occurs a spatial variation in the local electric field (represented as a step function on the logarithmic scale). The electric field profile across the semiconductor varies temporally (t_1, t_2, t_3, t_4), which corresponds to variation in the voltage and current. Figure adapted from Sze, 1981, and Thim, 1993.

As the electrons bunch near the leftmost interface, the local potential begins to drop. At the same time the electron bunch is moving towards the right, but as the local potential drops the velocity of the bunch is changing, as dictated by the slope of the *I–V* plot. When the local voltage decreases to $V = V_P$, the bunching process ceases because the electron velocities on both

[12] A depletion layer is a region in which the concentration of electrons is lower than its surroundings, see Cusack, 1957; Sze, 1981.

sides of the interface equalize. At this stage of the cycle the electron density begins to drop as V falls below V_P, and this sequence of events describes the formation of a half-wave (Figure 8).

The oscillatory waveform (current) is produced by the repetitive formation of bunched electrons and the crossing of these electron bunches through a non-ohmic region between two ohmic conductors. In his original experiments, Gunn (1963; see also Hilsum, 1962 and Hakki & Irvin, 1965) applied a short DC voltage pulse across the device (the pulse amplitude exceeded V_P) and observed that the peak voltage of the waveform was oscillatory as it passed the device. The dynamics of the electron bunch is unstable, however, and the frequency of the oscillation is determined by the transit time of the electrons through the depletion (non-ohmic) zone. The transit time, in turn, is determined by the length of the depletion zone and the materials that are used to fabricate the device.[13] Frequencies up to 100 GHz are readily attained at useful power levels, and signals up to 500 GHz are theoretically possible (Bosch & Engelmann, 1975; Thim, 1993).

The Gunn diode is not by itself a clean source of low distortion, high spectral purity sine waves. Its stability is affected by the load, the bias voltage, and the quality of its fabrication (Bosch & Engelmann, 1975). It is therefore possible to obtain many modes of operation from such a device. Stability and tunability are achieved by attaching a low impedance resonant circuit to the device. In other words, the Gunn oscillator is essentially a current amplifier coupled with an *RLC* resonator in much the same way as the Hartley and Colpitts oscillators.

Figure 9 illustrates several methods of stabilizing and tuning a Gunn diode. The diode may be incorporated into a cavity or stripline resonator that provides the inductive reactance and stabilization. As such, the device is considered free-running (*i.e.* no phase lock) and the frequency can be varied by means that affect either the bias voltage or the reactance of the stabilizing resonator. For example, a cavity resonator may be tuned by plungers (Figure 9A,B), but a broader tuning range may be had by incorporating varactors into the circuit.[14] The varactor can be inserted in series or parallel to the Gunn diode (Figure 9C) and independently biased so as to control the reactance of the resonator circuit. Finally, a current controlled tuning method for Gunn devices has been demonstrated by incorporating an Yttrium Indium Garnet (YIG) resonator. A YIG resonator is a sphere of ferromagnetic material, and when put into a magnetic field it oscillates at a frequency corrresponding to the precession frequency of the electron; a loop coupled to

[13] Unlike electron tubes, which are robust, semiconductors have low limits of voltages that may be applied prior to catastrophic breakdown, and there is only a limited amount of operating frequency control provided by the bias voltage.

[14] A varactor is a semiconductor device that exhibits a voltage-variable capacitance.

the resonant field of the YIG sphere (Figure 9D) contributes to the voltage bias of the Gunn diode (Bosch & Engelmann, 1975; Helszajn, 1985).

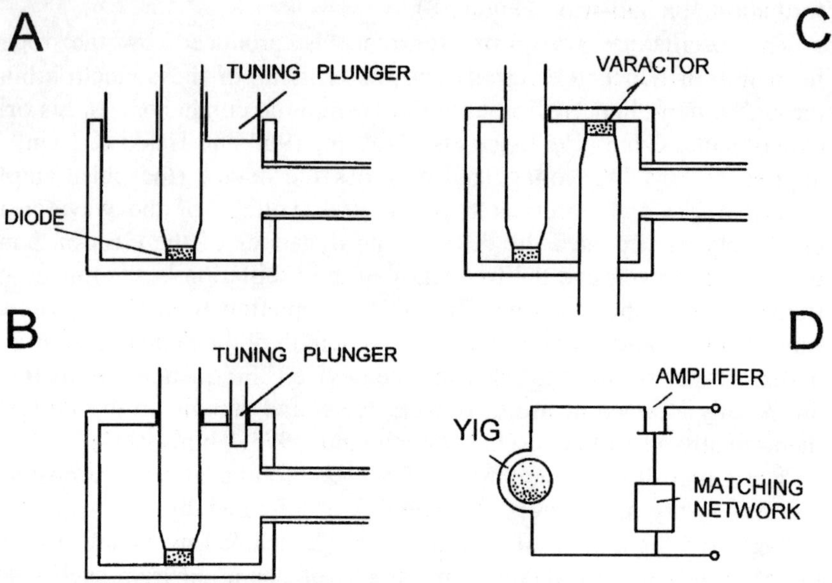

Figure 9. Methods of stabilizing a solid-state oscillator. The basic principles depicted in Figures 1 & 2 are repeated. Although transit time effects determine the waveform frequency, these are stabilized by coupling the device to an external *RLC* circuit, namely, a cavity resonator (A-C) or a YIG oscillator (D).

Until recently, most semiconductor devices for microwave generation and amplification have been fabricated from GaAs. InP is a likewise suitable material for microwave products, but it has been more problematic with regard to fabrication (Swaminathan & Macrander, 1991). As regards performance, however, differences between GaAs and InP devices are observed as a result of their differing band structure; InP has a slightly faster response time and is often warranted for devising high bandwidth devices.

2.3.2 Avalanche Diodes

Electron bunching and associated changes in the local voltages at a semiconductor interface is the basis of all semiconductor frequency synthesizers. The principal difference among the various types stems from the nature of the interface and the mechanism of electron transfer across that interface. The otherwise universal feature of all devices is the low conductance region through which the bunched electrons must travel in order to set up the voltage oscillation.

A second type of oscillator is constructed by replacing the leftmost interface of the Gunn device by a *p-n* junction. A *p-n* junction is formed at the interface between *p*-doped and *n*-doped semiconductors. When two such semiconductors are abutted, electrons diffuse from *n* to *p*, and 'holes' drift from *p* to *n*. As these mobile charges diffuse across the interface they leave behind fixed charges that generate an electric field. A resultant drift current counteracts the diffusion current until an equilibrium condition of no net current is reached (Sze, 1981). A carrier depletion region is therefore formed at the interface between the *n*- and *p*- bulk semiconductors.

Application of a voltage bias across the interface disrupts the current equilibrium. The depletion region is defined by the area at the junction through which the currents balance, and the bias voltage is taken to be defined across the depletion region. This voltage accelerates residual electrons whose kinetic energy is sufficient to ionize neutral atoms. Each electron-atom collision in the depletion region produces an electron-hole pair, and the process repeats itself so that a large current is rapidly generated: the so-called 'avalanche'.

The avalanche diode is a current-controlled device. As the avalanche current breakdown rises, the finite resistance of the circuit forces the voltage to drop and the depletion region expands. This expanding depletion region acts as a type of feedback that saturates the production of electron-hole pairs. At this point the bias voltage stops its decline and begins to increase again; the resultant *I–V* curve is S-shaped, and the requisite property of negative conductance is introduced.

An oscillator is produced if, instead of a DC bias and resistive load, the DC bias circuit includes a resonator circuit in series with the diode junction. An n^{+}-*p*-junction is reverse biased just below the avalanche breakdown voltage, which puts a high field across the junction. At this bias point, any additional microwave bias field will add to the existing DC bias and will momentarily exceed the breakdown condition. A relatively large current is therefore produced once per period of the oscillatory waveform. An (current) amplified microwave waveform is thereby generated via half-cycles, but the electron transit time through the drift region must match the period of the waveform or else destructive interference occurs.

The device itself is deceptively simple. Although Read (1958) described the principles of operation, it was Johnson, Loach and Cohen (1965) who first built a Read diode oscillator by *p*-doping a layer of a small piece of *n*-type silicon and then plating both ends with nickel. The chip was then inserted into a waveguide and reverse biased with a DC pulse at a low duty cycle. A spectrum analyzer at the waveguide output detected microwave power extending to K_U-band and at powers of tens of milliwatts. Further experimentation verified many of the theoretical predictions of Read and demonstrated the validity of the proposed mechanism of operation.

2.4 Photonic Methods of Generating Microwave Signals

The non-linear response of diodes may be used to modify the frequency of an incident sinusoidal waveform. For example, a mixer (see Section 4) is a device that consists of a bridge-like array of diodes and combines the frequencies of two waveforms. It is typically a three-port device and, if f_1 and f_2 are two frequencies passed to the device as input, then the combinations $f_1 \pm f_2$ are detected at the output, and therefore the device can be used as an up- or down-converter. Downconversion applications of mixers are familiar in the detection schemes of EMR (Section 4); both homo- and hetero-dyne detection are based on the recovery of the EMR signal as a DC or radio frequency (10–100 MHz). The Special Triple technique of ENDOR is another example of frequency conversion in which one of the two radio frequency signals is swept so that the symmetrically displaced ENDOR transitions are simultaneously driven.

This principle of signal generation via down-conversion may be applied to microwave and millimeter wave generation, where conventional oscillators are not available. A schematic diagram of the apparatus is shown in Figure 10. The output of two lasers operating at frequencies f_1 and f_2 is mixed by using a fiber optic coupler that yields a beat frequency $|f_1 - f_2|$, which would be the desired output frequency for a magnetic resonance experiment. One photodiode serves as the interface between the photonic and electronic circuits, and a second is used with a delay line and is the basis of the feedback loop that controls the tuning.

If one assumes, for example, that the first source operates at $\lambda_1 = 850.000$ nm, the second source must operate at $\lambda_2 = 850.7730$ nm in order to yield a downconverted output frequency of 330 GHz, or $\lambda_2 = 850.8492$ nm in order to yield 30 GHz. As one moves to laser sources that operate in the near infrared (e.g. $\lambda \sim 3000$ nm) the frequency that is obtained by downconversion becomes more of a significant fraction of the fundamental frequencies; 330 GHz is a beat frequency of 3000.0000 nm and 3009.9254 nm, and the requisite shift is approximately 10 nm, as opposed to 1 nm.

A photonic approach to microwave frequency generation is attractive for a variety of reasons. The first is the potentially enormous tuning range of cw frequencies that may be had by simply modifying the tuning characteristics of one of two laser sources. The preceding paragraph describes a simple example in which a laser source could span the entire microwave and far IR spectrum. But more importantly, the photonic approach to microwave and millimeter wave signals enables one to perform experiments with extremely short pulses. With electron tubes and conventional semiconductor technology one is constrained to operate with pulses whose width is measured on the tens of nanosecond time scale. Ultra-short (sub-picosecond) pulses are readily attained by mode-locking methods (New, 1983; Durling,

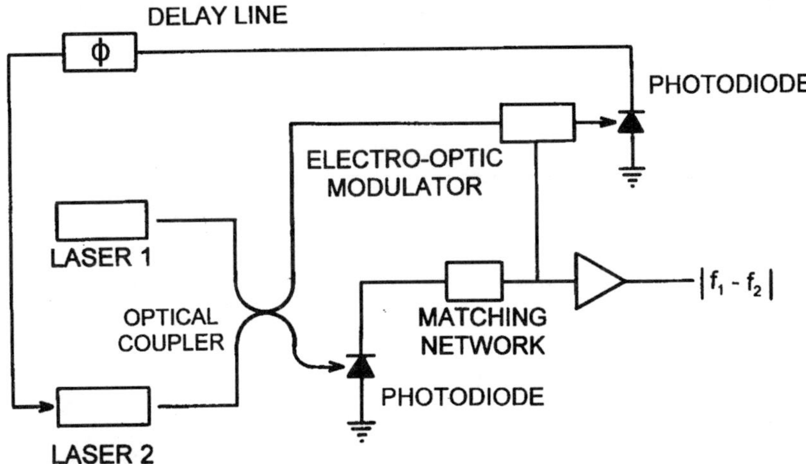

Figure 10. Schematic diagram for an apparatus designed for photonic generation of microwave signals. The output of two lasers is mixed and downconverted so that their difference $|f_2 - f_1|$, corresponds to the desired microwave frequency. Tunability is obtained by varying the frequency of one laser. Adapted from R. Logan, *NASA Tech. Briefs*, Nov. 1997, p54.

1995) and with semiconductor diode lasers. It is therefore feasible that one could generate ultrashort microwave pulses via the heterodyne method illustrated in Figure 10 by operating one of the two lasers in pulsed mode.

2.5 Spectral Purity & Frequency Control

All oscillators are stabilized by coupling to a resonant structure, such as an *RLC* network, whose time-domain response specifies an oscillatory waveform of discrete frequency (Section 2.1). In the absence of a resonant structure, for example, thermal emission or transit-time semiconductor diodes, one obtains a dispersion of output frequencies because the electrons do not all travel in phase (Pierce, 1951). In practical applications such as spectroscopy, one wishes to ensure that the source is generating near-monochromatic radiation, and that the frequency of this radiation remains constant. Ideally, one desires a radiation power spectrum that resembles Figure 11, upper left, in which a single narrow peak represents a very narrow dispersion of frequencies generated by an oscillator.

Frequency sources are stabilized by coupling an oscillator to a tuned resonant circuit, which is equivalent to a bandpass filter. Ideally, this filter would have a very narrow pass band that restricts the oscillator's output to a single, delta-function-like, frequency, but in practice a bandpass filter

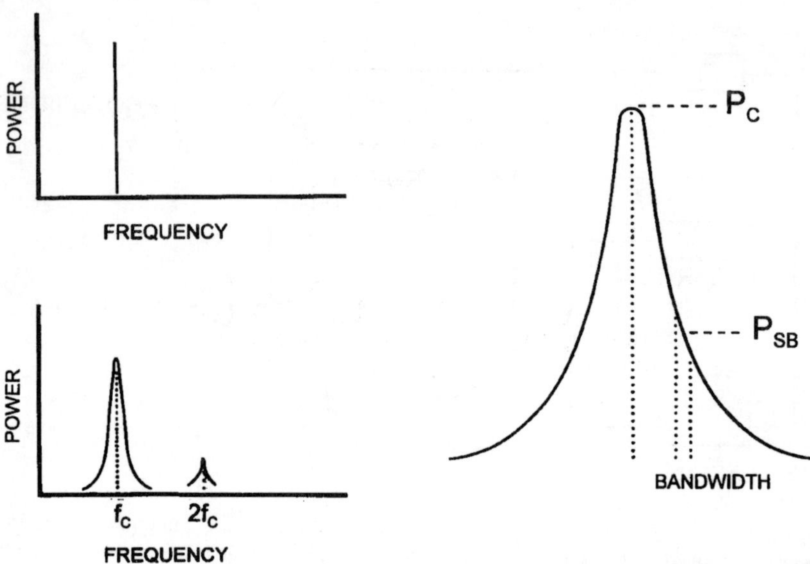

Figure 11. Representation of spectral purity. An ideal oscillator yields power at a single frequency as a δ-function power spectrum (top, left), but practical oscillators yield a dispersion of frequencies at a fundamental frequency f_c that satisfies some harmonic differential equation defined by the *RLC* parameters, plus some harmonics ($2f_c$, $3f_c$, etc.). The dispersion of frequencies corresponds to a phase shift in the harmonic equation, the magnitude of which corresponds to the dissipative behavior of the harmonic system. Phase noise is reported in terms of the ratio P_{ssb}/P_0, right.

transmits frequencies according to the properties of resonant circuits, for which there are two major sources of spurious frequencies. In section 2, resonant circuits were introduced as a network of elements (*e.g.* resistors, capacitors, inductors) that respond to a stimulus in an oscillatory manner. For example, the temporal flux of charge in a simple *LC* loop is described by a differential equation $L\, d^2x / dt^2 = -qC$, for which the solution is $q(t) = \frac{1}{2}q_0 [\exp(i\omega_0 t) + \exp(-i\omega_0 t)]$. It follows that $\omega_0 = (LC)^{-1/2}$, but an important point to be made is that multiples of ω_o, that is, harmonics, also satisfy the differential equation. One source of spectral impurity is therefore harmonics of the resonator that is coupled to the oscillating device, and the amplitude of the harmonics are typically specified relative to the fundamental (carrier) in dBc. For example, the specifications of the Gigatronics 12000A VXIbus microwave synthesizer read that spurious harmonics are ≤ 60 dBc.

Although a natural outcome of the resonant circuit, harmonics are usually much attenuated relative to the fundamental frequency. This can be rationalized on the basis of the distributed nature of high frequency circuits and the fact that geometry is an important factor in setting the frequency of the standing wave that may be supported by the structure (see Section 3). A resonator is commonly constructed of elements whose dimensions correspond to $\lambda/4$ or $\lambda/2$. Each multiple nf reduces the initial wavelength by a factor of $1/n$ (*i.e.* $\lambda' = \lambda/n$), and the ability of a guide to support a non-optimal mode is greatly diminished. One simple illustrative example is rectangular waveguide; a WR-90 guide that propagates a 9 GHz waveform will severely attenuate the 18 GHz harmonic.

In the region about the fundamental carrier frequency f_c, the spectral purity is affected by the quality of the linear network and the resonator. For example, the linear networks (amplifier and feedback, Section 2.1) are affected by physical variables. The feedback network must maintain a constant amplitude and phase at the amplifier input, and factors such as temperature and the load can alter the network's reactance and behavior. Similarly, the amplifier is affected by temperature and load; it is also affected by voltage fluctuations. All of these factors must be controlled in order to ensure a high spectral purity.

The power spectrum profile of a real source is represented by Figure 11, bottom left; there is a dispersion of frequencies about f_c. The breadth of this peak is commensurate with the dispersion of phase angles in the electron transit times (*cf.* Section 2.2.2). The variation in the phase angle is represented by a sinusoidal function, and therefore the otherwise 'pure' sine wave function of the carrier, $V(t) = V_0 \sin\omega_c t$, is modified to be $V(t) = V_0 \sin(\omega_c t + \Delta\phi)\sin(\omega_m t)$, where $\omega_c = 2\pi f_c$, $\Delta\phi$ is the maximum deviation of the phase angle, and ω_m is the modulation frequency (*i.e.* a noise function, $f_m \geq 0$, see Barber, 1971). Trigonometric identities allow one to rework the expression for $V(t)$ so that $V(t) = V_0 [\sin(\omega_c t) + \Delta\phi \sin(\omega_c t + \omega_m t) + \Delta\phi \sin(\omega_c t - \omega_m t)]$. In other words, the power spectrum of an oscillator features f_c plus sidebands at $f_c \pm f_m$, where f_m constitutes a set of frequencies weighted by $\Delta\phi$. In practice, one measures f_m in 1 Hz steps away from f_c, and measures the 'phase noise' from the ratio of the powers P_c and P_{sb} (Figure 11, right).

For a reactive element (*e.g.* an inductor or capacitor) in an *RLC* network, the impedance Z is a variable function that depends on the frequency. For any element, therefore, a plot of Z as a function of f is non-linear and typically features a critical point at which $dZ / df = 0$, indicating resonance (see Chapter 4 discussion of ENDOR coils). We know from AC circuit theory that the reactive devices shift the phase of the incident waveform by some amount $|\phi|$ whose sign will depend on the slope of the Z vs. f plot. A positive slope corresponds to a capacitive response in which the phase angle is positive; a negative slope signifies an inductive circuit and negative phase angle. At resonance (*i.e.* $dZ / df = 0$), $|\phi| = 0$. As the frequency deviates from f_c, $|\phi|$ increases, and therefore the phase angle can be used as a feedback test criterion. The response $\Delta\phi / \Delta f$ is proportional to the quality factor Q of the resonant circuit (see Kajfez, 1995).

This property of *RLC* circuits can be used to stabilize free-running oscillators. The underlying principle is to design a so-called discriminator circuit that produces an error voltage whose sign varies according to the relative magnitude of the desired and actual frequency, and whose amplitude is commensurate with the deviation (Figure 12). The error voltage is then

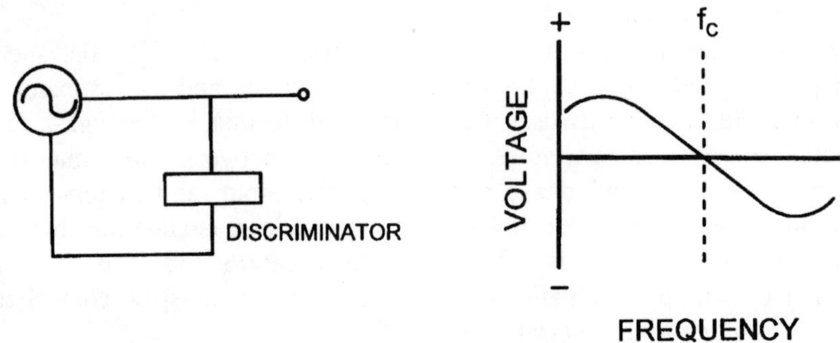

Figure 12. A general block diagram for a frequency stabilizer circuit. A sample of the oscillator output is passed to a 'discriminator' that yields a voltage whose sign and amplitude is determined by the deviation of the sampled frequency from some determined value. The characteristics of this discriminator voltage are indicated in the accompanying graph.

amplified in order to provide a feedback control to the oscillator, as depicted schematically in the figure. Frequency stabilization may be effected by coupling the oscillator to a high-*Q* cavity or a second high stability oscillator. In radar application, the reflex klystron serves as the high-stability local oscillator, and it is locked to the transmitter oscillator. Most EMR spectrometers, however, use the same oscillator for both sample perturbation and as the LO input on the receiver (see Section 4), and therefore a variation of the feedback loop (Figure 12) is used; such circuits typically yield stability of one part in 10^8.

The Pound stabilizer (Figure 13) illustrates a method of electronic automatic frequency control (AFC). A sample of the oscillator's waveform is passed to a magic tee (see Section 3.2.4 for description of 'tee' properties and applications) whose ports are terminated by a tunable high-*Q* reference cavity and crystal diodes. The signal at diode A is modulated by a stable low frequency source (the AFC modulation frequency), and diode B serves as a comparator of f_c and $f_c + f_{AFC}$; if $f_c = f_{ref}$, then the frequency output by the diode B is f_{AFC}, otherwise it is $f_{AFC} \pm f_{error}$. A phase sensitive detector is then use to compare the signal output by diode B and the original AFC waveform, and this error voltage is then amplified and used as a control voltage on the source oscillator (*e.g.* the reflector voltage on a reflex klystron). With the Pound stabilizer one sets the desired frequency on the tunable cavity, then tunes the oscillator to that frequency. In an EMR spectrometer, therefore, one would tune the sample resonator, tune the reference cavity (align the 'dips' in the cavity modes), and finally tune the klystron to this frequency.

A consistent feature of oscillators and the Pound stabilizer is the use of a filter, which has for the most part been an *RLC* circuit (*e.g.* cavity resonator). The Hartley and Colpitts oscillators that were introduced in Section 2.1 to illustrate this point are used to generate stable low frequencies by constructing the *RLC* from discrete components, and it is just as simple to

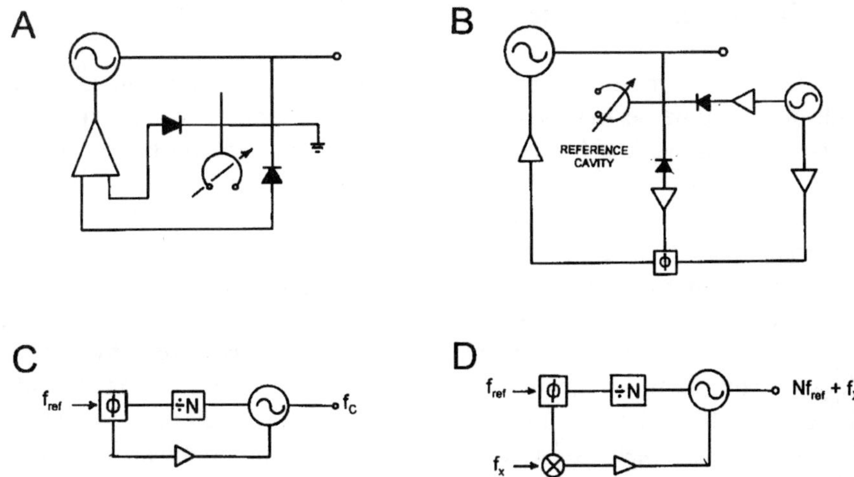

Figure 13. Practical frequency stabilizers. The Pound stabilizers (A & B) use hybrid tees to compare the frequency of the oscillator to the frequency to which a high-Q resonator is tuned. Frequency division (C & D) enables one to directly compare and lock the microwave oscillator to a high-Q quartz resonator.

use a piezoelectric element, such as a quartz crystal, as the filter element. One can similarly use low-frequency quartz oscillators that feature a high Q as part of the frequency stabilizing circuits for microwave oscillators. Again, a phase discriminator circuit is used, but a frequency divider and/or a second mixer stage is used to couple the high and low frequency stages of the oscillator unit (Figure 13C, D). Block diagrams of commercial microwave frequency synthesizer and how they are locked to 1 MHz reference standards are listed in a text by Manassewitsch (1987).

Finally, as regards spurious noise and modulation of the oscillator, the packaging of the device and its power supplies has some effect. In many of the microwave techniques books cited or listed in the bibliography, circuits are described in which the oscillator, its power supply, and frequency stabilizer circuits are housed within the same shielded enclosure, which at rf frequencies is typically a 1/4" aluminum box with tongue-in-groove joints. Air cooling is effected via ports that are fabricated from below-cutoff waveguide sections (Montgomery, 1947), although spectrometer bridges typically use a copper heat sink and cooling water (tapped from the magnet plumbing).

3. WAVE PROPAGATION & MANIPULATION

3.1 Transmission Lines

A functional circuit is designed by assembling discrete components that each impart some desired property to the network, and one can express the 'laws' of a network's behavior in terms of constant variables such as resistance, capacitance, and inductance. The simplicity of the analysis follows from the fact that one can ignore the properties of the electromagnetic field at low frequencies because the wavelength, $\lambda \propto f^{-1}$, is very large (approaching infinity in the limit of DC signals) and the dimensions of the circuit elements is comparatively small. Under such circumstances a network can be mathematically described by discrete elements, but as the frequency increases, the wavelength becomes smaller and eventually becomes comparable to the size of the circuit elements. At this point the geometry of a circuit element and the circuit layout become design factors.

Also problematic at high frequency is the fact that common parameters that are used to define a circuit's response are no longer applicable. For example, voltage is ordinarily defined as the difference in electrical potential between two points that may be arbitrarily defined. The formal definition of voltage, however, is a work function associated with traversing the region between two points, that is, a line integral $V = \int \mathrm{E} \cdot ds$. At low frequency this integral is path-independent and $V = E_B - E_A$, but at high frequency the integral is path-dependent and the ordinary means of defining and measuring V are invalid. At high frequency a circuit is therefore defined in terms of its wave properties, and the transition frequency between distributed and lumped circuits is often taken as 1 GHz, where $\lambda \approx 12$ in (30 cm). A summary of the properties of electrical circuits and their measurement at various frequency ranges is outlined as follows (after Ginzton, 1957):

Table 1. Electrical Measurements at Low and High Frequency

Frequency	Voltage	Current	Impedance
DC	voltmeter	ammeter	Wheatstone Bridge
Audio to 50 kHz	voltmeter, rectifier	thermocouple	Wheatstone Bridge
low rf	voltmeter, rectifier	thermocouple	Wheatstone Bridge
high rf	not accurate	thermocouple	bridge, SWR
microwave	not significant	not significant	bridge, SWR

A single conductor will transmit a DC current with losses that can be attributed to resistance, which is typically described as a scattering phenomenon between the conduction electrons and the atomic nuclei of the conductor material. An electromagnetic wave, by contrast, will likewise propagate along a single conductor, but its energy will be radiated away from the conductor. In other words, the single conductor device acts as an antenna and is subject to high radiative losses. A shield is required to prevent these radiative losses, and transmission lines are predominantly two-conductor structures such as parallel lines, coaxial lines, or hollow conductors (*i.e.* waveguides). In all cases the direction of propagation is along the axis of the transmission line, and one defines the propagating electromagnetic field by whatever component (electrical or magnetic vector) is perpendicular to the propagation vector.

An electromagnetic wave has two orthogonal components, the electric and magnetic fields, which are respectively analogous to the DC voltage and current. It follows that the propagating modes of an electromagnetic wave can be defined as transverse electrical (TE), transverse magnetic (TM), or transverse electromagnetic (both E and H, TEM). One or more of these modes may be suppressed by the geometry or the dielectric properties of the transmission line. For example, a waveguide is more restrictive with respect to allowed propagating modes than a coaxial line. The design and measurement of circuits at high frequency are therefore based upon wave phenomena, and circuit parameters R, L, and C are replaced by an all-encompassing impedance Z,[15] which can be related to the scattering behavior of waves.

Microwave circuits are commonly defined by a Voltage Standing Wave Ratio (VSWR) coefficient, and this is best explained by beginning with a mechanical model such as a wave in a fluid medium. The ideal scenario for wave propagation is one in which waves are propagated through a medium with minimal loss of amplitude, that is, energy, and this energy is delivered to a terminus with an efficiency of unity. An example would be water waves that impinge on a gradually sloping shoreline and are therefore not reflected.

One can take as a starting example the case in which a microwave source delivers an oscillatory signal to an infinite line. The electromagnetic wave will propagate through this medium presumably ad infinitum, but one finds that the signal weakens as the distance from the source increases. An observer at some arbitrary distance from the source would record the oscillations of the traveling wave: $E(x) = E_0 \sin(\omega t - \beta x + \phi)$. In this equation $\omega = 2\pi f$, where f is the oscillator frequency, and β is the phase constant of the medium, $2\pi/\lambda$. This equation describes a traveling wave and

[15] Formally, circuit analysis at high frequency introduces imaginary variables to account for the time-domain response of circuit elements. Traditional DC variables constitute a real component, whereas the imaginary component is called a reactance.

applies equally to both infinitely long transmission lines and a transmission line that is terminated by a load that perfectly absorbs all the energy of the incident wave. Measurement of the time averaged signal intensity at all points along the line would yield a constant value in the absence of resistive losses. For a very long line in which losses become a factor, the signal intensity would decline linearly with distance. Losses in signal intensity vary among the types of transmission lines, being related to geometry and the material from which it is fabricated. For example, standard rectangular copper waveguide at X-band (WR-90) typically dissipates between 4.5 and 6.5 dB of power per 100 linear feet.

An infinitely long or perfectly absorbing line is one extreme case in describing the propagation of waves through a transmission line. The other extreme corresponds to the case in which the transmission line is terminated by a load that perfectly reflects the incident wave. A so-called stationary wave is set up in such a situation, and the important point is that, unlike the traveling wave whose time average amplitude is the same for all points, the stationary wave amplitude varies with position. Returning to the water wave analogy, the scenario is equivalent to the difference between waves impinging upon a vertical wall as opposed to a gradually sloping beach.

The amplitude of a stationary wave is characterized by peaks and troughs that are formed by the interference between the incident and reflected wave. As a simple illustrative example, consider the incident wave that completes a half cycle at the terminus, where it is then reflected (the amplitude is zero). This would correspond to a perfect short circuit in the transmission line and a perfectly reflecting load, and the reflected wave can have either positive or negative polarity (0° or 180°). Negative polarity nulls the incident wave, but positive polarity reinforces it, and the standing wave mirrors the incident wave with a boundary condition of zero amplitude; nodes (troughs with amplitude zero) occurs at $x = \frac{1}{2} n \lambda$.

A transmission line that is terminated by a partially reflecting load likewise supports a stationary wave, but true nodes are not created. In the case of the perfectly reflecting wave, nodes are created because the incident and reflected wave have equal amplitudes, and the out of phase waves null one another. When the amplitude of the reflected wave is less than the amplitude of the incident wave, the waves do not null one another, and even if the reflected wave is 180° out of phase with the incident wave, there is still some signal amplitude left at $x = \frac{1}{2} n \lambda$. The measured signal intensity at some point x is the sum of all possible interference effects, and a partially reflecting terminus therefore creates troughs whose minima differ from zero. It follows that the relative magnitude of the stationary wave maximum and minimum provides a measure of the wave reflection. We define a reflection coefficient $\Gamma = E_{max} / E_{min}$, where $1.0 \leq \Gamma \leq \infty$; a perfectly reflecting wave nulls the incident wave $E_{max} / E_{min} \rightarrow \infty$, and a perfectly absorbing termination reflects no wave, creating no nodes, so that $E_{max} / E_{min} = 1.0$.

Complete reflection represents an infinite impedance to wave propagation, whereas the perfect absorber 'matches' the transmission line by absorbing all the incident power. The ratio of the wave maximum and minimum, which has been defined in the preceding paragraph and is called the Voltage Standing Wave Ratio (VSWR), therefore provides a measure of the match between a microwave circuit device and the transmission line source of electromagnetic radiation. Impedance is therefore only defined with respect to the transmission line; one does not need to be concerned with the absolute value of the impedance. The important design factor is the ratio Z / Z_0, where Z is the impedance of the device and Z_0 the impedance of the transmission line feed.

3.2 Impedance Matching

The object of matching a device to a transmission line is to minimize losses in the wave amplitude as it propagates through a circuit. There is nothing inherently wrong with operating under the condition of mismatch.[16] In general, the need to carefully match a device is dictated by the experimental requisite for maximum power transfer and operation at or near the maximum breakdown strength of a device's constituent materials. As for measurements, a careful match is a necessity for comparing the properties of two or more loads; this latter criterion applies to EMR spectroscopy, and the familiar tuning procedure matches the cavity to the remainder of the so-called bridge circuit.

The 'standard' 50 Ω impedance that one sees specified on many devices is a compromise value between 30 and 70 Ω. Power loss (per unit linear dimension of transmission line) and power handling capability of a transmission line vary with the impedance, and 30 and 70 Ω represent optimal impedance values for minimizing losses and maximum power handling, respectively. The source of the electromagnetic radiation will possess a characteristic impedance that is derived from its *RLC* network parameters, and is treated as a two terminal unit to be matched with the load circuit.

Transmission lines vary in geometry, but certain attributes render some more competitive than others. For example, a one inch diameter copper tube is a universal line for microwave communications because its losses are very small, and it will propagate many modes and frequencies (Miller, 1954). But in some experiments one wishes to make measurements with a well-defined propagating mode; in EMR, the resonance condition imposes specific

[16] The performance of an oscillator, including its frequency stability, will depend on the load.

relative orientations on the DC and oscillatory magnetic fields (Bleaney & Stevens, 1953), and rectangular waveguide that operates in the TE mode is the most frequently used transmission line. A second advantage of rectangular guide is that departures from linearity or uniformity of cross section appear only as a reactance, which can easily be compensated by the addition of another reactance (Marcuvitz, 1951).

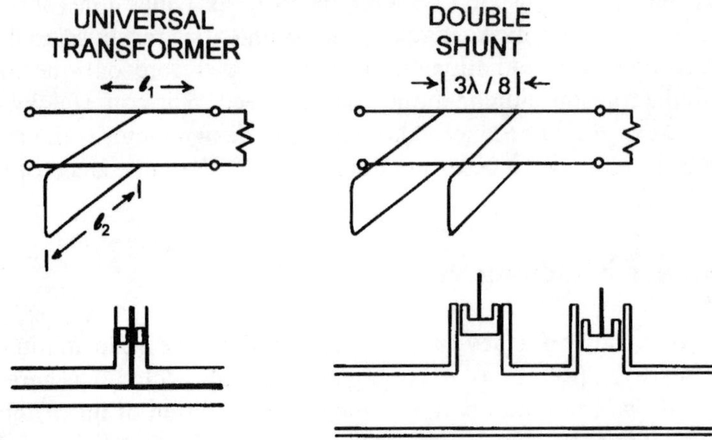

Figure 14. Impedance matching transformers. The 'universal' transformer consists of an adjustable short whose position along the line can be adjusted to a point at which the conductance equals the characteristic admittance. In practice, the device that best approximates a universal transformer is the slide screw tuner, whose variable reactance can be positioned at some point along the line where a 'sweet spot' repeats itself at some multiple nλ/2. Other practical devices cope with positional limitations along the line by adding one or more reactances (double and triple slug tuners, right).

Devices for impedance matching correct for a non-optimal reactance by introducing a variable reactance. A theoretical 'universal' transformer is merely a shunt on a transmission line (Figure 14). There are only two adjustable components on such a transformer: the length of the shunt, and the distance of the shunt from the mismatched load. The distance of the shunt from the mismatched load is selected so that the conductance equals the characteristic admittance (*i.e.* the resistance equals Z_0). At this point, therefore, all one has to do is introduce a pure reactance equal and opposite to the reactance of the line, and one can eliminate the reactive component of the impedance; since $R = Z_0$ at this point, the loaded line appears matched to the source. The length of the shunt controls the reactance of the transformer, which should be equal and opposite the line's reactance at the point of its attachment.

In EMR spectrometers, all of the components are matched in the sense that the bridge circuit is laid out to the optimal Z_0; this is true even of home-built spectrometers, which tend to be tinker-toyed together from discrete components that have connectors and are optimized for 50 Ω circuits. The only 'unknown' is the sample resonator, whose impedance will vary according to what sample material occupies the field lines. A cavity resonator tends to be coupled to a rectangular guide via an aperture and a variable post (the so-called iris), but other devices, such as the E-H tuner, the slide-screw tuner, and the double-stub tuner, may be used to match non-standard sample resonators (laboratory experiments for demonstrating the use of these devices are compiled in Reich *et al.*, 1957). E-H tuners are specified in some spectrometers that employ a helical sample resonator (Webb, 1962), and a slide-screw tuner was initially specified for matching the Bruker ENB250 TM_{110} ENDOR cavity when the coil was in place.[17]

3.2.1 Apertures and Irides

A transmission line has a characteristic resistance, capacitance, and inductance per unit length, and it is common to find theoretical treatments of transmission line behavior expressed in terms of an infinite or semi-infinite repetition of an *RLC* unit. The effective inductance and capacitance are parameters that depend on the geometry of the line and its dominant propagating mode. For example, in a rectangular guide the capacitance and inductance are determined by the dimensions of the narrow and broad walls, respectively. The analogy, based on the orientation of the respective field lines, to conventional capacitors and inductors is illustrated in Figure 15. Detailed analyses of the network properties of waveguide are found in texts by Slater (1942), Marcuvitz (1951), and Collin (1960).

[17] One would not ordinarily use two or more tuning devices in this manner, but each tuning device has only a finite range of impedances over which it can be varied. The slide screw supplies a greater range than the iris alone for matching the coil-containing ENDOR cavity; a similar strategy is used for an ESE-ENDOR cavity, in which the matching range of a Gordon Coupler is extended by adding an iris.

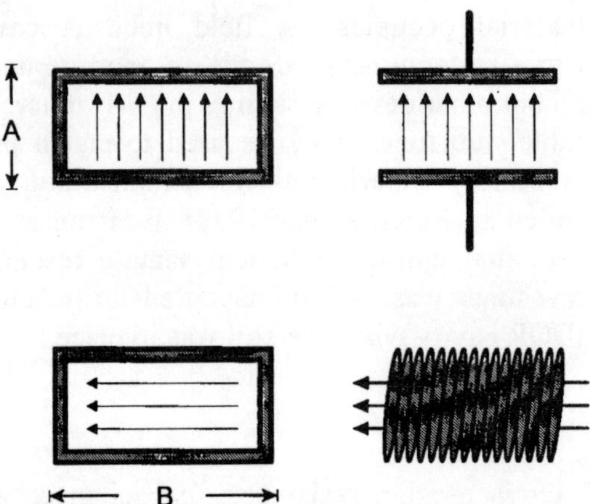

Figure 15. The analogy between the rectangular waveguide cross-sectional parameters and distributed parameter devices (capacitor and inductor).

A common method of correcting a mismatch is to insert a compensating inductance or capacitance into the circuit. One can therefore insert a very narrow diaphragm[18] through the broad or narrow wall to introduce

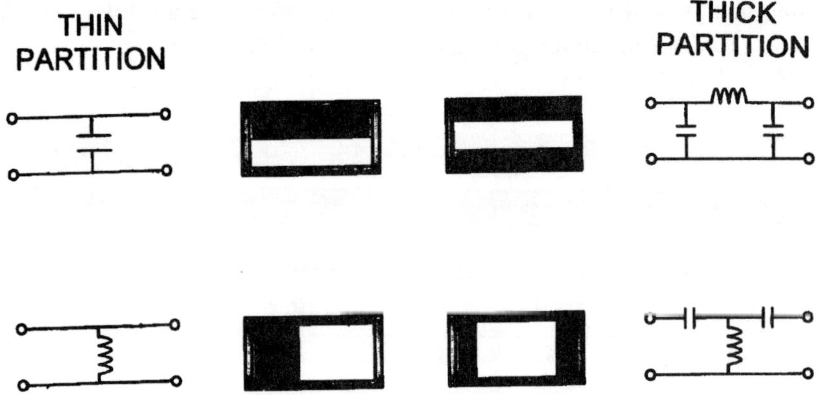

Figure 16. Waveguide obstructions and their equivalent circuit representations. Thin obstructions (left) and thick obstructions (right) are indicated.

capacitance or inductance (Figure 16). As appertains coupling to a cavity or

[18] Narrow diaphragms in waveguide are taken, in a practical sense, to be of thickness equal to or less than 1/32", see Marcuvitz, 1951.

another section of waveguide, an aperture may be viewed in much the same way as a thin diaphragm. Either circular holes or slots may be used, but the former couples via the electric field, whereas the latter couple via the magnetic field. One variant of the latter, the cruciform coupler, is useful because it allows one to impose field orientation constraints. It is important to recognize, however, that coupling apertures must be of a size so that they do not resonate.

Finite thickness obstructions scatter an incident electromagnetic wave and, through the analogy between wave scattering and impedance, introduce both series and shunt reactances. A post perpendicular to the electric field lines acts as a capacitive shunt, whereas a parallel post acts as an inductive shunt. The equivalent circuits of capacitive and inductive obstructions are indicated in Figure 16. Transformers are readily fabricated from fractional waveguide obstructions and optimally ‘tuned’ to the mismatch, coated with a thin layer of solder, slid into the optimal position, and heating the waveguide to make the solder joint. Bandpass filters often feature a step-like cross sectional profile of $\lambda / 4$ transformer sections (see Matthaei, Young, & Jones, 1980, Chapter 3).

3.2.2 The Slide-Screw Tuner

A vertical post inserted through the broad wall of a waveguide behaves as though it were an inductor, but if the post does not completely pass through the waveguide, there is a capacitance that arises between the post and the wall. A post whose position is variable may be used to introduce a reactance in the form of a series inductance and capacitance, which may then be used as a means to provide a nulling reactance so that a load may be matched. The slide-screw tuner is such a device that works on the principle of the universal transformer described in the introductory paragraphs of this section.

The adjustable post is mounted on a carriage so that it may be moved along the guide in the direction of propagation, and it follows that there are two adjustable parameters that optimize the impedance match. The optimal point along the transmission line, where $R = Z_0$, repeats itself at half wavelength intervals, and therefore the slide screw tuner device can be installed at some arbitrary point near the load (*i.e.* the tuner is a device that may be purchased as a waveguide section to be 'dropped in'). Since most applications of tuners correspond to resonator tuning, the slide screw is situated between the source and close to the resonator. The screw is withdrawn from the waveguide, and the resonator tuned as well as possible by using the iris, if one is available (presumably, one is monitoring coupling via the 'dip' in the source’s output). The screw is then inserted into the waveguide until it affects the tuning parameters, and then its position along the waveguide is adjusted in order to maximize the effect. Once the optimal position is found, the screw is inserted to optimize coupling.

3.2.3 The Double-Stub Tuner

The universal transformer that was described in the introduction to this section requires that one find a 'sweet spot' along the transmission line where $R = Z_0$. An adjustable short can, in principle, be placed anywhere along the line, but in practice this becomes problematic because one simply cannot move a waveguide or coaxial junction arbitrarily along the main transmission line. The double-stub tuners obviate the problem by introducing a second variable short that compensates for the immobility of the tuning plungers.

Placement of the stubs affects the performance of the tuner. For two stub device, the placement should be at some odd multiple of $\lambda/8$; the three stub tuner performs optimally when the distance separating the stubs is $\lambda/4$. The stub closest to the load to be matched provides an adjustable admittance so that the conductivity of the load matches that of the line; the second stub (furthest from the load) then provides an adjustable admittance to the transmission line.

3.2.4 Waveguide Tees and the E-H Tuner

A stub tuner is realized as a junction in a coaxial or waveguide transmission line (Figure 14), but because of the TE_{01} mode pattern in rectangular guide, the equivalent circuit of shunts varies according to which wall bears the junction. An E-plane junction puts all three branches in the electric plane, and therefore the circuit elements lie in series. By contrast, an H-plane junction puts all three branches in the magnetic plane; any circuit element (*e.g.* a load) attached to a given terminal will appear in parallel. Waveguide tees are therefore analogous to parallel and series junctions at low frequency and may be used to divide power among the other two ports. Power division is equal if the ports are identically matched.

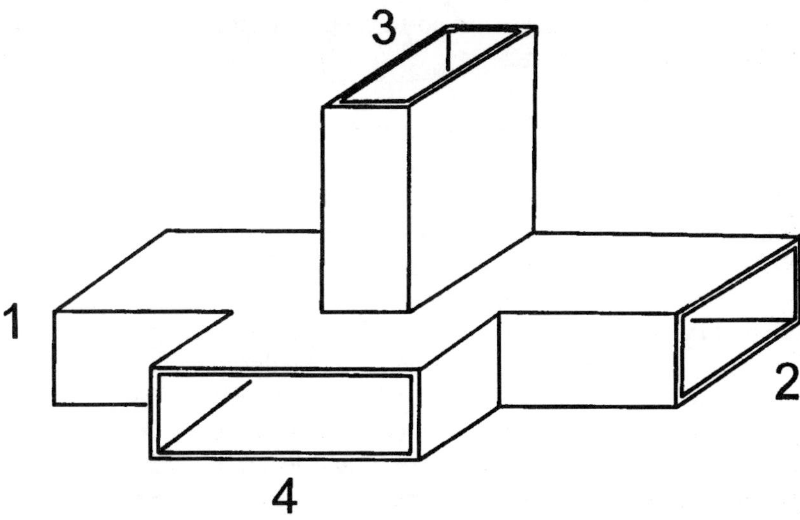

Figure 17. The hybrid tee. Branch 3 appears in series with the main waveguide section denoted by ports 1 and 2; branch 4 appears in parallel. An E-H tuner features adjustable shorts in both of the branches and therefore puts reactance in series and parallel to the main circuit that is connected to ports 1 and 2. Because of the symmetry and distribution of power delivered to the device, it is also use as a means of creating bridge circuits (see Sections 2.5 and 4)

A hybrid tee features both E-plane and H-plane junctions on the main guide, and the symmetry of the device renders certain pairs (*i.e.* 1-2, 3-4; see Figure 17) decoupled. If signals enter the tee from port 3, it will be divided equally among ports 1 and 2 if these two ports are matched loads. A signal appears at port 4 only if there is a mismatch between ports 1 and 2, and this is the bridge method of detection that was used in many early autodyne EMR spectrometers. Ports 1 and 2 are connected to the sample cavity and a standard 50 Ω load. With the sample off-resonance, the sample cavity is tuned to match the 50 Ω load, thereby nulling the signal at port 4. As the magnetic field is swept through the resonance condition the impedance of the sample resonator changes, creates a mismatch, and diverts power to port 4.

As impedance-matching devices, tees enable one to introduce reactive elements in series or parallel to the load because ports 3 and 4 are decoupled. A sliding short is situated in both the E- and H-plane at a common junction, and operates on the same principle as the double-stub tuner. This tuning device is more compact than the double-stub (i.e. series) tuner and may be used at frequencies up to 300 GHz. In this case, port 1 is the input and supplies power to port 2 (*i.e.* the load to be matched) only if the ports 3 and 4 are mismatched, which is the case because they are series and parallel shunts. Since the desired match condition (*i.e.* source to load) requires that no power be reflected back from port 2 to port 1, ports 3 and 4 are matched from the standpoint of port 2.

3.3 Attenuation and Attenuators

The concept of VSWR is related to electrical impedance and defined by the transmission of a wave across a junction between two devices. As such, VSWR represents a ratio of the amplitudes of the transmitted and reflected waves. A device at the end of a transmission line therefore receives a fraction of the wave's power that is incident on the junction between the line and device, and the term 'insertion loss' is defined as the power loss associated with a device in a transmission line. Formally, insertion loss, like all comparative measures of power, is defined as the log of a ratio, that is, $10\log(P_1 / P_2)$ (in units of dB), where P_1 and P_2 are powers delivered to the load when the device is absent and present in the circuit, respectively.

Insertion loss depends on the impedance parameters of the circuit and the operating power level. Using an illustrative example described by Wind & Rappaport (1955), consider a high frequency circuit driven by a 100 V AC signal. If the respective load and the source impedance are 20 and 30 Ω, the power ($I \cdot V$) delivered to the load is 120 W.[19] Inserting a device whose resistance is 100 Ω alters this power to 13.3 W, and the insertion loss is therefore 9.5 dB. It follows from this example that altering the source power or any of the resistance values will affect the computed insertion loss. For example, if the source resistance is 70 Ω instead of 20 Ω, then the insertion loss becomes 6.0 dB.

Any transmission line may be written as a repeating array of impedance that consists of resistive, capacitive, and inductive elements. At low frequencies we speak of an 'IR-drop' that is associated with losses in the power delivered by a source to a given load and defined by the resistance of the load; the lost power is often radiated as heat. With a high frequency circuit losses are defined in terms of the loss of amplitude of the electromagnetic wave (*i.e.* its power); but, in contrast to low frequency circuits, the losses may be radiative as well as dissipative (heat).

Resistance varies with the composition of the conductor; wires of different composition, for example copper vs. nichrome, vary with respect to conductivity and therefore resistive losses. This is true at both high and low frequencies, but a significant difference between conductors carrying low vs. high frequency waves concerns the depth of the penetrating current, the so-called 'skin depth'. At high frequency the skin depth is very thin and conduction occurs effectively within a thin layer at the conductor's surface

[19] Power delivered to the load $I \cdot V$ is written $I^2 \cdot R_L$, where $I = \{V / R_S + R_L\}$

(which is why the outer surface of a hollow conductor - waveguide - is at ground potential). Practical (coaxial) attenuators are therefore fabricated by depositing a metallic film onto a dielectric substrate and using this plated dielectric as an inner conductor. The thickness (and metal type) can be varied so that the wave penetrates into the lossy dielectric in a controlled manner and thereby attenuate the wave; the device is sometimes called a 'pad'. The waveguide attenuator likewise employs a film deposited onto a lossy substrate, but this 'resistive card' is mobile within the guide walls and may be moved manually so that the conducting layer intercepts the E-field lines to a greater or lesser extent.

Besides resistive cards, wire grates may be used to absorb or reflect incident microwaves, and such devices provide a precise means of controlling the attenuation in waveguides. A wire brought into an orientation that is parallel to the electric field lines will introduce large losses in the propagating wave by absorbing or reflecting specific modes in waveguide (Bronwell & Beam, 1947; p394). A device that consists of several independently rotating waveguide sections, each containing one transverse wire, will behave as a variable attenuator. A complete design would include rectangular-circular waveguide transitions and perhaps a grate that converts modes; precision rotary attenuators (*e.g.* the Hewlett-Packard HP382 series) operate in this manner.

Attenuators can also be active devices. A familiar example is the diode limiter, which is a diode inserted in the transmission line in such a way that a voltage dependent shunt is introduced. If the incident voltage (power) exceeds the breakdown voltage of the diode, the circuit is shunted and wave transmission terminated. The design is analogous to clipper circuits encountered at low frequency, and practical devices are easily realized by mounting diodes in the E-plane of rectangular guide.

The diode switch is an active device in the sense that an external DC bias controls the properties of the diode and its behavior. Fundamental to the *p-i-n* diode is the fact that, under certain conditions of forward and reverse bias, the diode will respond as a low impedance resistor or high impedance capacitor, respectively. This behavior also applies to GaAs FETs, although the corresponding values for C and R are higher. Like the limiter, switched diodes (acting as attenuators or on-off gates) are shunt devices and positioned parallel to the E-field; the DC bias and driver voltage is supplied to the diode via a connectorized contact.

3.4 Isolators

For a device having more than one port, VSWR can be expressed for both the forward and reverse direction. An isolator is a two-port device that is used to protect microwave power devices (*e.g.* oscillators or amplifiers) because it transmits radiation unidirectionally. The most common isolators are either short sections of loaded waveguide that are surrounded by a small permanent magnet or a microstrip equivalent. The waveguide or microstrip 'loads' are ferrites of specific geometric shape that are situated in a uniform magnetic field (of the permanent magnet). In all cases one exploits the time-dependent behavior of the microwave magnetic field and its interaction with the precessing spins of the ferromagnet in order to induce directionality of the propagating wave.

The basic ferrite isolator features a small U-shaped magnet into which a section of rectangular guide is inserted. In the TE_{01} mode of operation, the electric field vectors are directed perpendicular to the broad wall and the field intensity is at the guide center (see Poole, 1983, Chapter 2). A field displacement isolator leaves the electric field maximum near the center is one direction of propagation, but shifts it for a wave propagating in the opposite direction; by shifting it into a lossy material, one effectively cuts off the reverse wave. The ferrite material and its precessing spins afford the directionality because the 'sense' of the precessing spins with respect to the propagation direction of the forward and reverse wave differs.

The interaction between the precessing spins of a ferrite and incident electromagnetic waves can be used to alter the distribution of power among branches in multi-port junctions. For example, the power of a traveling wave is distributed evenly between the two branches of a Y-junction, but this symmetric power distribution can be broken by placing a ferrite at the junction (Chait & Curry, 1959). Using any of the three ports (*i.e.* 1, 2, or 3) as an input, the propagating wave encounters the ferrite at the junction and is preferentially diverted in one direction because of the relative 'sense' of the spins and the field vectors. A wave input at port 1 might be deflected preferentially to port 2; a wave input at port 2 would then be deflected to port 3; and a wave input at port 3 deflected to port 1. In other words, the three-port device is a circulator. If the port from which the ferrite deflects the wave is terminated with a non-reflecting load (*e.g.* Port 3 if Port 1 is input, as described above), the device behaves as an isolator.

3.5 Resonators

3.5.1 Overview

There are two ways of conceptualizing sample resonators for EMR when choosing a design for a specific application. For the most part, one can regard the sample resonator as a bandpass filter (Matthaei, Young, & Malherbe, 1979), but one can find equally useful designs among the antenna literature (Aharoni, 1946; Kraus, 1950; Silver, 1949; Watson, 1947). The principal concern, however, when adopting filter or resonator structures for EMR sample resonators is that one must exercise care for ensuring that the magnetic field lines of the resonator mode is perpendicular to the DC field (*i.e.* $H_0 \perp H_1$ because the resonance signal intensity goes as the sine of the angle between H_1 and H_0 (Bleaney & Stevens, 1953).

Like all forms of spectroscopy, the desired EMR sample resonator design maximizes signal-to-noise and sensitivity. The nature of the experiment dictates that the sample interact with an oscillatory high frequency magnetic field H_1, and therefore one desires that the resonator structure maximally concentrates H_1 within the sample volume. For example, in a rectangular cavity resonating in the TE_{102} mode, the H_1 field line runs in a double loop with a concentration at the center, which is well suited to an axial sample in a tube. The so-called filling factor is a measure of the H_1 concentration within the sample volume and assumes a numerical value between zero and one (see Poole, 1983, p156).

The other important sample resonator design factor is the quality, or *Q*, factor. In general, one wants a high *Q* for highest sensitivity, and this is preferred for continuous wave experiments, but with high *Q* comes a propensity towards 'ringing',[20] which leads to long deadtime in pulsed applications or experiments that record a transient response. A Mims transmission line resonator for pulsed EMR typically features a *Q* of approximately 100, whereas a cylindrical TE_{011} resonator for ENDOR might have a *Q* of 5000. Poole (1983, p124ff) outlines much of the relevant theory of *Q*, but, in general, *Q* is affected by surface finish and material of the resonator, the match of the resonator dimensions to the wave length, and the coupling arrangement to the transmission line.

3.5.2 Cavity Resonators

In Section 3.2 it was stated that an inductive iris can be fabricated by inserting thin strips through the wall of rectangular guide. A common form of filter can be designed by inserting a second inductor pair in series along the transmission line, and this filter can be 'tuned' to a specific frequency by adjusting the distance between these two strips so that $\ell = \lambda / 2$. One can introduce capacitance by closing off the aperture from the E-plane and forming an iris. This type of cavity/filter operates as a selective resonator because its geometry dictates what TE mode it will support. In other words, a standing wave cannot form within the confines delineated by the irides unless the field lines 'fit' (boundary conditions for the electric and magnetic fields met). Cavity operating modes refer to the number of half-cycles that span the enclosure.

[20] A time-dependent, damped, oscillation of the standing wave amplitude following a rapid change in the wave amplitude.

A cavity resonator is a section of hollow waveguide that is closed off so that a standing wave is set up and confined to the enclosure. A simple model is a section of shorted waveguide section, which acts as a perfect reflector (see Section 3.1) that is delineated by a thin diaphragm. An aperture in the thin diaphragm permits entry of the traveling wave, and the wave reflected at the shorted end sets up a standing wave. If the diaphragm is located at a nodal point in the standing wave, the structure traps the energy and is said to resonate.

3.5.3 Stripline & Loop-Gap Resonators

Like the hollow waveguide, a pair of discontinuities in a microstrip transmission line forms a filter (Matthaei, Young & Malherbe, 1979), and again the dimension of the partitioned section is $\lambda / 2$. With rectangular strips the H_1 field extends to either side and is a maximum along the short symmetry axis (*cf.* Edwards, 1981). The sample volume therefore has to be situated close to the strip and near the center, which makes these devices ideally suited for EMR studies of dielectric films (Wallace & Silsbee, 1991). Another prospective application is spectroelectrochemistry; a $\lambda / 2$ strip can be simultaneously operated as an electrode for standard electrosynthesis procedures because any small diameter conductor used to hook up the strip to the DC waveform generator will not support microwave transmission.

The so-called Mims (1965) cavity for pulsed EMR epitomizes the adaptation of stripline filter design for spectroscopy. A pair of waveguide sections run side-by-side and are terminated by a short. A rectangular $\lambda / 2$ strip is situated approximately $\lambda / 2$ away from the waveguide terminus and electrically couples to the standing waves in each of the parallel sections. The strip is coupled and matched to the waveguide sections only to the extent afforded by tapers in the waveguide and small indentations to the either side of the strip; there is no iris or coupling structure that would maximize Q. The sample sits in wells located to either side of the strip at the centerline.

The rectangular strip or linear conductor is a low inductance lumped parameter circuit element (see Terman, 1934; Young, 1977; Abrie, 1985) that will resonate when its length equals $2n\lambda / 4$ (open) or $(2n + 1)\, \lambda / 4$ (shunted), and the Q is commensurate with the strip's width. The H-field of the resonant TE mode resembles a half-wave along the longitudinal axis, and therefore the sample should be small and lie close to the strip's surface in order to maintain H_1 field homogeneity. One can, however, fold the rectangular strip around the short axis without drastically altering the resonant frequency, and this will improve the field homogeneity in the region that is now enclosed by the strip's inner surface. The correlation between the resonant frequency, as determined by $\lambda / 2$, and the actual length of the strip depends on the fringing capacitance and, as the radius of the folded strip decreases, the proximity of the two edges introduces a capacitance that will affect the equivalent *RLC* circuit and the resonant frequency of the device.

A folded rectangular strip is the most basic form of the so-called loop-gap resonator, which is a generic term that is applied to lumped-parameter sample resonators (*cf.* Fronciz & Hyde, 1982). Loop gap resonators combine any number of loops (inductors) with gaps (capacitors) as part of an *RLC* resonant network, and their principle advantage for EMR applications comes from their compact size. Rather than a large rectangular cavity and its disperse field lines, the loop gap offers a small structure whose geometry (*i.e.* the circular 'loop') lends itself well to sample tubes and concentrates the H-field precisely at the sample; there is hence an improvement in the filling factor of the loaded resonator.

3.5.4 Helical Resonators

Helical resonators (Webb, 1962; Poole, 1967, p562ff, 752ff) are attractive for EMR applications that require broadband operation (*e.g.* swept-frequency zero field EMR, Bramley & Strach, 1983) or low Q (*e.g.* pulsed EMR), but tend not to be widely used because there is a perception that they are not competitive with conventional resonant cavities. There is also the matter of radiating mode as it relates to the helices' diameter; helices large in comparison to λ (in other words, those suited to accommodate standard EMR sample tubes) radiate in the wrong mode for EMR applications.

One records EMR spectra as a change in the power level (*i.e.* the impedance) at the receiver as a material in a resonant circuit undergoes a change in permittivity. With resonant cavities the change in power is proportional to Q, that is $dP = \frac{1}{2}\eta\cdot\chi''\cdot P\cdot Q$, and therefore high-$Q$ resonators are preferred. By comparison, however, the analogous equation (Siegman, 1960) for the change in power in helices is $dP = 2\pi\cdot\chi''\cdot P\cdot S\cdot N$. In both equations P is the microwave power and χ'' is the complex susceptibility. The parameter η is the filling factor, S is defined as the slowing factor, and N is the number of carrier frequency wavelengths spanned by the helix. The slowing factor depends on the bandwidth of the helix, but may be nearly 1000 for a device of a few percent bandwidth, and the N-factor is approximately ⅓ at 10 GHz (Siegman, 1959; 1960). It follows that the helix is competitive with many cavities and lumped parameter devices such as the loop-gap resonator.

Siegman's analysis applies to small diameter helices, which radiate in the normal, as opposed to the axial, mode (see Kraus, 1950). The normal radiating mode is achieved with helices wound so that the diameter is approximately 0.1 λ and an interwinding spacing of 0.05 λ. The current distribution in these helices tends to be sinusoidal if the circumference c is less than $2\lambda/3$, but this is rendered uniform as $c \sim \lambda$ (Marsh, 1951). One must also use care in selecting the conductor material since its diameter affects normal mode radiation (Tice & Kraus, 1949). A large helix that radiates in the axial mode is typically 0.32 λ in diameter and has an interwinding spacing of 0.25 λ. In the axial mode, the diameter of the conductor has little effect on the helix characteristics.

3.6 Remarks on High Frequency Transmission Lines

The most commonly used transmission line is the rectangular pipe whose cross-sectional dimensions conform to the ratio 2:1. The actual cross-sectional dimensions are commensurate to the wavelength of the guided wave and therefore rectangular waveguides assume very small cross-sections as the carrier frequency increases. Such guides become miniscule at the en vogue spectrometer operating frequencies of 100 GHz and up. Oversized waveguides may be used as a transmission line for very high frequencies, but they suffer the disadvantage of propagating more than one mode. As an alternative to the oversize guides, one can use a waveguide cross-sectional ratio other than 2:1 at high frequencies; these waveguides are still manageable with standard machining techniques and retain mode selectivity.

So-called tall guides are rectangular cross-section guides whose width is elongated while keeping the height invariant. Ratios of 10:1 and 20:1 are commonly used formats whose properties vary. The number of modes that are supported in a guide increases as the ratio decreases, but the ease of fabrication is also greater (Griemsmann, 1963). One variation of the tall guide is the grooved guide, which concentrates the electric field lines in the cross-sectional center of the guide (Griemsmann, 1963).

H-guide is another variation of the tall or parallel plate wave guide in which a dielectric slab occupies the center portion of the guide, resembling a light pipe. The characteristics of these transmission lines (Tischer, 1959; Morse & Beam, 1956) mirror that of the tall guide, but TM modes cannot be supported. Attenuation is low, but the fabrication of components is difficult.

4. DETECTION METHODS AND RECEIVERS

4.1 Thermal EMF Methods

Microwave power measurements may be made by converting the heat generated by incident microwave power into a thermal emf. Terms such as bolometer, barretter, and thermistor are found in the literature and describe detection procedures in which microwave power is measured via the change in resistance in a bias circuit as heat alters the conductivity of some material. Spectroscopic applications of thermal detection methods are found in the older microwave spectroscopy literature (*cf.* Gordy *et al.*, 1953; Harvey, 1963), but were supplanted by diode detection methods. Bolometer methods are well suited to high-frequency (*i.e.* millimeter and sub-millimeter) EMR, however, and will therefore be briefly described here.

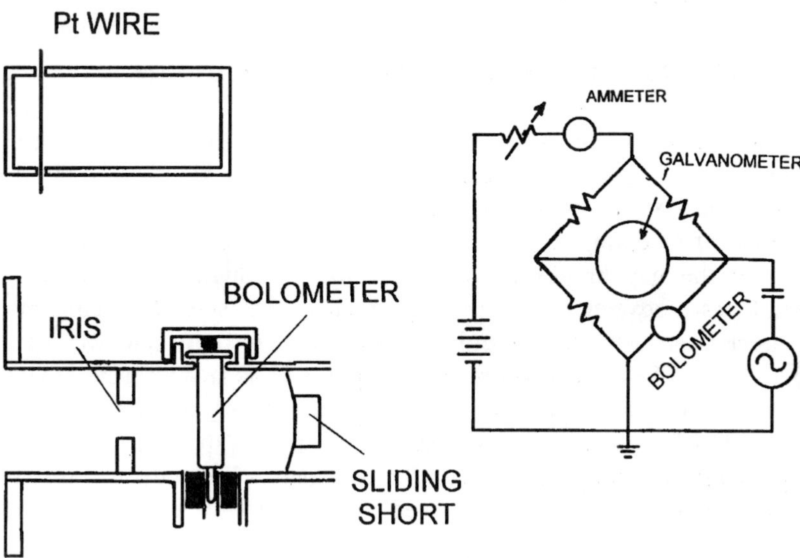

Figure 18. The bridge method of power detection by using bolometers. The bolometer is positioned in a waveguide (or coaxial) section where it absorbs power from the incident electromagnetic field. The power absorption (heating) alters the resistance of the device, which is then detected on a conventional impedance bridge. This variant of the Wheatstone Bridge uses a low frequency AC signal to compensate for ambient temperature fluctuations; other compensation techniques place matched temperature-sensitive resistors (those located immediately to the left of the bolometer in the figure) on the same mount as the bolometer, but not in the microwave field.

The simple bolometer is a thin platinum wire that terminates a transmission line and is mounted so that it maximally absorbs incident microwave radiation. A typical waveguide mount is shown in Figure 18, and the tuning plunger is used to optimize the microwave standing wave for maximal absorption by the wire. Other bolometer configurations are metal films deposited on a dielectric or grids, the latter of which are advantageous in the sense that the power may be partially transmitted. A thermistor is conceptually similar, but the device consists of two non-contacting wires that are bound within a bead of a conducting material such as a metal oxide matrix. Although bolometers work well at all frequencies, thermistors are contraindicated above 15 GHz.

The power measurement is made via bridge methods (Figure 18). The device is biased by a DC source and variations in its R are measured via the bridge. Typical specifications of the older platinum wire devices were a minimal power detection of 0.01 μW and a response of 50 μV/μW. Modern bolometers that are manufactured for use as detectors in radioastronomy are semiconductor composites (Ge-Si) and cryogen cooled, yielding much higher responsivities. Bolometer mixers and pulse detection are also feasible with current technology.

4.2 Crystal Detectors and Autodyne Detectors

Early detectors of electromagnetic waves consisted of a resonator (*i.e.* a loop antenna) and a spark gap, and reception of a signal was based on the observation of the spark or the deflection of a galvanometer. The spark gap method of reception requires that the incoming signal be of high amplitude so that the gap breaks down, and it was gradually replaced by contact interfaces between oxide-forming metals, the so-called 'coherers'. The coherer was fashioned as either a point contact between two metals, for example, a whisker contacting a metal plate, or as an aggregate of fine metal particles. The metal had to form an oxide layer at its surface, and the principle of operation was similar to the spark gap, that is, the ordinarily high resistance dry joint between the two metals became a low resistance to an rf signal; both receivers allowed for reception of pulse-code modulated carriers.

The principles of coherer operation, namely, the breakdown of the metal oxide interface, are suggestive of the point contact rectifying diode, which has been a standard detector for high frequency electromagnetic waves. An early version was fabricated at the interface between a phosphor bronze whisker and iron pyrite (Cleeton & Williams, 1934), and later devices adopted semiconductor materials such as boron-doped silicon (Scaff & Ohl, 1947; see also Levine, 1964). Modern crystal detectors rely on a metal layer that is evaporated directly onto the surface of the semiconductor, the latter of which is itself a more sophisticated device that is fabricated as an *n-i-p-i-n* GaAs rectifier.

The whisker contact to a semiconductor forms a Schottky barrier (Sze, 1981), and the mode of operation follows from the current-voltage relationship. Figure 19 depicts a typical *I-V* curve of a Schottky barrier diode. If the barrier is biased by an applied voltage, a current flows across the interface, and the magnitude of the current flux is specified by the conduction properties of the semiconductor. An oscillatory waveform is rectified, as indicated, because of the asymmetry of the *I-V* curve, and this rectified current signal can therefore be used as a video detector (i.e. a measure of the current amplitude) of microwave and rf signals. The response is linear insofar as the *I-V* curve is linear, and therefore the device is suitable for small signal detection only.

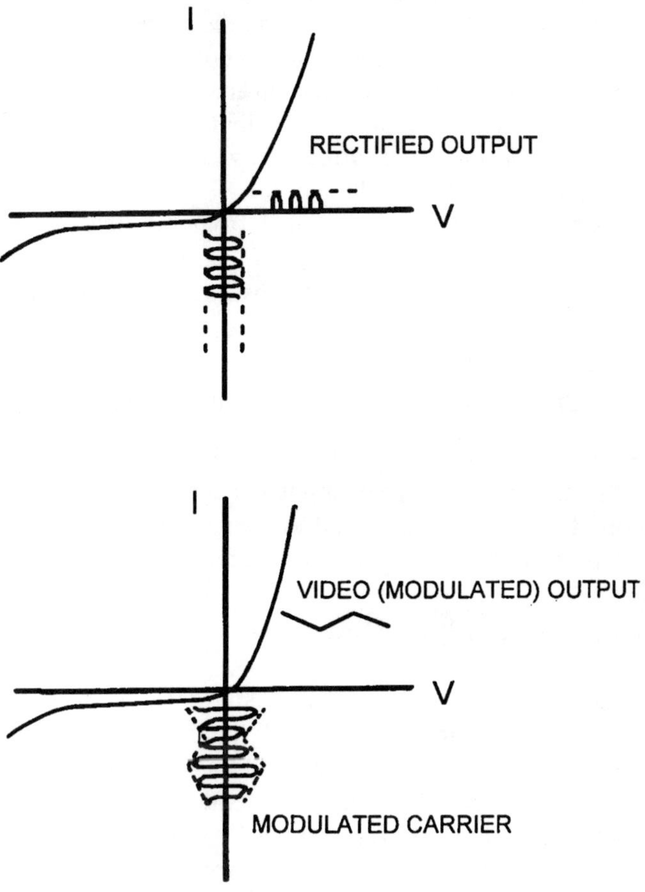

Figure 19. The current-voltage curves of a diode. Rectification of an AC signal is achieved due to the nonlinear profile of the curve (top), and if the AC signal is modulated (e.g. a sawtooth, as indicated in the bottom portion), this modulation, or 'information' is transferred to the diode output.

The barrier diode is essentially a building block of more complicated detectors. They can, for example, be combined in a bridge configuration to form a mixer (Section 4.3), but they can be used as stand-alone units in either waveguide or coaxial mounts. As a waveguide-based device, the diode is mounted in much the same way as has been described for limiters and attenuators (Section 3.3), the principal difference being that the diode terminus is outfitted with an output for metering the current amplitude (Figure 19). Most modern detectors, however, are packaged as coaxial devices, featuring an SMA connector for the high frequency input and a

BNC connector for the video output. Typical products are the Hewlett-Packard GaAs planar doped diode detectors (for high speed response) and traditional Schottky barrier detectors.[21] The crystal is therefore used as a video detector of rf and microwave signals. The response is linear in so far as the *I-V* curve is linear, which limits their application to small signal detection, although one tends to not operate in the linear region of the *I-V* curve because at this region the diode contributes the most noise. The absolute amplitude of the output is affected by the impedance of the rf source and the load characteristics of the whisker.

The detector output voltage is determined from the characteristic *I-V* curve, which is non-linear. The equation $V_{out} = IR$ is modified by writing I as a power series, $I = a_0 + a_1V + a_2V^2 \ldots$, and substituting the time-dependent expression for the rf voltage, $V(t) = V_0 \cos\omega t$. If one rewrites the expansion of I using the expression for $V(t)$ in place of V, and then uses the substitution $\cos^2\omega t = \frac{1}{2} + \frac{1}{2}\cos 2\omega t$, then the first two terms of the powers series expansion become $I = \frac{1}{2} a_2V_0^2 + a_1V_0 \cos\omega t$, and the DC output is proportional to the square of the peak input voltage. This is the so-called 'square-law' property of the detector, and one wishes to operate at voltage levels such that this term dominates the diode signal output.[22]

4.3 Homo- and Heterodyne Detection

Figure 20 illustrates the placement of a crystal detector in a section of waveguide where a single carrier signal will impinge upon it. In many detection circuits or receivers, however, one will find a configuration similar to those illustrated in the figure, in which a second signal is imposed upon the diode by a local oscillator (LO), for example, in radar and some similarly configured pulsed EMR spectrometers. In other words, there are two carrier signals $f_{c,1}$ and $f_{c,2}$. In this case the rf input voltages are $V_1(t) = V_1 \cos\omega_1 t$ and $V_2(t) = V_2 \cos\omega_2 t$, and the net voltage $V(t) = V_1(t) + V_2(t)$. Substituting into the series expansion for I yields two DC terms $a_2/2\,[V_1^2 + V_2^2]$ plus rf

[21] Part numbers of both diode types are similar. HP8474 series are GaAs planar diodes; HP8373B & C are Schottkys. Wiltron likewise markets high frequency coaxial detectors as a 75 series.

[22] See also Ishii, 1990, Chapter 7 for a more rigorous analysis of the square-law response.

signals $a_2/2\,V_1^2\cdot\cos 2\omega_1 t$, $a_2/2\,V_2^2\cdot\cos 2\omega_2 t$, $a_2 V_1\cdot V_2\cdot\cos(\omega_1 t + \omega_2 t)$, and

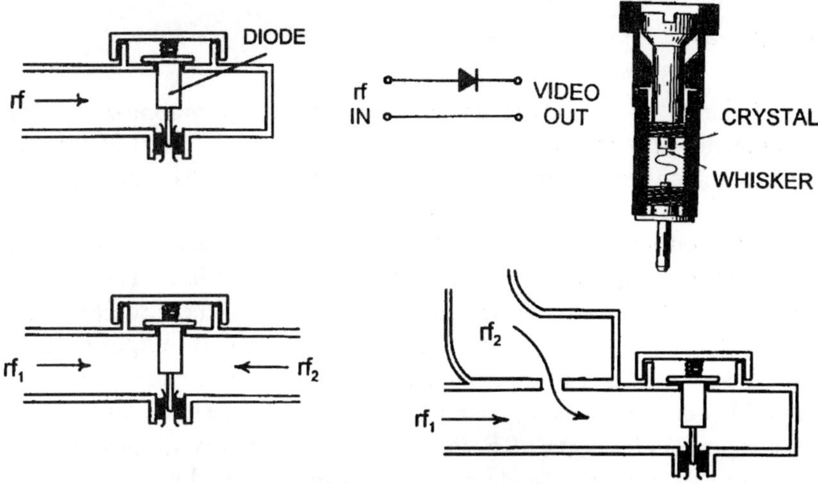

Figure 20. Diode detection schemes for auto-, homo-, and heterodyne detection. As with the bolometer, the diode is positioned in a transmission line so that the E-field lines intercept the device. A crystal diode detector configuration is depicted in the top left corner, whereas homo- and heterodyne arrangement are depicted on the bottom left and right diagrams. The equivalent circuit of the diode detector and 'anatomy' of a diode cartridge mount are also illustrated.

$a_2 V_1\cdot V_2\cdot\cos(\omega_1 t - \omega_2 t)$. For homodyne detection purposes one ensures that $\omega_1 = \omega_2$, and the DC signal therefore becomes $a_2\cdot(V_1/\surd 2 + V_2/\surd 2)^2$. One of the signals is a reference signal and does not vary in response to some change in impedance or external modulation, whereas the other signal is the information carrier. The effect of the second reference signal in this type of detection is that it supplies a bias to the diode and renders the current response of the diode proportional to $V_s^2 + V_s$, as opposed to simply V_s^2 (V_s denotes the signal voltage), and therefore there is a gain in sensitivity.

Heterodyne detection refers to the situation in which the two frequencies ω_1 and ω_2 are not identical, and the information is carried on an intermediate frequency $|\omega_1 - \omega_2|$. The rationale behind the superheterodyne approach to detection stems from the performance of amplifiers at DC, as opposed to higher frequencies. There tends to be more noise produced in the amplification of a DC signal than a moderate frequency AC signal, and there was at one time an impetus to use superheterodyne receivers in order to maximize sensitivity (see Feher, 1956). But like the once eschewed solid-state microwave sources, the noise figures of solid-state amplifiers has greatly improved with modern fabrication techniques, and homodyne detection is competitive with superheterodyne (and obviates the need to lock two sources to one another).

4.4 Mixers

The features of homodyne and heterodyne detection appear in a single semiconductor crystal because of its nonlinear response (*i.e.* I-V curve), but in most high-sensitivity applications two or more crystal diodes are used in tandem to improve performance (Pound, 1948;Tsui, 1983; Ishii, 1990). A so-called balanced mixer (*cf.* Pound, 1948, Ch. 6) combines two crystal diode mixers that are simultaneously driven by the local oscillator (*i.e.* reference) and received signals. The diodes may be inserted into the circuit by using a magic tee (or microstrip hybrid junction) so that two IF output signals are obtained. For example, by placing the two crystals on separate branches of the hybrid (*i.e.* Ports 1 & 2), one can ensure that the signal from the local oscillator port (LO – Port 3; RF- Port 4) arrives at the two diodes 180° out of phase, and therefore any random noise that modulates the source oscillator signal will be nulled as the output of the IF stages are combined. The symmetry of the device, namely the hybrid, is the determining factor in noise reduction by this detection device because the symmetry determines the extent to which each port (diode) is isolated from the others (*cf.* Section 3.2.4).

One may also build on this basic balanced mixer by adding diodes pairwise in a symmetric manner and thereby further reduce spurious noise because in so doing one introduces more isolated ports in the detector circuit. For example, one may combine two balanced (each containing two crystal diodes) mixers in parallel so that they are out of phase. Rendered as a pairing of two magic tees, the signal of the local oscillator is split between two branches by using either a simple tee or a magic tee. At the end of each branch is a balanced mixer of the type described in the preceding paragraph; the received signal must likewise be split and passed to the designated RF port (number 4) of the two balanced mixers. Figures 6-20, 6-21, and 6-22 in the book by Pound (1948) illustrate the construction details of the double balanced mixer comprised of two or more hybrid tees. The advantage of using such a configuration is that it increases the isolation between LO and RF ports and suppresses harmonics, and in so doing reduces the conversion loss from the rf to IF signal. The amount of isolation, however, depends upon the symmetry of the device and how well fabrication methods can ensure symmetry for the frequency range specified.

5. CONCLUDING REMARKS

Principles of select microwave devices have been described with the stated purpose of providing the reader with background information relevant to EMR spectrometer design and operation. Few spectrometers are now constructed in the laboratory because the quality of commercial instruments is high, and these instruments are sufficiently versatile that experimental needs may often be met. In concluding this chapter, however, I would like to address multi-frequency EMR experimentation as a technique for which the home-built spectrometer remains a necessity.

It is long recognized that the interpretation of EMR spectra is greatly facilitated by performing experiments at two or more spectrometer frequency/field combinations, and a number of laboratories have addressed this by operating the spectrometer at select operating bands (*e.g.* S-, C-, X-, K-, and Q-bands). The motivation for conducting experiments using this strategy is that one may exploit the field dependence of the spin Hamiltonian in order to better parameterize each of its contributing terms (*cf.* Hyde & Froncisz, 1982; Singel, 1989; Hoffman *et al.*, 1993). But in the majority of these cases the spectrometer operates at a fixed frequency within this band due to the fact that the microwave source is a device (*e.g.* reflex klystron or Gunn oscillator) with a limited tuning range. Only in 'classical' microwave (rotational) spectroscopy does one find octave and multi-octave microwave sources being used on a spectrometer (*cf.* Townes & Schawlow, 1955; Gordy *et al.*, 1953).

The reflex klystron has been an ideal source of microwave power because it supplies moderate powers and radiation of high spectral purity; the same reason it has been popular as a local oscillator in radar receivers applies to its choice as a microwave power source for EMR spectroscopy. As stated in Section 2, advances in solid-state device fabrication have yielded solid-state microwave sources, which in turn has led to many changes in the field of test & measurement engineering. Microwave test instruments (*e.g.* network and spectrum analyzers) are now typically solid-state instruments whose sources feature high spectral purity over multiple octaves; even higher spectral purity is available for specialized synthesized sources intended as local oscillators for receiver systems. With these changes in test & measurement engineering as the driving force, the EMR community now has the opportunity for performing true multi-frequency spectroscopy.

Modern synthesizers provide a very stable frequency source at adequate powers[23] in a compact package, and there are various options one can take in adopting solid-state devices for EMR spectrometers with a very favorable

[23] The power needs of a receiver LO are generally less than +10 dBm, and a 1 Watt reflex klystron provides more than ample power as a local oscillator.

recoup of reflex klystron based designs. Despite a lingering perception that solid-state devices are inferior to klystrons, the spectral purity of solid-state devices is very high, although specifications vary with the operating bandwidth. For example, Communications Technologies Inc. (CTI) manufactures narrow band synthesized sources (CTI 1056A series) whose output frequency may be stepped in 50 kHz increments over a 400-500 MHz tuning range; the phase noise of these units is –115 dBc/Hz @ 10 kHz.[24] By comparison, a CTI multi-octave source features a phase noise specification of –80 dBc/Hz @ 10 kHz. Racal-Dana likewise offers a narrow band synthesized source that provides a –115 dBc/Hz @ 1 kHz phase noise and +15 dBm power output.

The power output of the synthesized sources is admittedly lower than a reflex klystron, but this can be remedied by modifying the bridge design. In prototype studies of spectrometer design and operation, a Gigatronics 903 synthesized source that was tunable from 2–20 GHz was substituted for reflex klystrons on a pulsed EMR spectrometer.[25] Since the synthesizer is fully protected, its output was passed directly to the LO port of the receiver, whose LO power input was then optimized at the source output attenuator. A phase shifter and directional coupler (6 dB sample) were the only devices between the source's output and the LO input ports. The portion of the microwave signal that is sampled at the directional coupler is then amplified to spectroscopically useful power levels by a GaAs FET broadband amplifier (HP 83020A), which has a maximum power output of +30 dBm and noise figure of 10 dB. Single-octave low noise (< 6 dB) GaAs FETs can be substituted for the system amplifier if the 2–20 GHz range is not required.

In principle, the spectrometer design configuration that is outlined in the preceding paragraph could be adopted for cw-EMR, giving one a continuously tunable 2–20 GHz spectrometer. The power output of the GaAs FET amplifier is adequate to saturate most spin systems, and is certainly comparable to the 200 mW rated output of many commercial spectrometers. The spectral purity of the synthesized source was not detrimental to the performance of the receiver,[26] as compared to the reflex klystron stabilized to better than one part in 10^8. A direct measure of the phase noise of the amplified signal was not made, however, nor was a measure of narrow

[24] The specifications cited are obtained from the manufacturer's 1997 catalog.

[25] This spectrometer prototype was used to record the echo modulation data reported by Bender & Peisach (1998); the layout appears in Figure 17 of chapter 5, but a system amplifier was after the directional coupler and before the pulse-forming circuit. The Gigatronics 903 synthesizer has no provision for pulse modulation, and so the unit, configured as described in this passage, replaced the normally used Varian X-13 and X-12 reflex klystrons. Other synthesizers are outfitted for pulse modulation, but this introduces problems in designing the homodyne receiver system.

[26] Phase noise at the LO input of a receiver can, in principle, lead to a decline in performance.

linewidth (~10 mG) EMR spectra undertaken; the synthesized source and amplifier combination did yield cw-EMR spectra of strong and weak pitch standard samples that were exactly the same as those obtained with a conventional bridge.

Interestingly, the switch to solid-state signal sources has been accompanied by a widespread development of modular computer programmed test instruments, since the new sources may be packaged as plug-in modules, and this too has important implications for EMR spectroscopy. EMR spectrometers are fundamentally universal in design, and experiments for the most part simply differ with respect to the stimuli applied to the spin system and the subsequent data acquisition. In this sense, therefore, different EMR experiments are dictated by the computer software and, to a lesser extent, the hardware. This parallels the move among electronics engineers towards test instruments wherein the computer controls the measurement process via menus that drive modular instruments via a standardized communication bus. The current standards are VMIbus and VXIbus, which are the modern computer-driven answer to NIM and CAMAC modules.

Commercial VXIbus modules are currently available that may be used as the core of an EMR spectrometer. These include narrow- and broad-band sythesized sources for radio- and microwave frequencies, oscilloscopes, counters and meters, logic programmers and delay lines, and receivers. In principle, it is therefore possible to build a VXIbus EMR spectrometer with very little effort devoted to custom circuits. The major 'missing' device is the field control unit, but Bruker's BH-15 unit is a stand-alone device and may be linked to the VXIbus mainframe via IEEE-488 ports.

In concluding, therefore, we return to the double-edged nature of spectroscopy's link to the high frequency electronics marketplace. Although we have been forced to abandon traditional components and designs for EMR spectrometers, the advances in materials science, test and measurement, and quantum electronics provide us with the means to perform new types of experiments to get information about the spin system that was previously unobtainable or ambiguous.

6. ACKNOWLEDGMENTS

A Cherry B. Emerson Fellowship at the Emory University Emerson Center for Computational Chemistry (1995) is gratefully acknowledged, during which time this manuscript was originally researched and drafted. Thanks also to the Cooper Union for the Advancement of Science and Art (New York) for access to its computer and microwave engineering laboratories during the 1997-98 academic year, and thanks to Theodore Camenisch of the

National Biomedical ESR Center in Milwaukee for much useful advice on engineering over the years.

Symbols & Abbreviations

AFC	Automatic Frequency Control
C	capacitance
ENDOR	Electron Nuclear Double Resonance
EMR	Electron Magnetic Resonance
E	electric Field
f	frequency, in units of Hz
G	conductance
H	magnetic field
I	current
J	Bessel Function
L	inductance
P	power
R	resistance
VSWR	Voltage Standing Wave Ratio
X	susceptance
YIG	Yttrium Indium Garnet
Z	impedance
ω	frequency, in units of radians·sec^{-1}

7. REFERENCES

Abragam, A., 1961, *The Principles of Nuclear Magnetism*, Oxford University Press, Oxford.

Abrie, P.C.D, 1951, *The Design of Impedance-Matching Networks for Radio-Frequency and Microwave Amplifiers*, Artech House, Norwood.

Adler, R., 1946, `A Study of Locking Phenomena in Oscillators', *Proc. IRE*, **34**: 351 A.,

Aharoni, J., 1946, *Antennae*, Oxford University Press, Oxford.

Alger, R.S., 1968, *Electron Paramagnetic Resonance: Techniques and Applications*, Interscience, New York.

Angelo, E.J., 1958, *Electronic Circuits: A Unified Treatment of Vacuum Tubes and Transistors*, McGraw-Hill, New York.

Auld, B.A., 1952, *Applications of Group Theory in the Study of Symmetrical Wave Guide Junctions*, Microwave Laboratory Report 157, Stanford University, Palo Alto.

von Aulock, W.H., 1965, Handbook of Microwave Ferrite Materials, Academic Press, New York.

Baden Fuller, A.J., 1987, Ferrites at Microwave Frequencies, Peter Peregrinus, London.

Bailey, A.E., 1989, Microwave Measurements, 2nd ed., Peter Peregrinus, London.

Barber, R.E., 1971, `Short Term Frequency Stability of Precision Oscillators and Frequency Generators', *Bell System Tech. J.*, **50**: 881.

Barrow, W.L., and Meiher, W.W., 1940, `Natural Oscillations of Electrical Cavity Resonators', *Proc. IRE*, **28**: 184.

Beatty, R.W., 1973, *Applications of Waveguide and Circuit Theory to the Development of Accurate Microwave Measurements*, National Bureau of Standards Monograph 137, US Government Printing Office, Washington DC.
Beck, A.H.W., 1948, *Velocity Modulated Thermionic Tubes*, Cambridge University Press, Cambridge.
Beck, A.H.W., 1958, *Space Charge Waves and Slow Electromagnetic Waves*, Pergamon Press, New York.
Begovich, N.A., 1966, 'Frequency Scanning', In: *Microwave Scanning Antennas*, (R.C. Hansen, ed.), Volume 3: Array Systems, Academic Press, New York. p135.
Bell, D.A., 1985, *Noise in the Solid State*, Wiley, New York.
Bender, C.J., and Peisach, J., 1998, 'Electron Spin Echo Modulation Spectroscopic Study of the Type I Copper Center Associated with Stellacyanin from {\it Rhus vernicifera}', *J. Chem. Soc., Faraday Trans.*, **94**: 375.
Benham, W.E., 1957, *The Ultra-high Frequency Performance of Receiving Tubes*, McGraw-Hill, New York.
Berlin, H.M., 1978, *The Design of Phase-locked Loop Circuits with Experiments*, Howard Sams, Indianapolis.
Bevensee, R.M., 1964, *Electromagnetic Slow Wave Systems*, Wiley, New York.
Bhattacharyya, A.K., Shafai, L., and Garg, R., 1991, 'Microstrip Antenna: A Generalized Transmission Line', In: *Progress in Electromagnetic Research*, Vol. 4 (J.A. Kong, ed.), Elsevier, New York. p45.
Blanchard, A., 1976, *Phase Locked Loops*, Wiley, New York.
Bleaney, B., and Stevens, K.W.H., 1953, 'Paramagnetic Resonance', *Rep. Prog. Phys.*, **16**: 108.
Bosch, B.G., and Engelmann, R.W.H., 1975, *Gunn-Effect Electronics*, Wiley, New York.
Bradley, E.H., 1956, 'Design and Development of Strip Line Filters', *IRE Trans.*, **MTT-4**: 86.
Bramley, R., and Strach, S.J., 1983, 'Electron Paramagnetic Resonance at Zero Magnetic Field', *Chem. Rev.*, **83**: 49.
Bronwell, A.B., and Beam, R.E., 1947, *Theory and Application of Microwaves*, McGraw-Hill, New York.
Bruin, F., 1961, 'The Autodyne as Applied to Paramagnetic Resonance', In: *Advances in Electronics and Electron Physics*, Vol. 15 (L. Marton, ed.), Academic Press, New York. p327.
Carroll, J.E., 1970, *Hot Electron Microwave Generation*, Edward Arnold, London.
Casper, S., 1963, 'Bolometer Characteristics', *Microwaves*, **2**(12): 56.
Caulton, M., Hughes, J.J., and Sobol, H., 1966, 'Measurements on the Properties of Microstrip Transmission Lines for Microwave Integrated Circuits', *RCA Rev.*}, **27**: 377.
Chait, H.N., and Curry, T.R., 1959, 'Y-Circulators', *J. Appl. Phys.*, **30**: 15 **S**.

Chatterjee, R., 1988, *Advanced Microwave Engineering: Special Advanced Topics*, Ellis Horwood, Chichester.
Christen, P., Baude, A., and Günthard, Hs.H., 1972, 'Backward Diodes as Sensitive Detectors for Microwave Spectroscopy', *Rev. Sci. Instrum.*, **43**: 349.
Chronos Group, 1994, *Frequency Measurement and Control*, Chapman & Hall, London.
Clarricoats, P.J.B., 1961, *Microwave Ferrites*, Wiley, New York.
Cleeton, C.E., and Williams, N.H., 1934, 'Electromagnetic Waves of 1.1 cm Wavelength and the Absorption Spectrum of Ammonia', *Phys. Rev.*, **45**: 234.
Collin, R.E., 1966, *Foundations for Microwave Engineering*, McGraw-Hill, New York.
Cook, A.B., and Liff, A.A., 1968, *Frequency Modulation Receivers*, Prentice-Hall, Engelwood Cliffs.

Cook, J.S., Kaempfner, R., and Quate, C.F., 1956, 'Coupled Helices', *Bell System Tech. J.*, **35**: 127.

Crosby, M.G., 1937, `Frequency Modulation Noise Characteristics', *Proc. IRE*, **25**: 427.

Czoch, R., and Francik, A., 1989, *Instrumental Effects in Homodyne Electron Paramagnetic Resonance Spectrometers*, Ellis Harwood, Chichester.

Dammers, B.G., Haantjes, J., Otte, J., and van Suchtelen, V., 1950, *Applications of the Electron Valve in Radio Receivers and Amplifiers, Volume 1: RF and IF Amplification - Frequency Changing - Determining the Tracking Curve - Parasitic Effects and Distributions due to Curvature of Valve Characteristics - Detection*, N.V. Phillips, Eindoven.

Dammers, B.G., Haantjes, J., Otte, J., and van Suchtelen, V., 1950, *Application of the Electron Valve in Radio Receivers and Amplifiers, Volume 2: AF Amplification - The Output Stage - Power Supplies*, N.V. Phillips, Eindoven.

Davis, E.E., 1948, *Locking Phenomena in Microwave Oscillators*}, MIT Research Laboratory Electrical Technology Report No. 63, Cambridge.

Dawirs, H.N., 1962, `How to Design Impedance Matching Transformers', *Microwaves*, **1**(3): 22.

Durling, I.N., 1995, *Compact Sources of Ultrashort Pulses*, Cambridge University Press, Cambridge.

Erst, R.R., 1984, *Receiving Systems Design*, Artech House, Norwood.

Freymann, R., and Soutif, M., 1960, *La Spectroscopie Hertzienne Applique a la Chimie: Absorption Dipolaire, Rotation Moleculaire, Resonances Magnetique*, Dunod, Paris.

Gewartowski, J.W., and Watson, H.A., 1965, *Principles of Electron Tubes, Including Grid-Controlled Tubes, Microwave Tubes, and Gas Tubes*, van Nostrand, Princeton.

Gilmour, A.S., 1986, *Microwave Tubes*, Artech House, Norwood.

Ginzton, E.L., 1957, *Microwave Measurements*, McGraw-Hill, New York.

Gittens, J.F., 1965, *Power Traveling Wave Tubes*, Elsevier, New York.

Golding, E.W., and Widdis, F.C., 1963, *Electrical Measurements and Measuring Instruments*, 5th ed., Pitman, London.

Goubau, G., 1961, *Electromagnetic Waveguides and Cavities*, Pergamon Press, New York.

Gordy, W., Smith, W.V., and Trambarulo, R.F., 1953, *Microwave Spectroscopy*, Wiley, New York.

Griemsmann, J.W.E., 1963, `Oversized Waveguides', *Microwaves*, **2**(12): 20.

Grivet, P., 1965, *La Resonance Paramagnetique Nucleaire: Dipolaires et Quadrupolaires*, Centre Nationale Recherche Scientifique, Paris.

Grivet, P., 1976, *Physics of Transmission Lines at High Frequency*, Academic Press, London.

Groll, H., 1969, *Mikrowellennesstechnik*, Vieweg & Sohn, Braunschweig.

Groszkowski, J., 1964, *Frequency of Self-Oscillation*, MacMillan Co., New York.

Gunn, J.M., 1963, `Microwave Oscillations of Current in III-V Semiconductors', *Solid State Comm.*, **11**: 63.

Gunston, M.A.R., 1972, *Microwave Transmission-Line Impedance Data*, van Nostrand, London.

Gupta, M.S., 1971, `Noise in Avalanche Transit-Time Devices', *Proc. IEEE*, **59**: 1674.

Haantjes, J.C., 1960, *Classification of Electron Tubes*, MacMillan, New York.

Häggblom, H., 1959, `The Spectral Density of AM Noise in Reflex Klystrons', *Proc. IEE*, **106**B: 497.

Hague, B., 1971, *Alternating Current Bridge Methods*, 6th ed., Pitman, London.

Hakki, B.W., and Irwin, J.C., 1965, `cw-Microwave Oscillations in GaAs', *Proc. IEEE*, **53**: 80.

Harrison, A.E., 1947, *Klystron Tubes*, McGraw-Hill, New York.

Harvey, A.F., 1959, `Optical Techniques at Microwave Frequencies', *Proc. IEE*, **106**B: 141.
Harvey, A.F., 1963, *Microwave Engineering*, Academic Press, New York.
Helszajn, J., 1969, *Principles of Microwave Ferrite Engineering*, Wiley, London.
Helszajn, J., 1985, *YIG Resonators and Filters*, Wiley, Chichester.
Hilsum, C., 1962, `Transferred Electron Amplifiers and Oscillators', *Proc. IRE*, **50**: 185.
Horowitz, P., and Hill, W., 1989, *The Art of Electronics*, 2nd ed., Cambridge University Press, New York.
Howe, H., 1974, *Stripline Circuit Design*, Artech House, Norwood.
Howes, M.J., 1976, *Microwave Devices: Device-Circuit Interactions*, Wiley, New York.
Hund, A., 1933, *High Frequency Measurements*, McGraw-Hill, New York.
Huntoon, R.D., and Weiss, A., 1947, `Synchronization of Oscillators', *Proc. IRE*, **35**: 1415.
Hyde, J.S., and Froncisz, W., 1989, `Loop Gap Resonators', In: *Advanced EPR: Applications in Biology and Biochemistry*, (A.J. Hoff, ed.), Elsevier, Amsterdam. p277.
Ishii, T.K., 1990, *Practical Microwave Electron Devices*, Academic Press, San Diego.
Johns, E.M.T., and Bolljahn, J.T., 1956, `Coupled Strip Transmission Line Filters and Directional Couplers', *IRE Trans.*, **MTT-4**: 75.
Kajfez, D., 1995, *Q Factor*, Vector Fields, Urbana.
Kenney, J.M., 1980, *Semiconductor Measurement Technology: Modulation Measurements and Microwave Mixers*, National Bureau of Standards Document NBS 400-16, US Government Printing Office, Washington DC.
Kharkovich, A.A., 1962, *Nonlinear and Parametric Phenomena in Radio Engineering*, Rider, New York.
King, R.J., 1978, *Microwave Homodyne Systems*, Peter Peregrinus, London.
Kochelaev, B.I., and Yablokov, Y.V., 1995, *The Beginning of Paramagnetic Resonance*, World Scientific, Singapore.
Kraus, J.D., 1950, *Antennas*, McGraw-Hill, New York.
Lancaster, G., 1980, *DC and AC Circuits*, 2nd ed., Clarendon Press, Oxford.
Lancaster, G., 1992, *Introduction to Fields and Circuits*, Oxford University Press, Oxford.
Laverghetta, T., 1991, *Microwave Materials and Fabrication Techniques*, Artech House, Norwood.
Lax, B., and Button, K.J., 1962, *Microwave Ferrites and Ferrimagnetics*, McGraw-Hill, New York.
Leibrecht, K., 1968, `New Method for the Measurement of the Q Factor of Cavities and its Application to Research in Solid State Physics', *Rev. Sci. Instrum.* **39**: 1919.
Lewin, L., 1959, `Phase Measurements Through Tapered Junctions', *Proc. IEE*, **106**B: 495.
Lubkin, Y.J., 1963, `Diode Mixers', *Microwaves*, **2**(6): 34.
Malherbe, J.A.G., 1979, *Microwave Transmission Line Filters*, Artech House, Norwood.
Manassewitch, V., 1987, *Frequency Synthesizers*, 3rd ed., Wiley, New York.
Marcuvitz, N., 1951, *Waveguide Handbook*, McGraw-Hill, New York.
Marsh, J.A., 1951, `Measured Current Distributions on Helical Antennas', *Proc. IRE*, **39**: 668.
Matarè, H.F., 1960, `Recent Developments in the Field of Semiconductor Devices for High Frequency', In: *Fortschritte der Hochfrequenztechik*, (M. Strutt and F. Vilbig, eds.), Bande 5, Akademische Verlagsgesellschaft, Frankfurt. p347.
Matthaei, G., Young, L., and Jones, E.M.T., 1980, *Microwave Filters, Impedance Matching Networks, and Coupling Structures*, Artech House, Dedham.
Matthews, P.A., and Stephenson, I.M., 1968, *Microwave Components*, Chapman & Hall, London.
Messinger, G.C., and McCoy, C.T., 1957, `Theory and Operation of Crystal Diodes as Mixers', *Proc. IRE*, **45**: 1269.
Montgomery, F., 1947, *Techniques of Microwave Measurements*, McGraw-Hill, New York.

Morrison, R., 1992, *Noise and Other Interfering Signals*, Wiley, New York.

Morse, R.A., and Beam, R.E., 1956, *Proceedings of the National Electronics Conference, Inc.*, Chicago.

Nilson, J.W., 1983, *Electric Circuits*, Addison-Wesley, Reading.

Pettai, P., 1984, *Noise in Receiving Systems*, Wiley, New York.

Penfield, P., and Rafuse, R.P., 1962, Varactor Applications}, MIT Press, Cambridge.

Pengelley, R.S., 1986, Microwave Field Effect Transistors - Theory, Design, and Applications, 2nd ed., Research Studies Press, Letchfield.

Pierce, J.R., and Shepard, W.G., 1947, `Reflex Oscillators', *Bell System Tech. J.*, **26**: 460.

Pierce, J.R., 1951, `Waves in Electron Streams and Circuits', *Bell System Tech. J.*, **30**: 626.

Poole, C.P., 1967, *Electron Spin Resonance. A Comprehensive Treatise on Experimental Techniques*, Wiley, New York.

Poole, C.P., 1983, *Electron Spin Resonance. A Comprehensive Treatise on Experimental Technique*, 2nd ed., Wiley, New York.

Pound, R.V., 1948, *Microwave Mixers*, McGraw-Hill, New York.

Pritchard, W.L., 1961, `Loading Effects', In: *Crossed-Field Microwave Devices*, (E. Okress, ed.), Vol. 2, Academic Press, New York. p423.

Ragan, G.L., 1948, *Microwave Transmission Circuits*, McGraw-Hill, NewYork.

Read, W.T., 1958, `A Proposed High-Frequency Negative-Resistance Diode', *Bell System Tech. J.*, **37**: 401.

Reich, H.J., Skalnik, J.G., Ordung, P.F., and Kraus, H.L., 1957, *Microwave Principles*, van Nostrand, Princeton.

Reich, H.J., 1961, *Functional Circuits and Oscillators*, van Nostrand, Princeton.

Reintjes, J.F., and Coate, G.T., 1952, *Principles of Radar*, McGraw-Hill, New York.

Rhea, R.W., 1994, *HF Filter Design and Computer Simulation*, Noble Publishing, Atlanta.

Ridenour, L.N., 1964, *Radar System Engineering*, McGraw-Hill, New York.

Rider, J.F., 1937, *Automatic Frequency Control*, Rider, New York.

Robins, W.P., 1982, *Phase Noise in Signal Systems: Theory and Applications*, Peter Peregrinus, London.

Robinson, B.H., Thomann, H., Beth, A.H., Fajer, P., and Dalton, L.R., 1985, `Experimental Considerations: Instrumentation and Methodology', In: *EPR and Advanced EPR Studies of Biological Systems* (L.R. Dalton, ed.), CRC Press, Boca Raton. p111.

Robson, P.N., 1993, `Microwave Receivers', In: *Handbook on Semiconductors*, Completely Revised Edition (T.S. Moss, ed.), Vol. 4: Device Physics (C. Hilsum, ed.), Elsevier, Amsterdam. p 545.

Schneider, F., and Plato, M., 1971, *Elektronenspin-resonanz, Experimentalle Technik*, K. Thiemig, München.

Schneider, M.V., 1969, `Microstrip Lines for Microwave Integrated
Circuits', *Bell System Tech. J.*, **48**: 1421.

Schwartz, M., 1959, *Information, Transmission, Modulation, and Noise. A Unified Approach to Communication Systems*, McGraw-Hill, New York.

Sevin, L.J., 1965, *Field Effect Transistors*, McGraw-Hill, New York.

Shevichik, V.N., 1963, *Fundamentals of Microwave Electronics*, Pergamon Press, Oxford.

Siegman, A.E., 1959, *Traveling Wave Techniques for Microwave Resonance Measurements*, Technical Report 155-3, Signal Corps. Contract DA 36(039) SC73178, Electron Devices Laboratory, Stanford University, Palo Alto.

Siegman, A.E., 1960, `Traveling Wave Techniques for Microwave Resonance Measurements', In: *Quantum Electronics* (C.H. Townes, ed.), Columbia University Press, New York.

Silvan, L., 1996, *Microwave Tube Transmitters}*, Chapman & Hall, New York.

Silver, S., 1949, *Microwave Antenna Theory and Design}*, McGraw-Hill, New York.
Sinclair, G., 1948, 'Patterns of Slotted-Cylinder Antennas', *Proc. IRE*, **36**: 1487.
Skilling, H.H., 1952, *Transient Electric Currents*, McGraw-Hill, New York.
Slater, J.C., 1942, *Microwave Transmission*, McGraw-Hill, New York.
Slater, J.C., 1946, 'Microwave Electronics', *Rev. Mod. Phys.*, **18**: 441.
Slater, J.C., 1950, *Microwave Electronics*, van Nostrand, Princeton.
Smith, D.C., 1993, *High Frequency Measurements and Noise in Electronic Ciruits*, van Nostrand, New York.
Snoller, T.C., Brown, E.R., and Lee, H.Q., 1988, 'Microwave and Millimeter Wave Resonant Tunneling Diodes', *Lincoln Lab. J.*, **1**: 89.
Soohoo, R.F., 1960, *Theory and Applications of Ferrites*, Prentice-Hall, Engelwood Cliffs.
Sporleder, F., and Unger, H.-G., 1979, *Waveguide Tapers, Transitions, and Couplers*, Peter Peregrinus, London.
Strandberg, M.W.P., 1972, 'Sensitivity of Radio Frequency Measurements in the Presence of Oscillator Noise', *Rev. Sci. Instrum.*, **43}**: 307.
Strangeway, R.A., Ishii, T.K., and Hyde, J.S., 1988, 'Low Phase-Noise Gunn Diode Oscillator Design', *IEEE Trans. Microwave Theory Tech.*, **36**: 792.
Sugiura, T., and Sugimoto, S., 1969, 'FM Noise Reduction of Gunn Effect Oscillators by Injection Locking', *Proc. IEEE*, **57**: 77.
Swaminathan, V., and Macrander, A.T., 1991, *Materials Aspects of GaAS and InP Based Structures*, Prentice-Hall, Engelwood Cliffs.
Sze, S.M., 1981, *Physics of Semiconductor Devices*, 2nd ed., Wiley, New York.
Sze, S.M., 1991, *Semiconducting Devices: Pioneering Papers*, World Scientific, Singapore.
Terman, F.E., 1934, 'Resonant Lines in Radio Circuits', *Elect. Eng.*, **53**: 1046.
Thim, H.W., 1993, 'Microwave Sources', In: *Handbook on Semiconductors*, Completely Revised Edition (T.S. Moss, ed.), Vol. 4, Device Physics (C. Hilsum, ed.), Elsevier, Amsterdam. p475.
Thomas, R.L., 1976, *Practical Introduction to Impedance Matching*, Artech House, Dedham.
de Thomasson, M., 1962, 'Nature of mm-Tubes', *Microwaves*, **1**(5): 38.
Thourel, L., 1964, *The Use of Ferrites at Microwave Frequencies* (translation from the French by J.B. Arthur), Pergamon Press, Oxford.
Tice, T.E., and Kraus, J.D., 1949, 'The Influence of Conductor Size on the Properties of Helical Beam Antennas', *Proc. IRE*, **37**: 1296.
Tien, P.J., 1953, 'Traveling Wave Tube Helix Impedance', *Proc. IRE*, **41**: 1617.
Tischer, F.J., 1958, *Mikrowellen-Messtechnik*, Springer, Berlin.
Tischer, F.J., 1959, 'Properties of the H-Guide at Microwave and Millimetre Wave Regions', *Proc. IEE*, **106**(B, Suppl. 13): 47.
Townes, C.H., and Schawlow, A.L., 1955, *Microwave Spectroscopy*, McGraw-Hill, New York.
Tralle, G.E., 1962, 'Guide to Noise Figure', *Microwaves*, **1**(1): 22.
Tsui, J.B.-Y., 1983, *Microwave Receivers and Related Devices*, US Department of Commerce, National Technical Information Service, PB84-108711, Washington DC.
Vizmuller, P., 1987, *Filters With Helical and Folded Helical Resonators*, Artech House, Norwood.
Volino, F., Csakvary, F., and Servoz-Gavin, P., 1968, 'Resonant Helices and Their Application to Magnetic Resonance', *Rev. Sci. Instrum.*, **39**: 1660.
Waldron, R.A., 1961, *Ferrites: An Introduction for Microwave Engineers*, van Nostrand, London.
Walter, C.H., 1965, *Traveling Wave Antennas*, McGraw-Hill, New York.

Watson, H.A., 1968, *Microwave Semiconductor Devices and Their Circuit Application*, McGraw-Hill, New York.
Welsby, V.G., 1960, *The Theory and Design of Induction Coils*, 2nd ed., MacDonald, London.
Wilmhurst, T.H., 1968, *Electron Spin Resonance Spectrometers*, Plenum Press, New York.
Young, L., 1972, *Microwave Filters Using Parallel Coupled Lines*, Artech House, Dedham.
Yunik, M., 1973, *Design of Modern Transistor Circuits*, Prentice-Hall, Engelwood Cliffs.
Zverev, A.I., 1967, *Handbook of Filter Synthesis*, Wiley, New York.

Chapter 2

EPR Spectrometers at Frequencies Below X-band

Gareth R. Eaton and Sandra S. Eaton
Department of Chemistry and Biochemistry, University of Denver, Denver, Colorado 80208

Abstract: The majority of EPR spectra are obtained at X-band (*ca.* 9 GHz) EPR. However, there are many incentives for obtaining EPR spectra at frequencies below X-band. This review focuses primarily on the hardware aspects of EPR experiments at lower frequencies including tabulations of papers concerning low-frequency spectrometers, magnets, field gradients, radiofrequency sources, bridges, and resonators.

1. SCOPE OF THIS REVIEW

Literally since the first observation of EPR, measurements have been done at frequencies lower than the one (9-10 GHz) that has become most common. After a brief review of this early history, this review focuses on the instrumental aspects of performing EPR at frequencies lower than X-band. As explained herein, these will be generically called "low-frequency" in spite of the imprecision of the term. The literature is covered through early 2000 and was updated in late 2002. Papers that apply low-frequency EPR are not cited unless they provide some insight into the instrumentation, methodology, or sensitivity of the measurements. Swartz and Halpern (1998) recently reviewed *in vivo* EPR, comprehensively covering the low-frequency literature, and Eaton and Eaton have published a series of reviews of the EPR imaging literature (Eaton, Eaton and Ohno, 1991; Eaton and Eaton, 1988, 1990, 1991, 1993b, 1999c, 2000). Greenslade *et al.* (1996) critically reviewed low-frequency EPR and discussed relative benefits of various approaches. Zero-field EPR was reviewed by Bramley and Strach (1983). EPR at higher frequencies, with a focus on use of high magnetic fields, has been reviewed separately (Eaton, 1992; Eaton and Eaton, 1993a, 1999a,b).

Chapters in Biol. Magn. Res. Vol. 18 discuss bridges (Koscielniak), resonators (Rubinson), and time-domain spectrometer design and imaging methodology (Subramanian *et al.*) with an emphasis on L-band and RF EPR. Two recent reviews of EPR imaging provide many references to low-field EPR spectrometers and their applications (Eaton and Eaton, 2000; Lurie, 2002).

2. EARLY HISTORY

Initial attempts by Gorter, and the later success by Zavoisky, in detecting electron paramagnetic resonance were conducted in the MHz frequency range (Gorter, 1936a,b,c, 1947, 1967; Zavoisky, 1945, 1946; Hauser, 1978). A definite absorption maximum at 133 MHz and 47.6 gauss was the first demonstration of a paramagnetic resonance absorption maximum (Zavoisky, 1945). Less than a year after he reported the observation of paramagnetic resonance, Zavoisky (1946) extended the measurements to the microwave region. Zavoisky (1946) observed well-defined absorption peaks for $CrCl_3$ at 2.19 and 2.76 GHz, and for $MnSO_4$ at 2.75 GHz. Stimulated by Zavoisky's studies at 133 MHz and by the method of Purcell, Torrey, and Pound (1946) for the study of nuclear magnetic moments, Cummerow and Halliday (1946) extended the study of paramagnetic resonance to higher frequencies independently of, and simultaneously, with Zavoisky. Using a klystron as a frequency source they observed resonance at 2.93 GHz and 1100 gauss with 173 grams of $MnSO_4 \cdot 4H_2O$. In the full paper submitted a year later they extended the work to 9.375 GHz (Cummerow *et al.*, 1947). The spectrometer they used incorporated the technique of modulating the microwaves together with ac amplification of the detected signal. Good signal to noise was obtained with much smaller samples than their first report. Structure was observed in some of the spectra, but it was not understood. Comprehensive citation of the earliest literature is provided in chapters in Eaton *et al.* (1998b).

Inspired by the Zavoisky and Cummerow and Halliday papers, two more research groups entered the field. Bagguley and Griffiths (1947) investigated paramagnetic absorption in a single crystal of chrome alum at wavelengths between 3 and 10 cm. The figures in this paper and in (Cummerow and Halliday, 1947) are the first X-band spectra, and the first to show several absorption peaks. Theoretical treatment was given by Bagguley and Griffiths in terms of zero-field splittings (ZFS). The full papers published the following year reported that measurements had been made on over a hundred salts (Bagguley *et al.*, 1948a,b). It is interesting to note that this paper proposed using pulsed EPR to measure relaxation times. Independently, and nearly simultaneously, Weiss and coworkers (1947, 1948) studied powdered and single crystal samples of $NH_4Cr(SO_4)_2 \cdot 12H_2O$. Their spectrometer introduced the important innovation of using a magic T bridge and slide screw tuner coupling of the cavity. The structure in the spectrum was interpreted in terms of the ZFS of the Cr(III).

Halliday and Wheatley (1948) reported the first X-band absorption by a paramagnetic species in solution. At 9.375 GHz they obtained the spectrum of aqueous $MnSO_4$ in a thin capillary along the axis of a cylindrical cavity. Copper sulfate also received attention from several research groups (Wheatley and Halliday, 1949; Arnold and Kip, 1949; Bagguley and Griffiths, 1948). The spectrometer used by Arnold and Kip included a TE_{102} cavity coupled to the waveguide by an iris of adjustable aperture. A recording EPR spectrometer, incorporating many features that have evolved into the modern EPR spectrometer, such as AFC system, reflection cavity with coupling iris, magnetic field modulation, and lock-in detection, was described by Strandberg *et al.* (1956). This spectrometer could operate at 1.25, 3, and 10 cm wavelength.

This brief historical survey illustrates that within a few years of the first observation of paramagnetic resonance several research groups were exploring important magnetic interactions in solids with X-band spectrometers, and the groundwork had been laid for extension to species in solution. The many advantages of X-band wavelengths for EPR led to the dominance of X-band for EPR studies over the next four decades. While most studies were being done at X-band, with occasional studies at Q-band (*ca.* 35 GHz), a few researchers continued to explore the information that could be obtained at higher and lower frequencies.

3. NOMENCLATURE

Among most EPR spectroscopists, it would be generally agreed that the frequencies being explored for *in vivo* spectroscopy (a few tens of MHz to *ca.* 1 GHz) are "very low frequency." Unfortunately, the IEEE designation for 30 to 300 MHz is VHF (very high frequency), which arises out of an historical perspective. Hence, low = high. We believe that for unambiguous communication, when a broad range of frequencies is described, EPR spectroscopists should agree to use specified frequency ranges (*e.g.*, 100 to 500 MHz) or phrases such as "in the hundreds of MHz range." When describing a particular spectrometer or experimental result, use of the specific frequency, *e.g.*, 257 MHz, avoids the low = high issue, and communicates more precisely than using either VHF or "very low frequency." In this review we will use "low-frequency" as a non-specific term to refer to EPR at frequencies lower than X-band.

To complicate the picture even more, there can be multiple "standard" nomenclatures resulting in labels that are not unique. For example, C band is 4-8 GHz in coax, but 0.5 – 1 GHz in the military designation, and C is not used to label a waveguide band. In coax, S is 2 – 4 GHz, but it is 2.6 – 3.95 GHz in waveguide. L-band is 1 –2 GHz in coax, but designates 40-60 GHz in the military nomenclature. There is a handy pictorial representation of band designations on page 186 of the HP RF & Microwave Test Accessories 1999/2000 Catalog. Another summary is in Table 1 of the review by Eaton and Eaton (1993a). The column labeled "new military" in Eaton and Eaton (1993a) is the same nomenclature as the spectrum labeled "electronic warfare bands" in the HP catalog.

For ease of comparisons, we have converted all reported unloaded Q values to loaded (matched) Q values by dividing by 2 to report them in this review.

4. FACTORS INFLUENCING SELECTION OF FREQUENCIES BELOW X-BAND

Much of the renewed emphasis on EPR at frequencies below X-band (*ca.* 9 GHz) is stimulated by the desire to perform *in vivo* measurements, both spectroscopy and imaging. So far, there has been no optimization of the frequency for *in vivo* studies. Hutchinson's (1971) estimate that 500 to 750 MHz would be an optimum frequency for *in vivo* studies, based on a simplified dielectric model of a mouse, stimulated further spectroscopy in this frequency range. Resonator structures can be made to accommodate larger samples (*e.g.*, mice) at lower frequencies. The tradeoffs between sample volume, resonator filling factor, and microwave loss factor can be optimized at frequencies below X-band for some lossy samples, including biological (aqueous) samples.

The half-field transitions that result from electron-electron spin-spin interactions (Eaton and Eaton, 1982, 1989) are more intense, relative to the g ~ 2 signal, proportional to the inverse square of the Zeeman frequency. Very intense g ~ 4 signals were observed by Yoshioka (1977) in 135 MHz EPR spectra of crystalline organic radicals, and have been observed in our lab in 250 MHz spectra of ultramarine blue.

The tradeoffs between g-anisotropy and a-anisotropy can result in the minimum linewidth, and hence the best hyperfine resolution, occurring at frequencies below X-band (Hyde and Froncisz, 1982; Abdrachmanov and

Ivanova, 1973; Rothenberger *et al.*, 1986). For example, Ag(II) is best studied between 2.0 and 3.5 GHz (Abdrachmanov and Ivanova, 1976). Hyde and coworkers demonstrated the ability to observe nitrogen hyperfine coupling on the S = -1/2 line of Cu(II) in frozen solution at 2.3 GHz. This has resulted in a large literature of S-band EPR of Cu^{2+}, of which we cite only leading references (Hyde and Froncisz, 1982; Yuan *et al.*, 1999).

Measurements of relaxation times at a wide range of magnetic fields is needed to provide a test of mechanism, because some electron spin relaxation mechanisms are magnetic-field-dependent.

An enabling technology for low-frequency EPR is effective lumped circuit microwave resonators that permit a higher filling factor and higher B_1 per watt than cavity resonators at these frequencies. Rectangular cavity resonators or cylindrical cavity resonators would be extremely large at these low RF frequencies (see the review by Hyde and Froncisz, 1986, and many of the references cited in the tables below).

For many years, X-band (*ca.* 9-10 GHz) CW spectrometers have dominated the commercial EPR market (*e.g.*, all of the commercial spectrometers listed by Alger (1968) operated at X-band or higher). 34-35 GHz (called Q-band by most EPR spectroscopists) has been the second-most-common commercial EPR frequency. Recently, JEOL has marketed an L-band spectrometer and Bruker has marketed L-band, S-band and W-band spectrometers. In this commercial market it is easy to forget that long ago very low frequency EPR spectrometers were commercially available. For example, Singer (1962) reported 30, 65, and 320 MHz spectra obtained on Alpha Scientific Labs AL55 and AL340 ESR spectrometers.

4.1 RF Penetration as a Function of Frequency

Animals are lossy, but not as lossy as "physiological saline" (0.15 M NaCl). The penetration depth is greater than that of saline solution, because the animal is heterogeneous. Halpern *et al.* (1989) estimated 6 cm skin depth in tissue at 250 MHz. There is a long history of predicting that penetration depth and power deposition would limit MRI to much lower fields/frequencies than have proved to be practical. So far, MRI has successfully exploited higher frequencies than each "limit" that is predicted to be the highest practical. In part, this success is due to post-processing of data to correct for phase and amplitude variations due to the dielectric constant and loss of the body. Recently, Robitaille and coworkers have extended human MRI to 8T, and in the process once again have shown that RF losses (power deposition) are less than had been anticipated from lower frequency data (Robitaille *et al.*, 1998, 1999; Kangarlu *et al.*, 1999). Among the important observations is that conductivity suppresses dielectric resonances, and that the usual skin depth formula is valid only within very good conductors, and should not be applied to low-conductivity media and dielectrics.

Stevens (1994) reported that loaded Q values for a 250 MHz 27.5 mm i.d. resonator decreased from 250 for the empty resonator to 200 with H_2O, to 120 with an equal volume of 0.15 M NaCl solution, and to *ca.* 100 when resting on a human head (acting as a surface coil). Brivati *et al.* (1991) made similar measurements with a higher-Q LGR. The unloaded Q of the empty resonator was 5000. The Q decreased to 2000 with 200 mL of H_2O in a 4.2 cm i.d. quartz tube and to 300 with 200 mL of 0.1 M NaCl solution.

4.2 Frequency Dependence of Signal-to-Noise

A discussion of the frequency dependence of signal/noise is given in Halpern and Bowman (1991) and in our articles (Eaton *et al.*, 1998a; Rinard *et al.*, 1999a, b, c) and in another chapter in this book (Rinard *et al.*, 2003). A key conclusion is that if the sample is unlimited, and if the sample lossiness is not dominant, the S/N can actually increase at lower frequency (Eaton *et al.*, 1998a). Practical animal studies will not be described by the limiting cases in the published tables (Eaton *et al.*, 1998a; Rinard *et al.* 1999, 2003). EPR for *in vivo* physiology inherently focuses on a small region, so the filling factor for the region of interest will be small, and the volume will be a size defined by the animal physiology. Consequently, the tabulated cases are only a guide to optimism for low-frequency EPR. Depending on the region of interest and the study to be made, a surface coil or other local resonator, may give better S/N than a resonator that could contain the entire animal. This is as has been found in MRI. The use of a cavity or loop-gap resonator to study the tail of a mouse is one example of this selectivity.

Willer *et al.* (2000) stated that the echo intensity at 4 GHz was an order of magnitude lower than at X-band for the same overall gain, bandwidth, and resonator losses. Romanelli *et al.* (1994) found S/N 30-50 times lower at S-band than at X-band. Comparing spin echoes at 2.68 and 9.52 GHz, we have shown agreement between experiment (X/S = 9.5) and theory (X/S = 7.6) in a case where the resonator was well-characterized and all gains and losses in the signal path were taken into account (Rinard *et al.*, 1999a,b; 2002d,e). Experimental CW EPR signal intensities at 250 MHz, 1.5 GHz, and 9.1 GHz agree within experimental error with predictions from first principles that the signal at the lower frequency will be larger than the signal at the higher frequency by the ratio 1.57 (Rinard *et al.*, 2002d,e, 2003).

4.3 Animal Motion

The low filling factor is not the only challenge for EPR of living animals. Physiological motion (heart, breathing, other muscle movement, twitching, *etc.*) changes the interaction of the animal with the electromagnetic field of the microwaves or RF. When a reflection resonator is used, the changes in Q and in resonant frequency that result from physiological motion cause noise in the EPR signal. Various automatic frequency control (AFC) and automatic coupling control (ACC) systems have been implemented to decrease the detrimental effect of animal motion (Halpern *et al.*, 1989; Brivati *et al.*, 1991).

PEDRI is less sensitive to animal motion, because the RF frequency used for NMR detection is much lower, and the changes in the Q and frequency of the resonator used to excite the electron spins have effect only through changes in the amount of power reflected from the resonator, and hence do not impact the electron spins that produce the PEDRI effect.

Two aspects of LODESR make it different from "standard" EPR with a reflection resonator. First, detection of the longitudinal magnetization occurs with a sensitivity that is strictly linearly dependent on the magnitude of the Zeeman field for a constant number of spins. Second, the LODESR measurement usually uses a pickup coil assembly orthogonal to the exciting coil or resonator. The RF at the Zeeman frequency is modulated at a lower frequency. The EPR signal detection system operates at a frequency much lower than the Zeeman frequency, often hundreds of KHz, so animal motion does not affect the Q of the detection system very much.

5. SPECTROMETER DESIGN

Following the heroic early period that was briefly summarized above, most low-frequency EPR spectrometers have been of the CW reflection resonator type. The PEDRI and LODESR spectrometers have rather different construction. Table 1 provides references to the papers that have given the most comprehensive descriptions of the overall spectrometer systems. Tables 2-6 list papers that describe one or more of the functional units of a spectrometer in some detail. The tables are organized by topic, and within each table by RF frequency. Consistent with the primary focus of current practitioners on improving the instrumentation and methodology for low-frequency EPR, the topics selected for emphasis in the tables are largely instrumental. Few details are given in the tables – their purpose is to efficiently guide the reader to the original literature. Tables 7 and 8 list a few examples of low frequency EPR studies.

The most complete descriptions of EPR spectrometers in the 250-300 MHz range are by Halpern, *et al.* (1989; Halpern and Bowman, 1991), Symons and coworkers (Brivati *et al.*, 1991; Stevens and Brivati, 1994; Stevens 1994), Murali Krishna and coworkers (Bourg *et al.*, 1993; Murugesan *et al.*, 1997, 1998; Subramanian *et al.*, 1999; Devasahayam *et al.*, 2000, 2002; Koscielniak *et al.*, 2000; Yamada *et al.*, 2000; Afeworki *et al.*, 2000) and the Denver group (Rinard *et al.*, 2002 a,b,c; Quine *et al.*, 2002). These spectrometers were constructed for the study of biological specimens. Sotgiu and coworkers incorporated multipole magnets, which can produce magnetic field gradients, and reentrant resonators in low-frequency EPR imaging systems (Sotgiu, 1986; Colacicchi *et al.*, 1996; Alecci *et al.*, 1992a,b; Sotgiu *et al.*, 1995).

5.1 CW Spectrometers

The 250-MHz Halpern spectrometer used a strip-line-type resonator (*ca.* 1.25 cm radius) that could hold a small mouse, a hybrid coupler to direct RF to and from the resonator, an air-core Helmholtz magnet, and Anderson-type gradient coils (Halpern *et al.*, 1989). The coils for the main magnet were splayed to create the gradient in the z-direction. This spectrometer is described in considerable detail, including the structural and electronic aspects of the automatic coupling control (ACC), automatic frequency control (AFC), and the strip-line resonator. The bridge design uses high-dynamic range double-balanced mixers to accommodate animal-induced changes in reflected power.

Symons (1995) gives a brief description of a 250 MHz spectrometer intended for whole-body human imaging. The overall design is similar to that of the smaller 300 MHz system (Brivati *et al.*, 1991). One of the resonators is a 45 cm diameter 3-loop-4-gap LGR (the 2 outer loops are 10 cm). The Q of *ca.* 300 reduced to *ca.* 100 upon inserting a human head. The Symons spectrometer was designed for larger samples, up to about 5 cm radius at 300 MHz. The signal to noise was about 1/10 that of a standard X-band spectrometer.

The 300 MHz spectrometer described by Brivati *et al.* (1991) incorporated AFC, ACC, and automatic phase control to compensate for animal motion artifacts. Smith and Stevens (1994) reported that sloping baselines were due to interaction of stray magnetic fields with the directional coupler and 3-port circulator in the bridge, even though these components are well-shielded and at a distance of 4 m from the magnet coils. Similar sloping baselines have been observed in prototype systems in Denver, but seemed more sensitive to the cabling for the magnetic field modulation system than to the circulator or other bridge components.

The 300 MHz animal imaging spectrometer at NCI (Koscielniak *et al.*, 2000) uses a "magic-T" bridge, a commercial lock-in amplifier, and an AFC system. The field of the water-cooled air-core magnet could be swept up to 50 G in 4 seconds. A 3 cm diameter spherical volume had ±200 ppm field homogeneity. The field of the water-cooled air-core magnet could be swept up to 50 G in 4 seconds. A 3 cm diameter spherical volume had ±200 ppm field homogeneity.

A 650 MHz spectrometer that uses both automatic frequency control (AFC) and automatic coupling control (ACC) circuits could compensate for the breathing motions of a living rat (Yokoyama *et al.*, 2001). A varactor positioned one-quarter wavelength from a single-turn coupling loop was used to fine-tune the coupling to a LGR. The impact of the ACC circuit was greater when measurements were made in the chest region of the rat than when the head was measured. Stability analysis was performed for an AFC system designed for an L-band CW EPR spectrometer (Hirata and Luo, 2001).

Sato *et al.* (1997) constructed a spectrometer in which EPR at 680 MHz and NMR at 27.7 MHz could be performed using as many of the same components as possible. The loop-gap transmission line resonator was resonant at both frequencies, and an iron-core electromagnet was used to produce the required field strengths.

Ono *et al.* (1998) demonstrated frequency modulation as an alternative to magnetic field modulation for a 700 MHz spectrometer, in order to avoid microphonics due to the modulation field interacting with the resonator. In addition, it is possible to scan the frequency more rapidly than the magnetic field can be scanned. The 700 MHz signal from an Anritsu MG3633A RF signal source was modulated at 100 KHz to a depth of 1 MHz. The modulation signal coherently controlled the oscillator, the lock-in amplifier, and the resonator frequency. The resonator Q was 200. Spectra of DPPH powder were compared using frequency modulation and field modulation. Alecci *et al.* (1998a) used frequency modulation to obtain CW spectra with a pulsed spectrometer whose resonator shield was too thick to use magnetic field modulation.

The CW L-band spectrometer in the Berliner lab has been described (Nishikawa *et al.*, 1985, Berliner and Fujii, 1985; Fujii and Berliner, 1985; Berliner and Koscielniak, 1991). There have been many reports of low-frequency EPR spectroscopy from the Swartz lab, whose 1 GHz spectrometer was briefly described in Nilges *et al.* (1989).

Since few EPR spectra have spectral extents and relaxation times that make it feasible to use FT-EPR to enhance S/N per unit time, there are also efforts to acquire spectra with rapid-scan CW EPR (Oikawa *et al.*, 1995). The rate with which the magnetic field can be scanned is limited by hysteresis of an iron-core magnet or eddy currents in other metallic components. Consequently, Oikawa *et al.* (1995) and Ogata (1995) constructed a rapid-scan system for 700 MHz EPR using an air-core magnet. This system could achieve 15 mT/s, at which rate there was a field "delay" that could be corrected for. With this system 82 spectra for an image were obtained in 150 s (Ogata, 1995) and 81 spectra in 1.5 min (Oikawa *et al.*, 1996).

5.2 Pulsed Spectrometers

A 220 MHz pulsed spectrometer described by the Sotgiu group in L'Aquila (Alecci *et al.*, 1998a,b) used orthogonal resonators to excite the spins and detect the FID. Two orthogonal saddle coils for the receiver were inside and orthogonal to the LGR used to excite the spins (22-40 dB isolation). The empty Q values were 85 and 50 for the LGR and saddle coils, respectively. The Q values dropped to 45 and 18 when the resonator contained 55 mL physiological saline solution. Use of a class C amplifier avoided the need for the diode circuit used in most pulsed systems. The central concept in this system is that by having two receiver coils, one perpendicular to B_0 to detect the FID and the other parallel to B_0, with identical ringing times, detection of their signals 90° out of phase cancels the ringing and shortens the dead time. A 90° flip angle was achieved with a 300 ns 2.4 W pulse. By using a composite pulse to quench the ringing of the transmit resonator, and combining the two receiver signals out of phase, reduced the dead time to 300 ns (Alecci *et al.*, 1998b). The combination of the signals from the two receiver coils increases the noise by 3 dB (Alecci *et al.*, 1998a).

A series of four papers provide extensive detail about the pulsed 250 MHz EPR imaging spectrometer in Denver (Rinard *et al.*, 2002a,b,c; Quine *et al.*, 2002). This spectrometer is designed as an engineering test facility and incorporates multiple excitation and signal paths to facilitate testing of a variety of resonators for both CW and pulsed EPR. Innovations relative to prior low-frequency EPR spectrometers include a four-coil, air-core magnet and gradient coils, a crossed loop resonator, dynamic Q-switching to reduce dead time in pulsed EPR, and a narrow-band bridge based on circulators. The four-coil magnet achieved about 40 ppm magnetic field homogeneity over a 15 cm diameter working volume. The crossed loop resonator included Q-switching capability. A dead time after the RF pulse could be reduced to a few hundred nanoseconds (depending upon signal strength) while maintaining a Q of 200 during data acquisition for high sensitivity (Rinard *et al.*, 2002a). FID, spin echo, and echo-detected saturation recovery were demonstrated.

The 300 MHz pulsed EPR spectrometer in Murali Krishna's lab at the NIH National Cancer Institute has been described in detail (Bourg *et al.*, 1993; Murugesan *et al.*, 1998), as have the fast (1 Gs/s 2-channel) digitizer (Pohida, 1994; Subramanian *et al.*, 1999; Afeworki *et al.*, 2000) and the resonators (Rubinson *et al.*, 1998; Devasahayam *et al.*, 2000). To use fast (20-70 ns) pulses, the resonators were overcoupled to Q ~ 20-30 (Murugesan *et al.*, 1998, Devasahayam *et al.*, 2000; Afeworki *et al.*, 2000). Instrument dead times were in the range 300-450 ns (Murugesan *et al.*, 1998; Afeworki *et al.*, 2000). A 500 MS/s signal acquisition and summing computer board was compared (Devasahayam *et al.*, 2002) with the previously described Analytek data acquisition system (Subramanian *et al.*, 1999) and found to have sufficient capability that the imaging time was about one-third of the time previously required. Hardware implementation was described, and examples of phantom and mouse images were demonstrated. Quadrature detection of FIDs was implemented with a quadrature mixer and post-acquisition processing for phase cycling (Subramanian *et al.*, 2002). This was used to obtain FID intensity data at a constant time after the pulse, as has been done in NMR. The images obtained were free of many of the artifacts that occur in filtered back projection images using the entire FID.

An imaging spectrometer was constructed to perform both NMR (at 1.52 MHz and EPR (at 1-GHz), in a field of 357 G (Giuseppe *et al.*, 2001). The EPR portion of the bridge incorporated a narrow-band circulator to isolate the receiver by 25 dB from the transmitter. The pulsed EPR imaging used FID data collection following 750 ns 90° pulses with a 15 μs repetition.

The L-band (Quine *et al.*, 1996) and S-band (Rinard *et al.*, 1999a) pulsed spectrometers in Denver have been described. Some details of 250 MHz, L-band, and S band spectrometers were reported in a series of papers that confirmed predictions of frequency dependence of sensitivity (Rinard *et al.*, 1999a,b, 2002d,e).

Clarkson and coworkers have also described a pulsed S-band spectrometer (Belford *et al.*, 1987; Clarkson *et al.*, 1989, 1992), as has the Zurich group (Willer *et al.*, 2000). A *ca.* 2 GHz spectrometer described by Adkins and Nolle (1966) had PIN diode attenuators that permit saturation recovery measurements in addition to CW spectroscopy.

A standard commercial pulsed NMR spectrometer can be used for EPR by placing the sample in the fringe field outside the solenoid magnet. Dormann *et al.* (1983) demonstrated FID detection of the EPR signal of a crystalline fluoranthenyl radical using an NMR spectrometer at *ca.* 4.4, 9, and 27.4 MHz. Similarly, Callaghan *et al.* (1994) used a Bruker 300 MHz NMR spectrometer and placed the probe 0.9 m from the magnet, yielding a field of 10.5 mT. A spin-echo sequence and a single-pulse FID were used to study a crystalline fluoranthenyl radical, which has $T_1 \sim T_2 \sim 6$ μs (Coy *et al.*, 1996). Dead time before data collection was *ca.* 3-4 μs, so one can study only samples with long relaxation times, such as the one chosen for the demonstration (Callaghan *et al.*, 1994; Coy *et al.*, 1996; Feintuch *et al.*, 2000; Alexandrowicz *et al.*, 2000). Additional pulsed EPR measurements have been made in the fringe field of NMR spectrometer magnets (see references in Eaton and Eaton, 2000; Lurie, 2002).

These recent efforts to build pulsed EPR spectrometers at frequencies below X-band should be viewed in the historical perspective of the Bowers and Mims (1959) 3.8-4.4 GHz saturation recovery spectrometer and the Rannestad and Wagner (1963) pulsed 1.8 GHz spectrometer.

A long-awaited comparison of pulsed and CW low-frequency EPR imaging (Yamada *et al.*, 2002) shows that each has benefits for particular applications. Spectrometer dead time, which is a greater instrumental problem the lower the RF frequency and the weaker the EPR signal, limits pulsed EPR to samples with phase memory times long enough to have significant signal to observe after the dead time. Yamada *et al.* (2002) estimated that the phase memory time needed to be longer than 275 ns to use pulse EPR for imaging at 300 MHz. The gradient magnitudes also are limited, because the interference patterns of FIDs from multiple magnetic field positions damp the signal rapidly. There is hope that the spin echo capability demonstrated at 250 MHz by Rinard *et al.* (2002) may extend the range of applications of pulsed low-frequency EPR. Yamada *et al.* (2002) found that for narrow-line triarylmethyl spin probes, pulsed, FID-detected EPR yielded better sensitivity and better temporal and spatial resolution compared with CW EPR, but that CW EPR does not restrict the choice of spin probes to these narrow-line triarylmethyl radicals.

5.3 Proton Electron Double Resonance Imaging (PEDRI)

Proton Electron Double Resonance Imaging (PEDRI), also called Overhauser effect imaging, uses the dynamic nuclear polarization (DNP) effect of paramagnetic species on nuclear spins of the solvent to enhance the sensitivity of the NMR (MRI) measurement (Grucker, 1990; Grucker and Chambron, 1993; Lurie *et al.*, 1988, 1989; Lurie 1995; Alecci and Lurie, 1999; Puwanich *et al.*, 1999; Mülsch *et al.*, 1999; Youngdee *et al.*, 2001; Lurie *et al.*, 2002). This effect can be exploited to obtain EPR spectra and EPR images at very low frequency (Lurie *et al.*, 1991a,b,c, 1992, 1998; Nicholson *et al.*, 1994). For example, Guiberteau and Grucker (1996) used field-cycling to observe NMR at 68 G, but switched to different fields for the duration of EPR irradiation at 62, 66, 72, 74 MHz. The relaxation time of the free radical has been identified as the most important parameter in optimizing the S/N in PEDRI (Konijnenburg and Mehlkopf, 1996). Lurie and co-workers demonstrated PEDRI with constant magnetic field (Lurie *et al.*, 1988) and with field cycled to various values for the electron spin irradiation period (Lurie *et al.*, 1989; Puwanich *et al.*, 1999; Mülsch *et al.*, 1999).

A 0.38 T whole-body MRI system was converted to a 20.1 mT small-animal PEDRI imager with an NMR frequency of 856 kHz and an EPR frequency of 564 MHz (Lurie *et al.*, 2002). A 4 cm diameter by 6 cm long NMR solenoid was inside an Alderman-Grant type EPR resonator that was 7 cm diameter and 7 cm long. These were surrounded by a 14 cm diameter, 26 cm long cylindrical shield. The Overhauser enhancement of protons in the presence of triarylmethyl radicals was measured as a function of power at the EPR frequency (Li *et al.*, 2002). PEDRI images of a mouse enabled measurement of the distribution and clearance of the paramagnetic probe.

5.4 Longitudinally-Detected ESR (LODESR)

LODESR at 300 MHz (Nicholson *et al.*, 1994b, 1996; Yokoyama *et al.*, 1998) and 710 MHz (Yokoyama *et al.*, 1997a,b) has been used for imaging. Yokoyama *et al.* (1997a) used 600 KHz on/off modulation of the 710 MHz frequency, and saddle-type pickup coils (30 mm o.d.) inside the LGR, which was oriented perpendicular to B_0. Signal detection was with a lock-in amplifier at the modulation frequency. The LODESR signal intensity was proportional to applied RF power in the range up to 15.8 W. The LODESR signal was less sensitive to perturbation of the resonator (by inserting a tube of saline solution) than was normal CW EPR. In principle, the LODESR signal should be linearly proportional to B_0, since the Zeeman population difference is measured directly. Yokoyama *et al.* (1997b) used resonators and pickup coils of similar size at 296, 663, and 883 MHz, with 1 MHz modulation frequency. The Q of the resonator with pickup coil in place was 160, 150, and 230 for the three frequencies. Nitroxyl radical dissolved in benzene was used as the non-dielectric loss sample and the same nitroxyl dissolved in physiological saline solution was used as the high dielectric loss sample. The resulting Q values for these samples at the three frequencies were 130, 150, 220 and 65, 55, 30, respectively. After correcting for the differences in Q, the S/N values for the benzene solution were 4, 8, and 30, whereas for the saline solution the S/N values were 50, 50, and 30. Similar comparisons were made with nitroxide in a rat head at 300, 678, and 955 MHz. The Q values were 60, 60, and 20, and the S/N ratios were 50, 50, and 30. It was concluded that the dielectric loss of the saline solution or rat head increased with increasing frequency more strongly than the increase in Zeeman population, so it was better to use lower frequency. The power deposition in the rat head was 4, 5, and 7 W at 300, 700, and 900 MHz.

The Aberdeen groups of Lurie, Hutchison, and coworkers are developing instrumentation for longitudinal detected EPR (LODESR) (Panagiotelis *et al.*, 2001a,b). A birdcage coil tuned to 280 MHz for electron spin excitation, and a solenoid NMR detection coil were used for relaxation-sensitive LODESR maps of triarylmethyl radicals in the abdominal region of a rat (Panagiotelis *et al.*, 2001). Relaxation times of nitroxyl radicals and triarylmethyl radicals in fluid solution were measured by LODESR methods (Panagiotelis *et al.*, 2001a). EPR imaging was compared with NMR imaging in Davies *et al.* (2001). Methods of combining field-cycled PEDRI and snapshot NMR imaging to improve signal-to-noise ratio and resolution were presented in Youngdee *et al.* (2002). An Alderman-Grant type resonator was used for EPR irradiation at 120 MHz, and the 2.5 MHz NMR solenoid was inside the EPR resonator. Images were obtained of the abdomen of a rat.

5.5 Other Spectrometer Types

An induction EPR spectrometer, operating at 900 MHz, described by Duncan (1967), had a single-turn coil in a coaxial cavity. The coil was positioned such that no (-80 dB) microwave flux cuts the coil, except at EPR resonance.

An electrically-detected magnetic resonance spectrometer, detecting EPR at 900 MHz (Sato *et al.*, 2000), was designed to study semiconductors. Up to 50 W could be incident on the sample in a 4.4 cm diameter, 1 cm long, bridged LGR whose Q was 510.

Frequency modulation has been used only rarely in EPR because of the high Q of the EPR resonator. Hirata *et al.* (2002) combined frequency modulation with an electronically tunable 1.1 GHz surface-coil resonator and an automatic tuning control (ATC) system. Approximately the same signal-to-noise ratio was obtained with frequency modulation of 28 kHz and magnetic field modulation of 90 kHz.

Ultra-low-field (earth's magnetic field) EPR spectrometers have been developed by Moussavi and coworkers (Duret *et al.*, 1991, 1992; Kernevez *et al.*, 1992) and by Gebhardt and Dormann (1989) for use as magnetometers.

References to these and other overall spectrometer designs are summarized in Table 1.

Table 1. Spectrometer Systems

Frequency, Spectrometer type	Comment	Reference
variable	for measuring weak magnetic fields in the 0.4-21 Oersted range	Chirikov, 1959
0-285 Gauss	$\lambda/4$ coaxial line oscillator-detector circuit	Matheson and Smaller, 1955
1.845 MHz	earth's magnetic field	Duret *et al.*, 1991
1-300 MHz	magnetometer	Gebhardt and Dormann, 1989
1.2-15 MHz	used 3 g sample of DPPH	Garstens *et al.*, 1954
4.4, 9, and 27.4 MHz	FID detection with NMR spectrometer, Gaussmeter applications	Dormann *et al.*, 1983
5 to 450 MHz	hybrid junction, DBM, saturation recovery, no details given	Kume and Mizoguchi, 1985; Mizoguchi and Kume, 1985
10-70 MHz ultrasonic	for sound absorption in paramagnetic substances	Golenishchev-Kutuzov and Kharakhash'yan, 1965
10-120 MHz	magic T, DBM, lock-in amplifier	Hatch and Kreilick, 1972
19.3 MHz	no details given	Misra *et al.*, 1973
20 MHz	Clapp-oscillator, tank circuit in Dewar, Helmholtz coils, 50 Hz magnetic field modulation	Bruin and Bruin, 1956
30-100 MHz	similar to Decorps and Fric (1972) spectrometer, sample can be at 77 K	Fric and Mignot, 1975
30 MHz	dispersion spectrometer; no details given	Grobet *et al.*, 1971
30, 65, 320 MHz, 9 GHz	Alpha Scientific Labs spectrometers	Singer, 1962
40, 80, 210 MHz	circulator at 210 MHz; hybrid junction at 40 and 80 MHz	Decorps and Fric, 1972
56 MHz	magnetometer	Duret *et al.*, 1992
57 MHz	Broadline NMR spectrometer, observed half-field transition for DPPH	Verdelin *et al.*, 1974
60 MHz	Helmholtz magnet, Colpitts oscillator, 30 Hz magnetic field modulation and phase-sensitive detection	Lloyd and Pake, 1954
77 MHz	acoustic paramagnetic resonance	Antokol'skii *et al.*, 1977
80 MHz	marginal oscillator, 0.5 mL samples	Kent and Mallard, 1965
80 MHz	no details given	Barbarin and Germain, 1975

Frequency, Spectrometer type	Comment	Reference
100 MHz	crystal-controlled oscillator feeds the Helmholtz coil resonator through a 10 dB directional coupler	Hutchinson and Mallard, 1971
100-450 MHz	metals in glasses	N. S. Garifyanov, doctoral dissertation, Kazan University, 1965, cited by Abdrachmanov and Ivanova, 1973
100-1000 MHz	crossed coil resonator, Helmholtz coil magnet	Borel and Manus, 1957
135 MHz	sample coil and transmission line make up part of the oscillator	Benedek and Kushida, 1960
193.3 MHz	1 or 4 KHz modulation, sample cooled in N_2 gas	Matsui *et al.*, 1993; Terakado *et al.*, 1998
ca. 200 MHz	reflection resonator, magic T, quadrature detection, imaging system	Bolas *et al.*, 1996
220 MHz pulsed	samples up to 50 mL	Alecci *et al.*, 1998a,b
237 MHz	combined PEDRI and CW EPR, hybrid coupler, DBM, LGR	McCallum *et al.*, 1996b
240-360 MHz	homodyne RF circuit, LGR, 400 mL samples	McCallum *et al.*, 1996a
250 MHz	imaging system	Halpern *et al.*, 1989
250 MHz	sized for human whole-body imaging	Symons, 1995
250 MHz	Pulsed imaging	Quine *et al.*, 2002; Rinard *et al.*, 2002a,b c
280 MHz	AFC via movable diaphragm that changes the length of the cavity; superheterodyne detection	Hill and Wyard, 1967
280 MHz	imaging system	Alecci *et al.*, 1992b
280 MHz	quartz oscillator feeds a Helmholtz excitation coil orthogonal to a solenoidal detection coil	Dijret *et al.*, 1994
280 MHz	pulsed DNP, LGR	Alecci and Lurie, 1999
300 MHz	EPR of metals at 4 K	Feher and Kip, 1955
300 MHz	push-pull oscillator induction-coupled to sample coil; electromagnet modulated at 6 KHz	El'sting, 1960
300 MHz	transmission mode	Cook and Stoodley, 1963

Frequency, Spectrometer type	Comment	Reference
300, 700, 900 MHz	LODESR	Yokoyama *et al.*, 1997a, 1997b
300 MHz pulsed	50-70 ns pulses, 350-400 ns recovery time, FID detection, imaging system	Bourg *et al.*, 1993; Murugesan *et al.*, 1997, 1998; Afeworki *et al.*, 2000; Subramanian *et al.*, 2002
300 MHz CW	imaging system	Brivati *et al.*, 1991; Stevens and Brivati, 1994
300, 600 MHz	similar to spectrometer of Feher and Kip, 1955	Alger *et al.*, 1959
302 MHz	operation down to 0.3 K	Medvedev *et al.*, 1976
310, 930 MHz	coaxial hybrid ring, superheterodyne detection	Duncan and Schneider, 1965
315 MHz	DPPH sample in oscillator-detector assembly	Marcley, 1961
680 MHz EPR, 27.7 MHz NMR	combined CW EPR and CW NMR spectrometer	Sato *et al.*, 1997
700 MHz	rapid scan imaging	Oikawa *et al.*, 1995; Ogata, 1995
750 MHz	imaging system	He *et al.*, 1999
900 MHz	induction spectrometer	Duncan, 1967
900 MHz	electrically detected magnetic resonance, BLGR	Sato *et al.*, 2000
1 GHz	designed for aqueous samples	Brown, 1974
1 GHz	circulator, surface coil	Nilges *et al.*, 1989
1 GHz	Pulsed, for *in vivo* imaging	Giuseppe *et al.*, 2001
1-2 GHz	solid state source, circulator, reflection resonator	Giordano *et al.*, 1976
1-1.8 GHz	tunable, homodyne	Dahlberg and Dodds, 1981
L-band	circulator, stub tuner, BLGR added to commercial EPR	Ono *et al.*, 1986
L-band	used iron-core electromagnet and console of a Varian E112	Nishikawa *et al.*, 1985; Fujii and Berliner, 1985; Berliner and Fujii, 1985; Berliner and Koscielniak, 1991
1-2 GHz	imaging system	Zweier and Kuppusamy, 1988, 1994; Kuppusamy *et al.*, 1994; Kuppusamy *et al.*, 2001
1.3 GHz	3D gradients	Colacicchi *et al.*, 1988; Alecci *et al.*, 1992b
1.8 GHz pulsed	pulsed microwaves and rapid field scan for fast-passage recovery measurement of relaxation times	Rannestad and Wagner, 1963
2 GHz	circulator, helix, 360 Hz field modulation; PIN modulator	Adkins and Nolle, 1966

Frequency, Spectrometer type	Comment	Reference
	permitted saturation recovery measurements	
S-band pulsed	circulator, LGR, 1 KW TWT, DBM detector	Clarkson *et al.*, 1989, 1992
S-band	for process control	Hyde and Froncisz, 1981
S-band, pulsed	ESE, circulator, LGR, 1 KW TWT, 4.2 K	Hankiewicz *et al.*, 1993; Romanelli *et al.*, 1994
3 GHz	for determination of the sign of the Lande g-factor	Charru, 1956
10 cm wavelength (3 GHz)	hybrid ring in place of the magic T that was used at 3 cm wavelength; AFC	Strandberg *et al.*, 1956
3.8-4.4 GHz	klystron and pulsed TWT, rectangular TE_{011} cavity, for saturation recovery	Bowers and Mims, 1959
1001 MHz, zero-field	for study of organic triplet states at 1.2 K with 1 W microwave pulses, circulator, DBM	Schmidt, 1972
PEDRI, Overhauser effect	69, 74, 197, 198, 208 MHz	Grucker and Chambron, 1989; Grucker, 1990; Grucker and Chambron, 1993; Guiberteau and Grucker, 1997
PEDRI, Overhauser effect	field cycling	Lurie *et al.*, 1988, 1998
EPR by DNP	EPR at 62, 66, 69, 72, 74, 198 MHz	Guiberteau and Grucker, 1996; Grucker *et al.*, 1996
LODESR	300 MHz	Nicholson *et al.*, 1994b, 1996
EPR and DNP; CW EPR, pulsed NMR	9.5 mT, 267 MHz for EPR, 404 KHz for NMR	Ardenkjaer-Larsen *et al.*, 1998

6. MAGNETS

For much of the EPR spectroscopy in the *ca.* 1 to 9 GHz range, standard EPR iron-core magnets have been used. The air-gap between the pole faces of such magnets is usually less than about 60-70 mm, which limits the size of the resonator and sample that can be used. For imaging with magnetic field gradients, there is a problem concerning where to locate the magnetic field sensing device in iron-core magnets. If the magnet has tapered pole faces, some labs put a Hall probe on the tapered face. In other cases, researchers iteratively find a location that is unaffected by the field gradients. In the Bruker E540 system the effects of the gradients on the Hall probe are calibrated and the required corrections are included in the software.

For EPR below *ca.* 1 GHz, an air-core magnet is a practical way to produce the Zeeman field. An air-core magnet can use current control rather than field control, and avoid the problem of sensor location. Air-core magnets also generally provide a larger experiment volume in which one can use larger resonators, and in which larger gradient coils can be located. Some of the magnets described have used a pair of coils in the Helmholtz configuration (coil separation equals coil radius). A solenoid, as is commonly used in NMR, is less attractive in EPR because of the restriction on orientation of the resonator and restriction on experimental access perpendicular to the axis of the solenoid. A four-coil magnet design is a reasonable compromise between ease of experimental access and achieving a large volume of magnetic field homogeneity (Garrett, 1967). A four-coil design has shown high homogeneity in Denver and Chicago (Rinard *et al.*, 2002c).

The Halpern magnet has been described in some detail (Halpern *et al.*, 1989; Halpern and Bowman, 1991). For 1 GHz EPR a Helmholtz pair, 13 cm i.d., 3.6 cm apart, was wound with 5 cm wide anodized aluminum tape (Brown, 1974). 37.8 mT was produced with 7.5 A. A magnet support structure for a Helmholtz magnet and gradient coils is shown in Ogata (1995) and Yokoyama *et al.* (1996). The use of Mumetal discs (16 inch diameter, 1/8 inch thick) on Helmholtz-type coils (18 inch mean diameter) increased the spacing between the coils to about 6 inches. This provided space for variable temperature apparatus. The field was homogeneous to 3 parts in 10^5 over a 2 cm diameter sphere (Collingwood and White, 1967). To scan 15 mT in 1 s, a Helmholtz coil pair was constructed (Oikawa *et al.*, 1995; Ogata, 1995). The coil was 272 mm i.d., 333 mm o.d., and 4 mm thick, constructed of 32 4-turn layers, with a total resistance of 5 Ω (Yokoyama *et al.*, 1996).

Sotgiu designed a multipolar magnet with which it was possible to rotate the direction of the magnetic field and field gradients by controlling the currents on each pole (Sotgiu, 1986). A 16-pole implementation for low-frequency EPR imaging was described in detail (Alecci *et al.*, 1991). It achieved 30 mT fields and 50 mTm^{-1} gradients. A finite element analysis has been performed (Chiricozzi *et al.*, 1998). Calculations were presented of the generation of axial fields of arbitrary profile by multiple cylindrical coils (Sotgiu *et al.*, 1987b) and of generalized Anderson coils (Momo *et al.*, 1988). Methods of analyzing and correcting multipolar magnets were published (Alecci *et al.*, 1992c; Cirio *et al.*, 2001). A magnet for 1 GHz imaging system had a homogeneity of 100 ppm in a 10 cm diameter sphere (Placidi *et al.*, 2002).

A unique yokeless iron-core electromagnet described by Strandberg *et al.* (1956) provided good access to the experiment in the magnet gap, but had very large stray fields.

References to these and other magnet designs are summarized in Table 2.

Table 2 Magnet

Type of magnet	Comment	Reference
yokeless iron-core	11,400 Gauss with a 2 5/8 inch gap	Strandberg *et al.*, 1956
solenoid	6 cm diameter, 50 cm long, water cooled	Feher and Kip, 1955
solenoid	10 inches long by 2 inch diameter, dimensions chosen to fit gamma source	Cook and Stoodley, 1963
solenoid	30 cm diameter, 45 cm long, 800 A T^{-1}, 80 ppm homogeneity over 10 mL sample volume	Hornak *et al.*, 1991
shaped Helmholtz coils	equivalent turns on each coil are spaced one radius apart to increase homogeneity	Marcley, 1961
Helmholtz-type coils and Mumetal pole pieces	18 inch diameter, 6 inch apart, with 16 inch diameter, 1/8 inch thick Mumetal discs	Collingwood and White, 1967
Helmholtz	0.09 T	Halpern *et al.*, 1989
Helmholtz, similar to Halpern design	homogeneous to 3 parts in 10,000 over a 4 cm diameter sphere; main magnet field was modulated at 1-4 KHz	Bolas *et al.*, 1996
Helmholtz	30 cm diameter, water cooled; 10 ppm field homogeneity in 5 cm^3 volume	Murugesan *et al.*, 1998
Helmholtz	0.5 m bore, ±0.05% homogeneity within central 0.1 m^3 region; 3 KW	Brivati *et al.*, 1991; Stevens and Brivati, 1994
2 Helmholtz pairs, one for rapid scan	for 700 MHz EPR, magnet coils were 407 mm i.d., 633 mm o.d., 62 mm thick, and 262 mm apart, 28 turns/layer x 62 layers, with supplementary Helmholtz pair for rapid scan 272 mm i.d., 333 mm o.d., 4 turns/layer x 32 layers	Oikawa *et al.*, 1995, 1996; Ogata, 1995; Yokoyama *et al.*, 1996
field-cycled	field compensation using one constant magnet and another magnet to switch fields in *ca.* 5-10 ms	Lurie *et al.*, 1989, 1991b

Type of magnet	Comment	Reference
4-coil	100 G, 0.1 G homogeneity over an oblate spheroid of 11.5 by 9.7 cm	Hill and Wyard, 1967
4-coil	8 cm diameter by 8 cm long; 1 cm homogeneous volume	Dijret *et al.*, 1994
4-coil resistive magnet with secondary Helmholtz pair	same magnet system was used for field-cycled PEDRI and LODESR	Lurie *et al.*, 1992; Nicholson *et al.*, 1994
4-coil	vertical field of up to 0.04 T	McCallum *et al.*, 1996a
4-coil	40 ppm homogeneity over 15 cm working diameter at 90 G	Rinard *et al.*, 2002c
8 Cu coils with iron structure	23 cm i.d., 48 cm long, 12 cm diameter homogeneous region, 0.1 T, 2.2 KW	Sciandrone *et al.*, 2000
Multipole	16 individually-controllable poles, 27 cm aperture, 60 cm long	Sotgiu, 1986; Alecci *et al.*, 1991, 1992a,b,c; Chiricozzi *et al.*, 1998
multipole, laminated, produces B_0 and 2 gradients	80 mT, 40 ppm homogeneity in a 10 cm sphere	Chiricozzi *et al.*, 1998
Helmholtz	50 cm diameter, two additional sets of 36 cm diameter Helmholtz coils were used to cancel the earth's field at the center of the main coils	Gerkin and Szerenyi, 1969
Helmholtz sweep coils	coils on resonator for fast-passage recovery measurements	Rannestad and Wagner, 1963

7. MAGNETIC FIELD GRADIENTS

Many methods of generating magnetic field gradients were reviewed in Eaton, Eaton, and Ohno (1991). It is common to use coils that follow the scheme set forth by Anderson (1961). Detailed dimensions are given by Brivati *et al.* (1991) and by Stevens and Brivati (1994). Two key points should be emphasized. First, a coil system for generating a linear magnetic field gradient has to be about the same size, power, and design precision as a magnet that would produce a field of the same magnitude and uniformity. For example, a z-gradient of 1 T/m (10 G/cm) highly linear over a sample 10 cm long would have to be a four-coil magnet system with the same size and power as a magnet producing a field of 0.1 T with the same uniformity over this sample. This is rarely achieved in practice. Second, the magnetic fields used for low-frequency EPR are not large relative to the gradient fields needed for high-resolution imaging, so one cannot ignore the off-axis components, as can usually be done in X-band EPR imaging. The effect of vectorial components of the gradient fields when the gradients are not small relative to the main magnetic field was addressed by Gillies *et al.* (1994) and references therein.

References to these and other magnet field gradient designs are summarized in Table 3.

Table 3. Magnetic Field Gradients

Gradient design or goal	Comments	Reference
2-dimensional imaging	3 G/cm z-gradient produced by splaying coils; x-gradient produced with Anderson coils	Halpern *et al.*, 1989
G_z produced by Maxwell pair and G_x and G_y produced by two sets of 4 rectangular Anderson coils	*ca.* 20-50 mT/m; 3 KW power supplies for each set of coils, which are water-cooled	Brivati *et al.*, 1991; Stevens and Brivati, 1994
G_z produced by Maxwell pair, G_x produced by rectangular Anderson coils	*ca.* 23 G/cm	Nishikawa *et al.*, 1985; Fujii and Berliner ,1985
Maxwell pair	up to 5 G/cm, 2.9 cm diameter separated by 2.5 cm	Bourg *et al.*, 1993
anti-Helmholtz and rectangular coils	79 mm between coils, 3D gradients, up to 4 mT/cm in the range of 15 mm from the center	Ishida *et al.*, 1992; Oikawa *et al.*, 1995, 1996; Ogata, 1995; Yokoyama *et al.*, 1996, 1997a
"butterfly" coil	produces magnetic field along the cylindrical axis and a 0.35 G/cm field gradient in the perpendicular direction with 5% uniformity	Di Luzio *et al.*, 1998
multipole, laminated, produces B_0 and 2 gradients	0.15 T/m radial gradient, G_z = 0.025 T/m	Alecci *et al.*, 1992b; Chiricozzi *et al.*, 1998
Multipole	Homogeneity of 100 ppm in a 10 cm diameter sphere	Placidi *et al.*, 2001

8. MICROWAVE/RF SOURCE

For EPR experiments, in addition to the obvious matters of frequency and power output, the selection of a microwave/RF source includes consideration of the phase noise and the ability to control the source frequency with an AFC system. The phase-noise dependence on frequency away from the center band differs depending on the nature of the frequency generator. Since most CW EPR experiments use magnetic field modulation of the order of tens of KHz to move the detection frequency away from the center band frequency, the source phase noise in this spectral region should be considered. One source may have superior phase noise at 100 KHz from center, but another might be superior at 5 KHz. Lower modulation frequencies, such as 5.12 KHz (Halpern *et al.*, 1989) are advantageous for achieving smaller eddy current losses compared with the traditional 100 KHz modulation. Some sources, such as the PTS synthesizers commonly used in NMR spectrometers, cannot be controlled by an analog AFC circuit. See Hornak *et al.* (1991) for digital AFC of a PTS synthesizer.

No EPR system recently described has used a klystron in the L- and S-bands. The Engelman Microwave CC-12 and CC-24 cavity-stabilized oscillators used in some L-band and S-band spectrometers (Hankiewicz at al., 1993; Romanelli *et al.*, 1994; Rinard *et al.*, 1999a,b) are no longer available.

References to papers that discuss RF sources for low-frequency EPR are summarized in Table 4.

Table 4. RF source

RF source	Comments	Reference
Superregenerative FET oscillator	29.7 MHz, for NQR, ESR, and NMR	Bruin and Khunaysir, 1970
HP 3200B	used for 10-120 MHz	Hatch and Kreilick ,1972
HP8640B signal generator	20 dBm output	Halpern *et al.*, 1989; Alecci *et al.*, 1992, 1998b
Fluke 6082A signal generator	SSB phase noise –140 dBm at 100 KHz from carrier at 1 GHz	Quine *et al.*, 1996
Magnum Microwave oscillator	730-780 MHz	He *et al.*, 1999; Chzhan *et al.*, 1999
transistor oscillator	1.06 GHz	Brown, 1974
tunnel diode in a cavity	*ca.* 1 GHz EPR	Anderegg *et al.*, 1963
Solid State Technology SSV-0113D oscillator	200 mW	Giordano *et al.*, 1976
phase-locked crystal oscillator	300 MHz	Brivati *et al.*, 1991
Engelman CC-12 cavity-stabilized oscillator	1-2 GHz	Nishikawa *et al.*, 1985; Fujii and Berliner, 1985; Berliner and Fujii, 1985; Berliner and Koscielniak, 1991; Nilges *et al.*, 1989
HP8644A signal generator	300 MHz	Bourg *et al.*, 1993
Gigatronics 610	300 MHz LODESR	Nicholson *et al.*, 1994b
klystron	3.8-4.4 GHz	Bowers and Mims, 1959
Polarad 1207M1 oscillator	*ca.* 4 GHz	Gerkin and Szerenyi, 1969
cavity-stabilized transistor oscillator	1-2 GHz	Zweier and Kuppusamy, 1988, 1994

9. BRIDGE DESIGN

Of the many aspects of EPR bridge design, we focus on the directional device and the detector, since these are most central to the overall bridge layout and performance. The detectors are usually crystals or double-balanced mixers. There are many tradeoffs in the selection of the directional device.

Kent and Mallard (1965) point out that a marginal oscillator has the advantage that an absorption signal free of dispersion is detected, but Hutchinson and Mallard (1971) argue that the marginal oscillator circuit is particularly sensitive to small changes in Q, so it would not be suitable for *in vivo* studies.

A circulator, such as is common in modern X-band spectrometers has the advantage of low insertion loss (of the order of 0.5 dB) (Brivati *et al.*, 1991; Stevens and Brivati, 1994), but below *ca.* 1 GHz the bandwidth becomes very small (*ca.* 10% of center band). This narrow bandwidth forces the resonator (+sample) to be designed to match the bridge frequency. Circulators in the VHF range are large, narrow-band, and expensive. A circulator was used in a 250 MHz CW and pulse spectrometer (Quine *et al.*, 2002) and in a 1 GHz spectrometer (Giuseppe *et al.*, 2001).

A directional coupler can be used to direct power to and signal from the resonator (Rannestad and Wagner, 1963). One advantage is that the bandwidth of a coupler is usually greater than that of a circulator, especially at low frequencies. Collingwood and White (1967) found adequate directivity at 85 MHz with a 10 dB coupler of nominal bandwidth 200-400 MHz. It is desired to have the signal pass through without attenuation except a small insertion loss (*ca.* 0.2 dB) and the coupling loss. This requires that the source be injected through the coupled port, thereby losing source power in accordance with the coupling value of the directional coupler. To keep source power from reaching the detector requires a coupler of high value. For example, a 20 dB coupler would isolate the detector from the source by about 20 dB, and lose very little of the signal, but would require that almost all of the source power be discarded (into the load on port 4 of the directional coupler). A 3-dB coupler would be functionally equivalent to a hybrid coupler (see below) in this application. A compromise of about 10 dB would discard less source power, but would not isolate the detector from source power by very much. Use of a directional coupler is practical only for low-power operation. Berliner demonstrated an L-band bridge based on a directional coupler (Koscielniak and Berliner, 1994).

A common approach is to use a hybrid coupler to direct the source power to and signal from the resonator (Klein and Phelps, 1967). Power into one port of a hybrid (also called a hybrid junction) divides equally between two other ports. The phase relation between these two other ports is either 90 degrees (quadrature hybrid) or 180 degrees. In an EPR bridge, a 50 Ω load terminates one port, and half of the source power and half of the EPR signal are absorbed in this load. These losses are compensated by the convenience of much larger bandwidth than can be achieved with a circulator at low frequency. The hybrid coupler is analogous to the waveguide magic-T that was used in early X-band EPR spectrometers. The Halpern and Hornak spectrometers use a 180° hybrid coupler, and their bridge schematics show how the hybrid is implemented (Halpern *et al.*, 1989; Hornak *et al.*, 1991). Decorps and Fric (1972) used a circulator at 210 MHz and a hybrid junction at 40 and 80 MHz. Alecci *et al.* (1992b) used a hybrid junction that provided 40 dB decoupling between source and detector. These researchers compared a simple bridge design with 10-1000 MHz bandwidth and a reference arm bridge with a preamplifier and two mixers. There was no improvement in S/N with the reference arm bridge, and only a small improvement by placing a preamplifier in the detection path, indicating that detector noise did not dominate in this system (Alecci *et al.*, 1992b).

Ono *et al.* (1998) used a standing wave ratio (SWR) bridge (Wiltron 60N50) instead of a circulator. The SWR bridge is a precision directional device with good isolation. A 300 MHz pulsed EPR spectrometer (Bourg *et al.* 1993; Murugesan *et al.*, 1997, 1998) used a diplexer transmit/receive gate analogous to those in pulsed NMR spectrometers instead of a circulator or hybrid. Alecci *et al.* (1998b) used separate transmit and receive resonators to help isolate the FID from the incident pulse.

A homodyne configuration had lower noise than a heterodyne configuration in a 200 MHz spectrometer, but it was suggested that this was not an inherent benefit but merely a difficulty in locking all of the heterodyne frequencies together to have low effective FM noise figure (Bolas *et al.*, 1996).

Considerable effort has been applied to the task of designing an automatic frequency control (AFC) circuit for low-frequency spectrometers with low-Q resonators (Hyde and Gajdzinski, 1988; Alecci *et al.*, 1995; McCallum *et al.*, 1996b; Bolas *et al.*, 1996; Willer *et al.*, 2000). In addition to controlling frequency to compensate for the long-term thermal drift of the resonator, and give good lock to absorption and not dispersion, low-frequency systems intended for *in vivo* studies have to compensate for animal motion, and do so on the time scale of physiological motion.

References to these and other bridge designs are summarized in Table 5.

Table 5. Bridge

Operating frequency	Comments	Reference
1-400 MHz	magic T and reference delay line, DBM	Gebhardt and Dormann, 1989
85 MHz	directional coupler used below specified range	Collingwood and White, 1967
100 MHz	10 dB directional coupler, automatic tuning control to adjust for mouse physiology	Hutchinson and Mallard, 1971
100-1000 MHz	crossed coil resonator, rotate coils to change isolation, observe absorption or dispersion	Borel and Manus, 1957
ca. 200 MHz	reflection homodyne, quadrature detection	Bolas *et al.*, 1996
200 MHz	180-degree hybrid T; DBM	Hornak *et al.*, 1991
250 MHz	used high-level mixers, AFC, ACC	Halpern *et al.*, 1989
280 MHz CW	hybrid junction	Alecci *et al.*, 1992b
280 MHz	LODESR, bird cage for EPR excitation and solenoid for NMR	Panagiotelis *et al.*, 2001
300 MHz	pulsed	Bourg *et al.*, 1993
300 MHz	homodyne reflection type, with AFC, ACC, and APC	Brivati *et al.*, 1991; Stevens and Brivati, 1994
300 MHz	AFC	Alecci *et al.*, 1995
300 MHz	parallel coil, for pulsed EPR	Devasahayam *et al.*, 2000
310, 930 MHz	hybrid ring operates at 1, 3, 5 times 310 MHz	Duncan and Schneider, 1965
700 MHz	surface coil, analysis and impedance matching	Hirata *et al.*, 1995, Hirata and Ono, 1997; Tada *et al.*, 2000
750 MHz	narrow-band RF bridge	He *et al.*, 1999
760-820 MHz	circulator, 3-stub tuner, DBM	Ishida, *et al.*, 1989
774 MHz	dual-diode detector for homodyne bridge	Koscielniak and Berliner, 1994
1 GHz	circulator, AFC	Brown, 1974
1.2 GHz	fixed-frequency, compared with standard bridge	Chzhan *et al.*, 1995
1.8 GHz	directional coupler	Rannestad and Wagner, 1963
1-2 GHz	5-port high-isolation circulator	Giordano *et al.*, 1976
1-1.8 GHz	circulator, reference arm	Dahlberg and Dodds, 1981
1-2 GHz	circulator, Schottky diode detector, AFC	Zweier and Kuppusamy, 1994
L-band	circulator, stub tuner	Nishikawa *et al.*, 1985; Fujii and Berliner, 1985; Berliner and Fujii, 1985; Berliner and Koscielniak, 1991
L-band	pulsed ESE and SR	Quine *et al.*, 1996
L-,S-band	AFC system	Hyde and Gajdzinski, 1988
S-band	pulsed ESE	Hankiewicz *et al.*, 1993; Romanelli *et al.*, 1994
S-band	pulsed ESE and SR	Rinard *et al.*, 1999a
2.2 GHz	Bridged LGR	Petryakov *et al.*, 2001

10. RESONATOR

Over the years a wide variety of microwave structures have been used for EPR, especially at frequencies below X-band. Several of them are pictured in a review of (largely) Soviet research in this area (Abdrachamov and Ivanova, 1973). Since rectangular and cylindrical cavities are very large at low frequencies, interest has focused on more compact devices, such as coaxial cavities, helices, coaxial cavity containing a helix, "chink opening," *etc.*

Decorps and Fric (1969) compared several types of resonators at 210 MHz, and subsequently (1972) used a capacitively coupled strip-line (LGR) resonator at 210 MHz, and helices at 40, 80, and 210 MHz.

To avoid the large size of a cavity resonator at 1 GHz, Brown (1974) constructed a stripline resonator in which the center conductor was 13.2 cm by 2 cm, and the ground planes were 20 by 4.68 cm. The sample was placed near the center of the conductor, perpendicular to the long dimension, and between the center conductor and one of the ground planes. Near the ends of the center conductor were two quartz rods that were used to tune the frequency. The nodal E-plane is produced at the center of this assembly.

A 1-2 GHz lumped-circuit resonator (8.5 mm diameter) described by Giordano *et al.* (1976) had such good separation of E and H fields that the Q decreased less than 20% when an aqueous sample was inserted. Also, conducting materials, such as electrodes or other coils could be inserted with minimal effect.

Dahlberg and Dodds (1981) constructed 1-1.8 GHz stripline resonators from brass and from polystyrene wrapped with 20 μm thick "household" aluminum foil. Modulation frequencies up to 200 Hz could be used with the brass resonator, and up to 500 Hz with the aluminum foil resonator.

The Schneider and Dullenkopf slotted tube (1977), Hardy and Whitehead split-ring (1981), and Hyde and Froncisz loop-gap resonator (1982) papers stimulated many labs to design resonators to fit the experiment, especially at frequencies below X-band. Since 1982 (Hyde and Froncisz, 1982) the resonator for low frequency EPR usually has been a loop-gap-resonator (LGR) or related lumped-circuit type device. The "strip-line" resonator used in the Halpern 250 MHz spectrometer is similar to a LGR (Halpern *et al.*, 1989). Although many resonators are made of solid metal for convenience of construction, to achieve good modulation penetration, Halpern *et al.* (1995) fabricate resonators by electroplating ABS plastic with Cu and Au. Ono *et al.* (1986) introduced the important innovation of using a metal shield over the gap inside the LGR to minimize the fringing electric field in the sample.

A series of papers by Zweier and coworkers (Zweier and Kuppusamy, 1988; Chzhan *et al.*, 1993, 1995, 1999; Kuppusamy *et al.*, 1994, 1998a,b; He *et al.* 1999) describe L-band resonators designed for EPR imaging of perfused hearts and living mice. Since the heart preparations and animals involve lossy aqueous solutions, particular attention was paid to minimizing the E field at the sample. For example, the inner edge of the capacitive gap in a 1-loop-2-gap resonator was recessed to decrease the E fringe field (Zweier and Kuppusamy, 1988).

Nitroxide radicals in human skin were imaged using an S-band system (He *et al.*, 2001a). A bridged LGR, which resonated at 2.2 GHz when loaded with the human forearm skin, was used to image nitroxyl radical distribution and metabolism. A 7 mm diameter region of the skin was positioned with a special holder at the end of a 7 mm diameter LGR. Details of this resonator were described by Petryakov *et al.* (2001). Spectra were obtained of 10^{-8} moles of nitroxide applied to the surface of the skin of a human forearm. 1 mm from the end of the resonator, the microwave magnetic field B_1 was 0.62 $G/\sqrt{W}$, and significant EPR signal could be measured up to 6 mm from the end of the resonator. An analogous resonator was used for L-band (1.32 GHz) measurement of a mouse tumor in contact with the resonator, which could sample a volume 10 mm in diameter and 5 mm deep (Ilangovan, *et al.*, 2002). He *et al.* (2001b) described the use of varactor diodes for automatic tuning and automatic coupling of a 700 MHz resonator of a type previously described as transverse electric reentrant (Chzhan *et al.*, 1999).

To avoid the radiation losses inherent in some implementations of the LGR for animals, Sotgiu and coworkers (Momo and Sotgiu, 1984; Sotgiu and Gualtieri, 1985; Sotgiu, 1985; Alecci *et al.*, 1989) created a reentrant form of the LGR in which the return flux was contained within the resonator except for the cylindrical region via which the sample was inserted into the resonator. Essentially, an inductive loop, which may have several branches, surrounds one or more capacitances. This is a flexible design, which presages designs developed in our laboratory for the crossed-loop resonator (Rinard *et al.*, 1996a,b; 2000) and also used in other resonators where it is desirable to avoid a separate shield (Rinard *et al.*, 1999a). Sample regions with 50 mm diameter, sufficient for the study of mice, were produced at 1.34 GHz and 38 mm diameter at 1.68 GHz (Sotgiu, 1985).

Hornak *et al.* (1991) built a 200 MHz resonator analogous to a version of a LGR that had been used for ^{31}P NMR. Called a single turn solenoid, it was made of copper foil (30x82 mm, 0.08 mm thick) on a polyvinyl chloride form and three 20 pF chip capacitors were soldered across the gap. This was placed in a 38 mm diameter, 76 mm long, 0.5 mm thick silver-plated brass shield.

Brivati *et al.* (1991) tuned the frequency of a *ca.* 300 MHz LGR by moving a dielectric slab in the gap. This type mechanism was used as part of an automatic coupling control (ACC) circuit.

An innovative crossed-loop resonator (CLR) isolates the EPR signal from source noise and from pulse power in pulsed EPR. At L-band and S-band a CLR has been demonstrated to give results superior to a reflection resonator for CW dispersion spectroscopy, superheterodyne spectroscopy and ESE (Rinard *et al.*, 1996a,b, 2000). The CLR also has been implemented at 250 MHz with a 2.5 cm diameter sample loop (Rinard *et al.* 2002a).

A 300 MHz resonator used for pulsed EPR (Murugesan *et al.*, 1998; Devasahayam *et al.*, 2000) was 25 mm diameter and 25 mm long, being constructed of 11 parallel loops spaced 2.5 mm apart and connected in parallel. The resonator incorporated a parallel resistance to help lower the Q to decrease the ring-down time. Another resonator used for pulsed EPR in the 200-400 MHz range was a short-circuited coaxial line (Rubinson *et al.*, 1998), analogous to that used at 883 MHz (Rubinson *et al.*, 1995). At 300 MHz 10 W incident power yielded B_1 = 0.9 G. No tuning of the probe was needed to perform relaxation time measurements at 200, 250, 300, and 350 MHz.

To reduce the resonator Q for pulsed 286 MHz EPR, a two-turn coil-shaped polyethylene tube was placed inside a LGR, and water was flowed through the tubing, reducing the Q from 570 to 70 (Yokoyama *et al.*, 1999). Using 3 or 5 turns lowered the Q to 44 or 30. This resonator was used to obtain LODESR spectra of a triarylmethyl radical in water with 200 ns 100 W pulses repeated every 2000 ns (average power 10 W) (Yokoyama *et al.*, 1999). Although there might be situations in which it is desirable to lower the Q by adding loss to the resonator, the signal is larger using an overcoupled high-Q resonator than using a resonator with inherently lower Q (Rinard *et al.*, 1994).

An active resonator system was developed for 700 MHz CW EPR (Sato *et al.*, 2002). This is an example of what is called regeneration, as was used in early radio receivers to narrow the bandwidth and block interfering radio signals. Sato *et al.* showed that it increased the resonator Q, and hence the EPR signal, but it also increased the noise, such that the observed signal-to-noise ratio increased only 50% even though the signal increased a factor of seven.

Giuseppe *et al.* (2001) designed a one-loop two-gap bridged LGR tuned to 1 GHz for EPR with a coaxial solenoid outside the LGR tuned to 1.52 MHz for NMR detection. The homogeneous RF region was 3 cm diameter and 4 cm long. A shield surrounded the composite resonator.

A method for estimating the frequency of a bridged loop-gap resonator was described and compared with measurements on prototype resonators in the *ca.* 1.2-1.5 GHz range (Hirata and Ono, 1996). Microwave field distributions were calculated for an S-band BLGR (Willer *et al.*, 2000).

Multiple LGRs, magnetically coupled, yielded a 1.3 GHz resonator with wide bandwidth designed for pulsed EPR (Sakamoto *et al.*, 1995). Design equations and test data of a prototype were presented.

There are advantages to using surface coil resonators (Bacic *et al.*, 1989; Hirata *et al.*, 1995, 2000) for localized *in vivo* spectroscopy. Although some surface coil probes are a single loop of wire or a flat trace, an LGR or a dielectric resonator (Jiang *et al.*, 1996) can also be used as a surface coil. Varactor diodes were used to electronically tune the frequency and match of an L-band surface coil resonator (Hirata *et al.*, 2000). Flexible leads to the loop of a surface-coil-type resonator permit it to be used as an endoscope for *in vivo* spectroscopy (Ono *et al.*, 1994; Hirata and Ono, 1997; Lin *et al.*, 1997). A surface coil resonator for localized EPR was constructed by cutting a toroidal resonator along a chord to obtain a flat surface (Sotgiu *et al.*, 1987a).

Analysis (Hirata *et al*, 1995) and impedance matching (Hirata and Ono, 1997) of surface-coil-type EPR resonators have been described in detail, and their application to studying organs of nitroxide-treated rats was reported (Tada *et al.*, 2000). At 700 MHz, loop diameters of 6 - 18 mm (Hirata *et al.*, 1995) were tested, and the signal strength was largest for a diameter of *ca.* 8 mm. Impedance matching with varicap diodes provided a tuning range of 50 MHz (Hirata and Ono, 1997). The Q was slightly lowered relative to mechanical matching. Coils with 3, 4, and 10 mm diameter, resonating at about 720 MHz, were used to measure nitroxides in rats (Tada *et al.*, 2000). The B_1 field yielded measurable EPR signal up to about 2 mm from the surface of the coil. An exposed rat kidney was inserted into a 10 mm diameter surface coil to image nitroxyl radicals at 700 MHz (Ueda *et al.*, 2002). The loaded Q of the coil was about 70 under these conditions.

Some EPR systems also use a birdcage resonator (Bolas *et al.*, 1996), as is common in MRI systems. The LGR yields $\sqrt{3}$ larger B_1 for the same incident power as a birdcage resonator, but the axis of the LGR has to be arranged perpendicular to the axis of the birdcage resonator (Bolas *et al.*, 1996). A birdcage is usually capacitively coupled to the transmission line, but alternatively can be coupled with an inductive loop at some loss of B_1 homogeneity (Bolas *et al.*, 1996).

A 1.1 GHz "minimal cavity" (a ring resonator 30 mm diameter by 4.5 mm long), without a shield for ease of animal handling, was used to select a limited volume. The absence of a shield also was stated to avoid pickup of low-frequency modulation (Colacicchi *et al.*, 1996).

Resonant inductive coupling was proposed as an improved way of coupling resonators containing lossy electrically conducting samples, and demonstrated at 200 MHz. (Diodato *et al.*, 1998). In this scheme the inductive coupling loop is itself a resonator with tuning and matching capacitors, and the high-frequency combination of the two resonators was used. This was further elaborated in a resonator in which the resonant coupling loop is between two sections of a LGR (Diodato *et al.*, 1999).

References to these and other resonator designs are summarized in Table 6.

Table 6. Resonator

Type of resonator	Comments	Reference
1-400 MHz	lumped parameter delay line, untuned π network of coil and 2 capacitors	Gebhardt and Dormann, 1989
10-120 MHz coil of the Klein and Phelps, 1967 design	can be tuned *ca.* 20 MHz; obtained spectrum of 0.4 mM nitroxyl in water at 100 MHz	Hatch and Kreilick, 1972
27.7 MHz, 680 MHz	loop-gap and transmission line resonator, 48.5 mm i.d., Q=222 at 678.6 MHz, Q=96 at 27.66 MHz	Sato *et al.*, 1997
29-331 MHz	PEDRI: LGR, birdcage, or Alderman-Grant coil outside an NMR coil	Lurie *et al.*, 1990, 1991a,b,c, 1992; Nicholson *et al.*, 1994a; Mülsch *et al.*, 1999
51 MHz	8-leg, high-pass birdcage, 20 cm diameter, 20 cm long, Q=173	Lurie *et al.*, 1998
60 MHz	coil wound of flat Cu strip	Lloyd and Pake, 1954
85 MHz	helical, transmission and reflection	Collingwood and White, 1967
100 MHz coil	3 cm diameter single-turn Helmholtz pair	Hutchinson and Mallard, 1971
160, 288 MHz, 1.12 GHz	PEDRI, split-solenoid NMR coil 85 mm diameter, 67 mm long, and EPR coil of 20 loops connected in parallel 10 mm diameter, 20 mm long	Lurie *et al.*, 1988, 1989
200 MHz	single-turn solenoid, 12 mL volume	Hornak *et al.*, 1991
ca. 200 MHz	several birdcage resonators	Bolas *et al.*, 1996
200 – 400 MHz	LGR, multiply-tuned; can be set to irradiate simultaneously multiple transitions in a nitroxyl for PEDRI	Alecci *et al.*, 1996
210 MHz	several resonator types compared	Decorps and Fric, 1969
220 MHz LGR for transmit pulse and 2 pair of saddle coils for receiver	FT EPR; 59 mm i.d. by 20 mm long 1-loop 4-gap LGR, $\Lambda=19\mu T/\sqrt{W}$	Alecci *et al.*, 1998a,b
225 MHz	tunable LC circuit, 6 turns of #16 wire, ¾ inch diameter and 1 inch long	Strandberg *et al.*, 1956

Type of resonator	Comments	Reference
226 MHz	PEDRI with 625 kHz NMR transmit saddle coil and diameter solenoidal receive coil, 5.2 μT/√W.	Krishna *et al.*, 2002
237 MHz	LGR, 37 mm diameter, 37 mm long	McCallum *et al.*, 1996b
238-250 MHz	double split ring, 5.5 cm diameter, 10 cm long, Q=200 with 100 mL H_2O in 42 mm o.d. tube	Smith and Stevens, 1994
250 MHz	strip-line, 2.58 cm diameter, 3.2 cm long; Q = 550	Halpern *et al.*, 1989
250 MHz	strip-line, 1.6 cm diameter, 1.5 cm long, 0.27 G B_1 for 15 mW	Halpern *et al.*, 1995
250 MHz	Crossed loop resonator	Rinard *et al.*, 2002a
267 MHz	LGR, 2 cm diameter, 4.5 cm long, 4 cm diameter shield, Q=380	Ardenkjaer-Larsen *et al.*, 1998
268 MHz	low-temperature NMR coil probe with 4 capacitors for tuning and matching was proven with EPR of DPPH	Kim *et al.*, 1996
280 MHz	λ/4 coaxial line, 2.6 cm diameter, reflection cavity, Q=600	Hill and Wyard, 1967
280 MHz	orthogonal coils, 5.2 mm o.d. sample tube, solenoidal excitation coil and 2 cm diameter Helmholtz detection coil decoupled geometrically	Dijret *et al.*, 1994
280 MHz	LGR, 4.9 cm i.d., 10 cm long, 1-loop, 2-gap, Q=400 empty, 135 with rat	Quaresima *et al.*, 1992 Alecci *et al.*, 1992b
280 MHz	Cu tape on PVC form, 70 mm diameter, 120 mm long, Q=1300 empty, 25 with rat	McCallum *et al.*, 1996a
280 MHz LGR	for pulsed DNP, Cu tape on Perspex, 38 mm diameter, Q=173	Alecci and Lurie, 1999
280 MHz LGR	45 mm diameter Cu loop, 10 mm long, Q=570, Q was decreased to 70 with water in polyethylene tube inside LGR	Yokoyama *et al.*, 1999
200-400 MHz	short-circuited, coaxial line resonator	Rubinson *et al.*, 1998
300 MHz	LODESR; birdcage, 3.8 cm diameter, 3 cm long Q=70; NMR coil inside the birdcage	Nicholson *et al.*, 1994b, 1996
300 MHz	4-turn solenoid, 8 mm diameter	Bourg *et al.*, 1993
300 MHz	parallel coil resonator, 25 mm diameter, 25 mm long, Q=20-25 for pulsed EPR	Devasahayam *et al.*, 2000; Afeworki *et al.*, 2000
300 MHz	λ/2 coaxial transmission cavity, Q=1000-2000	Feher and Kip, 1955
300 MHz	half-wave coaxial line, transmission	Cook and Stoodley, 1963

Type of resonator	Comments	Reference
	mode, Q=450	
300 MHz	1-loop, 1-gap LGR, frequency tuned by moving dielectric slab in gap	Brivati *et al.*, 1991; Stevens and Brivati, 1994
300, 700, 900 MHz	LODESR, bridged LGRs with 4, 2, or 1 gap, with saddle-type pickup coils inside the LGR	Yokoyama *et al.*, 1997a,b
saddle-type pickup coils	analysis of sensitivity	Yokoyama *et al.*, 1997a,b, 1998
302 MHz	coaxial resonator with retarding helical system, Q=300 at room temperature	Medvedev *et al.*, 1976
310 MHz	λ/4 coaxial cavity, matched with double-stub tuner; Q=1000 at RT, 5000 at liq. He	Duncan and Schneider, 1965
300, 600 MHz	coaxial cavity, up to 10 mL of sample	Alger *et al.*, 1959
530 to 950 MHz	LGR frequency changed in 20 MHz steps by changing capacitance	Treiguts and Cugunov, 1995
549, 680 MHz	reentrant and split-ring	Alecci *et al.*, 1989
600-1200 MHz	λ/4 coaxial cavity with helical internal conductor, also used TE_{10n}, n=2-4, rectangular cavities	Abdrachmanov and Ivanova, 1973
700 MHz	flexible surface coil, 6 mm diameter, Q≈100	Hirata *et al.*, 1995
700 MHz	flexible surface-coil-type resonator, uses triaxial cable, 3.8 mm diameter loop, Q=82	Hirata and Ono, 1997
700 MHz	flexible surface-coil-type resonator, 5 mm diameter	Lin *et al.*, 1997
700 MHz	active resonator	Sato *et al.*, 2002
707 MHz	surface-coil-type, 7.5 mm diameter, Q=391	Ono *et al.*, 1994
750 MHz	rectangular reentrant resonator, transversely oriented electric field, Q=550	Chzhan *et al.*, 1999
760-820 MHz	bridged LGR, 43 mm diameter, 30 mm long, Q=1200	Ishida, *et al.*, 1989
883 MHz	short-circuited, coaxial line resonator, Q=69	Rubinson *et al.*, 1995
883 MHz,	reentrant LGR: 2-loop, 1-gap, 36 mm diameter, for mice, Q=525	Sotgiu and Gaultieri, 1985
890 MHz	BLGR, 4.4 cm diameter, 1 cm long, 0.5 mm Teflon spacers between loop and bridged shields, Q=510	Sato *et al.*, 2000
900 MHz induction	coil in coaxial cavity	Duncan, 1967
1.0 GHz	BLGR with coaxial solenoid for simultaneous EPR and NMR, B_1	Giuseppe *et al.*, 2001; Alfonsetti *et al.*, 2001

Type of resonator	Comments	Reference
	was uniform over about 25 mm	
1001 MHz	reentrant coaxial cavity, Q reduced to 250 by introducing an absorber into a region of high electric field; $B_1 \approx 0.3$ G with 1 W incident	Schmidt, 1972
1.06 GHz stripline	Q=909 empty, 604 with tube of water	Brown, 1974
1.07 GHz minimal cavity	30 mm diameter, 4.5 mm long, Q=90 empty, 40 with 15 mm diameter physiological saline, 20-30 with a 20-30 g mouse	Colacicchi *et al.*, 1996
1.1 GHz	double split-ring resonator surface coil, Cu on quartz tube, 8 mm diameter, Q≈200	Bacic *et al.*, 1989
1.1-1.2 GHz	LGR, 3.25 cm diameter, 2.5 cm long, two 1 mm bridged gaps	Jiang *et al.*, 1995
1.2 GHz	dielectric resonator surface probe	Jiang *et al.*, 1996
1.1-1.3 GHz	LGR, 2.5 cm diameter, 15 mL sample volume	Lukiewicz and Lukiewicz, 1984
1-1.8 GHz	quarter-wavelength stripline conductor, frequency tunable with capacitive element	Dahlberg and Dodds, 1981
1.2 GHz ceramic reentrant	3-loop-2-gap, 2 cm diameter sample loop, 0.14 G per square root watt	Chzhan *et al.*, 1993
1.2 GHz	electronically tunable 3-loop-2-gap LGR, piezoelectric adjustment of capacitance, 2 cm diameter sample loop	Chzhan *et al.*, 1995
1.2 GHz	toroidal surface coil	Sotgiu *et al.*, 1987a
1.25 GHz	surface coil BLGR built with Ag foil on quartz tubes, could sample a region 10 mm diameter and 5 mm deep	Kuppusamy *et al.*, 1998a,b
1.25 GHz	bridged LGR surface coil, 10 mm diameter, 5 mm deep sample volume	Kuppusamy *et al.*, 1998a,b
1.34 GHz	double reentrant LGR: 3-loop, 2-gap resonator, for mice, 50 mm diameter, Q=440; 1.68 GHz, 38 mm diameter, Q=515	Sotgiu, 1985
1.83 GHz	helix, various sizes were tested	Nishikawa *et al.*, 1985; Fujii and Berliner, 1985; Berliner and Fujii, 1985; Berliner and Koscielniak, 1991
1.86 GHz	flat loop surface coil, 4.1 cm long, 0.7 cm diameter, 0.8 mm gap	Nishikawa *et al.*, 1985; Fujii and Berliner ,1985;

Type of resonator	Comments	Reference
		Berliner and Fujii, 1985
L-band	interior Al foil shield added to LGR to restrain electric fringe field	Ono *et al.*, 1986
1-2 GHz	lumped LC circuit, Q=250 at 1 GHz and Q=150 at 2 GHz, sample in 3 parallel loops, 8.5 mm diameter and 3 mm apart; useful volume is 0.9 cm^3; the capacitive term is provided by stub tuners	Giordano *et al.*, 1976
L-band	reentrant, 4.2 mm diameter, 15 mm long sample loop	Quine *et al.*, 1996
L-band	electronically tunable surface coil	Hirata *et al.*, 2000
1-2 GHz recessed LGR	designed to minimize E fringe fields; 26 mm diameter, 25 mm long, Q=1000	Zweier and Kuppusamy, 1988
2 GHz	quarter wavelength coaxial cavity resonator with spiral internal conductor; Q=200	Denisov and Kalinichenko, 1965
L-band, S-band	crossed-loop resonator, dispersion, superheterodyne, ESE at L-band and S-band	Rinard *et al.*, 1996a,b, 2000
1-4 GHz reentrant cavity	Q≈2000, continuously tunable over 1-4 GHz	Shing and Buckmaster, 1976
2.1, 3.4, 6.4, and 9.4 GHz	re-entrant resonators, 4.9 to 8 mm diameter, Q=1650-1800	Momo and Sotgiu, 1984
2.5-2.9 and 2.9-3.4 GHz	cylindrical cavity with device for rotating sample at 4.2 K	Antipin, 1966
3.0-6.3 GHz	LGR	Hyde and Froncisz, 1981
3.8-4.4 GHz	rectangular TE_{011} cavity, for saturation recovery, single mode and dual-mode cavity for two-frequency SR measurements	Bowers and Mims, 1959
S-band LGR	inductive coupling	Hankiewicz at al., 1993; Romanelli *et al.*, 1994
ca. 4 GHz	cylindrical reentrant cavity made of steatite and silvered internally	Gerkin and Szerenyi, 1969

11. SENSITIVITY

Very few reports quantitatively assess sensitivity of low-frequency spectrometers. A detailed description of measuring the absolute signal intensity for an S-band pulsed spectrometer is available (Rinard *et al.*, 1999a,b). Sensitivities reported prior to 1967 were tabulated by Hill and

Wyard (1967), who also compared their sensitivity of $7x10^{14}$ and $2.3x10^{14}$ spins for two resonators with the theoretical values of $2.4x10^{14}$ and $5.6x10^{13}$ spins, respectively. Hutchinson and Mallard (1971) measured a sensitivity of $5x10^{15}$ spins per Gauss at 100 MHz, and estimated they should be able to detect $5x10^{14}$ spins per Gauss in a 10 mL sample. Kent and Mallard (1965) achieved a sensitivity of 10^{15} ΔH spins in *ca.* 0.5 mL samples using a marginal oscillator circuit.

11.1 < 500 MHz

Collingwood and White (1967) found a limiting sensitivity of 10^{16} spins in a mm^3 sample volume, using a helical resonator in a low-frequency (*ca.* 85 MHz) spectrometer. For aqueous samples, Decorps and Fric (1972) measured a sensitivity (defined as S/N = 1) of 10^{-7} M in 4 cm^3 at 210 MHz and 10^{-6} M at 40 MHz. At 200 MHz, Hornak *et al.* (1991) detected a 30 μM di-t-butyl nitroxide solution in a 10 mL ethanol sample, and stated that this corresponded to a sensitivity of $8x10^{21}$ spins T^{-1}. Bolas *et al.* (1996) showed that source (PTS310 synthesizer) phase noise was the limiting noise in a 200 MHz imaging system. Halpern *et al.* (1989) estimated a concentration sensitivity of $3x10^{-8}$ M for a line width of 1 G and S/N of unity at 250 MHz. Smith and Stevens (1994) reported $4x10^{-7}$ M sensitivity for a 100 mL volume sample at 250 MHz. Quaresima *et al.* (1992) extrapolated to a minimum detectable nitroxide concentration of 44 μM with S/N=2 at 280 MHz. At 280 MHz, Dijret *et al.* (1994) measured a sensitivity of $3.4x10^{17}$ spins T^{-1}. They also obtained a spectrum of 0.1 mL 30 μM degassed di-t-butyl nitroxide in ethanol. El'sting (1960) used synchronous detection to achieve S/N = 2 with $1.4x10^{-8}$ mole of DPPH at 300 MHz. Duncan and Schneider (1965) estimated $2x10^{14}$ spins per Gauss at S/N=1 for a 310 MHz spectrometer, which they calculated had a theoretical sensitivity of $3x10^{13}$ spins per Gauss. Matheson and Smaller (1955) estimated a limiting sensitivity of 10^{14} spins for a low-frequency spectrometer system that was only briefly described. Medvedev *et al.* (1976) found that a S/N ~ 4 could be obtained with 10^{-4} mole of DPPH at room temperature with 10^{-9} W from the 302 MHz source, in a system designed for operation down to 0.3 K. At 300 MHz, Cook and Stoodley (1963) observed a sensitivity of $5x10^{15}$ spins. The 300 MHz CW EPR spectrometer built by Brivati *et al.* (1991) could detect about $2x10^{-7}$ M nitroxyl in a 200 mL aqueous sample. In a 300 MHz animal imaging system, the minimum detectable number of spins for a nitroxyl spin probe was found to be $6x10^{17}$ (Koscielniak *et al.*, 2000).

11.2 500 MHz to <1 GHz

A 700 MHz surface-coil-type resonator, 7.5 mm diameter, 1 mm away from a quartz tube containing10 mL of physiological saline solution was able to detect 10^{-5} M tempol with S/N = 3.5 (Ono *et al.*, 1994). Rubinson *et al.* (1995) extrapolated to a limit of detection of 1.7×10^{14} spins per Gauss linewidth at 883 MHz for a short-circuited coaxial-line resonator. An induction spectrometer at 900 MHz (Duncan, 1967) detected 2.4×10^{13} spins per Gauss, compared with a theoretical 6×10^{12} spins per gauss.

11.3 1-2 GHz

A 1-GHz image of 0.7 g of DPPH had a signal-to-noise of 18 (Giuseppe *et al.*, 2001). Brown's 1 GHz spectrometer designed for aqueous samples could detect 1.2×10^{14} spins per cm^3 of 0.1 mT line width at room temperature (Brown, 1974). It was argued that 1 GHz was the optimum frequency for aqueous samples, taking into account the dielectric properties of water. The same aqueous solution of Fremy's salt was used at X-band in a spectrometer whose electronic circuitry was similar to that of the 1 GHz spectrometer, and the X-band sensitivity was measured to be 1.3×10^{15} spins per cm^3 (Brown, 1974). Giordano *et al.* (1976) measured a sensitivity of 1.3×10^{13} spins per cm^3 per Gauss at 1.037 GHz using 0.6 mW incident power. They observed that the sensitivity was higher at lower frequency, which they attributed to a decrease in Q with increase in frequency for the style of resonator used. Zweier and Kuppusamy (1988) obtained S/N~5 for 2 μM nitroxyl filling a 13 mm cylindrical tube in a LGR at 1.085 GHz, suggesting a detection limit (S/N=1) of 0.4 μM. Chzhan *et al.* (1993) extended this sensitivity to 0.25 μM with S/N=2. Colacicchi *et al.* (1996) were able to detect 10 μM aqueous CTPO in a 10 mm diameter tube in a simple 1.07 GHz ring resonator whose Q was reduced from 90 to 38 by also inserting a 15 mm diameter tube containing physiological saline. The weak pitch S/N check, standard at X-band, is usually not appropriate at lower frequency because the lower-frequency resonators are usually designed for larger samples. However, with a 4.2 mm sample loop, reentrant-type, LGR at 1.75 GHz S/N = 35 was measured using a standard weak pitch sample (Quine *et al.*, 1996). Nishikawa *et al.* (1985) found that the sensitivity of a L-band (1.83 GHz) "surface" coil immersed in 10^{-6} M tempol solution was equal to or better than that for the same solution in a flat cell in an X-band cavity. Adkins and Nolle (1966) obtained a sensitivity of $10^{15} \Delta H$ spins with a helix resonator at *ca.* 2 GHz.

11.4 PEDRI and LODESR

The sensitivity of PEDRI has been examined by Konijnenburg and Mehlkopf (1996). Based upon Röschmann's estimate that power deposition in the body depends on $\omega^{2.15}$, a simplified formula was derived that S/N is proportional to the product of the field at which the electron spins are saturated and the field at which the nuclear spins are detected, and of a factor that depends on relaxation times. If the same field is used for both saturation and detection, RF penetration depth results in optimum S/N at about 10 mT. If field cycling is used, the optimum field for detection is *ca.* 0.4 T (since losses in the body dominate the noise at detection fields higher than about 0.4 T) and the optimum field for electron spin saturation increases from *ca.* 1 to 10 mT with increasing electron spin relaxation times (see figure 2 in Konijnenburg and Mehlkopf, 1996). To avoid excessive temperature increases, DNP imaging has to be performed with EPR saturation at less than about 100 MHz (Guiberteau and Grucker, 1997). The increase in power deposition with frequency of electron spin irradiation is the incentive for field-cycled PEDRI (Lurie *et al.*, 1989, 1998).

RF LODESR has lower sensitivity than standard RF ESR with transverse detection. "The dominant noise source in RF-LODESR is the thermal noise of the detection coil…" (Panagiotelis *et al.*, 2001a)

12. APPLICATIONS OF LOW-FREQUENCY EPR

Although the focus of the chapter is on spectrometer hardware, some examples of applications are given in Tables 7 and 8. See also the reviews by Basosi *et al.* (1993), Swartz and Halpern (1998), Eaton and Eaton (1991, 1999c, 2000), Lurie (2002), and volume 18 in this series.

Table 7. Examples of *in vivo* studies

Topic	Comments	Reference
CTPO in mouse	250 MHz, 30 s acquisition time	Halpern *et al.*, 1989
trapped OH radicals	250 MHz, formed by gamma radiation and measured in a living mouse	Halpern *et al.*, 1995
nitroxyl in rat	280 MHz 2D images, 3 and 8 mm resolution	Quaresima *et al.*, 1992; Alecci *et al.*, 1994
nitroxyl in rat head	290 MHz LODESR, 16 W average power	Yokoyama *et al.*, 1998
nitroxyl in rat head	680 MHz EPR and 27.7 MHz NMR	Sato *et al.*, 1997
nitroxyl in rat head	760-820 MHz	Ishida, *et al.*, 1989, 1992
nitroxyl in rat head	700 MHz	Ogata, 1995; Oikawa *et al.*, 1996; Yokoyama *et al.*, 1996
nitroxyl in rat rectum	700 MHz flexible surface-coil-type resonator	Lin *et al.*, 1997
nitroxyl in rat head	720 MHz LODESR, 7.92 W average power	Yokoyama *et al.*, 1997a
nitroxyl in mouse	883 MHz, L-band	Sotgiu and Guatieri, 1985; Sotgiu, 1985
nitroxyl in mouse	1.1-1.3 GHz LGR	Lukiewicz and Lukiewicz, 1984
nitroxyl radicals	L-band	Nishikawa *et al.*, 1985; Fujii and Berliner, 1985; Berliner and Fujii, 1985
trityl	O_2 measured by relaxation times agree with Clark electrode	Subramanian *et al.*, 2002
trityl	Overhauser imaging (PEDRI). in mice, observed heterogeneity of a tumor.	Krishna *et al.* 2002
Trityl	^{19}F PEDRI at 628 kHz in perfused rat kidneys by	Murugesan *et al.*, 2002
Free radicals *in vivo* at 240 MHz EPR frequency	PEDRI *in vivo* at 240 MHz EPR frequency	(Puwanich *et al.*, 1999)

There are many low-frequency *in vivo* measurements that are not included in this table. See, for example, the review by Swartz and Halpern (1998) for further references.

Table 8. Samples other than *in vivo*

Topic	Comments	Reference
DPPH recrystallized from various solvents	30, 65, 320 MHz and 9 GHz	Singer, 1962
DPPH	28 MHz, 1.5 to 300 K	Van Gerven *et al.*, 1962
monochloroporphyrexide, porphyrexide, and nitroxyl	135 MHz	Yoshioka, 1977
$CrOF_5^{2-}$, $MoOF_5^{2-}$	600-1200 MHz, 4.2, 77 K	Abdrakhmanov *et al.*, 1968
Cr(V), Mo(V), Os(III), Cu(II), Ag(II), Ce(III), and Nd(III)	600-1200 MHz, 77 K	Abdrachmanov and Ivanova, 1973
CTPO	250 MHz, resolved 0.1 G line width	Halpern *et al.*, 1989
crystals of fluoranthenyl radical	pulsed imaging at 300 MHz, using NMR spectrometer	Callaghan *et al.*, 1994; Coy *et al.*, 1996; Feintuch *et al.*, 2000; Alexandrowicz *et al.*, 2000
crystals of fluoranthenyl radical	*ca.* 15 mG line width, independent of frequency for 15 MHz to 9.4 GHz	Dormann *et al.*, 1983
p-benzosemiquinone radical ion	384 MHz	Reynolds and Shanholtzer, 1967
asphaltine from crude oil	384 MHz	Reynolds and Shanholtzer, 1967
Fremy's salt	9.214 MHz (*ca.* 4 - 22 G)	Pake *et al.*, 1952
Fremy's salt	60 MHz	Lloyd and Pake, 1954
Fremy's salt	384 MHz	Reynolds and Shanholtzer, 1967
Fremy's salt	linewidth does not change from 267 MHz to X-band	Ardenkjaer-Larsen *et al.*, 1998
Trityl	$T_1 = T_2 = 8.2\ \mu s$ at infinite dilution in water at 250 MHz	Ardenkjaer-Larsen *et al.*, 1998
Li in methylamine	193.3 MHz	Matsui *et al.*, 1993
Na in ammonia	193.3 MHz	Terakado *et al.*, 1998
Li phthalocyanide	FID at 301 MHz	Bourg *et al.*, 1993
Li phthalocyanide	FID at 220 MHz	Alecci *et al.*, 1998a,b
Light-illuminated crystalline silicon	EDMR at 300-900 MHz	Fukui *et al.*, 2002.
solvated electrons from Na in liquid NH_3	FID at 301 MHz	Bourg *et al.*, 1993
Fremy's salt and DPPH	40, 80, 210, 9000 MHz	Decorps and Fric, 1972
nitroxyl radical in ethylene glycol at 77 K	80, 100 MHz	Fric and Mignot, 1975
triphenylene triplet state	67-90 K	Gerkin and Szerenyi, 1969
nitroxyl radicals	80 – 790 MHz	Gillies *et al.*, 1994
Mn^{2+}	530-950 MHz	Treiguts and Cugunov, 1995
VO^{2+} and Mn^{2+}	1 GHz, JEOL RE-3L with LGR	Utsumi *et al.*, 1992

Topic	Comments	Reference
$VO(acac)_2$	595 MHz	Belford *et al.*, 1987
trans-polyacetylene	5 to 450 MHz	Kume and Mizoguchi, 1985; Mizoguchi and Kume, 1985
Fe in $K_3Co(CN)_6$	1.8 GHz	Rannestad and Wagner, 1963
Cu^{2+} complexes	S-band	Hyde and Froncisz ,1982; Yuan *et al.*, 1999
Cu^{2+}	560 MHz, 1.4 GHz, 2.4 GHz	Belford *et al.*, 1987

There are many low-frequency EPR measurements that are not included in this table. See, for example, the extensive studies of Cu(II) species at S-band by Hyde, Antholine, and coworkers.

13. ABBREVIATIONS USED:

AFC automatic frequency control
ACC automatic coupling control
CLR crossed-loop resonator
CW continuous wave
CTPO 3-carbamoyl-2,2,5,5-tetramethylpyrolin-1-yloxy
DBM double-balanced mixer
DNP dynamic nuclear polarization
EPR or ESR electron paramagnetic resonance
ESE electron spin echo
LGR loop gap resonator
MRI magnetic resonance imaging
NMR nuclear magnetic resonance
PEDRI Proton Electron Double Resonance Imaging; also called Overhauser effect imaging
Q resonator quality factor
S/N signal to noise
SR saturation recovery

14. ACKNOWLEDGMENTS

This research was supported in part by NIH grants RR12183, GM21556, GM57577, and P41 RR12257. We thank George A. Rinard and Richard W. Quine for helpful background explanations and for comments on drafts of this review.

15. REFERENCES

Abdrakhmanov, R. S., Garif'yanov, N. S., and Semenova, E. I. (1968). EPR Study of the Structure of Some Oxyfluorides. *J. Struct. Chem.* **9**, 453-455 (pages 530-531 in Russian).

Abdrachmanov, R. S. and Ivanova, T. A. (1973). The EPR study of glasses and frozen solutions in the L and S bands, *J. Molec. Structure* **19**, 683-692. This paper contains citations to early Russian work.

Abdrachmanov, R. S. and Ivanova, T. A. (1976). Computer Analysis of Frequency Dependence of the ESR Spectra of nd^9-ions in Vitreous Media, *J. Appl. Spectros.* **25**, 1195.

Adkins, L. R. and Nolle, A. W. (1966). EPR Spectrometer Utilizing Traveling Wave Helix for 2 Gc Region, *Rev. Sci. Instrum.* **37**, 1404-1405.

Afeworki, M., van Dam, G. M., Devasahayam, N., Murugesan, R., Cook, J., Coffin, D., A-Larsen, J. H., Mitchell, J. B., Subramanian, S., and Krishna, M. C. (2000). Three-dimensional whole body imaging of spin probes in mice by time-domain radiofrequency electron paramagnetic resonance, *Mag. Reson. Med.* **43**, 375-382.

Alecci, M., and Lurie, D. J. (1999). Low Field (10 mT) Pulsed Dynamic Nuclear Polarization, *J. Magn. Reson.* **138**, 313-319.

Alecci, M., Gualtieri, G., and Sotgiu, A. (1989). Lumped parameters description of RF losses in ESR experiments on electrically conducting samples, *J. Phys. E: Sci. Instrum.* **22**, 354-359.

Alecci, M., Gualtieri, G., Sotgiu, A., Testa, L., and Varoli, V. (1991). Multipolar magnet for low-frequency ESR imaging, *Meas. Sci. Technol.* **2**, 32-37

Alecci, M., Della Penna, S., Sotgiu, A., and Testa, L., (1992a). R.F. (280 MHz) EPR Imaging of Extended Samples: Apparatus and Preliminary Results, *Appl. Magn. Reson.* **3**, 909-915.

Alecci, M., Della Penna, S., Sotgiu, A., and Testa, L., and Vannucci, I. (1992b). Electron paramagnetic resonance spectrometer for three-dimensional *in vivo* imaging at very low frequency, *Rev. Sci. Instrum.* **63**, 4263-4270.

Alecci, M., Colaicchi, S., Sotgiu, A., Testa, L., and Varoli, V. (1992c). Automatic trimming technique for multipolar magnets, *J. Appl. Phys.* **71**, 3053-3055.

Alecci, M., Farrari, M., Quaresima, V., Sotgiu, A., and Ursini, C. L. (1994). Simultaneous 280 MHz EPR Imaging of Rat Organs During Nitroxide Free Radical Clearance, *Biophys. J.* **67**, 1274-1279.

Alecci, M., McCallum, S. J., and Lurie, D. J. (1995). Design and Optimization of an Automatic Frequency Control System for a Radiofrequency Electron Paramagnetic Resonance Spectrometer, *J. Magn. Reson. A*, **117**, 272-277.

Alecci, M., Nicholson, I., and Lurie, D. J. (1996). A Novel Multiple-Tune Radiofrequency Loop-Gap Resonator for Use in PEDRI, *J. Magn. Reson. B* **110**, 82-86.

Alecci, M., Brivati, J. A., Placidi, G., and Sotgiu, A. (1998a). A Radiofrequency (220-MHz) Fourier Transform EPR Spectrometer, *J. Magn. Reson.* **130**, 272-280.

Alecci, M., Brivati, J. A., Placidi, G., Testa, L., Lurie, D. J., and Sotgiu, A. (1998b). A Submicrosecond Resonator and Receiver System for Pulsed Magnetic Resonance with Large Samples, *J. Magn. Reson.* **132**, 162-166.

Alexandrowicz, G., Tashma, T., Feintuch, A., Grayevsky, A., Dormann, E., and Kaplan, N. (2000). Spatial mapping of mobility and density of the conduction electrons in $(FA)_2PF_6$, *Phys. Rev. Lett.* **84**, 2973-2976.

Alfonsetti, M., Vecchio, C. D., Giuseppe, S. D., Placidi, G., and Sotgiu, A., 2001, A composite resonator for simultaneous NMR and EPR imaging experiments. *Meas. Sci. Technol.* **12**, 1325-1329.

Alger, R. S., Anderson, T. H., and Webb, L. A. (1959). Irradiation Effects in Simple Organic Solids, *J. Chem. Phys.* **30**, 695-706.

Alger, R. S., *Electron Paramagnetic Resonance: Techniques and Applications.* Wiley-Interscience, New York, 1968, pages 531ff.

Anderegg, M., Cornaz, P., and Borel, J.-P. (1963). Spectroscope de résonance paramagnétique dans la bande des 1000 MHz utilisant une diode tunnel, *Z. Angew. Meth. Phys.* **14**, 201-206.

Anderson, W. A. (1961). Electrical Current Shims for Correcting Magnetic Fields. *Rev. Sci. Instrum.* **32**, 241-250.

Antipin, A. A. (1966). Cavity for 10 cm waveband investigations, *Instrum. Exptl. Tech.* 366-367.

Antokol'skii, G. L., Baranov, V. S., and Iolin, E. M. (1977). Acoustic paramagnetic resonance in an aqueous solution of Fremy's salt, *JETP Lett.* **23**, 629-632.

Ardenkjaer-Larsen, J. H., Laursen, I., Leunbach, I., Ehnholm, G., Wistrand, L.-G., Petersson, J. S., and Golman, K. (1998). EPR and DNP Properties of Certain Novel Single Electron Contrast Agents Intended for Oximetric Imaging, *J. Magn. Reson.* **133**, 1-12.

Arnold, R. D., and Kip, A. F. (1949). Paramagnetic Resonance Absorption in Two Sulfates of Copper, *Phys. Rev.* **75**, 1199-1205.

Bacic, G., Nilges, M. J., Magin, R. L., Walczak, T., and Swartz, H. M. (1989). *In Vivo* Localized ESR Spectroscopy Reflecting Metabolism, *Magn. Reson. Med.* **10**, 266-272.

Bagguley, D. M. S., and Griffiths, J. H. E. (1947). Paramagnetic Resonance and Magnetic Energy Levels in Chrome Alum, *Nature* **160**, 532-533.

Bagguley, D. M. S., and Griffiths, H. E. (1948). Paramagnetic Resonance in Copper Sulphate, *Nature* **162**, 538-539.

Bagguley, D. M. S., Bleaney, B., Griffiths, J. H. E., Penrose, R. P., and Plumpton, B. I. (1948a). Paramagnetic Resonance in Salts of the Iron Group - A Preliminary Survey: I. Theoretical Discussion, *Proc. Phys. Soc. (London)* **61**, 542-550.

Bagguley, D. M. S., Bleaney, B., Griffiths, J. H. E., Penrose, R. P., and Plumpton, B. I. (1948b). Paramagnetic Resonance in Salts of the Iron Group - A Preliminary Survey: II. Experimental Results, *Proc. Phys. Soc. (London)* **61**, 551-561.

Barbarin, F., and Germain, J. P. (1975). Étude par resonance paramagnétique électronique des effets de la solvation sur le temps de correlation rotationnel du radical anion p-benzosemiquinone en milieu liquide, *J. Phys.* **36**, 475-480.

Basosi, R., Antholine, W. E., Hyde, J. S. (1993). Multifrequency ESR of Copper: Biophysical Applications, *Biol. Magn. Reson.* **13**, 103-150.

Belford, R. L., Clarkson, R. B., Cornelius, J. B., Rothenberger, K. S., Nilges, M. J., and Timkin, M. D. (1987). EPR Over Three Decades of Frequency: Radiofrequency to Infrared. in *Electronic Magnetic Resonance of the Solid State.* J. A. Weil, ed., Canadian Society for Chemistry, Ottawa, Canada, 21-43.

Benedek, G. B., and Kushida, T. (1960). Nuclear Magnetic Resonance in Antiferromagnetic MnF_2 under Hydrostatic Pressure, *Phys. Rev.* **118**, 46-57.

Berliner, L. J. and Fujii, H. (1985). Magnetic Resonance Imaging of Biological Specimens by Electron Paramagnetic Resonance of Nitroxide Spin Labels, *Science* **227**, 517-519.

Berliner, L. J., and Koscielniak, J. (1991). Low-Frequency EPR Spectrometers: L-Band, in *EPR Imaging and* in vivo *EPR*, S. S. Eaton, G. R. Eaton, and K. Ohno, eds., CRC Press, Boca Raton, Florida USA, Ch. 7.

Bolas, N. M., Gillies, D. G., Sutcliffe, L. H., and Symms, M.. R. (1996). A radiofrequency ESR spectrometer/imager, *Res. Chem. Intermed.* **22**, 525-537.

Borel, J.-P., and Manus, C. (1957). Un nouveau type de spectrographe Hertzian pour l'observation de la resonance électronique dans le domaine des ondes métriques et décimétriques, *Helv. Phys. Acta* **30**, 254-257

Bourg, J., Krishna, M. C., Mitchell, J. B., Tschudin, R. G., Pohida, T. J., Friauf, W. S., Smith, P. D., Metcalfe, J., Harrington, F. and Subramanian, S. (1993). Radiofrequency FT EPR Spectroscopy and Imaging, *J. Magn. Reson. B* **102**, 112-115.

Bowers, K. D., and Mims, W. B. (1959). Paramagnetic Relaxation in Nickel Fluorosilicate, *Phys. Rev.* **115**, 285-295.

Bramley, R., and Strach, S. J. (1983). Electron Paramagnetic Resonance Spectroscopy at Zero Magnetic Field, *Chem. Rev.* **83**, 49-82.

Brivati, J. A., Stevens, A. D., and Symons, M. C. R. (1991). A Radiofrequency ESR Spectrometer for *in Vivo* Imaging, *J. Magn. Reson.* **92**, 480-489.

Brown, G. (1974). A 1 GHz ESR spectrometer for the study of aqueous samples. *J. Phys. E: Sci. Instrum.* **7**, 635-638.

Bruin, F., and Bruin, M. (1956). Some Measurements on the Spectral Line Shape and Width of a Paramagnetic Resonance Absorption Line, *Physica* **22**, 129-140.

Bruin, F., and Khunaysir, H., (1970). Superregenerative FET Oscillator for NQR, ESR, and NMR, *Amer. J. Phys.* **38**, 1480-1481.

Callaghan, P. T., Coy, A., Dormann, E., Ruf, R., and Kaplan, N. (1994). Pulsed-Gradient Spin-Echo ESR, *J. Magn. Reson. A* **111**, 127-131.

Charru, M. A. (1956). Montage à 3000 MHz, pour l'étude de la resonance paramagnétique électronique et du sugne du facteur de Landé, *Compes Rendus* **243**, 652-654

Chiricozzì, E., Masciovecchio, C., Villani, M., Sotgiu, A., and Testa, L. (1998). Multipolar Laminated Electromagnet for Low-Field Magnetic Resonance Imaging and Electron Paramagnetic Resonance Imaging, *IEEE Trans. Biomed. Eng.* **45**, 928-933.

Chirikov, A. K. (1959). Measurement of weak magnetic fields by the electron resonance method, *Instrum. Expt. Tech.* 211-213.

Chzhan, M., Shteynbuk, M., Kuppusamy, P., and Zweier, J. L. (1993). An Optimized L-Band Ceramic Resonator for EPR Imaging of Biological Samples, *J. Magn. Reson. A* **105**, 49-53.

Chzhan, M., Kuppusamy, P., and Zweier, J. L. (1995). Development of an Electronically Tunable L-Band Resonator for EPR Imaging of Biological Samples, *J. Magn. Reson. B* **108**, 67-72.

Chzhan, M., Kuppusamy, P., Samouilov, A., He, G., and Zweier, J. L. (1999). A Tunable Reentrant Resonator with Transverse Orientation of Electric Field for *in Vivo* EPR Spectroscopy, *J. Magn. Reson.* **137**, 373-378.

Clarkson, R. B., Timkin, M. D., Brown, D. R., Crookham, H. C., and Belford, R. L. (1989). Enhancement of Nuclear Modulation in Electron Spin Echoes at Low Magnetic Fields: S-Band ESE Spectrometer, *Chem. Phys. Let.* **163**, 277-281.

Clarkson, R. B., Brown, D. R., Cornelius, J. B., Crookham, H. C., Shi, W.-J., and Belford, R. L. (1992). S-band electron spin echo spectroscopy. *Pure & Appl. Chem.* **64**, 893-902.

Colacicchi, S., Indovina, P. L., Momo, F., and Sotgiu, A. (1988). Low-frequency three-dimensional ESR imaging of large samples, *J. Phys. E: Sci. Instrum.* **21**, 910-913.

Colacicchi, S., Brivati, J., Barattelli, G., Gualtieri, G., and Sotgiu, A. (1996). A Minimal Cavity and Its Capacitive Coupling for *in Vivo* EPR Measurements on Mice. *Res. Chem. Intermed.* **22**, 549-556.

Collingwood, J. C., and White, J. W. (1967). Helical resonators for spin resonance spectroscopy, *J. Sci. Instrum.* **44**, 509-513.

Cook, R. F., and Stoodley, L. G. (1963). An electron-spin-resonance spectrometer for use while irradiating wet biological systems, *Int. J. Rad. Biol.* **7**, 155-160.

Coy, A., Kaplan, N., and Callaghan, P. T. (1996). Three-dimensional pulsed ESR imaging, *J. Magn. Reson.* **A121**, 201-205.

Cummerow, R. L., and Halliday, D. (1946). Paramagnetic Losses in Two Manganese Salts, *Phys. Rev.* **70**, 433.

Cummerow, R. L., Halliday, D., and Moore, G. E. (1947). Paramagnetic Resonance Absorption in Salts of the Iron Group, *Phys. Rev.* **72**, 1233-1240.

Cirio, L., Lucidi, S., Placidi, G., and Sotgiu, A. (2001) Automatic optimization strategy for the design of circular multipolar magnets. *J. Phys. D. Appl. Phys.* **34**, 313-318.

Dahlberg, E. D., and Dodds, S. A. (1981). Low-frequency tunable (1-1.8 GHz) electron spin resonance resonator and spectrometer, *Rev. Sci. Instrum.* **52**, 472-474.

Davies, G. R., Lurie, D. J., Hutchison, J. M. S., McCallum, S. J., and Nicholson, I. (2001) Continuous-Wave Magnetic Resonance Imaging of Short T_2 Materials. *J. Magn. Reson.* **148**, 289-297.

Decorps, M., and Fric, C. (1969). Etude comparative de divers types de volumes resonnants pour spectrometers a resonance paramagnetique electronique, en ondes metriques, *J. Phys. E: Sci. Instrum.* **2**, 1036-1040.

Decorps, M., and Fric, C. (1972). Un spectromètre basse frèquence a haute sensibilitè pour l'ètude de la resonance des spins èlectroniques, *J. Phys. E: Sci. Instrum.* **5**, 337-342.

Denisov, Yu. I., and Kalinichenko, V. V. (1965). Cavity Resonator for Decimeter Waveband EPR, *Instrum. Expt. Tech.* 379-381.

Devasahayam, N., Subramanian, S., Murugesan, R., Cook, J. A., Afeworki, M., Tschudin, R. G., Mitchell, J. B., and Krishna, M. C. (2000). Parallel coil resonators for time-domain radiofrequency electron paramagnetic resonance imaging of biological objects, *J. Magn. Reson.* **142**, 168-176.

Devasahayam, N., Murugesan, R., Yamada, K., Reijnders, K., Mitchell, J. B., Subramanian, S., Krishna, M. C., and Cook, J. A. (2002) Evaluation of high-speed signal-averager for sensitivity enhancement in radio frequency Fourier transform electron paramagnetic resonance imaging. *Rev. Sci. Instrum.* **73**, 3920-3925.

Dijret, D., Beranger, M., Bernerd, A., Jeandey, C., and Moussavi, M. (1994). A new 280 MHz ESR spectrometer, *J. Chim. Phys.* **91**, 1862-1867.

Di Luzio, S., Placidi, G., Di Giuseppe, S., Alecci, M., and Sotgiu, A. (1998). A novel, cylindrical, transverse gradient coil design for magnetic resonance imaging of large samples, *Meas. Sci. Technol.* **9**, 1663-1671.

Diodato, R., Alecci, M., Brivati, J. A., and Sotgiu, A. (1998). Resonant inductive coupling of RF EPR resonators in the presence of electrically conducting samples, *Meas. Sci. Technol.* **9**, 832-837.

Diodato, R., Alecci, M., Brivati, J. A., Varoli, V., and Sotgiu, A. (1999). Optimization of axial RF field distribution in low-frequency EPR loop-gap resonators, *Phys. Med. Biol.* **44**, N69-N75.

Dormann, E., Sachs, G., Stöcklein, W., Bail, B., and Schwoerer, M. (1983). Gaussmeter Application of an Organic Conductor, *Appl. Phys. A* **30**, 227-231.

Duncan, W. (1967). A 900 MHz induction electron spin resonance spectrometer, *J. Sci. Instrum.* **44**, 437-439.

Duncan, W., and Schneider, E. E. (1965). A 300 Mc/s electron spin resonance spectrometer, *J. Sci. Instrum.* **42**, 395-398.

Duret, D., Beranger, M., Moussavi, M., Turek, P., and Andre, J. J. (1991). A new ultra low-field ESR spectrometer, *Rev. Sci. Instrum.* **62**, 685-694.

Duret, D., Beranger, M., and Moussavi, M. (1992). An Absolute Earth Field ESR Vectorial Magnetometer, *IEEE Trans. Magn.* **28**, 2187-2189.

Eaton, G. R. (1992). EPR Spectroscopy, in *Proceedings of National High Magnetic Field Laboratory Workshop on Nuclear Magnetic Resonance*, E. R. Andrew, T. Mareci, and N. S. Sullivan, eds., University of Florida, ch. 4.iii.

Eaton, S. S., and Eaton, G. R. (1982). Measurement of Spin-Spin Distances from the Intensity of the EPR Half-field Transition, *J. Amer. Chem. Soc.* **104**, 5002-5003.

Eaton, G. R., and Eaton, S. S. (1988). EPR Imaging: Progress and Prospects, *Bull. Magn. Reson.* **10**, 22-31.

Eaton, G. R., and Eaton, S. S. (1989). Resolved Electron-Electron Spin-Spin Splitting in EPR Spectra, *Biol. Magn. Reson.* **8**, 339-397.

Eaton, S. S., and Eaton, G. R. (1990). Electron Spin Resonance Imaging, in *Modern Pulsed and Continuous-Wave Electron Spin Resonance*, L. Kevan and M. K. Bowman, eds., Wiley, 405-435.

Eaton, S. S., and Eaton, G. R., (1991). EPR Imaging, *Specialist Periodical Report, Electron Paramagnetic Resonance* **12b**, 176-190.

Eaton, S. S., and Eaton, G. R. (1993). Applications of High Magnetic Fields in EPR Spectroscopy, *Magn. Reson. Rev.* **16**, 157-181.

Eaton, G. R., and Eaton, S. S. (1993). Electron Paramagnetic Resonance Imaging, in *Microscopic and Spectroscopic Imaging of the Chemical State*, M. D. Morris, ed., Marcel Dekker, New York, ch 11, p. 395-419.

Eaton, G. R., and Eaton, S. S. (1999a). High-Field and High-Frequency EPR, *Appl. Magn. Reson.* **16**, 161-166.

Eaton, S. S., and Eaton, G. R. (1999b). High Magnetic Fields and High Frequencies, in *Handbook of Electron Spin Resonance*, C. P. Poole, Jr., and H. A. Farach, eds, vol 2, AIP Press, 345-370.

Eaton, G. R., and Eaton, S. S. (1999c). ESR Imaging in *Handbook of Electron Spin Resonance*, C. P. Poole, Jr. and H. A Farach, vol. 2, 327-343.

Eaton, S. S., and Eaton, G. R. (2000). EPR imaging in *Specialist Periodical Report, Electron Paramagnetic Resonance* **17**, 109-129.

Eaton, G. R., Eaton, S. S., and Ohno, K., eds. (1991) *EPR Imaging and In-Vivo EPR*, CRC Press, Boca Raton, FL.

Eaton, G. R., Eaton, S. S., and Rinard, G. A. (1998a). Frequency dependence of EPR Sensitivity, in *Spatially Resolved Magnetic Resonance*, P. Blümler, B. Blümich, R. Botto and E. Fukushima, eds., Wiley-VCH, Weinheim, 65-74.

Eaton, G. R., Eaton, S. S., and Salikhov, K. M., eds., (1998b). *Foundations of Modern EPR*, World Scientific, Singapore.

El'sting, O. G. (1960). A Device for Investigating Paramagnetic Electronic Resonance at Meter-Band Frequencies. *Instrum. Exptl. Tech.* 753-755 (page 64-66 in Russian).

Feher, G., and Kip, A. F. (1955). Electron Spin Resonance in Metals. I. Experimental, *Phys. Rev.* **98**, 337-348.

Feintuch, A., Alexandrowicz, G., Tashma, T., Boasson, Y., Grayevsky, A., and Kaplan, N. (2000). Three-dimensional pulsed ESR Fourier imaging, *J. Magn. Reson.* **142**, 382-385.

Fric, C., and Mignot, P. (1975). Étude par RPE en champ faible d'un monoradical nitroxyde dilué dans un solide, *Rev. Phys. Appl.* **10**, 305-308.

Fujii, H., and Berliner, L. J. (1985). One- and Two-Dimensional EPR Imaging Studies on Phantoms and Plant Specimens, *Magn. Reson. Med.* **2**, 275-282.

Fukui, K., Sato, T., Yokoyama, H., Ohya, H., and Kamada, H. (2002) Resonance-Field Dependence in Electrically Detected Magnetic Resonance: Effects of Exchange Interaction. *J. Magn. Reson.* **149**, 13-21.

Garrett, M. W. (1967). Thick Cylindrical Coil Systems for Strong Magnetic Fields with Field or Gradient Homogeneities of the 6th to 20th Order, *J. Applied Phys.* **38**, 2563-2585

Garstens, M. A., Singer, L. S., and Ryan, A. H. (1954). Magnetic Resonance Absorption of Diphenyl-Picryl Hydrazyl at Low Magnetic Fields, *Phys. Rev.* **96**, 53-56

Gebhardt, H., and Dormann, E. (1989). ESR gaussmeter for low-field applications, *J. Phys. E: Sci. Instrum.* **22**, 321-324.

Gerkin, R. E., and Szerenyi, P. (1969). Electron Paramagnetic Resonance Absorption in Oriented Triphenylene in Its Phosphorescent State at Low Magnetic Fields, *J. Chem. Phys.* **50**, 4095-4106.

Gillies, D. G., Sutcliffe, L. H., and Symms, M. R. (1994). Effects encountered in EPR Spectroscopy and Imaging at Small Magnetic Fields, *J. Chem. Soc. Faraday Trans.* **90**, 2671-2675.

Giuseppe, S. D., Placidi, G., and Sotgiu, A. (2001) New experimental apparatus for multimodal resonance imaging: initial EPRI and NMRI experimental results. *Phys. Biol. Med.* **46**, 1003-1016.

Giordano, M., Martinelli, M., Pardi, L., and Santucci, S. (1976). Electron spin resonance spectrometer with a new wideband resonator for aqueous samples, *Rev. Sci. Instrum.* **47**, 1402-1404.

Golenishchev-Kutuzov, V. A., and Kharakhash'yan, E. G. (1965) Ultrasonic Paramagnetic Spectrometer, *Instrum. Exptl. Tech.* 371-374 (pages 126-130 in Russian).

Gorter, C. J. (1936a). Paramagnetic Relaxation, *Nature* **137**, 190.

Gorter, C. J. (1936b). Paramagnetic Relaxation, *Physica* **3**, 503-515.

Gorter, C. J. (1936c). Paramagnetic Relaxation in Transversal Magnetic Field, *Physica* **3**, 1006-1008.

Gorter, C. J. (1947). *Paramagnetic Relaxation*, Elsevier, New York.

Gorter, C. J. (1967). Bad Luck in Attempts to Make Scientific Discoveries, *Physics Today*, January, p. 76-81; errata April 1967, p. 13.

Greenslade, D. J., Koptyug, A. V., and Symons, M. C. R. (1996). Aspects of Low-frequency Low-field Electron Spin Resonance. *Ann. Rep. Prog. Chem.* **92**, 3-21.

Grobet, P., Accou, J., Verlinden, R., and Van Gerven, L. (1971). Demagnetizing effects in g-factor measurements in *Magnetic Resonance and Related Phenomena*, Ursu, I., ed., Publishing House of the Academy of the Socialist Republic of Romania, pp. 268-270.

Grucker, D. (1990). *In Vivo* Detection of Injected Free Radicals by Overhauser Effect Imaging, *Magn. Reson. Med.* **14**, 140-147.

Grucker, D. and Chambron, J. (1989). Sensitivity of Overhauser Effect Imaging to Oxygen, *Physica Medica* **5**, 329-335.

Grucker, D. and Chambron, J. (1993).Oxygen Imaging in Perfused Hearts by Dynamic Nuclear Polarization, *Magn. Reson. Im.* **11**, 691-696.

Grucker, D., Guiberteau, T., and Planinsic, G. (1996). Proton-Electron Double Resonance: Spectroscopy and Imaging in Very Low Magnetic Fields, *Res. Chem. Intermed.* **22**, 567-579.

Guiberteau, T., and Grucker, D. (1996). EPR Spectroscopy by Dynamic Nuclear Polarization in Low Magnetic Field, *J. Magn. Reson.* B **110**, 47-54.

Guiberteau, T., and Grucker, D. (1997). Dynamic nuclear polarization imaging in very low magnetic fields as a noninvasive technique for oximetry, *J. Magn. Reson.* **124**, 263-266.

Halliday, D., and Wheatley, J. (1948). Paramagnetic Resonance Absorption in Aqueous Solutions of Manganese Sulfate, *Phys. Rev.* **74**, 1724.

Halpern, H. J., and Bowman, M. K. (1991). Low-Frequency Electron Paramagnetic Resonance. in *EPR Imaging and* in vivo *EPR*, S. S. Eaton, G. R. Eaton, and K. Ohno, eds., CRC Press, Boca Raton, Florida USA, ch. 6.

Halpern, H. J., Spencer, D. P., van Polen, J., Bowman, M. K., Nelson, A. C., Dowey, E. M., and Teicher, B. A. (1989). Imaging radio frequency electron-spin-resonance spectrometer with high resolution and sensitivity for *in vivo* measurements, *Rev. Sci. Instrum.* **60**, 1040-1050

Halpern, H. J., Yu. C., Barth, E., Peric. M., and Rosen, G. (1995). In situ detection, by spin trapping, of hydroxyl radical markers produced from ionizing radiation in the tumor of a living mouse, *Proc. Natl. Acad. Sci. USA* **92**, 796-800.

Hankiewicz, J. H., Stenland, C., and Kevan, L. (1993). Pulsed S-band electron spin resonance spectrometer, *Rev. Sci. Instrum.* **64**, 2850-2856.

Hardy, W. N., and Whitehead, L. A. (1981). Split-ring resonator for use in magnetic resonance from 200-2000 MHz, *Rev. Sci. Instrum.* **52**, 213-216.

Hatch, G. F., and Kreilick, R. W. (1972). Design of a Broadband Low-Field ESR Spectrometer, *J. Magn. Reson.* **8**, 126-128

He, G., Shankar, R. A., Chzhan, M., Samouilov, A., Kuppusamy, P., and Zweier, J. L. (1999). Nonivasive measurement of anatomic structure and intraluminal oxygenation in the gastrointestinal tract of living mice with spatial and spectral EPR imaging, *Proc. Natl. Acad. Sci. USA* **96**, 4586-4591.

He, G., Samouilov, A., Kuppusamy, P., and Zweier, J. L. (2001a) *In Vivo* EPR Imaging of the Distribution and Metabolism of Nitroxide Radicals in Human Skin. *J. Magn. Reson.* **148**, 155-164.

He, G., Petryakov, S., Samouilov, A., Chzhan, M., Kuppusamy, P., and Zweier, J. L. (2001b) Development of a Resonator with Automatic Tuning and Coupling Capability to Minimize Sample Motion Noise for *in Vivo* EPR Spectroscopy. *J. Magn. Reson.* **149**, 218-227.

Hill, M. J., and Wyard, S. J. (1967). A 280 MHz electron spin resonance spectrometer, *J. Sci. Instrum.* **44**, 433-436.

Hirata, H., and Ono, M. (1996). Resonance frequency estimation of a bridged loop-gap resonator used for magnetic resonance measurements, *Rev. Sci. Instrum.* **67**, 73-78.

Hirata, H., and Ono, M. (1997). A flexible surface-coil-type resonator using triaxial cable, *Rev. Sci. Instrum.* **68**, 1-2.

Hirata, H., and Ono, M. (1997) Impedance-matching system for a flexible surface-coil-type resonator. *Rev. Sci. Instrum.* **68**, 3528-3532.

Hirata, H., Iwai, H, and Ono, M. (1995). Analysis of a flexible surface-coil-type resonator for magnetic resonance measurements, *Rev. Sci. Instrum.* **66**, 4529-4534.

Hirata, H., Walczak, T., and Swartz, H. M. (2000). Electronically Tunable Surface-Coil-Type Resonator for L-Band EPR Spectroscopy, *J. Magn. Reson.* **142**, 159-167.

Hirata, H., Ueda, M., Ono, M., and Shimoyama, Y. (2002) 1.1 GHz Continuous-Wave EPR Spectroscopy with a Frequency Modulation Method. *J. Magn. Reson.* **155**, 140-144

Hornak, J. P., Spacher, M., and Bryant, R. G. (1991). A modular low frequency ESR spectrometer, *Meas. Sci. Technol.* **2**, 520-522.

Hutchinson, J. M. S. (1971). Electron spin resonance spectrometry on the whole mouse *in vivo*; optimum frequency considerations, *J. Phys. E: Sci. Instrum.* **4**, 703-704.

Hutchinson, J. M. S., and Mallard, J. R. (1971). Electron spin resonance spectrometry on the whole mouse *in vivo*; a 100 MHz spectrometer, *J. Phys. E: Sci. Instrum.* **4**, 237-239.

Hyde, J. S., and Froncisz, W. (1981). ESR S-Band Microwave Spectrometer for Process Control, in *Proceedings National Electronics Conference, 1981*, Vol. 35, National Engineering Consortium, Oak Brook, IL, pp. 602-606.

Hyde, J. S., and Froncisz, W. (1982).The Role of Microwave Frequency in EPR Spectroscopy of Copper Complexes, *Ann. Rev. Biophys. Bioeng.* **11**, 391-417.

Hyde, J. S., and Froncisz, W. (1986). Loop-Gap Resonators, *Specialist Periodical Report: Electron Spin Resonance* **10A**, 175-184.

Hyde, J. S., and Gajdzinski, J. (1988). EPR automatic frequency control circuit with field effect transistor (FET) microwave amplification, *Rev. Sci. Instrum.* **59**, 1352-1356.

Ilangovan, G., Li, H., Zweier, J. L., and Kuppusamy, P. (2002) *In vivo* meaurement of tumor redox environment using EPR spectroscopy. *Molec. Cell. Biochem.* **234/235**, 393-398.

Ishida, S., Kumashiro, H., Tsuchihashi, N., Ogata, T., Ono M., Kamada, H., Yoshida, E. (1989). *In vivo* analysis of nitroxide radicals injected into small animals by L-band ESR technique, *Phys. Biol. Med.* **34**, 1317-1323.

Ishida, S., Matsumoto, S., Yokoyama, H., Mori, N., Kumashiro, H., Tsuchihashi, N., Ogata, T., Yamada, M., Ono, M., Kitajima, T., Kamada, H., and Yoshida, E. (1992). An ESR-CT Imaging of the Head of a Living Rat Receiving an Administration of a Nitroxide Radical. *Magn. Reson. Imag.* **10**, 109-114.

Jiang, J., Liu, K. J., Walczak, T., and Swartz, H. M. (1995). An Analysis of the Effects of Eddy Currents on L-Band EPR Spectra, *J. Magn. Reson. B* **106**, 220-226.

Jiang, J., Liu, K. J., and Swartz, H. M. (1996). Low Frequency ESR Surface Probe Based on Dielectric Resonator, *Res. Chem. Intermed.* **22**, 539-547.

Kangarlu, A., Baertlein, B. A., Lee, R., Ibrahim, T., Yang, L., Abduljalil, A. M., and Robitaille, P.-M. L. (1999). Dielectric Resonance Phenomena in Ultra High Field MRI, *J. Comp. Assist. Tomog.* **23**, 821-831.

Kent, M., and Mallard, J. R. (1965). An 80 Mc/s electron spin resonance spectrometer for aqueous samples, *J. Sci. Instrum.* **42**, 505-506.

Kernevez, N., Duret, D., Moussavi, M., and Leger, J.-M. (1992). Weak Field NMR and ESR Spectrometers and Magnetometers, *IEEE Trans. Magn.* **28**, 3054-3059.

Kim, K., Bodart, J. R., and Sullivan, N. S., High-Sensitivity CW NMR Probe for Low Temperatures and Ultrahigh Frequencies, (1996). *J. Magn. Res. A* **118**, 28-32.

Klein, M. P., and Phelps, D. E. (1967). Radiofrequency Hybrid Tees for Nuclear Magnetic Resonance, *Rev. Sci. Instrum.* **38**, 1545-1546.

Koscielniak, J. and Berliner, L. J. (1993). Optimization of L-band Microwave Bridges for Maximum Sensitivity in in-vivo Experiments, Rocky Mountain Conference, Denver, Colorado, abstract 154.

Koscielniak, J. (2000). CW EPR signal detection bridges, in *Biol. Magn. Reson.* **18**, Berliner, L. J., ed., ch. 4, in press.

Konijnenburg, H., and Mehlkopf, A. F. (1996). Optimal SNR for proton electron double resonance imaging, *Res. Chem. Intermed.* **22**, 557-561.

Koscielniak, J., and Berliner, L. J. (1994). Dual diode detector for homodyne EPR microwave bridges, *Rev. Sci. Instrum.* **65**, 2227-2230.

Koscielniak, J., Devasahayam, N., Moni, M. S., Kuppusamy, P., Yamada, K., Mitchell, J. B., Krishna, M. C., and Subramanian, S. (2000) 300 MHz continuous wave electron paramagnetic resonance spectrometer for small animal *in vivo* imaging. *Rev. Sci. Instrum.* **71**, 4273-4281.

Kume, K., and Mizoguchi, K. (1985). Frequency dependence of the ESR in pristine trans-polyacetylene: spin-lattice relaxation time T_1, *Mol. Cryst. Liq. Cryst.* **117**, 469-472.

Kuppusamy, P., Chzhan, M., Vij, K., Shteynbuk, M., Lefer, D. J., Giannella, E., and Zweier, J. L. (1994). Three-dimensional spectral-spatial EPR imaging of free radicals in the heart: A technique for imaging tissue metabolism, *Proc. Natl. Acad. Sci. USA* **91**, 3388-3392.

Kuppusamy, P., Afeworki, M., Shankar, R. A., Coffin, D., Krishna, M. C., Hahn, S. M., Mitchell, J. B., and Zweier, J. L. (1998a). *In vivo* electron paramagnetic resonance imaging of tumor heterogeneity and oxygen in a murine model, *Cancer Res.* **58**, 1562-1568.

Kuppusamy, P., Wang, P., Shankar, R. A., Ma, L., Trimble, C. E., Hsia, C. J. C., and Zweier, J. L. (1998b). *In Vivo* Topical EPR Spectroscopy and Imaging of Nitroxide Free Radicals and Polynitroxyl-Albumin, *Magn. Reson. Med.* **40**, 806-811.

Kuppusamy, P., Shankar, R. A., Roubaud, V. M., and Zweier, J. L. (2001) Whole Body Detection and Imaging of Nitric Oxide Generation in Mice Following Cardiopulmonary Arrest: Detection of Intrinsic Nitrosoheme Complexes. *Magn. Reson. Med.* **45**, 700-707.

Li, H., Deng, Y., He., G., Kuppusamy, P., Lurie, D. J., and Zweier, J. L. (2002) Proton Electron Double Resonance Imaging of the *In Vivo* Distribution and Clearance of a Triaryl Methyl Radical in Mice. *Magn. Reson. Med.*, **48**, 530-534.

Lin, Y., Yokoyama, H., Ishida, S., Tsuchihashi, N., and Ogata, T. (1997). *In vivo* electron spin resonance analysis of nitroxide radicals injected into a rat by a flexible surface-coil-type resonator as an endoscope- or a stethoscope-like device, *MAGMA* **5**, 99-103

Lloyd, J. P., and Pake, G. E. (1954). Spin Relaxation in Free Radical Solutions Exhibiting Hyperfine Structure, *Phys. Rev.* **94**, 579-591.

Lukiewicz, S. J., and Lukiewicz, S. G. (1984). *In vivo* spectroscopy of large biological objects, *Magn. Reson. Med.* **1**, 297-298.

Lurie, D. J. (1995). Imaging using the Electronic Overhauser Effect. *Encyclopedia of Nuclear Magnetic Resonance*, Grant, D. M., and Harris, R. K., eds., Wiley, page 2481-2486.

Lurie, D. J. (2002) Techniques and Applications of EPR Imaging, in *Specialist Periodical Report, Electron Paramagnetic Resonance*, **18**, 137-160.

Lurie, D. J., Bussell, D. M., Bell, L. H., and Mallard, J. R. (1988). Proton-Electron Double Magnetic Resonance Imaging of Free Radical Solutions, *J. Magn. Reson.* **76**, 366-370.

Lurie, D. J., Hutchison, J. M. S., Bell, L. H., Nicholson, I., Bussell, D. M., and Mallard, J. R. (1989). Field-Cycled Proton-Electron Double-Resonance Imaging of Free Radicals in Large Aqueous Samples, *J. Magn. Reson.* **84**, 431-437.

Lurie, D. J., Nicholson, I., Foster, M. A., and Mallard, J. R. (1990). Free radicals imaged *in vivo* in the rat by using proton-electron double-resonance imaging, *Phil. Trans. R. Soc. Lond. A* **333**, 453-456.

Lurie, D. J., McLay, J., Nicholson, I., and Mallard, J. R. (1991a). Production of UV-Generated Hydroxyl Free Radicals Imaged by Proton-Electron Double-Resonance Imaging with Spin Trapping, *J. Magn. Reson.* **95**, 191-195.

Lurie, D. J., Nicholson, I., and Mallard, J. R. (1991b). EPR Spectral Information Obtained from Field-Cycled Proton-Electron Double-Resonance Images, *J. Magn. Reson.* **94**, 197-203.

Lurie, D. J., Nicholson, I., and Mallard, J. R. (1991c). Low-Field EPR Measurements of Field-Cycled Dynamic Nuclear Polarization, *J. Magn. Reson.* **95**, 405-409.

Lurie, D. J., Nicholson, I., McLay, J. S., and Mallard, J. R. (1992). Spin-Trapped Hydroxyl Free Radicals Studied at Low Field by Field-Cycled Dynamic Nuclear Polarization. *Appl. Magn. Reson.* **3**, 917-925.

Lurie, D. J., Foster, M. A., Yeung, D., and Hutchison, J. M. S. (1998). Design, construction, and use of a large-sample field-cycled PEDRI imager, *Phys. Med. Biol.* **43**, 1877-1886.

Lurie, D. J., Li, H., Petryakov, S., and Zweier, J. L. (2002) Development of a PEDRI free-radical imager using a 0.38 T clinical MRI system. *Magn. Reson. Med.* **47**, 181-186.

McCallum, S. J., Alecci, M., and Lurie, D. J. (1996a). Modification of a whole-body NMR imager into a radio frequency EPR spectrometer suitable for *in vivo* measurements, *Meas. Sci. Technol.* **7**, 1012-1018.

McCallum, S. J., Nicholson, I., and Lurie, D. J. (1996b). A Combined PEDRI and CW-EPR Instrument for Detecting Free Radicals *in Vivo*, *J. Magn. Reson. B* **113**, 65-69.

Marcley, R. G. (1961). Apparatus for Electron Paramagnetic Resonance at Low Fields. *Am. J. Phys.* **29**, 492-497.

Matheson, M. S., and Smaller, B. (1955). Paramagnetic Species in Gamma-Irradiated Ice. *J. Chem. Phys.* **23**, 521-528.

Matsui, H., Yamada, E., Shimokawa, S., and Nakamura, Y. (1993). Excess Electrons in the Lithium-Methylamine System Studied by Electron Spin Resonance at 200 MHz, *J. Phys. Chem.* **97**, 4284-4287.

Medvedev, L. I., Suleimanov, N. M., Kharakhash'yan, E. G., and Khlebnikov, S. Ya. (1976). Superheterodyne Decimeter-Waveband ESR Spectrometer for Operation at Ultralow Temperatures, *Instrum. Exptl. Tech.* **19**, 1797-1799 (page 185-187 in Russian).

Misra, B. N., Gupta, S. K., and Sharma, S. D. (1973).Temperature Variation of Spin Lattice Relaxation Time in a Free Radical Complex. *Z. Physik. Chem. Neue Folge* **85**, 64-68.

Mizoguchi, K., and Kume, K. (1985). Frequency dependence of the ESR in pristine trans-polyacetylene: line width, *Mol. Cryst. Liq. Cryst.* **117**, 459-462.

Momo, F., and Sotgiu, A. (1984). Re-entrant resonators for ESR spectroscopy between 2 and 10 GHz, *J. Phys. E: Sci. Instrum.* **17**, 556-558.

Momo, F., Adriani, O., Gualtieri, G., and Sotgiu, A. (1988). Generalized Anderson coils for magnetic resonance imaging, *J. Phys. E: Sci. Instrum.* **21**, 565-568.

Mülsch, A., Lurie, D. J., Seimenis, I., Fichtlscherer, B., and Foster, M. A. (1999). Detection of Nitrosyl-Iron Complexes by Proton-Electron-Double-Resonance Imaging, *Free Rad. Biol. Med.* **27**, 636-646.

Murugesan, R., Cook, J. A., Devasahayam, N., Afeworki, M., Subramanian, S., Tschudin, R., Larsen, J. A., Mitchell, J. B., Russo, A., and Krishna, M. C. (1997). *In vivo* imaging of a stable paramagnetic probe by pulsed-radiofrequency electron paramagnetic resonance spectroscopy, *Magn. Reson. Med.* **38**, 409-414.

Murugesan, R., Afeworki, M., Cook, J. A., Devasahayam, N., Tschudin, R., Mitchell, J. B., Subramanian, S., and Krishna, M. C. (1998). A broadband pulsed radio frequency electron paramagnetic resonance spectrometer for biological applications, *Rev. Sci. Instrum.* **69**, 1869-1876.

Murugesan, R., English, S., Reijnders, K., Yamada, K., Cook, J. A., Mitchell, J. B., Subramanian, S., and Krishna, M. C. (2002) Fluorine Electron Double Resonance Imaging for ^{19}F MRI in Low Magnetic Fields. *Magn. Reson. Med.* **48**, 523-529.

Nicholson, I., Lurie, D. J., and Robb, F. J. L. (1994a). The Application of Proton-Electron Double-Resonance Imaging Techniques to Proton Mobility Studies, *J. Magn. Reson.* **B104**, 250-255.

Nicholson, I., Robb, F. J. L., and Lurie, D. J. (1994b). Imaging Paramagnetic Species Using Radiofrequency Longitudinally Detected ESR (LODESR Imaging), *J. Magn. Reson.* **B104**, 284-288.

Nicholson, I., Foster, M. A., Robb, F. J. L., Hutchison, J. M. S., and Lurie, D. J. (1996). *In vivo* imaging of nitroxide-free-radical clearance in the rat, using radiofrequency longitudinally detected ESR imaging, *J. Magn. Reson. B* **113**, 256-261.

Nilges, M. J., Walczak, T., and Swartz, H. M. (1989). 1 GHz *in vivo* ESR Spectrometer Operating with a Surface Probe. *Phys. Med.* **5**, 195-201.

Nishikawa, H., Fujii, H., and Berliner, L. J. (1985). Helices and Surface Coils for Low-Field *in Vivo* ESR and EPR Imaging Applications, *J. Magn. Reson.* **62**, 79-86.

Ogata, T. (1995). Development of the rapid field scan L-band ESR-CT system. In *Bioradicals Detected by ESR Spectroscopy*, Ohya-Nishiguchi, H., and Packer, L., eds., Birkhäuser Verlag, Basel, Switzerland, pp. 103-111.

Oikawa, K., Ogata, T., Lin, Y., Sato, T., Kudo, R., and Kamada, H. (1995). Rapid Field Scan L-Band Electron Spin Resonance Computer Tomography System Using an Air-Core Electromagnet, *Anal. Sci.* **11**, 885-888

Oikawa, K., Ogata, T., Togashi, H., Yokoyama, H., Ohya-Nishiguchi, H., and Kamada, H. (1996). A 3D- and 4D-ESR Imaging System for Small Animals, *Appl. Radiat. Isot.* **47**, 1605-1609.

Ono, M., Ogata, T., Hsieh, K. C., Suzuki, M., and Kamada, H. (1986). L-band ESR spectrometer using a loop-gap resonator for *in vivo* analysis, *Chem. Lett.* 491-494.

Ono, M., Ito, K., Kawamura, N., Hsieh, K.-C., Hirata, H., Tsuchihashi, N., and Kamada, H. (1994). A Surface-Coil-Type Resonator for *in vivo* ESR Measurements, *J. Magn. Reson. B* **104**, 180-182.

Ono, M., Asahi, Y., Hirata, H., and Shimoyama, Y. (1998). A New Frequency Modulation Scheme of Electron Paramagnetic Resonance Spectroscopy. Presented at the workshop on EPR Studies of Viable Biological Systems and Related Techniques, Sept 13-18, 1998, Hanover, NH.

Pake, G. E., Townsend, J., and Weissman, S. I. (1952). Hyperfine Structure in the paramagnetic Resonance of the Ion $(SO_3)_2NO^{-}$, *Phys. Rev.* **85**, 682-683.

Panagiotelis, I., Nicholson, I., Hutchison, J. M. S. (2001a) Electron Spin Relaxation Time Measurements Using Radiofrequency Longitudinally Detected ESR and Application to Oximetry. *J. Magn. Reson.* **149**, 74-84.

Panagiotelis, I., Nicholson, I., Foster, M. A., Hutchison, J. M. S. (2001b) T_{1e}* and T_{2e}* Maps Derived *In Vivo* From the Rat Using Longitudinally Detected Electron Spin Resonance Phase Imaging: Application to Abdominal Oxygen Mapping. *Magn. Reson. Med.* **46**, 1223-1232.

Petryakov, S., Chzhan, M., Samouilov, A., He, G., Kuppusamy, P., and Zweier, J. L. (2001) A Bridged Loop-Gap S-Band Surface Resonator for Topical EPR Spectroscopy. *J. Magn. Reson.* **151**, 124-128.

Placidi, G., M. Alecci, and Sotgiu, A. (2002) First imaging results obtained with a multimodal apparatus combining low-field (35.7 mT) MRI and pulsed EPRI. *Phys. Biol. Med.* **47**, N127-N132.

Pohida, T. J., Fredrickson, H. A., Tschudin, R. G., Fessler, J. F., Krishna, M. C., Bourg, J., Harrington, F., and Subramanian, S. (1994). High-speed digitizer/averager data-acquisition system for Fourier transform electron paramagnetic resonance spectroscopy, *Rev. Sci. Instrum.* **65**, 2500-2504.

Purcell, E. M., Torrey, H. C., and Pound, R. V. (1946). Resonance Absorption by Nuclear Magnetic Moments in a Solid, *Phys. Rev.* **69**, 37-38.

Puwanich, P., Lurie, D. J., and Foster, M. A. (1999) Rapid imaging of free radicals *in vivo* using field cycled PEDRI. *Phys. Med. Biol.* **44**, 2867-2877.

Quaresima, V., Alecci, M., Ferrari, M., and Sotgiu, A. (1992). Whole rat electron paramagnetic resonance imaging of a nitroxide free radical by radio frequency (280 MHz) spectrometer, *Biochem. Biophys. Res. Commun.* **183**, 829-835.

Quine, R. W., Rinard, G. A., Ghim, B. T., Eaton, S. S., and Eaton, G. R. (1996). A 1-2 GHz Pulsed and Continuous Wave Electron Paramagnetic Resonance Spectrometer, *Rev. Sci. Instrum.* **67**, 2514-2527.

Quine, R. W., Rinard, G. A., Eaton, S. S., and Eaton, G. R. (2002) A Pulsed and Continuous Wave 250 MHz Electron Paramagnetic Resonance Spectrometer, *Magn. Reson. Engineer.* **15**, 59-91.

Rannestad, A., and Wagner, P. A. (1963). Paramagnetic Relaxation in Dilute Potassium Ferricyanide, *Phys. Rev.* **131**, 1953-1960.

Reynolds, G. F., and Shanholtzer, W. L. (1967). Electron Spin Resonance At Radio Frequencies, *Proc. West Virginia Acad. Sci.* **39**, 387-391.

Rinard, G. A., Quine, R. W., Eaton, S. S., Eaton, G. R., and Froncisz, W. (1994). Relative Benefits of Overcoupled Resonators vs. Inherently Low-Q Resonators for Pulsed Magnetic Resonance, *J. Magn. Reson.* **A108**, 71-81.

Rinard, G. A., Quine, R. W., Ghim, B. T., Eaton, S. S., and Eaton, G. R. (1996a). Easily Tunable Crossed-Loop (Bimodal) EPR Resonator, *J. Magn. Reson.* **A122**, 50-57.

Rinard, G. A., Quine, R. W., Ghim, B. T., Eaton, S. S., and Eaton, G. R. (1996b). Dispersion and Superheterodyne EPR Using a Bimodal Resonator, *J. Magn. Reson.* **A122**, 58-63.

Rinard, G. A., Quine, R. W., Song, R., Eaton, G. R., and Eaton, S. S. (1999a). Absolute EPR Spin Echo and Noise Intensities, *J. Magn. Reson.* **140**, 69-83.

Rinard, G. A., Quine, R. W., Harbridge, J. R., Song, R., Eaton, G. R., and Eaton, S. S. (1999b). Frequency Dependence of EPR Signal-to-Noise, *J. Magn. Reson.* **140**, 218-227.

Rinard, G. A., Eaton, S. S., Eaton, G. R., Poole, C. P. Jr., and Farach, H. A. (1999c). Sensitivity, in *Handbook of Electron Spin Resonance*, C. P. Poole, Jr., and H. A. Farach, eds, vol 2, AIP Press, 1-23.

Rinard, G. A., Quine, R. W., and Eaton, G. R., (2000). An L-band Crossed-Loop (Bimodal) Resonator, *J. Magn. Reson.* **144**, 85-88.

Rinard, G. A., Quine, R. W., Eaton, G. R., and Eaton, S. S. (2002a) 250 MHz Crossed Loop Resonator for Pulsed Electron Paramagnetic Resonance, *Magn. Reson. Engineer.* **15**, 37-46.

Rinard, G. A., Quine, R. W., Eaton, G. R., and Eaton, S. S. (2002b) Adapting a Hall Probe Controller for Current Control of an Air-Core Magnet, *Magn. Reson. Engineer.* **15**, 47-50.

Rinard, G. A., Quine, R. W., Eaton, G. R., and Eaton, S. S., Barth, E. D., Pelizzari, C. A., and Halpern, H. H. (2002c) Magnet and Gradient Coil System for Low-Field EPR Imaging, *Magn. Reson. Engineer.* **15**, 51-58.

Rinard, G. A., Quine, R. W., Eaton, S. S., and Eaton, G. R. (2002d) Frequency Dependence of EPR Signal Intensity, 250 MHz to 9.1 GHz, *J. Magn. Reson.* **156**, 113-121.

Rinard, G. A., Quine, R. W., Eaton, S. S., and Eaton, G. R. (2002e) Frequency Dependence of EPR Signal Intensity, 248 MHz to 1.4 GHz, *J. Magn. Reson.* **154**, 80-84.

Rinard, G. A., Quine, R. W., Eaton, S. S., and Eaton, G. R. (2003). Frequency Dependence of EPR Sensitivity, in *Biol. Magn. Reson.*, L. J. Berliner, ed., this volume..

Robitaille, P.-M. L., Abduljalil, A. M., Kangarlu, A., Zhang, X., Yu. Y., Burgess, R., Bair, S., Noa, P., Yang, L., Zhu, H., Palmer, B., Jiang, Z., Chakeres, D. M., and Spigos, D. (1998). Human magnetic resonance imaging at 8 T, *NMR Biomed.* **11**, 263-265.

Robitaille, P.-M. L., Kangarlu, A., and Abduljalil, A. M. (1999). RF Penetration in Ultra High Field MRI: Challenges in Visualizing Details Within the Center of the Human Brain. *J. Comp. Assist. Tomog.* **23**, 845-849.

Romanelli, M., Kurshev, V., and Kevan, L. (1994). Comparative Analysis of Pulsed Electron Spin Resonance Spectrometers at X-Band and S-Band, *Appl. Magn. Reson.* **7**, 427-441.

Rothenberger, K. S., Nilges, M. J., Altman, T. E., Glab, K., Belford, R. L., Froncisz, W., and Hyde, J. S. (1986). L-Band Parallel Mode EPR. Measurement of Quadrupole Coupling Through Direct Observation of Secondary Transitions, *Chem. Phys. Lett.* **124**, 295-298

Rubinson, K. A., Koscielniak, J., and Berliner, L. J. (1995). Modified, Short-Circuited Coaxial-Line Resonators for CW-EPR. *J. Magn. Reson. A* **117**, 91-93.

Rubinson, K., Cook, J. A., Mitchell, J. B., Murugesan, R., Krishna, M. C., and Subramanian, S. (1998). FT-EPR with a nonresonant probe: use of a truncated coaxial line, *J. Magn. Reson.* **132**, 255-259.

Rubinson, K. A. (2000). Resonators for Low-Field *in Vivo* EPR, *Biol. Magn. Reson.* **18**, Berliner, L. J., ed., ch. 5, in press.

Sakamoto, Y., Hirata, H., and Ono, M. (1995) Design of a Multicoupled Loop-Gap Resonator for Pulsed Electron Paramagnetic Resonance Measurements. *IEEE Trans. Microwave Theory Techniques* **43**, 1840-1847.

Sato, T., Oikawa, K., Ohya-Nishiguchi, H., and Kamada, H. (1997). Development of an L-band electron spin resonance/proton nuclear magnetic resonance imaging instrument, *Rev. Sci. Instrum.* **68**, 2076-2081.

Sato, T., Yokoyama, H., Ohya, H., and Kamada, H. (2000). Development and evaluation of an electrically detected magnetic resonance spectrometer operating at 900 MHz. *Rev. Sci. Instrum.* **71**, 486-493.

Sato, T., Yokoyama, H., Ohya, H., and Kamada, H. (2002) An active resonator system for CW-ESR measurement operating at 700 MHz. *J. Magn. Reson.* **159**, 161-166.

Schmidt, J. (1972). Electron spin echoes in photo-excited triplet states of organic molecules in zero magnetic field, *Chem. Phys. Lett.* **14**, 411-414.

Schneider, H. J., and Dullenkopf, P. (1977). Slotted tube resonator: A new NMR probe head at high observing frequencies, *Rev. Sci. Instrum.* **48**, 68-73.

Sciandrone, M., Placidi, G., Testa, L., and Sotgiu, A. (2000). Compact low field magnetic resonance imaging magnet: Design and optimization. *Rev. Sci. Instrum.* **71**, 1534-1538.

Shing, Y. H., and Buckmaster, H. A. (1976). Anomalous Splitting of the Ground State of Gadolinium Ion in Lanthanum Ethyl Sulfate. *J. Magn. Reson.* **21**, 295-309.

Singer, J. R. (1962). Exchange Narrowed ESR Absorption Lines at Low and Intermediate Frequencies. in *Paramagnetic Resonance: Proceedings of the First International Conference held in Jerusalem, July 16-24, 1962*, W. Low, ed., Academic Press, N.Y., Vol. II, pp. 577-581.

Smith, C. M., and Stevens, A. D. (1994). Reconstruction of images from radiofrequency electron paramagnetic resonance spectra, *Brit. J. Radiol.* **67**, 1186-1195.

Sotgiu, A. (1985). Resonator Design for *in Vivo* ESR Spectroscopy, *J. Magn. Reson.* **65**, 206-214.

Sotgiu, A. (1986). Fields and gradients in multipolar magnets. *J. Appl. Phys.* **59**, 689-693.

Sotgiu, A., and Gualtieri, G., (1985). Cavity Resonator for *in vivo* ESR spectroscopy, *J. Phys. E: Sci. Instrum.* **18**, 899-901.

Sotgiu, A., Fujii, H., and Gualtieri, G. (1987a). Torroidal surface coil for topical ESR spectroscopy, *J. Phys. E: Sci. Instrum.* **20**, 1428-1429.

Sotgiu, A., Gualtieri, G., and Momo, F. (1987b).Generation of axial fields of arbitrary profile by air core coils, *J. Phys. E: Sci. Instrum.* **20**, 1574-1576.

Sotgiu, A., Alecci, M., Brivati, J., Placidi, G., and Testa, L. (1995). New experimental modalities of low frequency electron paramagnetic resonance imaging. In *Bioradicals Detected by ESR Spectroscopy*, Ohya-Nishiguchi, H., and Packer, L., eds., Birkhäuser Verlag, Basel, Switzerland, pp. 69-92.

Stevens, A. D. (1994). A moderately sized resonator/surface coil for radiofrequency electron paramagnetic resonance spectroscopy and imaging *in vivo*, *Brit. J. Radiol.* **67**, 1243-1248.

Stevens, A. D., and Brivati, J. A. (1994). A 250 MHz EPR spectrometer with rapid phase-error correction for imaging large biological specimens, *Meas. Sci. Technol.* **5**, 793-796.

Strandberg, M. W. P., Tinkham, M., Solt, I. H. Jr., and Davis, C. F. Jr. (1956). Recording Magnetic-Resonance Spectrometer, *Rev. Sci. Instrum.* **27**, 596-605.

Subramanian, S., Murugesan, R., Devasahayan, N., Cook, J. A., Afeworki, M., Pohida, T., Tschudin, R. T., Mitchell, J. B., and Krishna, M. C. (1999). High-speed data acquisition system and receiver configurations for time-domain radiofrequency electron paramagnetic resonance spectroscopy and imaging, *J. Magn. Reson.* **137**, 379-188.

Subramanian, S., Mitchell, J. B., and Krishna, M. C. (2000). Time-Domain Radio Frequency EPR Imaging, *Biol. Magn. Reson.* **18**, Berliner, L. J., ed., ch. 2., in press.

Subramanian, S., Devasahayam, N., Murugesan, R., Yamada, K., Cook, J., Taube, A., Mitchell, J. B., Lohman, J. A. B., and Krishna, M. C. (2002) Single-Point (Constant-Time) Imaging in Radiofrequency Fourier Transform Electron Paramagnetic Resonance. *Magn. Reson. Med.* **48**, 370-379.

Swartz, H. M., and Halpern, H. (1998). EPR studies of living animals and related model systems (*in vivo* EPR), *Biol. Magn. Reson.* **14**, 367-404.

Symons, M. C. R. (1995). Whole body electron spin resonance imaging spectrometer, in *Bioradicals Detected by ESR Spectroscopy*, H. Ohya-Nishiguchi and L. Packer, eds., Birkhauser Verlag, Basel, 93-102.

Tada, M., Yokoyama, H., Toyoda, Y., Ohya, H., Ito, T., and Ogata, T. (2000) Surface-Coil-Type Resonators for *in Vivo* Temporal ESR Measurements in Different Organs of Nitroxide-Treated Rats. *Appl. Magn. Reson.* **18**, 575-582.

Terakado, O., Kamiyama, T., and Kakamura, Y. (1998). Low-field EPR study of the metal-non-metal transition in sodium-ammonia solutions, *J. Chem. Soc., Faraday Trans.* **94**, 867-869.

Treiguts, E., and Çugunov, L. (1995). Frequency dependence of Mn^{2+} ion electron spin resonance spectra in the ultra-high-frequency band, *J. Phys. Condens. Matter* **7**, 6789-6795.

Utsumi, H., Tatebe, T., and Hamada, A. (1992). ESR Spectra of VO^{2+} and Mn^{2+} in Aqueous Solution at L-band, *Chem. Lett.* 277-280.

Van Gerven, L., Van Itterbeek, A., and De Laet, L. (1962). Low Field EPR Measurements in DPPH at Low Temperatures in *Paramagnetic Resonance: Proceedings of the First International Conference held in Jerusalem, July 16-24, 1962*, W. Low, ed., Academic Press, N.Y., vol. II, pp. 905-918.

Verlinden, R., Grobet, P., and Van Gerven, L. (1974) Low Temperature Magnetic Properties of DPPH.$(C_6D_6)_x$, Studied by NMR and EPR. *Chem. Phys. Lett.* **27**, 535-539.

Weiss, P. R., Whitmer, C. A., Torrey, H. C., Hsiang, J-S., (1947). Magnetic Resonance Absorption of Chromic Ammonium Alum, *Phys. Rev.* **72**, 975.

Weiss, P. R. (1948). The Microwave Spectroscopy of Paramagnetic Salts: the Spectrum of Chromic Alum, *Phys. Rev.* **73**, 470-476.

Wheatley, J., and Halliday, D. (1949). Paramagnetic Absorption in Single Crystals of Copper Sulfate Pentahydrate, *Phys. Rev.* **75**, 1412-1415.

Willer, M., Forrer, J., Keller, J., Van Doorslaer, S., Schweiger, A., Schuhmann, R., and Weiland, T. (2000). S-band (2-4 GHz) pulse electron paramagnetic resonance spectrometer: Construction, probe head design, and performance, *Rev. Sci. Instrum.* **71**, 2807-2817.

Yamada, K., Murugesan, R., Devasahayam, N., Cook, J. A., Mitchell, J. B., Subramanian, S., and Krishna, M. C. (2002) Evaluation and Comparison of Pulsed and Continuous Wave Radiofrequency Electron Paramagnetic Resonance Techniques for *in Vivo* Detection and Imaging of Free Radicals. *J. Magn. Reson.* **154**, 287-297.

Yokoyama, H., Ogata, T., Tsuchihashi, N., Hiramatsu, M., and Mori, N. (1996). A Spatiotemporal Study on the Distribution of Intraperitoneally Injected Nitroxide Radical in the Rat Head Using an *in Vivo* ESR Imaging System, *Magn. Reson. Imag.* **14**, 559-563.

Yokoyama, H., Sato, T., Tsuchihashi, N., Ogata, T., Ohya-Nishiguchi, H., and Kamada, H. (1997a). A CT Using Longitudinally Detected ESR (LODESR-CT) of Intraperitoneally Injected Nitroxide Radical in a Rat's Head, *Magn. Reson. Imag.* **15**, 701-708.

Yokoyama, H., Sato, T., Ogata, T., Ohya-Nishiguchi, H., and Kamada, H. (1997b). *In Vivo* Longitudinally Detected ESR Measurements at Microwave Regions of 300, 700, and 900 MHz in Rats Treated with a Nitroxide radical, *J. Magn. Reson.* **129**, 201-206.

Yokoyama, H., Sato, T., Ohya-Nishiguchi, H., and Kamada, H. (1998). *In vivo* 300 MHz longitudinally detected ESR-CT imaging in the head of a rat treated with a nitroxide radical, *Magn. Reson. Mater. Phys. Biol. Med.* **7**, 63-68.

Yokoyama, H., Sato, T., Fukui, K., Ohya-Nishiguchi, H., Kamada, H. (1999). High-power Low-Q RF ESR Resonator for Pulse Techniques. *Chem. Lett.* 919-920.

Yokoyama, H., Sato, T., Ogata, T., Ohya, H., and Kamada, H. (2001) Automatic Coupling Control of a Loop-Gap Resonator by a Variable Capacitor Attached Coupling Coil for EPR Measurements at 650 MHz. *J. Magn. Reson.* **149** 29-35.

Yoshioka, T. (1977). Magnetic Properties and Low-Field ESR of Organic Free Radicals, Monochloroporphyrexide and Porphyrexide, *Bull. Chem. Soc. Japan* **50**, 1372-1378.

Youngdee, W., Planinšič, G., and Lurie, D. J. (2001) Optimization of field-cycled PEDRI for *in vivo* imaging of free radicals. *Phys. Biol. Med.* **46**, 2531-2544.

Youngdee, W., Lurie, D. J., and Foster, M. A. (2002) Rapid imaging of free radicals *in vivo* using hybrid FISP field-cycled PEDRI. *Phys. Med. Biol.* **47**, 1091-1100.

Yuan, H., Collins, M. L. P., and Antholine, W. E (1999). Type 2 Cu^{2+} in pMMO from Methylomicrobium album BG8, *Biophys. J.* **76**, 2223-2229.

Zavoisky, E. (1945). Spin Magnetic Resonance in Paramagnetic Substances, *J. Phys. (USSR)* **9**, 245.

Zavoisky, E. (1946). Spin Magnetic Resonance in the Deci-Meter-Wave Region, *J. Phys. (USSR)* **10**,197-198.

Zweier, J. L., Kuppusamy, P. (1988). Electron paramagnetic resonance measurements of free radicals in the intact beating heart: A technique for detection and characterization of free radicals in whole biological tissues, *Proc. Natl. Acad. Sci. (US)* **85**, 5703-5707.

Zweier, J. L., and Kuppusamy, P. (1994). EPR Spectroscopy of Free Radicals in the Perfused Heart, *Curr. Topics Biophys.* **18**, 14-25.

Chapter 3

Frequency Dependence of EPR Sensitivity

George A. Rinard, Richard W. Quine, Sandra S. Eaton, and Gareth R. Eaton
Department of Engineering and Department of Chemistry and Biochemistry, University of Denver, Denver, Colorado 80208 USA

Abstract: Contrary to some prior derivations, it is shown that the sensitivity of EPR measurements is, as expected, the same as for NMR, and that in general comparisons of EPR sensitivity as a function of frequency have been pessimistic by one factor of ω. The sensitivity of EPR can increase at lower frequency if the sample size is scaled inversely with frequency.

1. INTRODUCTION

To obtain adequate depth penetration into large aqueous samples, including animals, EPR imaging is performed at the lowest feasible frequency - typically between 250 MHz and 2 GHz. The feasibility of imaging physiologically significant signals at these frequencies is dependent upon obtaining adequate signal-to-noise (S/N). A frequently-posed question is "what, in principle, should one expect to be able to achieve with an ideal spectrometer"? This chapter presents a basis for making such predictions. Biological EPR spectroscopy also encompasses multi-frequency and high-field/high-frequency EPR (HFEPR), CW and pulsed, and the study of samples at very low temperatures. This chapter consequently discusses the frequency dependence of EPR S/N for a variety of conditions, including at, e.g., liquid helium temperature.

The issue of EPR signal intensity as a function of frequency has been confused for many years because of an unfortunate error in Poole's book (1967). Experimental tests of the frequency dependence of S/N have typically compared performance on extensively engineered X-band systems with results obtained on low-frequency EPR systems that have been built with minimal budget and/or for specialized purposes and are unlikely to approach ultimate sensitivity. The predictions and experimental results in the literature have provided a rather pessimistic picture.

In this chapter we first present a general discussion of EPR signal intensity and the frequency dependence of the terms that contribute to the signal intensity. Next, we briefly show the relation of this approach to the expressions commonly used in similar discussions of NMR. NMR and EPR traditions, reflecting the historic differences in frequency range of interest and the associated technologies, express the frequency dependence of sensitivity differently. A translation between these traditions is presented. Following this, we present a step-by-step derivation of the fundamental equations for magnetic resonance signals, both CW and pulsed. After introducing noise, we formulate expressions for the signal-to-noise (S/N) ratio, and consider various contributions to noise in EPR. Detailed expressions are then presented for the frequency dependence of S/N for two limiting situations - noise dominated by resistive losses in the resonator, and noise dominated by loss in the sample. Results are gathered into two tables for ease of reference. After comparison of these predictions with the literature, limitations and extensions are briefly discussed. In the final section this background is applied to the design of *in vivo* EPR and imaging measurements. Derivations of some of the equations stated in this paper are presented in a general discussion of signal and noise (Rinard *et al.*, 1999a), which should be referred to for more detail.

Contrary to some prior derivations, it is shown that the sensitivity (defined as S/N =1 with all noise due to thermal noise of the resonator) of EPR measurements is, as expected, the same as for NMR, and that in general comparisons of EPR sensitivity as a function of frequency (ω) have been pessimistic by one factor of ω. The sensitivity of EPR can increase as frequency is decreased if the sample size is scaled inversely with frequency.

1.1 Major Contributions to Signal and Noise

In the following paragraphs each of the contributors to the frequency dependence of the EPR signal and the contributors to noise are considered in detail. Here, we provide a general perspective on signal and noise. The frequency dependence of the EPR signal is inherent in the spin magnetization term in the susceptibility (Eqs. 1 and 2 below), and in the dependence on frequency of the filling factor, η, and the resonator Q. Both of these depend on geometric details of the sample and the resonator, such that the net dependence of S/N on frequency can range from $\omega^{-1/2}$ to $\omega^{11/4}$ for cases in which thermal noise of the resonator or sample losses dominate the noise. There are many contributions to noise in an EPR spectrometer. In the ideal case that the dominate noise is the thermal noise of the resonator, the noise can be predicted quantitatively. For the case in which sample losses dominate, quantitative prediction of noise is possible within the limitations of knowledge of the frequency dependence of sample losses, which must be determined empirically. In all cases, components of the spectrometer between the signal in the resonator and the final display add noise. Knowing the gain or loss, bandwidth, and noise figure of each component permits quantitative prediction of the impact of each component on the final S/N of the spectrometer. There are experimentally realizable cases (indeed they exist in many commercial and home-built spectrometers under various conditions of operation) in which the limiting noise in the spectrometer can be microwave source phase noise, microphonics in the resonator due to vibrations in the environment, microphonics in the resonator due to eddy currents and Lorentz forces resulting from the magnetic field modulation, 1/f noise of the detector, high noise figure of the first stage amplifier, high noise figure of the detector, or even, in extreme cases, power supply noise and ground loops. Most of these are identifiable (sometimes with considerable effort), can be avoided with good engineering, and are not fundamental to the EPR experiment per se. For *in vivo* EPR, animal physiology (e.g., breathing, heartbeat) may dominate the noise. In this chapter we focus on noise sources that are fundamental to the EPR experiment - thermal noise in the resonator or losses in the sample. Noise voltage results from a current in a resistance, so we will formulate the problem in terms of the relevant resistances. For generality, and precision in the arguments, we start with the case in which the relevant noise is that generated in the resistance of the resonator. Following this, we replace this resistance with the specific case important for *in vivo* EPR in which the dominant resistance is that of the organism being studied.

2. FREQUENCY DEPENDENCE OF EPR SIGNAL INTENSITY

A general expression for EPR signal intensity is Eq. (1) (Rinard *et al.*, 1999a),

$$V_S = \chi'' \eta Q \sqrt{P Z_0} \tag{1}$$

where V_S is the signal voltage at the end of the transmission line connected to the resonator, η (dimensionless) is the resonator filling factor (see Poole, 1967, p. 291), Q is the loaded quality factor of the resonator, Z_0 is the characteristic impedance of the transmission line, and P is the microwave power to the resonator produced by the external microwave source. The magnetic susceptibility of the sample, χ'', is the imaginary component of the effective RF susceptibility.

For a Lorentzian line with linewidth $\Delta\omega$ at resonance frequency ω, substituting for the susceptibility in (1) yields (2):

$$V_S = \frac{N_S g^2 \beta^2 \mu_0}{4k_B T_S}\left(\frac{\omega}{\Delta\omega}\right)\eta Q_L \sqrt{Z_0 P} \tag{2}$$

where N_S is the number of spins per unit volume, k_B is Boltzmann's constant, β is the Bohr magneton, $\mu_0 = 4\pi x 10^{-7}\ T^2J^{-1}m^3$ is the permeability of vacuum, T_S is the temperature of the sample, and we have introduced numerical coefficients for the case of S = 1/2.

To see relationships between some of the terms in this equation, consider the effect of increasing the size of the resonator and sample while keeping both the sample concentration, N_S, and the filling factor, η, constant. The number of spins increases proportional to the increase in volume of sample. The increase in EPR signal is not directly proportional to this increase in number of spins, because, as will be shown below, the increase in size also affects Q through the change in resistance R and inductance L of the resonator. However, if the resonator size and frequency are kept constant, and the sample size is increased, thus changing the filling factor η, the EPR signal would increase in proportion to the increase in volume of the sample. These statements assume that the sample is non-lossy and does not have a dielectric constant large enough to distort the RF B_1 distribution.

Prediction of the frequency dependence of EPR signal-to-noise involves examination of the frequency dependence of each term in Eq. (1). Equation (2) explicitly shows the frequency dependence of the magnetic susceptibility. We now seek the frequency dependence of the filling factor, η, and of Q. Since an experiment may be done at constant incident power or constant B_1, we also derive the frequency dependence of these terms. Without loss of generality we can assume a sample and resonator geometry such that we do not have to include the physical description of end effects of coils or samples. In practical cases these effects and/or special geometries are encompassed in the filling factor or coil efficiency factor.

The results for three cases are summarized in Table 1: case 1 - the size (linear dimensions) of the sample and resonator are constant, case 2 - the size of the sample and resonator are scaled with $1/\omega_0$, and case 3 - the size of the sample is constant and the size of the resonator is scaled with $1/\omega$.

2.1 Frequency Dependence of Q for LGR

To provide specificity and coherence to the presentation, we assume the resonator has the topology of a loop-gap-resonator (LGR). Any other resonator type whose frequency can be changed without changing the overall size of the resonator could be used to obtain the results for case 1 (constant sample size and constant resonator size). Resonators whose size scales with frequency, such as cavity resonators, could yield cases 2 and 3. To consider other resonator, the only change in the calculations would be to change the specific formulae for the inductance, resistance and capacitance of the resonator. As long as the scaling of size and frequency meets the criteria of the entries in the Tables, the trends displayed there should apply, regardless of the type of resonator used.

The loaded Q, Q_L, for a loop gap resonator (LGR), is given by $Q_L = \omega L/2R$. The inductance, L, is related to the length z and the cross-sectional area of the resonator loop, $A_R = \pi d^2/4$, (d is the diameter of the loop) by Eq. (3).

$$L = \mu_0 \frac{A_R}{z} \tag{3}$$

The resistance, including skin effect, (skin depth $\delta = [\omega\mu_0\sigma/2]^{-1/2}$) of the resonator loop is

$$R = \frac{\pi d}{z}\sqrt{\omega\frac{\mu_0}{2\sigma}} \tag{4}$$

where σ is the conductivity of the material of the resonator. Substitution of (3) and (4) into the definition of Q, gives the frequency dependence of Q, (5):

$$Q_L = \frac{d\sqrt{2}}{8}\sqrt{\omega}\sqrt{\mu_0}\sqrt{\sigma} \tag{5}$$

Since σ can be assumed to be independent of frequency, the frequency dependence of Q is determined by $\omega^{1/2}d$, which is $\omega^{1/2}$ if the resonator size is kept constant, and is $\omega^{-1/2}$ if the size is scaled proportional to ω^{-1}.

This expression for Q inherently includes the resistance of the resonator. If there were no other sources of noise in the spectrometer system, this resistance would establish the noise floor of the spectrometer. Other noise sources will be added in subsequent development below.

2.2 Frequency Dependence of Filling Factor

The filling factor, η, must be determined for each sample and resonator combination, and requires a detailed analysis of the electromagnetic fields in the resonator. However, we often can make an educated guess for η. We will assume, as a first approximation, that the filling factor for LGRs as a function of frequency will scale as if the RF magnetic field were uniform and confined to the loop. For resonators with a constant loop size and sample size, this approximation will overestimate η at higher frequency because the gap spacing must increase with frequency for a constant loop size, and as the gap increases, the magnetic field will no longer be completely confined to the loop.

2.2.1 Filling Factor for Lossy Samples

When a dielectric sample is placed in a resonator the resonant frequency decreases (Harrington, 1961, p.323). A familiar example is the decrease in frequency when a quartz Dewar in placed in a cavity resonator. In the opposite case of a highly conducting sample (e.g., a metal), the resonant frequency increases if the material is in a region of large magnetic field, and decreases if the material is in a region of large electric field (Harrington, 1961, p. 320). Most biological samples involve both dielectric and conductive interactions with the microwaves.

If a sample is "lossy," microwave energy is converted to heat within the sample as a result of interaction of the microwave magnetic and electric vectors with polar and charged species in the sample. In the case of living organisms, water and ions are the primary causes of loss. To a first approximation, the microwave B_1 decreases exponentially into the organism due to the loss. The real case is somewhat more complicated due to the heterogeneity of the organism (Halpern 1989; Jiang, 1995; Sueki, 1996). Halpern and Bowman (1989) summarized available information as showing about 5 cm RF penetration in animals at 200-250 MHz. The filling factor of a lossy organism in a resonant cavity is not simply the ratio of the volume of the sample to that of the resonator, as can be approximated for a lossless dielectric (Poole, 1967, p. 291). This is due to the fact that in the calculation of the filling factor B_1 varies over the organism due to the effects of the loss. In the extreme case, a portion of the interior of the organism is beyond the depth to which microwaves penetrate to any significant degree.

2.3 Frequency Dependence of RF Field B_1

The frequency dependence of B_1 is given by Eq. (6) (Rinard *et al.*, 1999a) for the special case in which only the resistance of the materials of construction of the resonator contribute to the power loss:

$$B_1 = K_b \frac{1}{\sqrt{zd}} \frac{\sqrt{P}}{\omega^{1/4}} \tag{6}$$

where: $K_b = k_1 \sqrt[4]{2\mu_0^3 \sigma / \pi^2}$ and k_1 is a constant of proportionality between square root of power and B_1, the rms magnitude of the circularly polarized component of the microwave magnetic field that is perpendicular to the external magnetic field and in phase with the Larmor precession. Typically, one would seek to maintain the same B_1 at the sample, unless the spin relaxation time changed with frequency. To keep B_1 constant it is necessary to scale the incident power as $\omega^{1/2}zd$. If the resonator size is scaled with frequency (i.e., d, z $\propto 1/\omega$), then the power required scales as $\omega^{-3/2}$.

Table 1. Predicted Frequency Dependence of EPR Sensitivity When Resonator Resistance Dominates[a]

		Case 1 const. sample size const. LGR size	**Case 2** sample size $\propto 1/\omega$ LGR size $\propto 1/\omega$	**Case 3** const. sample size LGR size $\propto 1/\omega$
1	L	1 (3)	ω^{-1} (3)	ω^{-1} (3)
2	R (resonator resistance)	$\omega^{1/2}$ (4)	$\omega^{1/2}$ (4)	$\omega^{1/2}$ (4)
3	Q	$\omega^{1/2}$ (5)	$\omega^{-1/2}$ (5, 6)	$\omega^{-1/2}$ (5)
4	η	1	1	ω^3 (44)
5	EPR S/N at constant P	$\omega^{3/2}$ (41)	$\omega^{1/2}$ (42)	$\omega^{7/2}$ (45)
6	$B_1/\sqrt{P}$	$\omega^{-1/4}$ (6)	$\omega^{3/4}$ (43)	$\omega^{3/4}$ (43)
7	P to maintain constant B_1	$\omega^{1/2}$ (6)	$\omega^{-3/2}$ (43)	$\omega^{-3/2}$ (43)
8	EPR S/N at constant B_1	$\omega^{7/4}$ (6, 40, 41)	$\omega^{-1/4}$ (40, 42, 43)	$\omega^{11/4}$ (40, 43, 45)[b]

[a]The equation numbers that are the bases for the various statements are given in parentheses.

[b] Note that in Rinard et al. (1999a) table 1, Case 3, there is a typographical error: the EPR signal is proportional to $\omega^{11/4}$, not $\omega^{5/4}$ as stated in that reference.

3. COMPARISON WITH EXPRESSIONS FOR FREQUENCY DEPENDENCE OF NMR

Due to the different paths in the historical developments of the fields, NMR (Makovski, 1996; Andrew, 1988; Abragam, 1961; Hoult, 1996; Hoult and Richards, 1976; Hoult and Lauterbur, 1979; Chen and Hoult, 1989; Vlaardingerbroek and den Boer, 1996) and EPR (Poole, 1967; Rinard *et al.*, 1999a) have expressed sensitivity, or signal-to-noise (S/N), from such different points of view that at first glance the expressions seem to show different dependences on fundamental parameters. Predictions for EPR are consistent with the commonly-cited expressions for NMR when the same assumptions are made, since the fundamental spin resonance is the same. In NMR, the signal typically is represented as a current induced in a coil by precession of the spins. The typical NMR case would be case 1, i.e., a constant sample size and constant loop size, for which the inductance L is constant. For the case in which coil resistance dominates the noise, NMR S/N depends on frequency to the 7/4 power (Hoult and Richards, 1976), in agreement with the results in Table 1.

4. DERIVATION OF THE FUNDAMENTAL EQUATIONS FOR A MAGNETIC RESONANCE SIGNAL

A key point in relating the traditional NMR and EPR expressions for S/N is that the expressions for NMR conventionally are written for the resonator side of an impedance matching network, and the S/N expressions for EPR usually are written for the detector side of the impedance matching network (Fig. 1).

In Fig. 1, R, L, and C are the resistance, inductance, and capacitance of the NMR or EPR resonator that may be, for example, a coil, a cavity, a birdcage resonator, or an LGR. In the case of an EPR cavity resonator, the impedance matching network is the iris assembly. V_E is the spin system voltage, and V_1 is the resultant voltage at the input to the impedance matching network. V_S is the magnetic resonance signal voltage sensed at the detector side of the impedance matching network. At resonance, with critical coupling the coil or cavity is a pure resistance R. The impedance matching network transforms R into Z_0, the impedance of the detector, and similarly transforms V_1 into V_S. Note that Eqs. (1) and (2) contain $\sqrt{PZ_0}$. They are written from the EPR point of view.

We now proceed by following the NMR convention of deriving equations in terms of quantities on the resonator side of the impedance matching network. Whether we discuss NMR or EPR, the absorptive part of the RF (we will use RF to denote any frequency of interest) magnetization, M_A, is proportional to B_1.

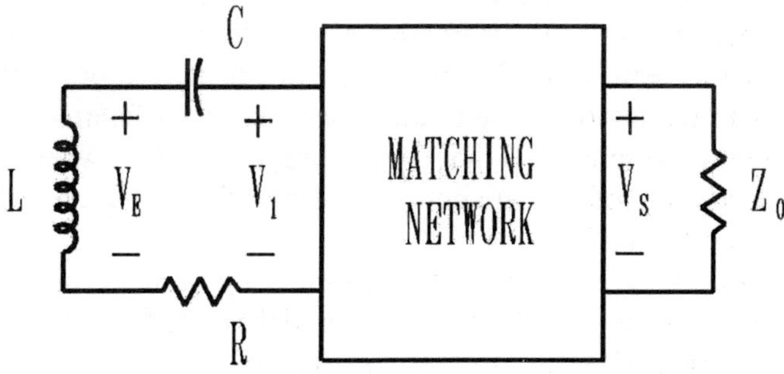

Figure 1. This sketch represents the transformation of V_1 and R on the resistor side into V_S and Z_0 on the detector side by an impedance matching network.

$$M_A = \frac{\chi''}{\mu_0} B_1 = \chi'' H_1 \tag{7}$$

Using for the moment the language of a coil with N turns and cross-sectional area A_R, the RF magnetic flux, ϕ_1, produced in the resonator is

$$\phi_1 = \frac{LI}{N} \tag{8}$$

where I is the current in the resonator. From the fundamental definition of B_1 as flux per unit area, and Eq. (8),

$$B_1 = \frac{\phi_1}{A_R} = \frac{LI}{NA_R} \tag{9}$$

Since $B_1 = \mu_0 H_1$, and $I = \sqrt{\frac{P}{R}}$, Eq. (9) can be rewritten, yielding

$$H_1 = \frac{L}{\mu_0 N A_R}\sqrt{\frac{P}{R}} \tag{10}$$

where P is the power at the resonator from the RF source. Note that the resistance R of the resonator is introduced at this point in the derivation. Anticipating future derivations, note that the resistance R in this equation relating incident power to H_1 represents all loss in the resonator, not just the resistance of the materials of construction of the resonator. Later we will generalize R to include loss due to the properties of the sample as well. The derivation to this point reveals that to maintain constant B_1 when R increases (hence Q decreases, other parameters held constant), the power has to be increased proportionately.

In the following we make the simplifying assumption that B_1 is homogeneous over the sample. We can then define the flux over the sample, ϕ_A, due to the magnetization of the spins, M_A, as

$$\phi_A = \mu_0 M_A A_S = \mu_0 \chi'' A_S H_1 \tag{11}$$

where A_S is the cross sectional area of the sample, and we have used Eq. (7). This flux actually varies as cos(ωt) in the laboratory frame. Substituting the expression (10) for H_1 into (11)

$$\phi_A = \mu_0 \chi'' A_S \frac{L}{\mu_0 N A_R}\sqrt{\frac{P}{R}} \tag{12}$$

The voltage V_E induced by M_A precessing at the Larmor frequency ω in the resonator (coil) is

$$V_E = N\frac{d\phi_A}{dt} = -\omega N \phi_A \sin(\omega t) \tag{13}$$

In the rotating frame we can drop the -sin(ωt) term.
Substituting (12) into (13), we can write for V_1,

$$V_1 = \frac{\chi'' \omega L \eta}{2}\sqrt{\frac{P}{R}} \tag{14}$$

$V_1 = V_E/2$ since the resonator is matched to the transmission line by the impedance matching network. We introduce the filling factor $\eta = A_S/A_R$ to subsume the various geometric terms. N does not appear explicitly in Eq.

(14), but remains in L ($L \propto N^2$), so the voltage still depends on the number of turns.

Assuming that the matching network is lossless, the signal power into the network, $\frac{V_1^2}{R}$, is equal to that delivered to the detector load, $\frac{V_S^2}{Z_0}$. Thus, to find the detected voltage V_S we multiply V_1 by $\sqrt{\frac{Z_0}{R}}$. Hence,

$$V_S = \frac{\chi''\omega L\eta}{2R}\sqrt{PZ_0} \tag{15}$$

Now if we use the definition $Q = \frac{\omega L}{2R}$, we can write this equation as

$$V_S = \chi''\eta Q\sqrt{PZ_0} \tag{16}$$

which is the same as (1), so this derivation is fully consistent with common expressions for EPR signal intensity. In spite of the fact that in EPR this expression is derived in terms of a reflection of the source RF power from the resonator, it is valid for EPR or NMR, CW or pulsed. If the resonator is overcoupled, appropriate adjustment of the B_1 per square root of incident power must be made.

5. COMPARISON OF CW AND PULSED EPR SIGNAL INTENSITIES

For most pulsed EPR measurements the detected S/N is lower than the optimal CW EPR signal for the same sample. In this section we briefly outline the nature of the CW and pulsed EPR signals, and discuss the different impact on their relative S/N due to the fact that the noise in actual spectrometers is not entirely "white" thermal noise. As discussed in the next section, magnetic field modulation is used in CW EPR to suppress low-frequency noise, whereas rapid signal acquisition and averaging techniques (such as analog boxcar or digital summation) are used in pulsed EPR to suppress noise. Because of the differences in time response, pulsed EPR uses a much larger detector bandwidth, which, by Eq. (24) in section 6, decreases the S/N. If averaging techniques are performed long enough to make the effective time constants the same, pulsed EPR can have larger S/N than a CW spectrum when the modulation amplitude is small enough to avoid distortion of the derivative line width. Relaxation times are also important to the comparison. If the relaxation time is long relative to the reciprocal of the modulation frequency, it is not possible to obtain an undistorted field-modulated CW EPR spectrum. If the spectrum is narrow enough that it can be fully excited in pulsed EPR, then in the same total time better S/N can be achieved in pulsed than in CW EPR. Most examples of this advantage for pulsed EPR occur for defect centers in solids (such as the irradiated fused SiO_2 standard sample available from Wilmad Glass, Buena, NJ, USA). It will also be true for some very narrow EPR signals, such as for the Nycomed radical discussed below, which are becoming useful in biological EPR.

5.1 Magnetic Field Modulation in CW EPR

Magnetic field modulation (often at 100 KHz) is used in CW EPR to encode the EPR signal at the modulation frequency so that subsequent phase-sensitive detection discards most of the noise except that in a narrow band near the modulation frequency. It is inherent in the properties of the phase-sensitive detector that the first-harmonic output signal is the first derivative of the magnetic resonance absorption. The principles are sketched in Figure D.6 on page 482 in Weil *et al.*, 1994. In order to obtain a faithful representation of the first derivative of the absorption signal, it is necessary that the magnetic field modulation amplitude be small relative to the absorption line width so that the portion of the line traversed by the modulation approximates the tangent to the absorption curve. As a practical compromise between fidelity and S/N, it is common to use a modulation amplitude approximately 1/10 of the line width. For detection of weak signals, modulation amplitude approximately equal to the line width is used, resulting in some distortion of the line shape. If a very large modulation amplitude is used, the EPR signal splits into two peaks of opposite polarity, each with an amplitude equal to the total absorption signal amplitude (see Figure E.6 on page 505 in Weil *et al.*, 1994). When the modulation amplitude is chosen to minimize distortion of the derivative line shape, the EPR derivative signal amplitude is less than the maximum possible, and is proportional to the modulation amplitude. (There are further complications when the reciprocal of the modulation frequency is of the order of spin relaxation times, but these passage effects are beyond the scope of this chapter.) The principal advantage of magnetic field modulation is that it suppresses noise at frequencies substantially lower than the modulation frequency. There are many contributions to the spectrometer response at low frequencies, including building vibrations, power supply ripple, power line frequencies, temperature variations, cooling fan noise, and even microphonics caused by the magnetic field modulation itself. Magnetic field modulation and phase-sensitive detection yields a smaller than maximal signal, but very much smaller noise than direct detection.

5.2 Pulsed EPR

Precessing electron spin magnetization induces a current in the walls of the resonator. The task of calculating the resultant signal level encompasses four major steps. First, the relation between magnetization and signal in the resonator is calculated from first principles, using the inductance and resistance of the resonator. The relation between EPR line shape and microwave B_1, as described by Bloom (1955) and Mims (1965, 1972), is used to calculate the echo amplitude. Then the signal in the resonator is transformed to the other side of the resonator coupling device. Gains and losses from this point to the detector are used in the calculation of the predicted echo.

The electron spin echo voltage induced in the resonator, as in Eq. (13), is given by

$$V_E = N\frac{d\phi_0}{dt} \tag{17}$$

where N is the number of turns in the resonator and ϕ_0 is the magnetic flux produced by the spin magnetization, M_0. Since the flux density produced by M_0 is $\mu_0 M_0$, ϕ_0 is given by

$$\phi_0 = \mu_0 \eta \vec{A} \bullet \vec{M}_0 \tag{18}$$

where $\vec{A}$ is the cross sectional area of the resonator sample loop, η is the filling factor, and $\mu_0 = 4\pi 10^{-7}$. M_0 varies sinusoidally at the resonant frequency ω, and if the magnetization is fully turned to the xy plane by the microwave pulse, the peak voltage for a single-turn coil (N=1 for a LGR) is

$$V_E = \mu_0 A_R \eta \omega M_0 \tag{19}$$

in agreement with Bloembergen and Pound (1954). Then, from Eq. (22) of Rinard *et al.*, (1994), the output voltage of the resonator coupling structure, $V_{E\beta}$, is given by

$$V_{E\beta} = \frac{\sqrt{\beta}}{1+\beta}\sqrt{\frac{Z_0}{R}}V_E \tag{20}$$

where R is the resistance of the resonator and Z_0 is the impedance of the transmission line (usually 50 Ω). The coupling parameter β can be calculated from the overcoupled Q and the critically-coupled Q, Q_H, by

$$\beta = \frac{2Q_H}{Q} - 1 \tag{21}$$

Combining Eqs. (19) and (20), $V_{E\beta}$ can be written as

$$V_{E\beta} = \frac{\sqrt{\beta}}{1+\beta}\sqrt{\frac{Z_0}{R}}A_R\omega\eta M_0 \tag{22}$$

5.3 Ratio of CW to Pulsed EPR Signal Intensities

The ratio of CW EPR signal intensity to electron spin echo intensity for the same sample is the ratio of Eq. (16) to Eq. (22). For clarity, set β = 1, which is always experimentally possible if the relaxation time is long enough. The algebra simplifies if it is noted that one can use the substitutions $B = \mu_0 H$, $\omega = \gamma B$, $M_0 = \frac{B\chi_0}{\mu_0}$, $Q_L = \frac{\omega L}{2R}$, $\chi'' = \chi_0 \frac{\omega}{\Delta\omega}$, $L = \mu_0 \frac{A_R}{z}$, where z is the length of the loop-gap resonator, and $\frac{\sqrt{P}}{\sqrt{R}}\frac{1}{z} = H_1$, with which it can be shown that

$$\frac{CW}{Echo} = \frac{\gamma B_1}{\Delta\omega} = \frac{B_1}{\Delta B} \tag{23}$$

For convenience, we have written the ratio in both frequency and field units. This ratio implies that if the echo is formed by all of the spins in the sample, the unsaturated CW spectral intensity is equal to the CW microwave B_1 divided by the EPR line width times the echo intensity. Most commercial EPR spectrometers have an output microwave power of 200 mW. For a standard rectangular resonator (loaded Q ~ 3600), this corresponds to a B_1 at the sample of ca. 0.5 G. If the EPR line is about 2.5 G wide, which could be fully excited by a microwave pulse, then the unsaturated CW EPR intensity at 200 mW would be ca. 0.2 times the intensity of the echo. In practice, most CW spectra are obtained with magnetic field modulation. If the magnetic field modulation were approximately equal to line width, this ratio would still hold. Such a large magnetic field modulation would distort the signal, so in practice a smaller modulation amplitude is usually used, resulting in proportionately smaller CW signal relative to echo signal.

6. THERMAL NOISE

Thermal noise in the resonator can be expressed as a noise voltage V_N (Lee and Dalman, 1994)

$$V_N = \sqrt{k_B TBR} \tag{24}$$

where T is the absolute temperature, k_B is Boltzmann's constant, and B is the effective noise bandwidth. Often in the literature the noise voltage is written $\sqrt{4k_B TBR}$. This is the voltage generated in the resistor (cf. the voltage V_E generated in the spin system). The output voltage will be ½ this value for the critically coupled case for the same reason that $V_1 = ½\ V_E$ in Eq. (14).

When the thermal noise is dominated by the resistance R of the materials of construction of the resonator, this is the same R as was assumed in Section 2. In common NMR notation, this is called the coil resistance R_c, which is given by Eq. (25).

$$R_c = \frac{\ell}{p}\sqrt{\frac{\mu_0 \omega \rho_c}{2}} \tag{25}$$

ℓ is the length and p the circumference of the wire, and ρ_c is the resonator resistivity at temperature T_c. The apparent resistance of a typical solenoid is increased about a factor of 5 due to proximity effects between conductors (Hoult and Lauterbur, 1979).

In the next section we will assume that the limiting noise is thermal noise. Later we will add other sources of noise.

7. SIGNAL-TO-NOISE IN MAGNETIC RESONANCE

Combining the results of the above sections, we can divide (14) by (24) and hence write the S/N as

$$S/N = \frac{V_1}{V_N} = \frac{\frac{\chi'' \omega L \eta}{2}\sqrt{\frac{P}{R}}}{\sqrt{k_B TBR}} \tag{26}$$

If the two terms in $\sqrt{R}$ are combined, and the expression for Q ($=\frac{\omega L}{2R}$) is used, the S/N can be written as proportional to $\frac{\omega Q}{\sqrt{k_B TB}}$. Since Q is proportional to $\omega^{1/2}$, the product ωQ yields the $\omega^{3/2}$ dependence at constant power in agreement with case 1 of Table 1.

The noise voltage and the signal voltage transform the same way in the impedance matching network, so if we take the ratio of signal to noise on the detector side of the impedance matching network, both V_1 and V_N are multiplied by $\sqrt{\frac{Z_0}{R}}$, yielding

$$S/N = \frac{V_S}{V_{Ns}} = \frac{\frac{\chi''\omega L\eta}{2R}\sqrt{PZ_0}}{\sqrt{k_B TBZ_0}} \tag{27}$$

$V_{Ns} = V_N\sqrt{\frac{Z_0}{R}}$ is the noise voltage on the detector side of the impedance matching network. Equation (27) is a general expression for S/N in EPR. Cancellation of Z_0 in the numerator and denominator of Eq. (27) gives Eq. (26). The terms in Z_0 are retained in Eq. (27) to facilitate subsequent substitutions. Equation (1) can be converted to Eq. (27) by dividing by $\sqrt{k_B TBZ_0}$ and using $Q = \frac{\omega L}{2R}$. Expressions (26) and (27) for S/N are equivalent. After transformation the resistance term occurs only in the numerator where terms can be gathered to replace $\frac{\omega L}{2R}$ with Q. Since conversion of Eq. (1) to (27) requires only division by $\sqrt{k_B TBZ_0}$, which is independent of frequency, Eq. (1) can be used to describe the frequency dependence of EPR S/N (Rinard *et al.*, 1999a).

A Lorentzian line shape function can be assumed without loss of generality, which changes (26) into (28), as in the conversion of (1) to (2).

$$S/N = \frac{\chi_0 \frac{\omega^2}{\Delta\omega}\eta\frac{L}{2}\sqrt{\frac{P}{R}}}{\sqrt{k_B TBR}} \tag{28}$$

We will use Eq. (28) in the following discussion of the frequency dependence of S/N.

The usual derivation of S/N for NMR (see section 3) also yields an ω^2 dependence, where one factor of ω comes from χ'' (as in Eqs. (1) and (2)) and the second factor of ω comes from Lenz's law (Hoult and Richards, 1976).

7.1 Absolute S/N Numerical Examples

In this section we provide numerical examples of S/N for CW and pulsed EPR for specific cases. A key message from these examples will be that sensitivity (S/N) differences between CW and pulsed EPR are a strong function of detector bandwidth and modulation amplitude.

For the pulsed EPR example we take results from a very detailed measurement and calculation at S-band (Rinard *et al.*, 1999b). For an irradiated fused quartz sample, the echo signal at the output of the resonator was calculated to be 190 μV. This value is prior to amplification in the detection path of the spectrometer. The noise power available at the same point is -174 + 10log(bandwidth) dBm. For the ESE measurement the detector bandwidth was 25.7 MHz. Hence, the thermal noise voltage is 2.2 μV in a 50 Ω load. Another way of saying this is that if all of the active devices had NF = 0 dB, the equivalent input noise voltage would be 2.2 μV, so a 190 μV signal would have S/N = 86, and a 2.2 μV signal would be detectable with S/N = 1. The actual experimental S/N was somewhat less, due to the noise added by loss and gain stages in the detection path. The original paper gives details. The sample contained ca. $9.4\text{x}10^{15}$ spins. The extrapolated ultimate S-band sensitivity then is ca. $1.1\text{x}10^{14}$ spins with S/N = 1 if the only noise is thermal noise.

The number of spins detectable with S/N = 1 decreases dramatically if the bandwidth is narrower, since the noise is proportional to square root of bandwidth. One way to narrow the effective bandwidth is to signal average (Wilmshurst, 1990), in which case the effective noise bandwidth decreases with the square root of the number of scans averaged. Thus, it is not totally artificial to consider a pulse experiment with a 1 Hz bandwidth due to signal averaging, and we consider the hypothetical case in which the ESE detection system has 1 Hz bandwidth in order to make a rough comparison with CW EPR sensitivity specifications. The thermal noise voltage in a 50 Ω load at 290 K detected with 1 Hz bandwidth would be $4.5\text{x}10^{-10}$ V, and one could observe $2.2\text{x}10^{10}$ spins with S/N = 1 at S-band.

It is well-known that state-of-the-art X-band EPR spectrometers are stated to have a CW sensitivity (S/N = 1) equivalent to $0.8\text{x}10^{10}$ spins/G at 200 mW for non-saturable, non-lossy sample, extending through a TE_{102} cavity, assuming an S = ½ system with a single Lorentzian line, with 1 s time constant and optimum magnetic field modulation. Note that the standard commercial definition of noise for sensitivity tests is peak-to-peak divided by 2.5, whereas the standard deviation noise we use is more nearly equal to peak-to-peak divided by 5. Note also that in some conventions various numbers of noise spikes are ignored.

We use Eq. (1) for V_S with best estimates of η (ca. 1%) and Q (ca. 3600). Assuming the stated X-band sensitivity of 0.8×10^{10} spins/G, and 200 mW maximum available power, we calculate a signal voltage of ca. 6×10^{-10} V prior to amplification. This compares with the noise voltage of 4.5×10^{-10} V in 1 Hz bandwidth. The standard S/N tests on commercial spectrometers use a "1 second" filter time constant. To compare our calculation with these standard conditions, we have to relate filter time constant to equivalent noise bandwidth (ENBW). On modern spectrometers the filter in the signal detection circuit has a roll off of 12 dB/octave, and hence ENBW = 1/8 times the reciprocal of the filter time constant. Consequently, for a nominal 1 second time constant the bandwidth used in the noise calculation should be 0.125 Hz, which decreases the predicted noise by 2.8 to 1.6×10^{-10} V. Comparing this with the predicted signal voltage, state-of-the-art spectrometers are about a factor of two-to-four from the thermal limit. One cannot be more precise than this because traditions of using peak-to-peak instead of standard deviation noise and of ignoring some noise spikes obscure the mathematical relation between standard deviation noise and the experimental S/N measurements.

Current X-band CW spectrometers probably have noise contributions from microphonics (including that due to use of high modulation amplitude), source noise (especially at high microwave power), and detector preamplifier noise (most spectrometers do not use a low-noise microwave preamplifier). The standard weak pitch S/N measurement is performed with magnetic field modulation larger than line width to maximize the signal amplitude (although distorting the line shape), and thus approximates the assumptions used in our treatment of the CW signal. Thus, both the CW and echo experiments measure approximately the total signal voltage. Note that in a field-modulated CW measurement in which the line shape is to be preserved, the modulation amplitude should be less than about 1/10 of the line width, so the signal voltage is substantially reduced from the maximum possible.

8. FREQUENCY AND SIZE DEPENDENCE OF S/N FOR LOSSY SAMPLES

If some other part of the spectrometer contributes noise, the denominator in Eqs. (19-21) will contain this noise contribution in addition to the thermal noise discussed here. The noise powers (noise voltages squared) sum under the square root sign in the denominator. Various gains, losses, and noise figures in the system are combined in the standard Friis equation (Lee and Dalman, 1994).

What about other contributions to R in the resonator? Anything that increases R will increase the noise and decrease the signal, as is evident from Eq. (28). Thus, when a lossy sample is put in the resonator it absorbs power and is seen as an increase in the noise, and will be accompanied by a decrease in the signal unless the incident power is increased to restore B_1 to its original value. This is the NMR point of view and is the origin of the concept of "sample noise." This portion of R due to sample loss increases as ω^2 (Hoult and Lauterbur, 1979; Chen and Hoult, 1989; Vlaardingerbroek and den Boer, 1996; Hoult et al., 1986). From the EPR point of view, when a sample adds loss to the resonator, increasing R, then the coupling to the resonator must be changed to maintain critical coupling to the transmission line. The signal and noise voltages are both transformed by the new coupling ratio $\sqrt{Z_0 / R}$. As a result, it appears that the signal is inversely proportional to R and the noise is constant (see the conversion of (26) into (27)). Equivalently, this is a decrease in Q. The resultant impact on S/N is the same, regardless of the point of view - it is reduced by the factor 1/R.

We now turn our attention to showing how sample loss affects the S/N, since MRI and physiological EPR imaging usually are performed on lossy samples. Following Hoult and Lauterbur (1979) we use generic dimensions a and b to represent the dimensions of the resonator and of the sample, respectively, and use ρ to represent sample resistivity. Thus, since we are focusing on overall trends, we can take parameter a as representing a linear dimension of the resonator and not make further distinction between, say, the diameter d and the length z as used in Eqs. (3) - (5). Similarly, we ignore the actual shape of the sample and merely use parameter b as representing a generic linear dimension of the sample.

Some of the loss terms of concern when studying magnetic resonance (NMR or EPR) in animals include the dielectric loss and the inductive (or magnetic) loss. Hoult and Lauterbur (1979) expressed these as effective resistances, R_e and R_m, respectively, which add to the "coil" resistance R used above, which we distinguished as R_c in (25).

$$R_e = \tau\omega^3 L^2 C_d \tag{29}$$

where τ is the dielectric loss factor, L is the coil inductance, and C_d is the distributed capacitance of the resonator (Hoult and Lauterbur, 1979). One strives to minimize this contribution to the loss by designing resonators to minimize the electric field in the sample (Röschmann, 1987). Gadian and

Robinson provided quantitative examples (1979). How well dielectric loss has been avoided in the context of biological MRI can be judged by the critically coupled frequency shift upon insertion of the sample (Röschmann, 1987). It should be noted that more generally, the insertion of a dielectric will shift the frequency of a resonator even if it is not lossy.

$$R_m = \frac{\pi\omega^2\mu_0^2 N^2 b^5}{30\rho(a^2 + g^2)} \tag{30}$$

In this expression ρ is the resistivity of the sample with radius b and the coil has N turns of radius a. "Hidden" within this expression is a factor in B_1, since it can be shown that for unit current the field in a coil of half-length g is (Hoult and Lauterbur, 1979)

$$B_1 = \frac{N\mu_0}{2\sqrt{a^2 + g^2}} \tag{31}$$

Hence the R_m equation can also be written as

$$R_m \propto B_1^2\omega^2 b^5 \tag{32}$$

We now combine in the denominator to Eq. (28) the contributions to noise due to the resistances R_c, R_e, and R_m. Since, in the design of the resonator one seeks to minimize R_e relative to the other terms, we now drop it as relatively small. In the following we seek the dependence on frequency and on resonator and sample dimensions, so we represent the relative contributions of R_c and R_m by coefficients α and β. Recalling the frequency dependence of R_c, Eq. (4), we can write the R in Eq. (28) as $\alpha\omega^{1/2} + \beta\omega^2 b^5$. This is similar to the formula Hoult and Lauterbur (1979) provided for the resistance as a function of frequency and dimensions if the coil resistance and the magnetic losses were the most important noise sources. Foster (1992) has pointed out that tissue resistivity is frequency-dependent over the range of frequencies of interest, so we modify the Hoult and Lauterbur expression by including the sample resistivity ρ, giving (33).

$$R \propto \alpha\omega^{1/2} + \frac{\beta\omega^2 b^5}{\rho a^2} \tag{33}$$

where α and β are coefficients that reflect the physical system. Any other loss terms, such as the dielectric loss R_e, would be included in the sum.

8.1 Constant Power Predictions

As stated in section 4, the resistance R in the above equations includes all loss in the resonator. Two limiting cases of primary interest are the case in which the dominant contribution to R is the resistive losses in the materials of which the resonator is constructed, and the case in which the dominant loss is due to the sample. The same expression for the resistance (all loss terms to be considered) has to be used in place of R in the numerator and in the denominator in Eq. (28). Hence, the general expression for S/N for this case becomes

$$S/N = \frac{\chi_0 \dfrac{\omega^2}{\Delta\omega} \eta \dfrac{L}{2} \sqrt{\dfrac{P}{\left(\alpha\omega^{1/2} + \dfrac{\beta\omega^2 b^5}{\rho a^2}\right)}}}{\sqrt{k_B T B \left(\alpha\omega^{1/2} + \dfrac{\beta\omega^2 b^5}{\rho a^2}\right)}} \tag{34}$$

Which readily simplifies to

$$S/N = \frac{\omega^2}{\Delta\omega} \frac{\chi_0 L \eta}{2\left(\alpha\omega^{1/2} + \dfrac{\beta\omega^2 b^5}{\rho a^2}\right)} \frac{\sqrt{P}}{\sqrt{k_B T B}} \tag{35}$$

This is the same expression as would have been obtained had we substituted the full expression for R in Eq. (27). Retaining just the terms in (35) that depend on dimensions or frequency, and using $\eta \propto \dfrac{V_S}{V_R} \propto \dfrac{b^3}{a^3}$ and $L \propto \dfrac{A_R}{z} \propto \dfrac{a^2}{a}$, we find that the functional dependence of S/N is given by

$$S/N \propto \omega^2 \frac{V_S}{V_R} \frac{A_R}{z} \frac{1}{\alpha\omega^{1/2} + \dfrac{\beta\omega^2 b^5}{\rho a^2}} = \omega^2 \frac{b^3}{a^3} \frac{a^2}{a} \frac{1}{\alpha\omega^{1/2} + \dfrac{\beta\omega^2 b^5}{\rho a^2}}$$

$$= \frac{\omega^2 b^3}{a^2\left(\alpha\omega^{1/2} + \dfrac{\beta\omega^2 b^5}{\rho a^2}\right)} \quad \text{(constant incident power)} \tag{36}$$

If incident power is kept constant, all dimensions are kept constant, and the sample loss dominates, Eq. (36) predicts no change in S/N with frequency, except for the frequency dependence of ρ. If all linear dimensions are scaled as $1/\omega$, and the coil resistance dominates, this predicts $\omega^{1/2}$ as in Table 1 case 2, row 5. When sample losses dominate, note the dramatic change in S/N dependence on frequency due to the large effect of b^5 in the denominator.

Under conditions in which sample resistivity dominates the noise, the S/N at higher frequencies is lower than would be predicted based on ignoring the frequency dependence of ρ (Hoult *et al.*, 1986; Foster, 1992), because the conductivity increases at higher frequency. Conductivity of biological tissues increases as about the 0.4 power of frequency in the frequency region commonly used for MRI, and then increases more slowly with frequency up to about L-band (Johnson and Guy, 1972). In the region where both R_c and R_m are important, their frequency dependencies are different ($\omega^{1/2}$ and ω^2, respectively). The combination of the effects of the larger contribution of R_c at lower frequency, and the frequency dependence of sample resistivity ρ may result in an apparent linear dependence of sensitivity on frequency for a limited range of experimental measurements (Edelstein *et al.*, 1986).

8.2 Constant B_1 Predictions

The case of constant B_1 requires recognizing that in Eq. (28) the power would be increased to compensate for the change in resistance. Returning to fundamental definitions, as in Eqs. (8) and (9),

$$B_1 = \frac{LI}{A_R} = \frac{L}{A_R}\sqrt{\frac{P}{R}} \tag{37}$$

where R is the general expression for resistance. We dropped the number of turns, N, from Eq. (9) since we seek only frequency dependence.

To keep B_1 constant, the term $\frac{L}{A_R}\sqrt{\frac{P}{R}}$ has to be kept constant by adjusting incident power. Therefore, this term can be factored out of the numerator of Eq. (28). Using the same dimension conventions as in (36), (28) becomes

$$S/N = \frac{\chi_0 \frac{\omega^2}{\Delta\omega}\eta \frac{A_R}{A_R}\frac{L}{2}\sqrt{\frac{P}{R}}}{\sqrt{k_B TBR}} \Rightarrow \frac{\chi_0\omega^2\eta A_R B_1}{\Delta\omega 2\sqrt{R}\sqrt{k_B TB}} \tag{38}$$

$$S/N \propto \frac{\omega^2}{\Delta\omega} \frac{\chi_0 \eta A_R}{2\sqrt{\alpha\omega^{1/2} + \frac{\beta\omega^2 b^5}{\rho a^2}}} \frac{1}{\sqrt{k_B TB}} \quad (B_1 \text{ is constant}) \tag{39}$$

The functional dependence of S/N on frequency for this case becomes:

$$S/N \propto \omega^2 \frac{\eta A_R}{\sqrt{\alpha\omega^{1/2} + \frac{\beta\omega^2 b^5}{\rho a^2}}} = \frac{b^3 a^2}{a^3} \frac{\omega^2}{\sqrt{\alpha\omega^{1/2} + \frac{\beta\omega^2 b^5}{\rho a^2}}}$$

$$= \frac{b^3}{a} \frac{\omega^2}{\sqrt{\alpha\omega^{1/2} + \frac{\beta\omega^2 b^5}{\rho a^2}}} \quad (B_1 \text{ is constant}) \tag{40}$$

This gives the $\omega^{7/4}$ dependence of Table 1, row 8, case 1 if coil resistance dominates, and linear dependence on ω if sample loss dominates (Table 2, row 4).

Using Eq. (36) or (40) with variation in parameters a or b, or both, while ω is varied result in the entries in Tables 1 and 2. This approach is quite general, and can be elaborated for other than these limiting cases by incorporating, for example, explicit functional dependences of dimensions or other expressions for R_m.

Table 2. Predicted Frequency Dependence of EPR Sensitivity When Sample Loss Dominates[a]

		const. sample size const. LGR size	sample size $\propto 1/\omega$ LGR size $\propto 1/\omega$	const. sample size LGR size $\propto 1/\omega$
1	S/N at constant P	ρ (35, 36)	$\omega^2\rho$ (35, 36)	ρ (35, 36)
2	$B_1/\sqrt{P}$	$\omega^{-1}\rho^{1/2}$	$\omega^{3/2}\rho^{1/2}$	$\omega^{-1}\rho^{1/2}$
3	P to maintain constant B_1	$\omega^2\rho^{-1}$	$\omega^{-3}\rho^{-1}$	$\omega^2\rho^{-1}$
4	S/N at constant B_1	$\omega\rho^{1/2}$ (39, 40)	$\omega^{1/2}\rho^{1/2}$ (39, 40)	$\omega\rho^{1/2}$ (39, 40)

[a]The equation numbers that are the bases for the various statements are given in parentheses. The tables list overall frequency dependence and do not include proportionality constants. In Table 2 ρ represents sample resistivity.

Equation (36) contains the same functional form of the coil and sample resistance as Foster's (1992) modification of the Hoult and Lauterbur (1979) expression, except that Eq. (36) also includes the dimensional dependence of ηL.

It is also useful to consider some of the results derived here from a more heuristic approach, starting with Eq. (2). For example, to calculate S/N at constant incident power, substitute (5) into (2) and divide by $V_{Ns} = \sqrt{k_B T B Z_0}$ yielding

$$S/N = K_e \eta d \omega^{3/2} \sqrt{P} \tag{41}$$

where terms that do not depend on frequency are combined in $K_e = \frac{\chi_0}{4\sqrt{2}\Delta\omega}\sqrt{\frac{\mu_0 \sigma}{k_B T B}}$, and $\chi_0 = N g^2 \beta^2 S(S+1)/3k_B T_s$. Equation (41) indicates that the frequency dependence of the ESR signal is a function of filling factor η and diameter d as well as $\omega^{3/2}$. For example, if η is constant and the dimension d scales proportional to $1/\omega$, and we consider only loss due to the materials of which the resonator is constructed, then the signal is proportional to $\omega^{1/2}$. This is Table 1, row 5, case 2.

Similarly, we can predict the frequency dependence if B_1 is kept constant, and again consider only the loss due to the materials of which the resonator is constructed. In case 1 we assume constant sample size and constant loop size. For example, one could have a LGR of a given loop size, with the frequency varied by changing the capacitance gap dimension. In this case, the variation with frequency is that given in Eqs. (5), (6), and (41) with filling factor and resonator dimensions held constant. This results in a dependence of the EPR signal on $\omega^{7/4}$ (Table 1, row 8, case 1).

In case 2 ("unlimited sample") we assume that sample and resonator linear dimensions scale as $1/\omega$. With these assumptions Eqs. (6) and (41) give expressions (42) and (43).

$$S/N \propto \eta \sqrt{\omega P} \tag{42}$$

$$B_1 \propto \omega^{3/4} \sqrt{P} \tag{43}$$

Hence, P has to change as $\omega^{-3/2}$ to keep B_1 the same (Table 1, row 7, case 2). Introducing this dependence into (42), the S/N is proportional to $\omega^{1/2}\sqrt{\omega^{-3/2}} = \omega^{-1/4}$. That is, the S/N gets larger as the frequency decreases. This is Table 1, row 8, case 2.

In case 3 ("limited sample") we assume that the sample size is constant and the linear dimension of the resonator scales as $1/\omega$. Since the sample volume is constant and the resonator volume varies as $1/\omega^3$, the frequency dependence of the filling factor is,

$$\eta \propto \omega^3 \tag{44}$$

Substituting (44) into (41) gives,

$$S/N \propto \omega^{7/2}\sqrt{P} \tag{45}$$

The relation between B_1 and P in this case is the same as for case 2 (Eq. (43)). The result is that the EPR signal varies as $\omega^{11/4}$. If the sample is very small and non-lossy, the best signal is obtained at the highest frequency at which the resonator matches the sample size.

8.3 Comparison with Literature Results

Consider first the predictions when resonator noise dominates. For the unlimited sample case, the dependence on $\omega^{1/2}$ agrees with the predictions of Abragam and Bleaney (1970), Wilmshurst (1968), and Fraenkel (1960). For the limited sample case the dependence on $\omega^{7/2}$ agree with the results of many others, including Feher (1957), Abragam and Bleaney (1970), and Fraenkel (1960). The dependence of signal at constant power (Abragam and Bleaney, 1970; Fraenkel, 1960) (unlimited sample, case 2) on $\omega^{1/2}$, of signal intensity for constant sample (case 3) at constant power (Abragam and Bleaney, 1970; Fraenkel, 1960; Feher, 1957; Varian) on $\omega_o^{7/2}$, and of constant sample (Varian) (case 3) at constant B_1 on $\omega^{11/4}$ agree with predictions in the references cited, among others. In the frequently cited discussion of N_{min} in (1), an additional factor of $1/\omega$ was included based upon a set of assumptions concerning the frequency dependence of the relationship between incident microwave power in the waveguide outside a cavity and B_1 in the cavity. The relationships in Table 1 do not bear out this assumption.

The range of exponents in Tables 1 and 2 indicates that one needs to consider carefully the experimental conditions in predicting frequency dependence of S/N for a particular situation. One also needs to consider practical realities. Scaling a resonator design over a wide frequency range may not be possible because of machining tolerances, or because gaps become too small to prevent arcing for high-power and high Q conditions. In addition, resonator dimensions may become so much smaller at higher frequencies that it is not possible to maintain a constant sample size. Thus the practical need for different resonators at different frequencies may prevent one from taking advantage of theoretically predicted advantages.

Based on early work by Hoult and coworkers (1979, 1989, 1986), Andrew (1988) and others (Vlaardingerbroek and ven Boer, 1996) it is now generally agreed (Halpern and Bowman, 1989) that the S/N in MRI and spectroscopy in living systems follows a $\omega^{7/4}$ to ω^1 frequency dependence. This is due to the fact that noise voltage due to resistance in the radiofrequency coil probe circuit is proportional to $\omega^{1/4}$, while noise due to losses in the patient's body is proportional to ω. At low frequencies where losses are small, the S/N varies as $\omega^{7/4}$ (Table 1, case 1). At higher frequencies, where the coil losses are less important relative to losses in the sample, the S/N depends linearly on ω (Table 2, case 1). Measurements at low frequency have confirmed the linear dependence of S/N on frequency in EPR (Stoodley, 1963) and NMR (Hoult *et al.*, 1986; Edelstein *et al.*, 1986).

For the conditions to which the results in Table 2 are applicable, we predict that if both sample size and resonator size scale with frequency, for constant B_1 the S/N scales as $\omega^{1/2}$, not as ω. Thus, if the spectrometer is optimized to the sample, there is only a weak dependence of S/N on frequency.

Willer *et al.* (2000) reported that "At 4 GHz the echo intensity is found to be an order of magnitude lower than in our home-built X-band spectrometer (same overall receiver gain, bandwidth and resonator losses). However, echo intensities obtained with larger sample volumes have been found to be up to three times higher than with standard 4 mm o.d. quartz tubes." Weber *et al.* (2002) reported that the sensitivity of their "pulsed S-band spectrometer was measured to be three times smaller than at X-band and in agreement with theory."

Direct measurements of electron spin echo signal and noise in well-characterized X-band and S-band spectrometers agree with our predicted frequency dependence (Rinard *et al.*, 1999c). For the particular spectrometers compared, the echo at 9.52 GHz was 9.5 times larger than the echo at 2.68 GHz, after scaling for differences in spectrometer gain. The calculated ratio was 7.6.

Experimental CW EPR signal intensities at 250 MHz, 1.5 GHz, and 9.1 GHz (Rinard *et al.*, 2002) agree within experimental error with predictions from first principles. When both the resonator size and the sample size are scaled with the inverse of RF/microwave frequency, ω, the EPR signal at constant B_1 scales as $\omega^{-1/4}$. Comparisons were made for three different samples in two pairs of loop gap resonators. Each pair was geometrically scaled by a factor of 6. One pair of resonators was scaled from 250 MHz to 1.5 GHz, and the other pair was scaled from 1.5 GHz to 9 GHz. All terms in the comparison were measured directly, and their uncertainties estimated. The theory predicts that the signal at the lower frequency will be larger than the signal at the higher frequency by the ratio 1.57. For 250 MHz to 1.5 GHz, the experimental ratio was 1.52 and for the 1.5 GHz to 9 GHz comparison the ratio was 1.14.

The electron paramagnetic resonance (EPR) pulsed free induction decay (FID) of a degassed solution of a triaryl methyl radical, methyl tris(8-carboxy-2,2,6,6-tetramethyl(-d_3)-benzo[1,2-d:4,5-d']bis(1,3)dithiol-4-yl) tripotassium salt, 0.2 mM in H_2O, was measured at VHF (247.5 MHz) and

L-band (1.40 GHz) (Rinard *et al.*, 2002b). The calculated and observed FID signal amplitudes (mV) agreed within 1 and 6%, and the ratio of the normalized FID signals at the two frequencies agreed within 5%. The FID decay time constant was 2.7 μs at both frequencies.

As we have emphasized in our papers (Rinard *et al.*, 1999a,b,c and 2002a,b), our experience confirms the truth of the statements made long ago that quantitative measures of absolute numbers of spins are extremely difficult. Our papers provide a model for the level of detailed measurement of the gains, bandwidths, and noise figure of the spectrometer needed to make valid comparisons between spectrometers.

8.4 Factors not Considered in Compiling Tables 1 and 2

In the derivations summarized in Tables 1 and 2 we assumed that the line width $\Delta\omega$ does not change with frequency. This would be a good assumption for the Lorentzian line in this model, but if a real line shape were dominated by g-anisotropy, the line would become narrower at lower frequency, and the S/N would improve somewhat at lower frequency as a result. Conversely, for broad lines whose width is determined by g-anisotropy, S/N will decrease at higher frequency by a factor of ω relative to the terms in the Tables due to the proportional decrease in spins per G as the frequency increases (Davoust *et al.*, 1996).

Omitted from this discussion are all effects that any filtering, such as Fourier filtering in FT measurements, would have on the effective bandwidth. Further, noise introduced after the resonator would reduce the S/N as a function of gain and noise figures for those stages of the spectrometer, per the standard Friis formula. The effects of relaxation times are discussed in the next section.

9. RELAXATION AS A FUNCTION OF FREQUENCY

9.1 Nitroxyl Radicals

There is very little information on relaxation times of nitroxyl spin probes as a function of frequency. Hyde and coworkers (1990) observed a decrease in T_1 from X-band to S-band for a spin-labeled stearic acid in a membrane at room temperature. This was described as "unexpected." Subsequently, Robinson and Mailer stated that the relaxation of nitroxyl radicals can be described in terms of rotational dynamics (spin-rotation interactions), electron-nuclear dipolar (END) interactions, spin diffusion interactions with solvent nuclei, and collisions with oxygen (Robinson *et al.*, 1994; Haas *et al.*, 1993):

$$\frac{1}{T_{1e}} = \frac{1}{T_{1e}^{SR}} + \frac{1}{T_{1e}^{END}} + \frac{1}{T_{1e}^{SD}} + \frac{1}{T_{1e}^{Ox}} \tag{46}$$

Hyde found that the oxygen relaxivity was the same at both X-band and S-band. The spin-rotational interaction is dependent on g-anisotropy and rotational correlation time, but not on microwave frequency. Lloyd and Pake (1954) reported that there was no observable difference between the 60 MHz CW power saturation measure of relaxation of peroxylamine disulphonate (Fremy's salt) in H_2O and D_2O. They proposed that the observed relaxation involves spin-orbit interaction. We propose that collisional modulation of spin-orbit coupling should be included in a description of nitroxyl spin-lattice relaxation, and may dominate under some conditions. This interaction effect does not have a frequency dependence.

9.2 Relaxation of the Nycomed Probe

The NCI lab (Murugesan *et al.*, 1997) observed that the room temperature relaxation time of a Nycomed probe is long enough at 300 MHz (> 1 μs) to obtain good FIDs. For radicals with very small nuclear couplings, no nitrogen nuclei, and small g-anisotropy, such as the Nycomed radicals, very little frequency dependence of T_1 is expected. Electron spin relaxation times of a Nycomed triarylmethyl radical (sym-trityl) in water, 1:1 water:glycerol, and 1:9 water:glycerol were measured at L-band, S-band, and X-band by pulsed EPR methods (Yong *et al.*, 2001). In H_2O solution, T_1 is 17 ± 1 μs at X-band at ambient temperature, is nearly independent of microwave frequency, and exhibits little dependence on viscosity. In H_2O solution, T_2 is 11 ± 1 μs at X-band, increasing to 13 ± 1 μs at L-band. For more viscous solvent mixtures, T_2 is much shorter than T_1 and weakly frequency dependent, which indicates that incomplete motional averaging of hyperfine anisotropy makes a significant contribution to T_2. In water and 1:1 water:glycerol solutions continuous wave EPR linewidths are not relaxation determined, but become relaxation determined in the higher-viscosity 1:9 water:glycerol solutions. The Lorentzian component of the 250 MHz linewidths as a function of viscosity is in good agreement with T_2-determined contributions to the linewidths at higher frequencies. In glassy aqueous glycerol solutions, the spin-lattice relaxation rates of the Nycomed radical at temperatures between about 20 and 160 K are about a factor of 5 slower than for nitroxyl radicals. The temperature dependence of T_1 in 1:1 water:glycerol is characteristic of domination by a Raman process between

20 and 80 K. The increased spin-lattice relaxation rates at higher temperatures, including room temperature, are attributed to a local vibrational mode that modulates spin-orbit coupling. The Nycomed radical does not exhibit the increase in spin echo dephasing rates at 100 to 150 K that are observed for nitroxyl radicals due to methyl group rotation (Zecevic *et al.*, 1998). An extensive discussion of relaxation times, including frequency dependence, is in volume 19 of this series (Eaton and Eaton, 2000).

9.3 Relaxation in Low Temperature Solids

The predicted dependence of relaxation mechanisms commonly encountered in the solid state are different in certain cases for even and odd numbers of unpaired electrons. We will focus on odd-spin systems. The rate of the direct process for spin lattice relaxation is proportional to the 4th power of magnetic field, so relaxation times are much longer at lower field. This prediction has been demonstrated (Davids and Wagner, 1964) and exploited to use higher observe power at higher frequency to improve S/N of otherwise too-slowly relaxing species (Muller *et al.*, 1989). There are limits on how far one should extrapolate such generalizations. For example, at high magnetic fields the density of phonon states may be low at the resonant frequency (Witowski *et al.*, 1997), cross relaxation is decreased at high frequency (Strutz *et al.*, 1992), and high polarization of the spins can suppress spin flip-flop probabilities, making T_2 longer at high fields (Kutter *et al.*, 1995). The direct process typically dominates only at temperatures below about 10 K.

The first-order Raman process can exhibit squared dependence on field (longer relaxation time at lower field), but the more common Raman process is predicted to be independent of magnetic field (Abragam and Bleaney, 1970; Witowski, 1991). The Orbach process is independent of magnetic field, except insofar as the field may alter the value of the energy gap between the coupled states. Davoust *et al.* (1996) reported that T_1 is shorter at Q-band than at X-band at 2-4 K. Prisner (1997) summarized relaxation times from the literature and his own measurements, comparing X-band and W-band (95 GHz) results for T_1 and T_2. In some cases relaxation times are longer at higher frequency, in some cases they are about the same, and in some cases relaxation times get shorter at higher frequency. The few data points span such widely different radicals that to find a pattern will require identifying the relaxation mechanisms for each type of sample. For example, if modulation of orientation-dependent parts of the Hamiltonian dominates relaxation, then relaxation times will be faster at high frequency.

These theoretical predictions and the few relevant experimental observations all argue that relaxation times of species in solids will not get shorter at lower microwave field (frequency), relative to X-band, and at low temperature are likely to get longer. Consequently, in most cases, relaxation times are not expected to strongly alter the predicted frequency dependence, and Table 1 can be applied to pulsed EPR as well as to CW EPR. At the highest fields and frequencies at which EPR is performed, numerical simulation is required of the effect of phonon density of states and of spin polarization. One other footnote that needs to be added is that echo envelope modulation (ESEEM) is strongly frequency dependent, being larger at lower frequencies, and would have to be accounted for in a comparison.

10. PHASE NOISE

Demodulation of source phase noise by the resonator contributes to noise in the detected signal in measurements for which microwave power is incident on the resonator during the signal measurement. This includes standard CW EPR detection and saturation recovery EPR detection. In an electron spin echo measurement microwaves are not incident on the resonator during the echo measurement, so the effect of the resonator on the source phase noise is not relevant to this case. However, source phase noise can always be a concern if the electrical paths of the signal side and the reference side of the detector crystal or mixer are not identical. Since most biological applications of EPR so far have used reflection resonators, source phase noise is an important consideration.

There are statements in the literature that demodulation by the resonator of frequency (phase) noise is linear in Q when the spectrometer is tuned to dispersion and quadratic in Q when tuned to absorption (Alger, 1968, p.80; Pfenninger *et al.*, 1995a, 1995b). Derivations have not been published. To a first approximation it would seem that phase noise in a spectrometer with a reflection resonator should be proportional to $Q\sqrt{P}$, just as is signal. In one set of L-band measurements with the critically-coupled resonator Q changed by loading with various aqueous salt solutions, we observed that phase noise increased proportional to Q and proportional to $\sqrt{P}$. Consequently, Table 3, Part 2, is based on the assumption that phase noise is proportional to Q. Hence, if the spectrometer noise is determined by RF source phase noise under the conditions of the measurement, the S/N will remain constant with increase in power. In addition, if the power is increased to maintain signal amplitude as the Q of the resonator is decreased, the phase noise also will be the same as it was before the Q was reduced. Consider the case in which the sample size and resonator are scaled with wavelength as frequency is lowered (case 2), while maintaining B_1 constant. To be specific, we consider lowering the frequency from 9.1 GHz to 200 MHz. From Table 1 Q varies as $\omega^{-1/2}$, P varies as $\omega^{-3/2}$, and the EPR signal varies as $\omega^{-1/4}$. The ratio of frequencies is 45.5, so Q is 6.75 times higher at 200 MHz than at 9.1 GHz. The power has to be increased 307 times to maintain constant B_1. The EPR signal is 2.6 times higher at the lower frequency, but the phase noise, which goes as $Q\sqrt{P}$ is 118 times higher at the lower frequency. This comparison assumes that the power used at the higher frequency was that where the phase noise just begins to dominate, and that the phase noise of the sources at both frequencies are the same. This assumption would require verification by experiment. However, if it is valid, the phase noise could increase substantially and become a problem at low frequency even though it would not be a problem at higher frequency. Experimental methods to eliminate the phase noise would substantially increase the low-frequency S/N for this case. One way to accomplish this would be with a resonator that isolated the signal from source phase noise. An example of this is the bimodal (Mailer *et al.*, 1980; Piaseki *et al.*, 1996) or crossed-loop resonator (Rinard *et al.*, 1996a, 1996b, 2000, 2002c). Depending on the relative importance of source phase noise, the S/N at 200 MHz could range from 2.6 to 0.022 times that at 9.1 GHz.

Table 3. S/N for *in vivo* EPR;
Part 1 – The three cases in Tables 1 and 2 when RF source phase noise is not important.

lower frequency from $\omega_1 \rightarrow \omega_2$	9.1 → 1 GHz	1 → 0.2 GHz	9.1 → 0.2 GHz
frequency ratio ω_2/ω_1	0.11	0.2	0.022
S/N ratio at constant B_1 if resonator resistance dominates			
Table 1 case 1 ($\omega^{7/4}$)	0.021	0.06	0.0013
Table 1 case 2 ($\omega^{-1/4}$)	1.74	1.5	2.6
Table 1 case 3 ($\omega^{11/4}$)	0.0023	0.012	0.000028
S/N ratio at constant B_1 if sample loss dominates			
Table 2 case 1 (ω)	0.11	0.2	0.022
Table 2 case 2 ($\omega^{1/2}$)	0.33	0.45	0.15
Table 2 case 3 (ω)	0.11	0.2	0.022

Part 2 - The three cases in Tables 1 and 2 when RF source phase noise is dominant. The power is adjusted to maintain constant B_1 at the sample.

S/N ratio $\propto \eta(\omega_2/\omega_1)$ $\eta = 1$ for cases 1 and 2, but $\eta \propto (\omega_2/\omega_1)^3$ for case 3	9.1 → 1 GHz	1 → 0.2 GHz	9.1 → 0.2 GHz
case 1	0.11	0.2	0.022
case 2	0.11	0.2	0.022
case 3	1.5×10^{-4}	1.6×10^{-3}	2.3×10^{-7}

These calculations (a) ignore frequency dependence of conductivity; (b) ignore g-dispersion; and (c) assume that the signal does not saturate at any of the powers compared.

11. CONDUCTIVITY

Another possible limitation in the predictions in Table 1 is whether Eq. (4) accurately represents the frequency dependence of conductivity of the materials of the resonator. Equation (4) is the standard text-book formula. However, as frequency increases the conductivity is influenced increasingly by the surface properties rather than the bulk properties of the conductor. Machined, buffed, and plated surfaces are rough relative to skin depth at microwave frequencies, and can contain inclusions from the finishing process. The machining and buffing processes result in work-hardening of the metal crystals, reducing their conductivity. These problems are discussed in (Alger, 1968; Benson, 1969; Goldsmith, 1982, 1998). Some measurements, summarized in (Alger, 1968; Benson, 1969; Goldsmith, 1982, 1998; Losee, 1997) reveal that the effective resistance at microwave frequencies is higher than the dc resistance, by as much as a factor of 2.5 or so, but typically that ratio is between 1.5 and 2, over the range 9 - 203 GHz. At 890 GHz the surface resistivity of evaporated gold was 2.2 times that expected from the dc resistivity (Batt *et al.*, 1977). At frequencies below 9 GHz, the effective resistance approached the dc resistance. Only a very rough, scattered correlation with frequency, and a stronger correlation with surface treatment and corrosion, was observed (Batt *et al.*, 1977). In general, plated specimens exhibited poor conductivity relative to theoretical. Recall that the resistance of the materials of which the resonator is built influence our predictions via the resonator Q, which is proportional to $\sqrt{\sigma}$. Since σ is the reciprocal of resistivity, a factor of 2 change in resistivity corresponds to a $\sqrt{2}$ decrease in Q and hence in EPR S/N by Eqs. (26) and (27). The conclusion from these observations is that the predictions of Eq. (15) may be as much as $\sqrt{2}$ optimistic for the improvement in signal intensity when the frequency is increased above X-band, but that if the resonator losses dominate the noise, the S/N should still increase as predicted in Table 1.

High-frequency (above X-band) spectrometers will require improvements in sources, resonators, and detectors to achieve the theoretical improvement in S/N with increasing frequency. The performance of low-frequency spectrometers should follow the predictions of Table 1 unless there are particularly resistive surfaces due to machining, plating, or corrosion.

12. DESIGN OF EPR *IN VIVO* IMAGING EXPERIMENTS

The frequency dependence of S/N at constant B_1 ranges (Table 1) from $\omega^{-1/4}$ to $\omega^{11/4}$ if resonator resistance dominates the noise, and depends on either ω or $\omega^{1/2}$ if sample loss dominates the noise. Typically, one might seek to match the resonator size to the portion of the animal to be studied, and then lower the frequency to get the desired depth of penetration. To make this quantitative, assume X-band (9.1 GHz) as the reference frequency, and assume the sample size allows one to keep the resonator size constant with frequency (case 1). Note that the filling factor is kept constant in this example. Because of interest in *in vivo* EPR at 1 GHz and 200 MHz, the results are presented for two steps: 9.1 to 1 GHz, and then 1 GHz to 200 MHz to achieve penetration depth. Further, we consider three possibilities for the dominant noise source, resonator resistance (Table 1), sample loss (Table 2) and RF source phase noise (section 5).

Source phase noise is proportional to $Q\sqrt{P}$, just as is the EPR signal (Eqs. (1) and (2)). To calculate the S/N when phase noise dominates, divide Eq. (2), by $Q\sqrt{P}$. The resulting frequency dependence of the S/N is $\eta\omega$. For cases 1 and 2, η is constant, so the S/N is proportional to ω. Note that for case 3 the filling factor depends on ω^3, which is a dominant effect, making S/N very poor at low frequency relative to higher frequency when the sample size is kept constant and RF source phase noise dominates. The effect of phase noise is calculated with the assumption that the power used at the higher frequency was that where the phase noise just began to dominate, and that the phase noise of the sources at both frequencies are the same. The predictions are summarized in Table 3.

The frequency dependence of the phase noise is strikingly different for case 3 relative to cases 1 and 2, because of the filling factor effect.

The numerical examples in Table 3 reveal orders of magnitude difference in S/N depending on the relative scaling of sizes of sample and resonator and on the dominant noise source. Not surprisingly, the best results are obtained if the sample size and the resonator size are scaled with inverse frequency. When the sample is limited in size, the best S/N could be achieved at the highest frequency compatible with the size and loss of the sample, because this will give the highest filling factor. If the resistance of the materials of which the resonator is constructed dominate the system noise, the EPR S/N can be larger at low frequency than at high frequency. If sample loss dominates the noise, the low-frequency EPR S/N decreases only by the square root of the frequency ratio, resulting in a relatively small S/N penalty for low-frequency *in vivo* EPR. If RF source phase noise dominates the spectrometer system noise, then the S/N scales linearly with frequency for cases 1 and 2, but S/N scales with ω^4 for case 3, because of the filling factor effect for a fixed sample size. Elimination of the source phase noise problem would substantially improve S/N. For example, compare case 2 for scaling of sample and resonator size with inverse of frequency from 9.1 GHz to 0.2 GHz. If source phase noise were eliminated the S/N would improve by

0.15/0.022 = 6.8 if the sample loss then became dominant, and by 2.6/0.022 = 118 if the resonator resistance then became dominant. An improvement of 118 in voltage S/N corresponds to about 41 dB, an isolation of RF source phase noise from the EPR signal, which is readily achievable with a crossed-loop or other bimodal resonator.

The real situation for animal samples will be somewhere between the extremes in the Tables. For example, there will be a limited range of application of the functional dependences in Table 2. At sufficiently high frequency, the depth of penetration will be limited by combinations of conductivity losses, dielectric concentration effects (lens effects), and for special geometries resonant effects (Röschmann, 1987; Gadian and Robinson, 1979; Foster, 1992; Zypman, 1996; Tofts, 1994; Petropoulos and Haacke, 1991; Mansfield and Morris, 1982). Heterogeneous tissue reduces the power loss in a given volume relative to the same mass in a homogeneous volume. All of these factors make the predictions of Table 2 very conservative over relatively small changes in frequency and inapplicable for large changes in frequency.

One might scale the resonator size with frequency (case 2) from X-band to L-band, in order to be able to study a mouse. But then there would be no further benefit in making the resonator large relative to the mouse, so then one would switch to keeping both the sample and resonator size constant (case 1) as the frequency is lowered by another factor of 5 to 200 MHz. For this, one would pay a signal amplitude price of about 17, ameliorated somewhat by the greater penetration depth of the RF at the lower frequency. If the signal came from the center of the mouse, one might not lose much signal by lowering frequency, since the skin depth is proportional to $\omega^{-1/2}$ and starts to become significant for an animal the size of a mouse at frequencies above a few hundred MHz. Noise due to losses in the sample also increases with frequency for animal samples in this frequency range. The net result is that in practical applications, sensitivity (S/N) appears to increase approximately linearly with frequency for animal imaging in the RF range. As measurements proceed to larger animals, it will be necessary to scale the resonator size to fit them, and the best results will be obtained if the wavelength is large compared with the size of the animal. If the organ of interest increases in size proportionately with the overall size of the animal, the situation is more like case 2, and S/N should not be much worse at lower frequency.

12.1 Dielectric and Conductivity Properties

Interaction of the dielectric and conductivity properties of the sample with the RF also affects S/N. Aqueous and conducting samples have frequency-dependent skin depths (Halpern and Bowman, 1989; Sueki *et al.*, 1996). The conductivity of biological media such as muscle and bone have been tabulated (Johnson and Guy, 1972). The change with frequency is small enough that over the frequency range 200 - 2000 MHz the changes in conductivity will not substantially alter the general trends as a function of ω displayed in Tables 1 and 2. The conductivity of biological media increases a factor of 1.8 from 40 to 200 MHz, which is a little slower than $\omega^{0.4}$ If we use this frequency dependence of conductivity in row 4 of Table 2, we predict for the three cases $\omega^{0.8}$, $\omega^{0.3}$, and $\omega^{0.8}$. The very mild $\omega^{0.3}$ dependence for case 2, where sample and resonator size scale inversely with frequency offers encouragement for low-frequency EPR imaging. On the other hand, for non-lossy and non-conductive samples, the $\omega^{11/4}$ dependence of Table 1, row 8, case 3, is a strong encouragement to high-frequency EPR for small samples.

The dielectric properties can produce phase changes that can null signals from certain locations in samples (Sueki *et al.*, 1996), and can cause the magnitude of the RF magnetic field to be larger at the center of a homogeneous cylinder of high dielectric material than at the surface (Zypman, 1996; Tofts, 1994). The much larger penetration depth at low frequencies can result in many more spins in a large sample being visible at low frequency than at high frequency. This effect can reverse the general trend toward greater sensitivity at higher frequency. Heterogeneous samples appear to have greater penetration depth than homogeneous samples of the same type of material. The physical interactions that result in the losses mentioned here also result in reduction of the resonator Q, and the detected ESR signal is proportional to resonator Q.

12.2 Skin Depth

One should be cautious about using the standard skin depth formula as given after Eq. (3) as anything more than a qualitative indicator when dealing with animals. The skin depth as defined there assumes electromagnetic radiation incident on a plane surface of a homogeneous conducting body. Animals are neither homogeneous nor planar. The formula for inductive (magnetic) loss assumes a homogeneous body. Vlaardingerbroek and den Boer (1996, page 248) point out that if a conductive spherical volume were divided into ten equal spheres with the same total volume, the power dissipated in the ten small spheres would be about 1/5 of the power dissipated in the larger sphere. Röschmann (1987) obtained good fit to experimental RF interaction data on humans with a tissue model in which small regions of high-conductivity (e.g. muscle) are separated by "quasi-insulating" layers of lower conductivity (e.g., adipose tissue).

Calculations on homogeneous bodies reveal that the value of the conductivity required to produce a certain current density (hence, power deposition) in the body increases from the planar body to the cylinder to the sphere (Petropoulos and Haacke, 1991). In terms of apparent skin depth, one could restate these results as implying that for the same conductivity the skin depth increases in the order planar to cylindrical to spherical.

13. APPLICATION AND EXTENSION OF THESE RESULTS

All of the above discussion is relevant to signal intensities in both CW and pulsed EPR, if the relaxation times are long enough, as they usually are in NMR. However, if electron spin relaxation times are short enough to be comparable to instrument dead-times, then new sensitivity considerations apply. If Q is constant, and the electron spin relaxation time is constant and equal to the dead time at one frequency, then the sensitivity for pulsed EPR decreases as $e^{-n\tau}$, where n is the proportional relative frequency. To avoid this strong dependence on frequency, it is necessary to reduce the dead time proportional to the reduction in frequency. This could be done with a combination of a crossed-loop resonator (Rinard *et al.*, 1996a, 1996b), active Q-spoiling (Pfenninger *et al.*, 1995b), and passive or active pulse cancellation (Davis and Mims, 1981; Narayana *et al.*, 1982). Pulsed EPR imaging of a radical injected into a mouse has been achieved (Murugesan *et al.*, 1997).

The reduction in Q due to a given lossy sample decreases with decrease in frequency, even for the same filling factor. This effect also tends to favor lower frequency EPR for lossy samples. All of these effects together could even reverse the expected trends of S/N with frequency, such that it is conceivable that the ultimate molar (spins per volume) sensitivity for biological samples and other lossy samples could increase as the frequency decreases. This remains to be demonstrated, but there is good basis for exploration in this direction.

14. ACKNOWLEDGMENTS

We appreciate discussions with Professor Howard J. Halpern (University of Chicago). GRE thanks Dr. Keith Earle for stimulating this examination of the frequency dependence of actual conductivities, and for pointing out the book by Goldsmith. Some background, and derivations of some of the relations discussed in this chapter, are in the Handbook chapter coauthored with Poole and Farach (Rinard *et al.*, 1999a). A brief treatment of part of this chapter appeared in (Eaton *et al.*, 1998). This work was supported in part by NSF grant BIR-9316827 (GRE), NIH grant RR12257 (H. Halpern, PI), and NIH grant GM57577 (GAR). Barnard Ghim helped with the unpublished L-band phase noise measurements discussed in section 9.

Literature references are comprehensive through mid-1988 and were updated in late 2002 to include recent experimental measurements of sensitivity and relaxation times.

15. LITERATURE CITED

Abragam, A., 1961, *The Principles of Nuclear Magnetism*, Oxford University Press, Oxford.

Abragam, A., and Bleaney, B., 1970, *Electron Paramagnetic Resonance of Transition Ions*, Oxford University Press, Oxford.

Alger, R. S., 1968, *Electron Paramagnetic Resonance: Techniques and Applications*, Wiley-Interscience, New York, p. 133ff.

Andrew, E. R., 1989, *Magnetic Resonance and Related Phenomena*, Elsevier (24th Ampere Congress, Poznan, 1988), p. 45-51.

Batt, R. J., Jones, G. D., and Harris, D. J., 1977, The Measurement of the Surface Resistivity of Evaporated Gold at 890 GHz, *IEEE Trans. Microwave Theory and Techniques* **MTT-25**:488-491.

Benson, F. A., 1969, Attenuation of rectangular waveguides, *in Millimetre and Submillimetre Waves*, F. A. Benson, ed., Iliffe Books Ltd., London.

Bloembergen, N., and Pound, R. V., 1954, Radiation Damping in Magnetic Resonance Experiments. *Phys. Rev.* **95**: 8-12.

Bloom, A. L., 1955, Nuclear Induction in Inhomogeneous Fields. *Phys. Rev.* **98**: 1105-1111.

Chen, C.-N. and Hoult, D. I., 1989, *Biomedical Magnetic Resonance Technology*, Adam Hilger, Bristol.

Davids, D. A., and Wagner, P. E., 1964, Magnetic Field Dependence of Paramagnetic Relaxation in a Kramers Salt, *Phys. Rev. Lett.* **12**:141-142.

Davis, J. L., and Mims, W. B., 1981, Use of a microwave delay line to reduce the dead time in electron spin echo envelope spectroscopy, *Rev. Sci. Instrum.* **52**:131-132.

Davoust, C. E., Doan, P. E., and Hoffman, B. M., 1996, Q-Band Pulsed Electron Spin-Echo Spectrometer and Its Application to ENDOR and ESEEM. *J. Magn. Reson. A* **119**:38-44.

Eaton, S. S. and Eaton, G. R., 2000, Relaxation Times of Organic Radicals and Transition Metal Ions, in *Distance Measurements in Biological Systems by EPR*, G. R. Eaton, S. S. Eaton, and L. J. Berliner, eds., *Biol. Magn. Reson.* **19**, 29-154.

Eaton, G. R., Eaton, S. S., and Rinard, G. A., 1998, Frequency Dependence of EPR Sensitivity, *in Spatially Resolved Magnetic Resonance*, P. Blümler, B. Blümich, R. E. Botto, and E. Fukushima, ed., Wiley-VCH Publ., pp. 65-74.

Edelstein, W. A., Glover, G. H., Hardy, C. J., and Redington, R. W., 1986, The Intrinsic Signal-to-Noise Ratio in NMR Imaging, *Magn. Reson. Med.* **3**:604-618.

Feher, G., 1957, Sensitivity Considerations in Microwave Paramagnetic resonance Absorption Techniques, *Bell System Technical Journal* **36**:449-484.

Foster, T. H., 1992, Tissue Conductivity Modifies the Magnetic Resonance Intrinsic Signal-to-Noise Ratio at High Frequencies. *Magn. Reson. Med.* **23**:383-385.

Fraenkel, G. K., 1960, Paramagnetic Resonance Absorption, *in Technique of Organic Chemistry, Vol. I - Part IV, Physical Methods of Organic Chemistry*, 3rd Ed., A. Weissberger, ed., Interscience Publishers, New York, ch. XLII.

Gadian, D. G., and Robinson, F. N. H., 1979, Radiofrequency Losses in NMR Experiments on Electrically Conducting Samples, *J. Magn. Reson.* **34**:449-455.

Goldsmith, P. F., 1982, Quasi-Optical Techniques at Millimeter and Submillimeter Wavelengths, *Infrared and Millimeter Waves* **6**:277-342.

Goldsmith, P. F., 1998, *Quasioptical Systems: Gaussian Beam Quasioptical Propagation and Applications*, IEEE Press, New York, p. 119ff, 303ff.

Haas, D. A., Sugano, T., Mailer, C., and Robinson, B. H., 1993, Motion in Nitroxide Spin Labels: Direct Measurement of Rotational Correlation Times by Pulsed Electron Double Resonance, *J. Phys. Chem.* **97**:2914-2921.

Halpern, H. J., and Bowman, M. K., 1989, Low frequency EPR imaging *in EPR Imaging and in vivo EPR*, G. R. Eaton, S. S. Eaton, and K. Ohno, eds., CRC Press, Boca Raton, FL, ch. 6.

Harrington, R. F., 1961, Time-Harmonic Electromagnetic Fields. McGraw-Hill Book Co., New York.

Hoult, D. I., 1996, Sensitivity of the NMR Experiment in *Encyclopedia of NMR*, D. M. Grant and R. K. Harris, eds., **7**:4256-4266.

Hoult, D. I., Chen, C.-N., and Sank, V. J., 1986, The Field Dependence of NMR Imaging II. Arguments Concerning an Optimal Field Strength, *Magn. Reson. Med.* **3**:730-746.

Hoult, D. I., and Lauterbur, P. C., 1979, The Sensitivity of the Zeugmatographic Experiment Involving Human Samples. *J. Magn. Reson.* **34**:425-433.

Hoult, D. I., and Richards, R. E., 1976, The Signal-to-Noise Ratio of the Nuclear Magnetic Resonance Experiment, *J. Magn. Reson.* **24**:71-85.

Hyde, J. S., Yin, J.-J., Feix, J. B., and Hubbell, W. L., 1990, Advances in spin label oximetry. *Pure & Applied Chem.* **62**:255-260.

Jiang, J., Liu, K. J., Walczak, T., and Swartz, H. M., 1995, An Analysis of the Effects of Eddy Currents on L-Band EPR Spectra. *J. Magn. Reson. B* **106**: 220-226.

Johnson, C. C., and Guy, A. W., 1972, Nonionizing Electromagnetic Wave Effects in Biological Materials and Systems, *Proceed. IEEE* **60**:692-718.

Kutter, C., Moll, H. P., van Tol, J., Zuckerman, H., Mann, J. C., and Wyder, P., 1995, Electron Spin Echoes at 604 GHz Using Far Infrared Lasers. *Phys. Rev. Lett.* **74**: 2925-2928.

Lee, C. A., and Dalman, G. C., 1994, *Microwave Devices, Circuits and Their Interaction*, Wiley, New York.

Lloyd, J. P., and Pake, G. E., Spin Relaxation in Free Radical Solutions Exhibiting Hyperfine Structure. *Phys. Rev.* **94**: 579-591 (1954).

Losee, F., 1997, *RF Systems, Components, and Circuits Handbook*, Artech House, Boston, p. 345.

Mailer, C., Thomann, H., Robinson, B. H., and Dalton, L. R., 1980, Crossed TM_{110} bimodal cavity for measurement of dispersion electron paramagnetic resonance and saturation transfer electron paramagnetic resonance signals for biological materials, *Rev. Sci. Instrum.* **51**:1714-1721.

Makovski, A., 1996, Noise in MRI, *Magn. Reson. Med.* **36**:494-497.

Mansfield, P., and Morris, P. G., 1982, *NMR Imaging in Biomedicine* (Supplement 2, Advances in Magnetic Resonance), Academic Press, pp. 310-330.

Mims, W. B., 1965, Electron echo Methods in Spin Resonance Spectrometry, *Rev. Sci. Instrum.* **36**: 1472-1479.

Mims, W. B., 1972, Electron Spin Echoes, in "Electron Paramagnetic Resonance," S. Geschwind, ed., Plenum Press, New York.

Muller, F., Hopkins, M. A., Coron, N., Grynberg, M., Brunel, L. C., and Martinez, G., 1989, A high magnetic field EPR spectrometer, *Rev. Sci. Instrum.* **60**:3681-3684.

Murugesan, R., Cook, J. A., Devasahayam, N., Afeworki, M., Subramanian, S., Tschudin, R., Larsen, J. A., Mitchell, J. B., Russo, A., and Krishna, M. C., 1997, *In Vivo* Imaging of a Stable Paramagnetic Probe by Pulsed-Radiofrequency Electron Paramagnetic Resonance Spectroscopy, *Magn. Reson. Med.* **38**: 409-414.

Narayana, P. A., Massoth, R. J., and Kevan, L., 1982, Active microwave delay line for reducing the dead time in electron spin-echo spectrometry, *Rev. Sci. Instrum.* **53**:624-626.

Petropoulos, L. S., and Haacke, E. M., 1991, Higher-Order Frequency Dependence of Radiofrequency Penetration in Planar, Cylindrical, and Spherical Models, *J. Magn. Reson.* **91**:466-474.

Pfenninger, S., Froncisz, W., and Hyde, J. S., 1995a, Noise Analysis of EPR Spectrometers with Cryogenic Microwave Preamplifiers. *J. Magn. Reson., A* 113, 32-39.

Pfenninger, S., Froncisz, W., Forrer, J., Luglio, J., and Hyde, J. S., 1995b, General method for adjusting the quality factor of EPR resonators, *Rev. Sci. Instrum.* **66**:4857-4865.

Piasecki, W., Froncisz, W., and Hyde, J. S., 1996, Bimodal loop-gap resonator, *Rev. Sci. Instrum.* **67**:1896-1904.

Poole, C. P. Jr., 1967, *Electron Spin Resonance*, Wiley, New York, ch. 14.

Prisner, T. F., 1997, Pulsed High-Frequency/High-Field EPR, *Adv. Magn. Optic. Reson.* **20**: 245-299.

Rinard, G. A., Quine, R. W., Eaton, S. S., Eaton, G. R., and Froncisz, W., 1994, Relative Benefits of Overcoupled Resonators vs. Inherently Low-Q Resonators for Pulsed Magnetic Resonance, *J. Magn. Reson. A* **108**: 71-81.

Rinard, G. A., Quine, R. W., Ghim, B. T., Eaton, S. S., and Eaton, G. R., 1996, Easily Tunable Crossed-Loop (Bimodal) EPR Resonator, *J. Magn. Reson. A* **122**:50-57.

Rinard, G. A., Quine, R. W., Ghim, B. T., Eaton, S. S., and Eaton, G. R., 1996, Dispersion and Superheterodyne EPR Using a Bimodal Resonator, *J. Magn. Reson. A* **122**:58-63.

Rinard, G. A., Eaton, S. S., Eaton, G. R., Poole, C. P., Jr., and Farach, H. A., 1999a, Sensitivity, in *Handbook of Electron Spin Resonance*, C. P. Poole, Jr. and H. A. Farach, eds, AIP Press, **2**:1-23.

Rinard, G. A., Quine, R. W., Song, R. Eaton, G. R., and Eaton, S. S., 1999b, Absolute EPR Spin Echo and Noise Intensities. *J. Magn. Reson.* **140**:69-83.

Rinard, G. A., Quine, R. W., Harbridge, J. R., Song, R., Eaton, G. R., and Eaton, S. S., 1999c, Frequency Dependence of EPR Signal-to-Noise, *J. Magn. Reson.* 140, 218-227.

Rinard, G. A., Quine, R. W., and Eaton, G. R., 2000, An L-band Crossed-Loop (Bimodal) Resonator, *J. Magn. Reson.* **144**, 85-88.

Rinard, G. A., Quine, R. W., Eaton, S. S., and Eaton, G. R., 2002a, Frequency Dependence of EPR Signal Intensity, 250 MHz to 9.1 GHz, *J. Magn. Reson.* **156**, 113-121.

Rinard, G. A., Quine, R. W., Eaton, S. S., and Eaton, G. R., 2002b, Frequency Dependence of EPR Signal Intensity, 248 MHz to 1.4 GHz, *J. Magn. Reson.* **154**, 80-84.

Rinard, G. A., Quine, R. W., Eaton, G. R., and Eaton, S. S., 2002c, 250 MHz Crossed Loop Resonator for Pulsed Electron Paramagnetic Resonance, *Magn. Reson. Engineer.* **15**, 37-46.

Robinson, B. H., Haas, D. A., and Mailer, C., 1994, Molecular Dynamics in Liquids: Spin-Lattice Relaxation of Nitroxide Spin Labels, *Science* **263**:490-493.

Röschmann, P., 1987, Radiofrequency penetration and absorption in the human body: Limitations to high-field whole-body nuclear magnetic resonance imaging, *Med. Phys.* **14**:922-937.

Stoodley, L. G. 1963, The Sensitivity of Microwave Electron Spin Resonance Spectrometers for use with Aqueous Solutions, *J. Elect. Control* **14**:531-546.

Strutz, T., Witowski, A. M., and Wyder, P., 1992, Spin-Lattice Relaxation at High Magnetic Fields. *Phys. Rev. Lett.* **68**: 3912-3915.

Sueki, M., Rinard, G. A., Eaton, S. S., and Eaton, G. R., 1996, Impact of High Dielectric Loss Materials on the Microwave Field in EPR Experiments, *J. Magn. Reson. A* **118**:173-188.

Tofts, P. S., 1994, Standing Waves in Uniform Water Phantoms, *J. Magn. Reson. B* **104**:143-147.

Varian spectrometer manual 87-125-052, page 3-8.

Vlaardingerbroek, M. T., and den Boer, J. A., 1996, *Magnetic Resonance Imaging*, Springer, New York.

Weber, A., Schliemann, O., Bode, B. and Prisner, T., 2002, PELDOR at S- and X-Band Frequencies and the Separation of Exchange Coupling from Dipolar Coupling. *J. Magn. Reson.* **157**, 277-285.

Weil, J. A., Bolton, J. R., and Wertz, J. E., Electron Paramagnetic Resonance: Elementary Theory and Practical Applications. Wiley, New York, 1994.

Willer, M., Forrer, J., Keller, J., Van Doorslaer, S., and Schweiger, A., 2000, S-band (2 - 4 GHz) pulse electron paramagnetic resonance spectrometer: Construction, probe head design, and performance. *Rev. Sci. Instrum.* **71**, 2807-2817.

Wilmshurst, T. M., 1968, *Electron Spin Resonance Spectrometers*, Plenum Press, New York.

Wilmshurst, T. H., 1990, *Signal Recovery from Noise in Electronic Instrumentation*, 2nd ed., Adam Hilger - IOP Publishing, page 87.

Witowski, A. M., 1991, The Two-Phonon Spin-Lattice Relaxation Processes in High Magnetic Fields. *Solid State Commun.* **77**: 23-27.

Witowski, A. M., Kutter, C., and Wyder, P., 1997, Spin-Lattic Relaxation at High Magnetic Fields: A Tool for Electron-Phonon Coupling Studies. *Phys. Rev. Lett.* **78**: 3951-3954.

Yong, L., Harbridge, J., Quine, R. W., Rinard, G. A., Eaton, S. S., Eaton, G. R., Mailer, C., Barth, E., and Halpern, H. J., 2001, Electron Spin Relaxation of Triarylmethyl Radicals in Fluid Solution. *J. Magn. Reson.* **152**, 156-161.

Zecevic, A., Eaton, G. R., Eaton, S. S., and Lindgren, M., 1998, Dephasing of Electron Spin Echoes for Nitroxyl Radicals in Glassy Solvents by Non-methyl and Methyl Protons, *Mol. Phys.* **95**: 1255-1263.

Zhou, Y., Bowler, B. E., Eaton, G. R., and Eaton, S. S., 1999, Electron Spin Relaxation Rates for S = ½ Molecular Species in Glassy Matrices or Magnetically-Dilute Solids at Temperatures Between 10 and 300 K. *J. Magn. Reson.* **139**:165-174.

Zypman, F. R., 1996, MRI Electromagnetic Field Penetration in Cylindrical Objects, *Comput. Biol. Med.* **26**:161-175.

Chapter 4

ENDOR Coils and Related Radiofrequency Circuits

Christopher J. Bender
Department of Chemistry, Fordham University, 441 E. Fordham Road, Bronx, New York 10458 USA

1. INTRODUCTION

Electron-Nuclear Double Resonance (ENDOR) is an advanced electron magnetic resonance (EMR) technique that provides a much higher level of spectral resolution than conventional microwave absorption methods. The ENDOR method was introduced by Feher (1956) as a variant of the Overhauser Effect (1953) in nuclear resonance, and it entails the observation of transitions between nuclear sublevels of the electron Zeeman Effect. The electron transitions are still used as a means of detection, however, because the sensitivity of the electron resonance measurement is greater than that of the nuclear resonance. In brief, an EMR transition ($\Delta m_S = \pm 1$, $\Delta m_I = 0$, Figure 1) is saturated, which leads to the collapse of the observed EMR signal as the corresponding state populations equalize. If one now irradiates the spin system so that transitions are induced between the nuclear sublevels (*i.e.* $\Delta m_I = \pm 1$, $\Delta m_S = 0$), the condition of saturation in the EMR transition is lifted as the nuclear sublevel populations shift, and there is a partial recovery of the EMR signal, the so-called 'ENDOR enhancement' (*cf.* Kevan & Kispert, 1976). It is common to view the pathways among the energy levels in the spin Hamiltonian state diagram (Figure 1) as being analogous to a resistive electrical network and the spin dynamics of the ENDOR effect as a short circuit phenomenon (*cf.* Dwek *et al.*, 1969; Kwiram, 1971; Möbius *et al.*, 1982; Kurreck *et al.*, 1988).

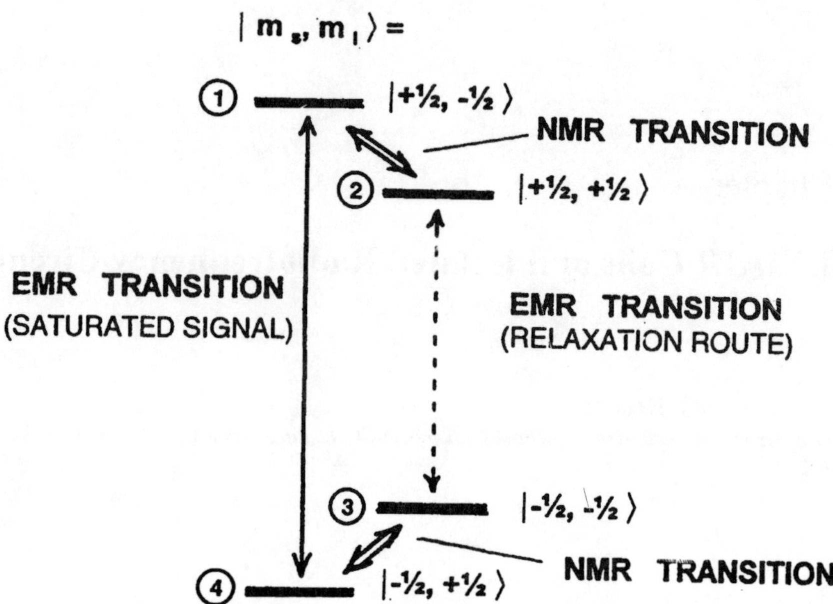

Figure 1. Four level state diagram ($m_S = \pm 1/2$; $m_I = \pm 1/2$) and the transitions of an ENDOR experiment. An allowed EMR transition is saturated, and the rf frequency is swept in order to detect NMR transitions.

The design of an ENDOR spectrometer system differs only slightly from the basic cw- or pulsed EMR spectrometer, and the subject has been exhaustively reviewed (Gothe, 1970; Kevan & Kispert, 1976; Box, 1977; Leniart, 1979; Schweiger, 1987; Kurreck *et al.*, 1988; Bender & Aisen, 1993; Thomann & Bernardo, 1993; Piekara-Sady & Kispert, 1994). But there is a gap in the ENDOR literature as concerns the so-called 'coil' that is used to generate the radiofrequency field that drives the NMR transitions, and this review is intended to meet that need. At issue is the need to irradiate the sample material by a secondary radiofrequency field (*i.e.* NMR transition), which entails the incorporation of a second 'resonator' without incurring serious deleterious effects on the primary microwave field (EMR transition). A loop or helical coil is used as a radiofrequency field generating element in a transmitter circuit that consists of a low power signal source or sweeper, a broadband power amplifier, and the coil network. The latter is typically two short sections of transmission line that are spanned by the coil and terminated by a power load (Figure 2). The manner in which the radiofrequency and microwave resonators are integrated affects both the method and the sensitivity of the spectroscopic measurement because one is combining two *RLC* structures in a distributed parameter network while trying to retain the behavior and properties of each device.

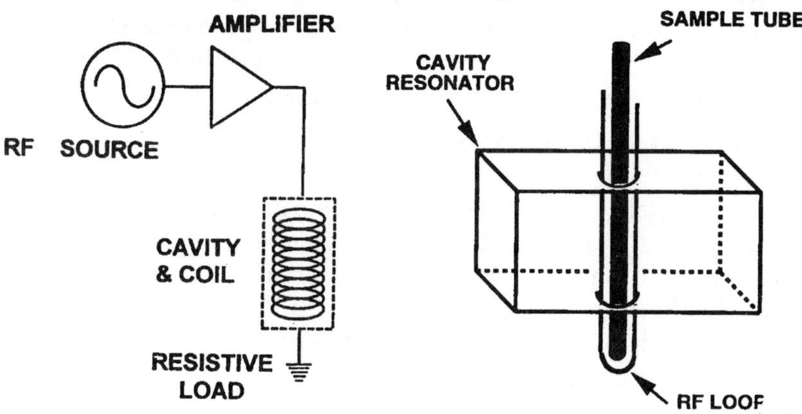

Figure 2. The basic configuration for simultaneously irradiating a sample material with radio- and microwave fields for ENDOR.

Methods of generating magnetic fields in the radiofrequency domain of the electromagnetic spectrum are analogous to those used in DC electricity theory. A single linear conductor generates a magnetic field in a circular pattern that is orthogonal to the conductor's axial direction. The magnetic field polarity is determined by the direction of current flow through the conductor according to the right hand rule, and the field strength can therefore be enhanced by modifying the geometry of the conductor so that fields reinforce one another. Two common conductor shapes are the loop and its extension, the multi-turn coil. Both of these geometries are commonly used for ENDOR spectroscopy, and the term coil will be used in this review as a synonymous reference to both.

2. THE PRINCIPLES OF COIL IMPLEMENTATION

2.1 The Coil as a Field-Generating Element

At its simplest, an ENDOR coil may be constructed by forming a wire loop or helix around the sample holder (Figure 2). A wire loop or multi-turn coil constitutes a device known as an inductor, wherein a magnetic field is generated by the current flow through the conductor. For direct current circuits, the magnetic field strength at the center of a loop (or helix) is given as $H = N{\cdot}I / 2\,r$ (and $B = \mu{\cdot}H$), where N, I, r, and μ are the number of turns in the loop, the magnitude of the current, the radius of the coil, and the magnetic permitivity of the medium (air and/or the sample material), respectively. The field is a vector quantity and has a direction that is determined by the direction of current flow. The loop therefore behaves as a magnetic dipole when alternating current is used. In the latter scenario, the energy stored in the magnetic field of the dipole is $P = \frac{1}{2} L \cdot I^2$, where L is the inductance of the device in units of Henrys, and power is in Watt·sec (Jordan & Balman, 1968).

The inductance of a conductor may be calculated by using its geometric parameters (*cf.* Terman, 1955). Besides the circular loop and helix (Section 3.2), common inductors that find application in ENDOR probeheads are parallel pairs of wire (Section 3.4) and parallel rectangular bars (Section 3.5). For each of these commonly used ENDOR coils the inductance is related to geometry as follows:

Straight Wire: $$L = 0.00508l(\ln\frac{4l}{d} - 0.75) \quad (1a)$$

Parallel Wire Pair: $$L = 0.01016l(\ln\frac{2D}{d} - \frac{D}{l} - \delta\mu) \quad (1b)$$

Parallel Rectangular Bars:
$$L = 0.01016l(\ln\frac{D}{b+c} + 1.5 - \frac{D}{l} + 0.2235\frac{b+c}{l}) \quad (1c)$$

Circular Loop: $$L = 0.01595D(\ln\frac{8D}{d} - 2 - \delta\mu) \quad (1d)$$

Single Layer Helix: $$L = FN^2D \quad (1e)$$

where in each case L is expressed in microhenrys and all dimensions are expressed in inches. The geometric variable D designates the diameter (loop/helix) or the spacing between the conductors (parallel wire/bar); d is the diameter of the wire; b and c are the width and breadth of the bar; and l is the length. The parameter F is a form factor (*cf.* Terman, 1955) that depends on the diameter to length ratio of the helix, and is assigned the value 0.005854 for the familiar 1:4 ratio that is used in the cylindrical TM_{110} cavity. The term $\delta\mu$ that appears in (1b) and (1d) is a numerical correction factor that is related to the skin depth and permeability of the conductor; it simply means that the numerical quantities within the parentheses are not as precise as would be inferred.

As an illustrative example, the multi-turn wire coil that is used in the commercial TM_{110} ENDOR cavity will be analyzed as follows. The diameter of the coil is approximately 1.0 cm, or 0.39 inches, which means that L will scale as $0.002305 \times N^2$. The estimated inductance of a 20-turn coil is therefore 9.22×10^{-7} Henrys.

2.2 The Effect of Reactance

Capacitors and inductors generate electric and magnetic fields that, in AC circuits, interact with the current and affect its time-dependent behavior by introducing a phase lag between the voltage and current. This so-called reactance is manifest as a frequency dependent complex valued term that in the case of the inductor is given by the formula $i\omega L$, where L is the value of the inductance. The AC current magnitude is related to the voltage in much the same way as the DC relation, except that the resistance is replaced by a complex-valued impedance, Z, which is equal to $i\omega L$ for the inductor. Since the rf field power is proportional to the square of the current, that is, $P = ½\, L{\cdot}I^2$, one desires to optimize the relation $I = Y{\cdot}V$, where Y is the admittance of the circuit, or Z^{-1}.

As depicted in Figure 2, the ENDOR coil circuit consists of a series pairing of the inductor/coil and a 50 Ω power resistor, often referred to as a load. The resistance R is real-valued, and the inductance supplies a complex-valued impedance $i\omega L$, so that one writes the impedance of the network as $Z = R + i\omega L$. As such, $V = Z{\cdot}R$, or $I = Y{\cdot}V$, and therefore the total power delivered to the network may be written as $P = Z{\cdot}I^2$ or $P = Y{\cdot}V^2$. The admittance $Y = (R + i\omega L)^{-1}$, and the power may be written:

$$P = \frac{V^2}{(R^2 + \omega^2 L^2)^{½}} \tag{2}$$

where the denominator represents the magnitude of the admittance. It follows from this equation that the term $\omega^2 L^2$ will limit the power delivered to the network since $\omega = 2\pi f$ and varies with the rf carrier frequency. For example, with $R = 50\ \Omega$ and $L = 9.22 \times 10^{-7}$ Henrys, as calculated in Section 2.1 for a 20-turn coil, parameters for the evaluation of $|Z|$ are:

f (MHz)	$\omega \times 10^{-6}$	$\omega^2 L^2$ ohms	$(R^2+\omega^2 L^2)^{1/2}$ ohms
1	6.28	33	50.3
5	31.4	839	57.8
10	62.8	3360	76.6
50	314	84000	294.1
100	630	335000	580.9

This simplified analysis of the ENDOR coil network, graphically depicted in Figure 3, illustrates why one seeks to minimize the inductance of the coil, which, as has been described in Section 2.1, can be controlled via the geometry of the device. For example, a pair of parallel rectangular strips (*cf.*

Section 3.5) of dimension 4 × 10 mm and separated by 4 mm (the approximate diameter of a standard sample tube)

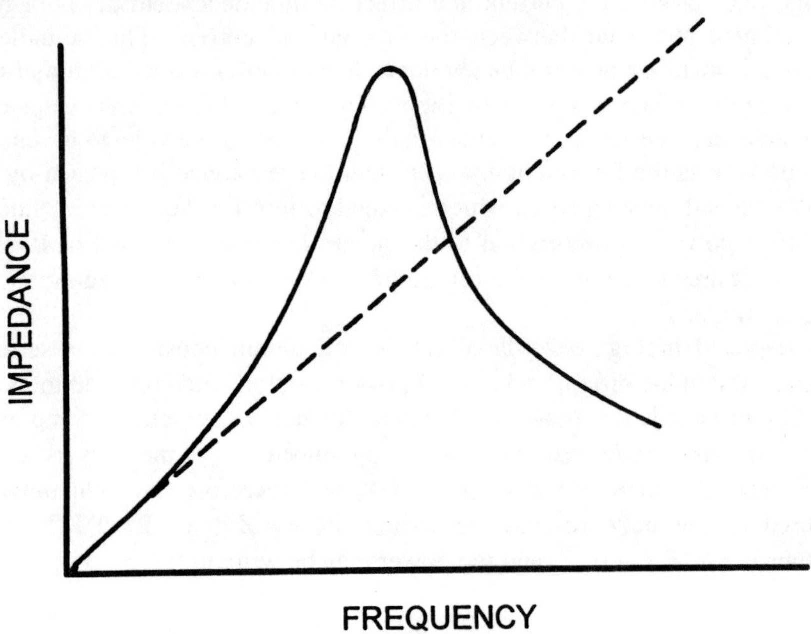

Figure 3. The variation of inductor impedance as a function of AC carrier frequency. Dotted line corresponds to ideal inductor (no parasitic capacitance). Solid line corresponds to real inductor and features resonance.

will have an approximate inductance (Equation 1c, where $b << D$) of $L = 4.8 \times 10^{-9}$ Henrys. The corresponding calculated parameters for evaluating $|Z|$ of the rectangular strips are:

f (MHz)	$\omega\times10^{-6}$	ω^2L^2 ohms	$(R^2+\omega^2L^2)^{1/2}$ ohms
1	6.28	0.001	50.0
5	31.4	0.022	50.0
10	62.8	0.089	50.0
50	314	2.234	50.0
100	630	9.000	50.1

2.3 Impedance-Frequency Behavior & Resonance

Another feature of AC circuits that can complicate ENDOR coil design is the fact that the size of the electrical circuit components approximate the electromagnetic wave length at radio- and microwave frequencies, and one must treat the circuit as a distributed network (*cf.* Murdoch, 1970). The idealized model of the ENDOR coil circuit that was described in the preceding paragraphs does not account for parasitic capacitances that arise between (a) the turns of the wire loop and (b) the loop and the wall of the microwave cavity resonator. These parasitic capacitances are a factor in the mathematical description of the ENDOR coil network, but, as the name suggests, these capacitances are not desired, and their manifestation in the ENDOR spectroscopy experiment is associated with artifacts.

The impedance of an inductor varies linearly as a function of the frequency according to $Z = R + i\omega L$, but the parasitic capacitance associated with a wire solenoid requires that the device be described as a parallel *RLC* network whose *Z-f* plot resembles Figure 3 (solid line). This plot is linear in the low frequency region, but deviates from the ideal behavior as some critical frequency f_c is approached. The impedance of a parallel *RLC* network is (Bowick, 1982; Nilson, 1983):

$$Z = \frac{R + i\omega L}{i\omega CR + (1 - \omega^2 LC)}. \tag{3}$$

As the frequency of the signal increases from, for example, DC to the rf or microwave region of the spectrum, the impedance, as defined by Equation (3), undergoes a change in the sense that the reactance (complex-valued term) is initially dominated by the inductance, but as the frequency increases the capacitance term becomes larger and the total impedance is then dominated by the capacitance terms. At some critical frequency f_c the imaginary part of the impedance vanishes, and this point demarcates the inductance-dominated and capacitance-dominated region of the *Z-f* plot. A series *RLC* network behaves in the inverse manner; the impedance achieves a minimum value at f_c. This critical frequency where the device, in effect, becomes purely resistive denotes a resonance.

As the impedance of an inductor increases, the power that is transmitted by the device declines, and therefore the radiofrequency field generated by the coil will likewise decline with increasing frequency. In other words, the coil is behaving as a low-pass filter. At resonance, the coil again

transmits

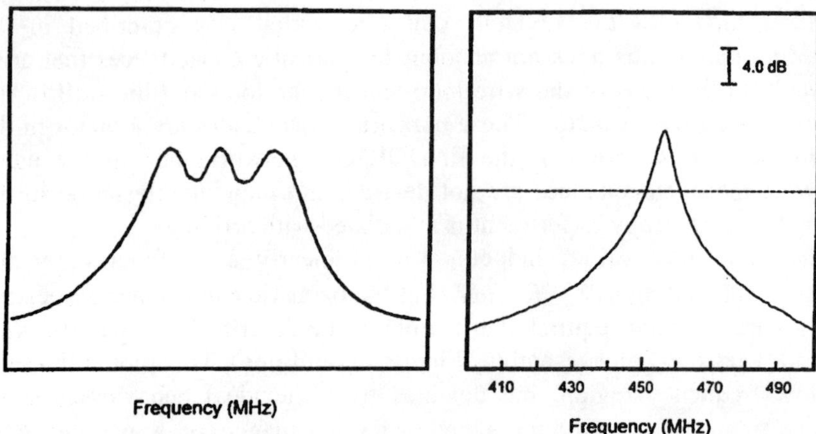

Figure 4. Power transmission profiles of a bandpass filter. On the left is a pass band modelled by a polynomial function. The center of the plateau region corresponds to the center of tuned frequency of the filter, and the pass band corresponds to the frequency region over which the power transmitted is ≤3 dB relative to the power transmitted at the center frequency. On the right is a power transmission profile through the resonance region of an 18-turn ENDOR coil.

power and behaves as a bandpass filter in a frequency 'window' near f_c. The ENDOR experiment is performed by sweeping through a frequency range, and one would ideally wish that the power and radiofrequency field strength remain very nearly constant during the course of the sweep. Since the pass band of the coil as filter is too narrow (see below), one typically operates the ENDOR coil in the low impedance region of the *Z-f* plot, where the coil's impedance is determined by the inductance *L* in Equation (3). In this region of the operating range of the coil, the slope of the *Z-f* plot is approximately proportional to the inductance, and it follows that low inductance coils are desired in order to flatten the response profile and extend the useful radiofrequency sweep range.

The ENDOR coil is not an isolated circuit element, and therefore, besides the self-capacitance, there is a second parallel capacitance that arises between the coil and the microwave cavity walls. The impedance-frequency profile of an ENDOR coil changes when it is placed inside a cavity, and direct measurements of the assembled ENDOR probe determined that an additional parallel capacitance is introduced by the interaction between the coil and the grounded cavity walls; it was found that a resonance condition existed in the 18-turn wire coil and a Bruker TM_{110} cavity at approximately 30 MHz (Bender & Babcock, 1992).

The ENDOR coil will behave both as a bandpass filter, owing to its characteristics as a parallel *RLC* network, and as a low pass filter. Typically, however, the coil is operated in the low frequency region of its operating range (as defined by the *Z-f* plot), that is, well below f_c, because the pass band of an ENDOR coil at resonance will not correspond to an rf sweep range of a typical spectrum. The power transmission profile of an ideal bandpass filter approximates a rectangular response that may be modelled by a polynomial expansion, as illustrated in Figure 4 (left panel), and the degree to which the pass band mimics a rectangle depends on the order and type of the polynomial function. Filters are typically defined by the polynomial that best approximates their behavior, examples being Bessel, Chebyshev (*cf.* Bowick, 1982; Rhea, 1994), and a pass band is designated by the center frequency plus and minus those frequencies that are attenuated by 3 dB or less.

The ideal bandpass filter power transmission profile may now be compared with that of a TM_{110} cylindrical ENDOR cavity apparatus outfitted with an 18-turn silver wire coil of length 1" (Figure 4, right, and 5). A Hewlett-Packard network analyzer and test set provided the swept frequency signal to the probehead and records the transmitted and reflected power through the device under test (DUT, see Appendix 2), which in this case is the ENDOR cavity assembly. The coil behaves as a classic inductor, as indicated by the steady decline in transmitted power as the frequency increases. Nearly all the power is attenuated at 300 MHz, but near 450 MHz the coil again transmits and behaves as though it were a bandpass filter with its 3 dB band spanning only 1.4 MHz. This narrow region in which the coil behaves as a bandpass filter is unsuitable for swept frequency ENDOR experiments despite some control over tuning the center frequency; if one spoils the coil's Q by altering the coil geometry (*e.g.* spacing between turns, height to diameter ratio, *etc.*) one will still not significantly broaden the pass band.[1]

The ENDOR coil is therefore operated in a frequency range well below resonance at frequencies where its impedance is low (optimally, near 50 Ω). The power fluctuation across the sweep range may be reduced by lowering the inductance of the coil (designs are reviewed in Section 3), and the optimum impedance attained by matching the transmission line near the center of the frequency sweep range (Section 4). Increasing the amplifier gain offsets the magnetic field strength losses accompanied by lowering L.

For the standard commercial cavity apparatus, that is, a 20-turn silver wire coil that is inserted into a cylindrical TM_{110} cavity, the power transmission

[1] Large coils denote ENDOR probeheads in which the wall of the microwave cavity resonator constitutes the rf coil. For example, wire-wound TE_{102} (Hyde, 1965) or spiral-cut TE_{112} (Gruber et al., 1974) cylindrical walls. Comparing the dimensions of these probeheads with coil design formulae suggest that the rf coil Q is high since the length:diameter ratio is nearly 1:1.

profile (Figure 5) shows that 5 to 15 MHz sweep is accompanied by only a 1

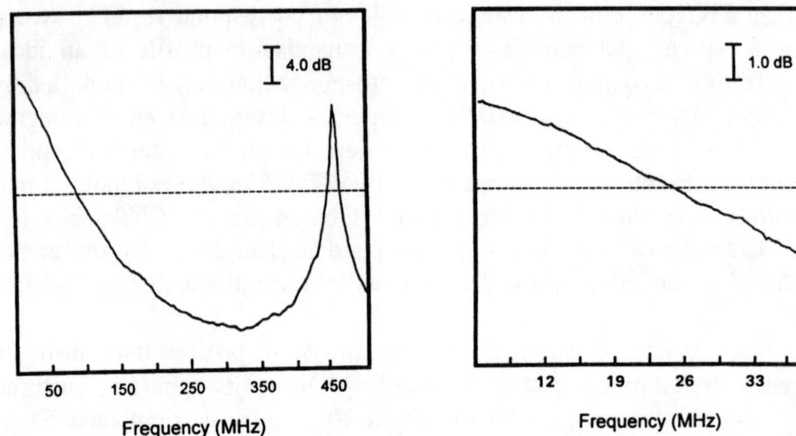

Figure 5. Power transmission profiles of a 20-turn silver wire helix that is inserted into a cylindrical TM_{110} cavity. On the left is a broad frequency sweep that shows resonance, whereas on the right is a narrow range sweep that demonstrates how power losses may be significant even on sweeps commonly used in an ENDOR experiment.

dB decline in power, but a 5 to 40 MHz sweep is subject to a 6 dB decline, which can affect the quality of the spectrum. The 5 to 15 MHz sweep is suitable for many ^{1}H-ENDOR spectra, and the 1 dB decline in power is sufficiently small that impedance matching is not necessary.[2] Beyond 20 MHz the transmitted power declines further, but with nearly the same slope, and therefore a corrective impedance match at 25 MHz will ensure that the radiofrequency field is optimal. At 30 MHz and beyond the slope of the power transmission plot steepens and there is a weak resonance because of the parallel capacitance to the cavity wall (Bender & Babcock, 1992). Beyond 30 MHz one should switch to a lower inductance coil or, if the frequency range of the sweep is small, use the same strategy as was employed between 20 and 30 MHz. This approach was used by Bender *et al.* (1989) for the ^{1}H-ENDOR study of the tyrosyl radical of ribonucleotide reductase; the spectrum spanned 40 MHz, but because the affected protons consisted of both α– and β– types, the power requirements for observing the ENDOR effect (*cf.* Hyde *et al.*, 1968; Allendoerfer & Maki, 1970) varied and the spectrum was assembled from 10 MHz portions. A full 2-50 MHz sweep was possible with a very low inductance coil, such as a wire loop, but

[2] Impedance will vary with design; in this example the impedance was approximately 12 Ω at 15 MHz and may be easily matched by using 1:4 transformers, see Section 4

for a given radiofrequency power level one would not have been able to optimize signal intensity of the α- and β-protons simultaneously.

2.4 S/N Factors Unrelated to Coil Design

2.4.1 Near-Crossing of Hyperfine States: Nuclear Cross Relaxation

The greatest challenge to the ENDOR experimentalist is optimizing the population transfer kinetics among the participating nuclear sublevels, which entails a systematic variation of the microwave and radiofrequency power levels, sample temperature, *etc*. All of these experimental parameters affect the so-called ENDOR enhancement (*cf.* Kevan & Kispert, 1976; Kurreck *et al*., 1988), and traditional problem areas for observing ENDOR have been the very low (≤ 5 MHz) and high (≥ 30 MHz) frequency regions. The low frequency region is associated with weakly coupled nuclei, such as ^{2}H and ^{14}N, where the electron-nuclear interaction is defined only by the nuclear Zeeman and quadrupole interaction terms of the spin Hamiltonian, whereas the high frequency region is associated with strongly coupled nuclei, which would include the electron-nuclear interactions of paramagnetic transition metal ions. In both cases the complaint is often heard that insufficient rf power is available to drive the spin dynamics and observe an ENDOR effect.

The power transmission profile of Figure 5 is obtained for very nearly all ENDOR coils, and power delivery to the sample is therefore certainly not due to a problem of poor coil design, since all the coil types possess an impedance that tends towards 50 Ω at $f \leq 5$ MHz. Instead, one must examine the spin population transfer kinetics, particularly in powder samples. In general, we draw the hyperfine state diagrams like Figure 1 with the assumption that both the individual states (*i.e.* levels) and the frequencies of the incident radiation are discrete, which is not the case. All rf and microwave sources are based on *RLC* networks, which are defined by a resonance curve and radiate a distribution of frequencies. Likewise, the energies that represent a given state $|\, m_S, m_I \rangle$ are distributed because of local field inhomogeneities. The ENDOR enhancement, which is based on the radiation-induced dynamics of spin populations among discretely selected hyperfine states, is at best an ideal because when these states can 'communicate' (via overlapping energy distributions) the spin populations tend to equalize and deleteriously affect the saturation of the desired EMR transition. This phenomenon is known as 'cross-relaxation' (*cf.* Grant, 1964a,b; Abragam & Bleaney, 1970, pp78, 243–51).

To cite a specific example, electron spin echo modulation (ESEEM) is often called a complement to ENDOR (*cf.* Tsvetkov & Dikanov, 1987) because it seems to perform better at detecting low frequency nuclear sublevel transitions (*e.g.* weakly coupled ^{2}H and ^{14}N). To be sure, nuclear quadrupole relaxation affects the system's spin dynamics (*cf.* Andrew, 1958; Abragam, 1961; 1962), but low frequency spectra are likely made problematic because of the proximity of nuclear sublevels involved. In each of these cases of weakly coupled nuclei the spin Hamiltonian (nuclear Zeeman, nuclear hyperfine, and nuclear quadrupole) terms are approximately equal in magnitude, and the nuclear sublevels merge in one m_S state (*cf.* Mims & Peisach, 1978; Schweiger *et al.*, 1979; Böttcher *et al.*, 1990). The proximity of the energy levels makes the conditions suitable for rapid cross-relaxation, hence the problematic nature of observing cw-ENDOR transitions of these weakly coupled nuclei. ESEEM, by contrast, owes its sensitivity to the interference among closely spaced energy levels (quantum beats) and a phase coherence of a population of spin states that can only occur as the state distributions merge.[3] Phenomenological scenarios that support this argument may be found in the ENDOR and ESEEM literature; cw-ENDOR of weakly coupled ^{14}N nuclei is readily observed in single crystals where linewidths are narrow (Böttcher *et al.*, 1990; Spaeth *et al.*, 1992), and deep modulation of the electron spin echo by strongly coupled ^{14}N in the copper site of azurin is observed at W-band, where a merging of hyperfine levels (exact cancellation) is imposed by the now commensurately large (relative to the nuclear hyperfine interaction) nuclear Zeeman energy (Coremans *et al.*, 1997).

2.4.2 The Effect of Magnetogyric Ratio and Spin Relaxation Times

The gist of the argument outlined in the preceding paragraph is that cross-relaxation destroys the requisite condition of saturation in the EMR transition and therefore the ENDOR enhancement. This inter-relationship between ENDOR enhancement and spin-relaxation rates has been examined by Allendoerfer & Maki (1970) and is expressed in general terms as

$$\frac{\gamma_n^2 H_2^2 T_i^2}{1+\gamma_n^2 H_2^2 T_i^2}, \tag{4}$$

where γ_n is the magnetogyric ratio of the nucleus, H_2 is the rf field strength, and T_i is the spin-lattice relaxation or cross-relaxation time (depending on mechanism of the transition).

[3] In the quantum electronics literature this phase coherence is attributed to cross-relaxation processes induced by interactions of the dipole-dipole type. The details of the mechanism are described by Grant (1964a,b), and the process is known as 'Hertzian Coherence', as distinct from 'Optical Coherence', which is associated with the simultaneous driving a several interfering states.

The magnetogyric ratio γ_n is equal to $g_n\, e/2m_p$, and representative values for nuclei whose hyperfine interactions are commonly measured via ENDOR are:[4]

Nucleus	g_n	γ_n ($s \cdot A \cdot kg^{-1} \times 10^{-6}$)
^1H	5.5857	267.54
^2H	0.8574	41.07
^{14}N	0.40376	19.34
^{15}N	−0.56638	27.13
^{17}O	−0.7575	36.28
^{23}Na	1.478	70.79
^{31}P	2.263	108.39
^{63}Cu	1.484	71.08

The variance of γ_n in the Equation (4) suggests that, all other factors remaining constant, one must increase the rf field strength (power) to compensate for low γ_n. For example, γ_n of ^1H and ^2H differ by an order of magnitude, which means that ^2H-ENDOR experiments will require a ten-fold increase in power over ^1H-ENDOR even before one considers differences among the nuclear spin relaxation rate; this is why power amplifiers rated at the hundreds or kilowatt power level are often specified in the descriptions of ENDOR spectrometers.

2.4.3 Remediation of Deleterious Magnetogyric and Relaxation Effects

Operating the spectrometer with higher rf powers is one method of compensating for low γ_n of the probe nucleus, and this is certainly the most effective since the energy stored in the magnetic field of the dipole (*i.e.* the wire loop) is $\frac{1}{2} L \cdot I^2$. But this power relation also indicates that the inductance of the loop is a factor to first order, and therefore increasing the inductance of the device can aid the driving conditions so that ENDOR may be observed. As described in preceding paragraphs, high inductance devices severely limit one's ability to sweep the frequency, but the nuclei in question possess a Larmor frequency that is small (at 0.3 T) in comparison to the typical hyperfine interaction, and the corresponding ENDOR transition frequencies will appear at approximately $\frac{1}{2} a \pm \nu_n$, where $\pm \nu_n$ is a narrow range (*e.g.* 3–5 MHz). One therefore does not need the capability of a broad sweep range and can afford to operate the spectrometer with high inductance coils; in fact, with such narrow sweep ranges one may be emboldened to use tuned coils that are spoiled.

[4] m_p is defined as proton mass and equals 1.6726×10^{-27} kg; $e = 1.60219 \times 10^{-19}$ s·A

2.4.4 ENDOR Spectra Above 50 MHz

One does not often see reports of ENDOR spectra above 50 MHz, and the reason for this may be inferred from Figure 5 or in practice by watching the amplifier power meter as one sweeps the rf carrier frequency upwards beyond 30 MHz; the power is dramatically attenuated beyond 50 MHz with the standard probehead consisting of a 20 turn wire coil and a cylindrical cavity because the impedance has simply increased to the point where the coil behaves as a cut-off filter. The remedy for this is to either use a low-inductance coil or incorporate an impedance matching network into the circuit so that the transmitter's impedance is close to the optimum 50 Ω (Section 4).

3. PERFORMANCE COMPARISON OF ENDOR COILS

3.1 Overview & General Strategy

In the preceding section coil optimization was described as flattening the *Z-f* response over the desired frequency range while ensuring that the radiofrequency field intensity was adequate to drive the spin population transfer kinetics. In experimental practice, however, the ENDOR signal intensity is affected more by how the coil integrates with the microwave resonator than by the impedance of the radiofrequency transmission line. The lossy metal of the coil will lower the microwave resonator quality factor, Q, if it enters the microwave field and couples to the electric field component. The experimenter's efforts are therefore best directed towards constructing the coil in such a way that minimal perturbation of the Q-factor is incurred.

The microwave resonator (*e.g.* cavity, lumped-parameter, slow-wave, etc.) behaves as a filter in the same manner as has been described in Section 2. The quality factor is formally defined as the energy stored in the system divided by the power absorbed (loss) by the system and is customarily estimated by the ratio of the resonance curve width Δf to the center frequency f_0. Prior to recording an EMR spectrum, one places a sample material within the resonant microwave field of an *RLC* structure, which alters the impedance of that structure, and then tunes the loaded microwave structure in order to match the bridge circuit (essentially nulling the power to the detector). As one sweeps the DC magnetic field through the resonance condition the permittivity of the sample material changes, thereby affecting the impedance of the loaded resonator and the match of the bridge circuit, which is detected as a power fluctuation at the detector. The power fluctuation is proportional to the loaded resonator's quality factor, that is,

$$dP = \tfrac{1}{2}\,\eta\chi''PQ, \tag{5}$$

and therefore the sensitivity of the experiment will be deleteriously affected if the Q-factor is lowered (*cf.* Poole, 1983; Talpe, 1971).

ENDOR coils were initially wound externally to high Q cavities (Feher & Gere, 1959; Hyde, 1965; Christidis & Heineken, 1973; Gruber *et al.*, 1974). This approach on the one hand has the advantage that the cavity Q is not spoiled, but, on the other hand, has the disadvantage that the volume enclosed by the coil is large. The generation of a radiofrequency field that is suitable for ENDOR spectroscopy therefore requires operation with very high powers because the field is inversely proportional to the coil diameter ($H = N{\cdot}I/r$, Section 2.1). External coils are used to best advantage with small resonators, such as high frequency cavities (Yagi *et al.*, 1970; Wang & Chasteen, 1995) and loop-gap resonators (Newton & Hyde, 1991).

The most efficient means of applying the radiofrequency field to the sample is by placing the coil inside the microwave cavity resonator and compensating for Q-factor spoilage by carefully selecting the cavity mode, sample orientation, and coil type. For example, solution ENDOR requires both high microwave and radiofrequency power in order to overcome the rapid relaxation rates of free radicals, and because many liquids also tend to be lossy, cavity Q is also lowered. The cylindrical TM_{110} cavity is well suited to these experiments because a capillary tube containing the sample material can be inserted along the cavity's cylindrical axis with minimum perturbation to the electric field lines of TM_{110} mode. The TM_{110} cavity is further suited to solution EMR by the fact that the mode's Q-factor increases with the cavity's length. At room temperature, where there is little concern for temperature gradients along the sample length, an elongated cavity will accommodate more sample material while at the same time feature a higher Q.

Modification of the TM_{110} cavity for ENDOR (Biehl *et al.*, 1977; Zweygart *et al.*, 1994) can be achieved by inserting a wire helix into the cavity with the coil's cylindrical axis colinear with the cavity's. The wire turns of the coil run very nearly perpendicular to the electric field lines of the TM_{110} mode and the resultant probehead geometry is therefore suited for optimum signal-to-noise. Silver wire helices wound onto quartz tubing of 1–2 mm thickness and outer diameter between 8 and 11 mm[5] have been used in ENDOR studies of free radicals in solution when used with the

[5] Poole (1983) and Alger (1968) list cavity factors that affect S/N. Likewise, the Varian operation manual features a section by Hyde on radiofrequency field concentration by quartz tubing inserts; for example, an optimum signal intensity was achieved in a rectangular TE cavity with a quartz tube $\phi_o = 10$ mm and $\phi_i = 6$ mm

commercial TM_{110} cavity (Bruker Instruments; part number ENB-250). Prototype TM_{110} cavities with slightly longer axial lengths (l = 5 cm *vs*. 4 cm; Bender, unpublished) yielded higher Q and signal-to-noise for identical sample concentrations. Leniart (1979) describes the coil arrangement of the now obsolete Bruker ER-420 ENDOR probe, which was the forerunner of the present version.

The elongated TM_{110} cavity can be used for ENDOR spectroscopy of solid samples and frozen solutions, but one must be concerned about the sample length in flow dewars.[6] A better cavity design tactic for ENDOR studies of frozen solutions with the TM_{110} cavity entails decreasing the cavity's axial length, which increases the filling factor η in Equation (5). There is some sacrifice in cavity Q, but the sample now occupies a greater portion of the cavity's axial length and thereby optimizes η. The combined decrease in Q and increase of η is a favorable design feature for electron spin echo detected ENDOR. TM_{110} cavity prototypes with axial lengths as short as 0.5" have been successfully tested by the author with stable radical-containing powders (γ-irradiated sucrose and citric acid); details of their construction are given in Appendix 1.

The cylindrical TE_{011} and TE_{112} cavities are also excellent sample resonators for ENDOR on solid samples that require temperature control. Both cavities have an inherently high Q-factor and can be used with two or four parallel posts that are arranged inside the cavity so that they are parallel to the cavity's central axis. The posts are externally joined and configured as a loop in order to generate the radiofrequency field, although the magnitude of the field produced by the loop is less than that of the helix for the same current. This method may also be applied to the rectangular TE_{102} cavity, and one variant (Castner & Doyle, 1968) orients the posts horizontally in a rectangular TE_{201} cavity.

The following section describes several coil types and their network characteristics that are derived from S-parameter measurements made by using a Hewlett-Packard 8753A Network Analyzer and S-Parameter Test Set. Graphical and numerical data are presented for such parameters as s_{11}, SWR, reactance, and the Smith Chart. For each case study the test coils were constructed so as to be compatible with the Bruker ENB-250 cylindrical TM_{110} (helix), the Varian or Bruker rectangular TE_{102} (posts, stripline), or Varian V4533 cylindrical TE_{011} (post, stripline) cavities. Wire loops and coils were wound onto quartz tubes of specified diameter and had a length of 1 inch. Parallel posts were made compatible with the cylindrical TE_{011} cavity

[6] Liquid immersion dewars (nitrogen or helium) can be used with the TM_{110} cavity, but one must use care to prevent bubbling of the cryogen. See Alger, 1968.

by using special modified end plates.[7] All coils were soldered directly to a 50 Ω Type-N bulkhead connector that mated a pair of 1 m cables that connected the device to the S-Parameter Test set and network analyzer. The test coils were constructed of commercial magnet wire (Belden Wire and Cable, Richmond, IN; MWS Wire Industries, Westlake Village, CA) that had an enamel coat; in spectroscopic experiments the coils are fabricated from silver wire (helices) or German silver[8] (posts). The network behavior of the standard magnet wire was identical to silver wire of the same gauge.

Table 1. Network parameters of a 20-turn wire coil (24 ga.) wound onto a 10mm o.d. quartz tube (1 mm wall).

f (MHz)	R (Ω)	X (Ω)	L or C	SWR (1:)
0.15	49.8	5.7	6.0 μH	1.12
0.25	50.2	5.2	3.3 μH	1.10
0.50	50.5	6.9	2.2 μH	1.15
1.00	51.6	12.1	1.9 μH	1.27
1.50	53.3	17.8	1.9 μH	1.42
2.00	55.7	23.6	1.9 μH	1.58
2.50	59.0	29.6	1.9 μH	1.76
3.00	63.2	35.8	1.9 μH	1.94
3.50	68.9	42.4	1.9 μH	2.16
4.00	75.8	48.9	1.9 μH	2.38
4.50	85.0	55.6	2.0 μH	2.63
5.00	96.9	62.0	2.0 μH	2.90
5.50	112	67.5	1.9 μH	3.19
6.00	132	71.0	1.9 μH	3.50
6.50	156	70.0	1.7 μH	3.80
7.00	187	61.0	1.9 μH	4.16
7.50	219	38.7	0.8 μH	4.53
8.00	245	0.0	25 nF	4.19
8.50	254	−53.0	350 pF	5.30
9.00	237	−110	170 pF	5.74
9.50	201	−144	120 pF	6.18
10.00	160	−160	100 pF	6.60

[7] The walls of the V4533 consist of a winding of silver wire that can be detached from the end plates, which are aluminum and were duplicated in the shop with fittings for mounting the ENDOR coil(s).

[8] An alloy of copper, nickel and zinc. Also known as nickel silver (Goodfellow Corp., Cambridge, UK and Berwyn, PA).

3.2 The Wire Helix

Besides formulae for predicting the inductance of wire helices, several practical guidelines accompany directions for optimizing coil performance (*cf.* Simpson, 1947; Terman, 1955; Welsby, 1960; Abrie, 1985; Vizmuller, 1987). Many of these guidelines are directed towards optimizing coil Q for operation as a bandpass filter near resonance and are therefore irrelevant to ENDOR systems. One guideline, however, that does seem to be useful, for reasons unrelated to the optimization of coil Q, concerns the winding pitch; Simpson (1947) and Medhurst (1947) recommend that the winding pitch of the coil be approximately 1.5d, where d is the diameter of the wire. This latter design factor is an empirical result that yields an optimal compromise between the coil inductance and self-capacitance as the turns per inch is

Table 2. Network parameters for a 20-turn wire coil (24 ga.) wound onto an 8 mm o.d. quartz tube (1 mm wall).

f (MHz)	R (Ω)	X (Ω)	L or C	SWR (1:)
0.15	49.8	5.3	5.6 μH	1.11
0.25	50.2	4.4	2.8 μH	1.09
0.50	50.4	5.4	1.7 μH	1.11
1.00	51.2	9.1	1.4 μH	1.20
1.50	52.5	13.0	1.4 μH	1.30
2.00	54.3	17.1	1.4 μH	1.40
2.50	56.6	21.3	1.4 μH	1.51
3.00	59.4	25.5	1.4 μH	1.64
3.50	63.1	29.8	1.4 μH	1.77
4.00	67.5	34.0	1.4 μH	1.91
4.50	73.1	38.2	1.4 μH	2.06
5.00	79.9	42.2	1.3 μH	2.22
5.50	88.2	45.8	1.3 μH	2.39
6.00	98.3	48.6	1.3 μH	2.57
6.50	110	50.1	1.2 μH	2.75
7.00	125	49.5	1.1 μH	2.95
7.50	142	45.4	0.9 μH	3.16
8.00	160	36.1	0.7 μH	3.38
8.50	178	20.3	0.4 μH	3.61
9.00	192	–3.2	5.5 nF	3.84
9.50	199	–32.0	510 pF	4.09
10.00	195	–64.7	250 pF	4.35

increased. Coincidently, the 1.5*d* winding pitch is a good compromise in the TM_{110} cavity with the requirement that the microwave mode E-field lines be oriented near-perpendicular to a conductor in order to avoid coupling losses.

Tables 1–4 list the network parameters for a series of 20-turn helical coils that are wound onto a quartz tube of wall thickness 1 mm. The axial length of each coil is 1 inch, and the data listed in the tables illustrate the effect of geometry of impedance. Tables 1–3 illustrate the effect of decreasing the coil diameter; as the coil becomes compact (all other factors, such as turn spacing etc., remaining constant), the impedance declines. Table 4 is analogous to Table 1 in the sense that the coil diameter is the same in both cases, but the coil that was used to generate the data listed in Table 4 was constructed of a narrower gauge wire, and the resulting increase in space between turns (the 20 turns per inch winding maintained) likewise lowers the impedance of the coil.

Table 3. Network parameters of a 20-turn wire coil (24 ga.) wound onto a 6 mm o.d. quartz tube (1 mm wall).

f (MHz)	*R* (Ω)	*X* (Ω)	*L* or *C*	SWR (1:)
0.15	49.9	4.9	5.2 μH	1.10
0.25	50.2	3.8	2.4 μH	1.08
0.50	50.4	4.2	1.3 μH	1.09
1.00	51.0	6.6	1.1 μH	1.14
1.50	51.9	9.3	1.0 μH	1.21
2.00	53.1	12.1	0.96 μH	1.27
2.50	54.7	14.9	0.95 μH	1.35
3.00	56.6	17.6	0.94 μH	1.42
3.50	59.1	20.4	0.93 μH	1.50
4.00	61.8	23.1	0.92 μH	1.59
4.50	65.1	25.7	0.91 μH	1.68
5.00	69.1	28.1	0.89 μH	1.77
5.50	73.8	30.3	0.88 μH	1.87
6.00	79.2	32.2	0.85 μH	1.97
6.50	85.4	33.5	0.82 μH	2.07
7.00	92.7	34.1	0.78 μH	2.19
7.50	101	33.6	0.71 μH	2.30
8.00	110	31.6	0.63 μH	2.42
8.50	120	27.7	0.52 μH	2.55
9.00	130	21.4	0.38 μH	2.68
9.50	140	12.0	0.20 μH	2.81
10.00	148	−3.7	43 nF	2.95

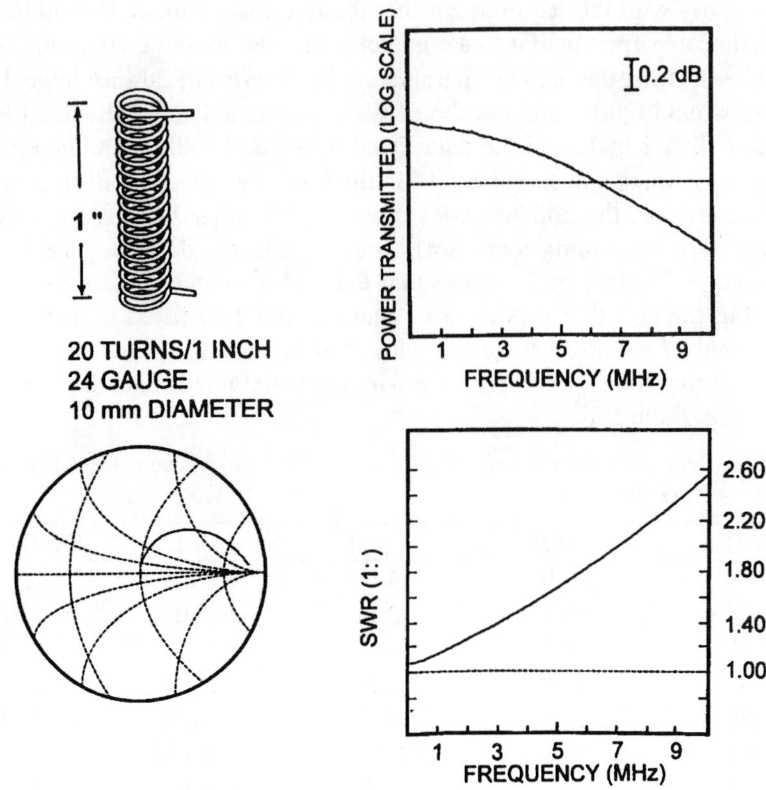

Figure 6. Representative network properties of a 20-turn copper wire helix wound onto a 10 mm diameter quartz tube (1 mm wall). Clockwise, from upper right: Power transmission (Log P; a broad sweep resembles Figure 5); SWR; and a Smith Chart. Corresponding numerical data are listed in Table 1.

Representative power transmission plots (*i.e.* s_{11}, see Appendix 2) of two wire helices are depicted in Figures 6 and 7. The extended frequency range plots (0.1–500 MHz) of these coils are similar to the plot illustrated in Figure 5, and therefore the data plotted in Figures 6 and 7 span a smaller range in order to best illustrate the relationship between network parameters and coil performance. The upper right hand panel is a standard logarithm plot of power transmitted as a function of carrier frequency, which corresponds to s_{11}. The lower right hand plot is the standing wave ratio (SWR), and the lower left hand plot is a Smith Chart representation of the same data. Any of these three formats may be used to characterize a given coil and are obtained from modern menu-driven network analyzers, such as the HP 8753A. Recalling that the coil spans two sections of 50 Ω cable that are terminated with a standard 50 Ω load resistor, the data show that the $Z = 50$ Ω optimum 'match' is met at low frequency, but becomes increasingly mismatched as the frequency increases. This deviation from match is manifest as a change in SWR (*i.e.* 1: n, where n > 1), and an outwardly spiraling trajectory on the Smith Chart (where the optimum match corresponds to the center point of the coordinate frame).

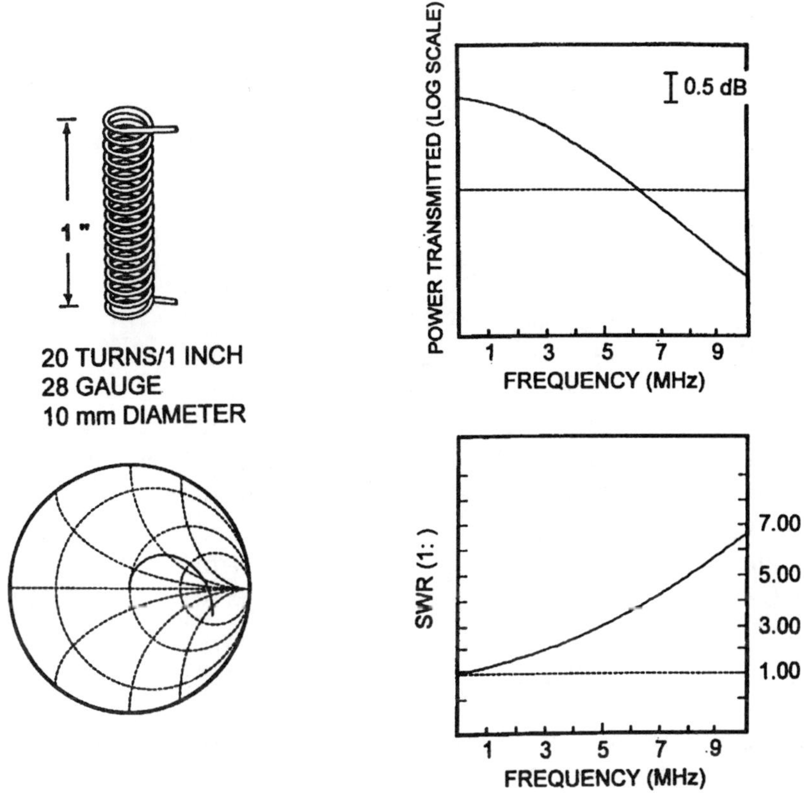

Figure 7. Representative network properties of a 20-turn copper wire helix wound onto a 10 mm diameter quartz tube (1 mm wall). The coil is identical to the one described in Figure 6, but the wire diameter is smaller and alters the network properties. Data as presented in Figure 6; Numerical data are found in Table 4.

Table 4. Network parameters of a 20-turn coil (28 ga) wound onto a 10 mm o.d. quartz tube (1 mm wall).

f (MHz)	R (Ω)	X (Ω)	L or C	SWR (1:)
0.15	49.8	4.7	5.0 μH	1.10
0.25	50.2	3.4	1.5 μH	1.07
0.50	50.3	3.7	1.1 μH	1.08
1.00	50.9	5.7	0.90 μH	1.12
1.50	51.6	7.5	0.84 μH	1.17
2.00	52.6	10.3	0.82 μH	1.22
2.50	53.5	12.6	0.80 μH	1.29
3.00	55.5	14.9	0.79 μH	1.35
3.50	57.5	17.1	0.78 μH	1.41
4.00	59.7	19.3	0.77 μH	1.48
4.50	62.4	21.4	0.76 μH	1.55
5.00	65.5	23.4	0.75 μH	1.63
5.50	69.2	25.2	0.73 μH	1.70
6.00	73.4	26.7	0.71 μH	1.78
6.50	78.1	27.8	0.68 μH	1.86
7.00	83.5	28.5	0.65 μH	1.95
7.50	89.5	28.4	0.60 μH	2.03
8.00	96.2	27.4	0.55 μH	2.13
8.50	103	25.4	0.48 μH	2.23
9.00	111	21.8	0.38 μH	2.32
9.50	118	16.4	0.38 μH	2.42
10.00	126	9.0	0.14 μH	2.53

The data of Tables 1–4 suggest that helical coils are best designed with inter-turn spacing greater than the wire diameter and a small cross-sectional diameter.

3.3 Variations of the Helical Coil

The simple wire helix is often altered in the construction of antenna for high-frequency transmission and reception (*cf.* Kraus, 1988), and some of these modified antenna designs appear as specialized NMR probes such as the 'birdcage' (*cf.* Rudin, 1992), which is a variant of the multifilar antenna. The wire helix data reported in Section 3.2 suggest that low impedance structures yield the flattest response over a broad frequency range of operation, and therefore many of the designs that are used in broadband antenna applications may be used as a starting point for prototyping novel ENDOR coils.

Zheng and Dismukes have used the parallel post arrangement of radiofrequency coil in the TM_{110} cavity by carefully placing the posts in the resonant microwave field (C. Dismukes, personal communication). A similar stratagem may be employed with half-turn (*i.e.* Kilgus, 1969, coils) helical antennas, which would otherwise seem to be prone to spoiling the cavity Q because of their near-vertical pitch. Two- and four-wire Kilgus coils were constructed for compatibility with the commercial TM_{110} ENDOR cavity and the standard ϕ_o = 10 mm quartz dewar insert. Electrical contact to the transmission line was likewise made via brass contact rings, and the results of the network transmission analysis for these two coils is illustrated in Figure 8. The near-vertical orientation of these coils did not seriously compromise cavity Q as long as a narrow wire ($\geq$ 28 ga.) was used.

The motivation behind experimentation with the Kilgus coil design and the TM_{110} cavity is a desire to perform circularly polarized ENDOR (*cf.* Schweiger & Günthard, 1981; Schweiger *et al.*, 1982) with the commercial probehead system. The circular polarization ENDOR method is used to suppress lines in the spectra that correspond to spin not precessing with the same sense of rotation as the circularly polarized rf field, which is generated by superimposing two orthogonal rf carriers at quadrature. The conventional method of performing this experiment is via four parallel posts that are operated as independent pairs; each loop carries the same frequency, but the traveling waves are 90° out of phase. By changing the sense of the circularly polarized field (*i.e.* left- or right-handedness) one can turn on or off different sets of ENDOR lines in a complicated spectrum. The quadrifilar Kilgus coil can be driven in the same manner provided that the brass end pieces (and launch connectors) are suitably modified.

Other modifications to the helical ENDOR coil follow from traveling wave engineering. Also known as slow-wave structures, these devices appear in the design of many broadband applications, for example, amplifier

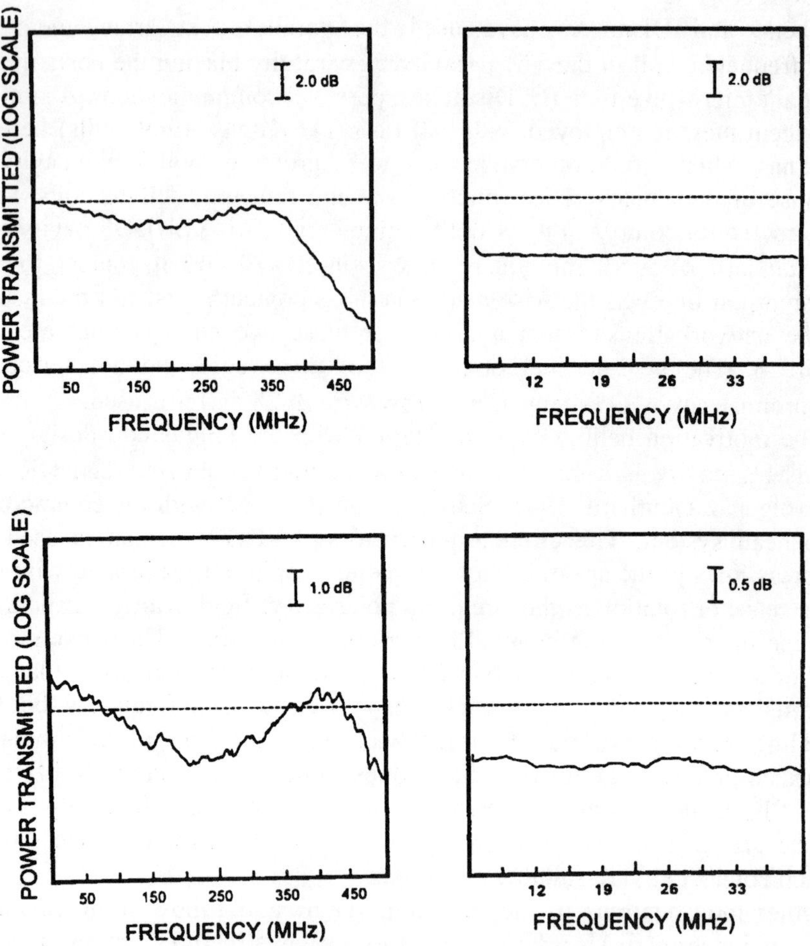

Figure 8. Power transmission plots (log P) of bifilar (top) and quadrifilar (bottom) Kilgus coils. The coils are wound onto a 10 mm o.d. quartz tube and inserted into a commercial TM_{110} ENDOR cavity.

tubes. One slow-wave structure that has characteristics that may be applicable to ENDOR is the cross-wound twin helix (*cf.* Chodorow & Chu, 1955). Two helices are co-wound but do not make electrical contact; prototypes for dynamic nuclear polarization and pulsed ENDOR have been constructed (Bender, unpublished).

The extent to which a modified helix can be adapted to a given ENDOR experiment will depend on the degree of compatibility between the coil and the microwave resonator. In both coil variants described in the preceding paragraphs, a very narrow wire was used in order to minimize perturbations to the cavity's TM_{110} mode. One also has to be careful with the helical coil's mode (*i.e.* axial *vs.* normal) of operation, which depends upon the coil dimension. In summary, therefore, the design of novel ENDOR probes is only warranted if the new design offers significantly better admittance characteristics over a desired frequency range or grants higher sensitivity in the EMR detection stage by perturbing the microwave field less. The principal design consideration for ENDOR probes is the manner by which the microwave circuit is affected.

3.4 Parallel Post Coils

The simplest example of a parallel post ENDOR coil is a wire loop that is cemented to the wall of a quartz tube or dewar and therefore requires no specialized connectors or cavity modifications. As such, the geometry of this coil does not permit one to generate as large a field for a given current as the multi-turn helix, but the performance is compensated by performing the experiment with higher powers while using the inherently lower inductance of the parallel post configuration. Parallel posts are most often used with TE mode cavity resonators, which tend to feature higher quality factors than the TM mode resonator. As stated in Section 3.1, cavity Q and Q-spoilage should be the primary concern of the experimenter, and one should start with probehead designs that optimize Q. There are, for example, reports of ENDOR cavities operating in the TE mode with Q-factors or 10000 or higher (Chacko, 1978).

Modified end plates for the Varian V4533 TE_{011} cavity were prepared by drilling holes in a breadboard fashion in order to configure various arrangements of parallel posts for ENDOR coil design optimization. This did not affect the performance of the microwave cavity design because the wall current of the cylindrical TE_{011} mode is essentially non-existent. Data are presented for three parallel post configurations: two symmetrically arranged pairs separated by 0.5" and 0.75", and a four-post arrangement.

The power transmission profile of the parallel post coil reflects the smaller inductance (than the multi-turn helix) and is flatter (Figures 9 & 10). Tables 6 and 7 list the relevant network parameters of the two parallel post prototypes and indicate that the inductance and SWR decrease as the spacing between the posts increases. The second paired data set corresponds to the four-post arrangement. Table 8 lists the network properties of a square arrangement with posts separated by approximately 0.5". A second four-post rectangular arrangement with sides $\frac{3}{8}$ " by $\frac{5}{8}$ " does not appreciably differ with respect to its network parameters.

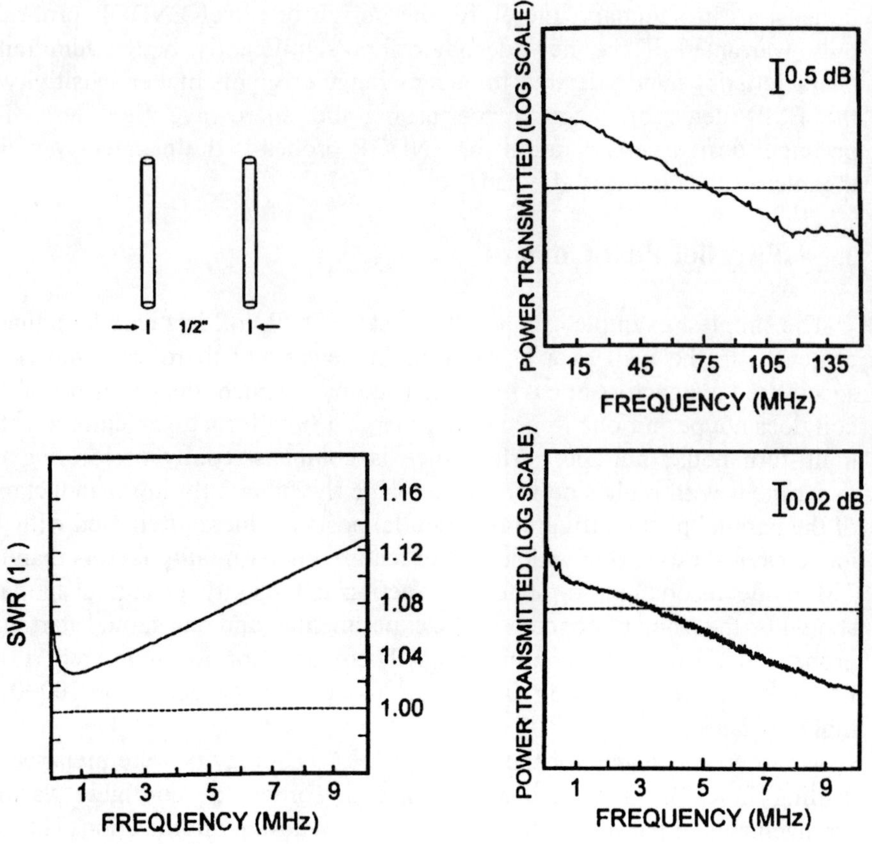

Figure 9. Representative network properties of a two-post ENDOR coil. Data are presented as in Figure 6 and correspond to numerical data in Table 5.

Table 5. Network parameters of parallel post (single loop) coil consisting of 20 ga copper wire spaced 0.5" (approximately 13mm) apart.

f (MHz)	*R* (Ω)	*X* (Ω)	*L* or *C*	SWR (1:)
0.15	49.8	4.0	4.3 μH	1.08
0.25	50.1	2.4	1.5 μH	1.05
0.50	50.2	1.5	0.49 μH	1.03
1.00	50.3	1.3	0.21 μH	1.03
1.50	50.5	1.5	0.15 μH	1.03
2.00	50.8	1.6	0.13 μH	1.04
2.50	51.0	1.8	0.11 μH	1.04
3.00	51.2	2.0	0.10 μH	1.05
3.50	51.4	2.1	0.10 μH	1.05
4.00	51.6	2.3	0.09 μH	1.06
4.50	51.9	2.4	0.09 μH	1.06
5.00	52.2	2.5	0.08 μH	1.07
5.50	52.5	2.7	0.08 μH	1.07
6.00	52.8	2.7	0.07 μH	1.08
6.50	53.1	2.8	0.07 μH	1.08
7.00	53.4	2.8	0.06 μH	1.09
7.50	53.8	2.9	0.06 μH	1.10
8.00	54.2	2.8	0.06 μH	1.10
8.50	55.3	2.6	0.04 μH	1.11
9.00	55.0	2.7	0.05 μH	1.11
9.50	55.3	2.6	0.04 μH	1.12
10.00	55.7	2.4	0.04 μH	1.12

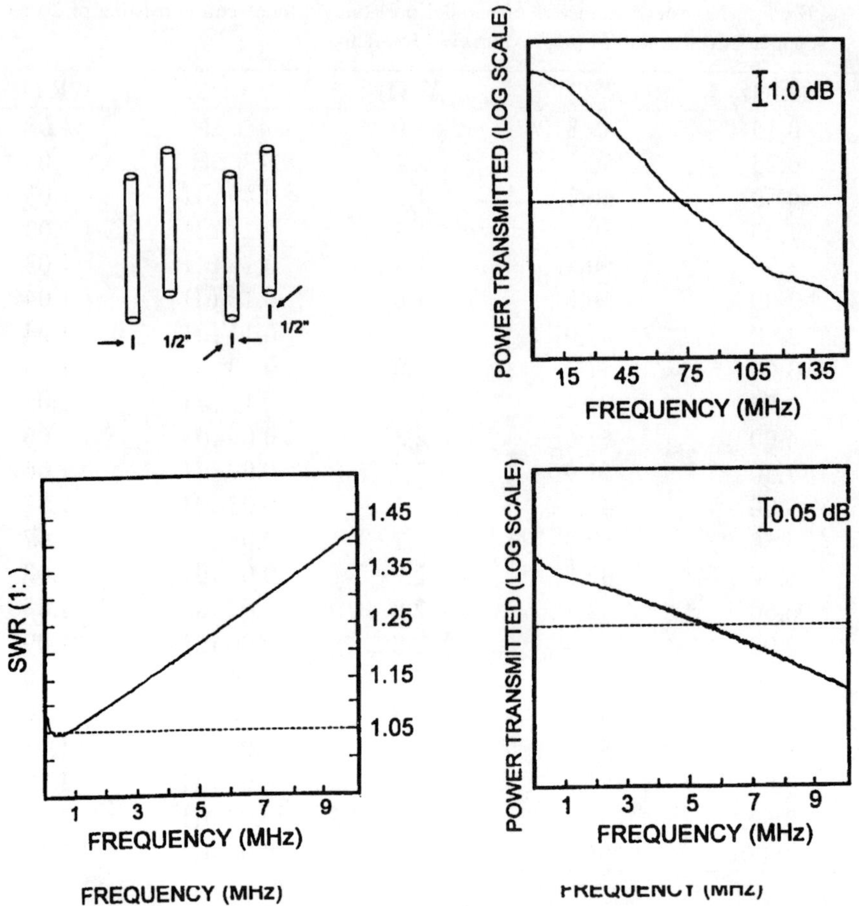

Figure 10. Representative network properties of a four post ENDOR coil arranged as a 1/2″ × 1/2″ array. Corresponding numerical data are listed in Table 6.

Table 6. Network parameters of a four parallel post coil consisting of 20 gauge wire posts occupying the corners of a rectangular layout 1/2″ × 1/2″

f (MHz)	R (Ω)	X (Ω)	L or C	SWR (1:)
0.15	49.9	4.2	4.5 μH	1.09
0.25	50.1	2.8	1.8 μH	1.06
0.50	50.3	2.2	0.70 μH	1.04
1.00	50.6	2.6	0.41 μH	1.05
1.50	51.0	3.3	0.35 μH	1.07
2.00	51.4	4.0	0.32 μH	1.09
2.50	51.9	4.7	0.30 μH	1.10
3.00	52.5	5.4	0.29 μH	1.12
3.50	53.2	6.1	0.28 μH	1.14
4.00	53.9	6.7	0.27 μH	1.16
4.50	54.7	7.3	0.26 μH	1.18
5.00	55.6	7.8	0.25 μH	1.20
5.50	56.6	8.3	0.24 μH	1.22
6.00	57.7	8.7	0.23 μH	1.24
6.50	58.8	9.0	0.22 μH	1.26
7.00	60.2	9.2	0.21 μH	1.29
7.50	61.5	9.3	0.20 μH	1.31
8.00	63.0	9.3	0.18 μH	1.33
8.50	64.4	9.1	0.17 μH	1.36
9.00	65.9	8.8	0.15 μH	1.38
9.50	67.5	8.3	0.14 μH	1.41
10.00	69.1	7.7	0.12 μH	1.44

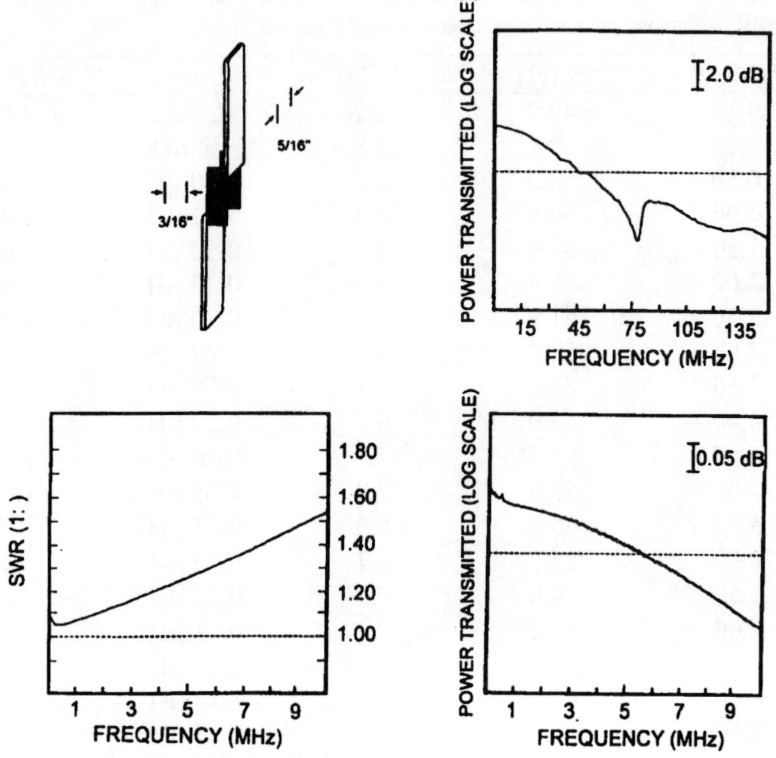

Figure 11. Representative network properties of a Colligani stripline ENDOR coil. Corresponding numerical data is listed in Table 7.

3.5 Stripline Coils

A novel low-inductance ENDOR coil for TE mode cavity resonators was introduced by Colligiani and is constructed of two stripline conductors that are bridged by a parallel pair of rectangular strips (Figures 11 & 12; *cf.* Colligiani *et al.*, 1979; Colligiani & Romanenghi, 1991; Romanenghi, 1985). Prototype stripline coils that were used for this comparative study were fabricated by milling 5 inch sections of rectangular brass tubing (Small Parts, Miami Lakes, FL) in such a manner that three walls were removed on the ends (so that a stripline conductor feed remained on either end) and two opposing walls were removed at the center (leaving the parallel rectangular strips/sample resonator). The length of the parallel strips section was 1 inch, and each of the 2 inch stripline sections was attached to a Type-N launch connector. The dimensions of the two prototype coils are listed in the corresponding figures describing their network behavior; the first prototype is compatible with standard 4 mm sample tubes, whereas the second is compatible with quartz capillary sample tubes.

Table 7. Network parameters of a stripline coil 3/16″ × 1″.

f (MHz)	R (Ω)	X (Ω)	L or C	SWR (1:)
0.15	49.8	4.3	4.5 μH	1.09
0.25	50.1	2.8	1.8 μH	1.06
0.50	50.2	2.3	0.74 μH	1.05
1.00	50.5	2.9	0.47 μH	1.06
1.50	50.9	3.8	0.41 μH	1.08
2.00	51.4	4.8	0.38 μH	1.10
2.50	52.0	5.7	0.36 μH	1.13
3.00	52.7	6.6	0.35 μH	1.15
3.50	53.5	7.5	0.34 μH	1.17
4.00	54.4	8.3	0.33 μH	1.20
4.50	55.5	9.1	0.32 μH	1.22
5.00	56.6	9.8	0.31 μH	1.25
5.50	57.9	10.4	0.30 μH	1.28
6.00	59.3	11.0	0.29 μH	1.30
6.50	60.8	11.4	0.28 μH	1.33
7.00	62.5	11.7	0.27 μH	1.36
7.50	64.2	11.9	0.25 μH	1.39
8.00	66.2	11.8	0.24 μH	1.41
8.50	68.1	11.6	0.22 μH	1.44
9.00	70.2	11.1	0.20 μH	1.47
9.50	72.4	10.4	0.17 μH	1.50
10.00	74.6	9.4	0.15 μH	1.53

The stripline coils have a very low inductance and flat response over a broad frequency range, and in this respect the stripline coil's performance is similar to that of the parallel posts that are described in Section 3.4. The stripline design is, however, more conducive to operation with high power amplifiers because of the greater surface area of the strips. The one disadvantage of the stripline *vs*. parallel posts was that the crudely assembled prototypes, which are not intended by the author to represent optimum geometries, were difficult to load into the Varian V4533 TE_{011} cavity without Q-spoilage. Colligiani *et al*. (1979) use smaller stripline segments.

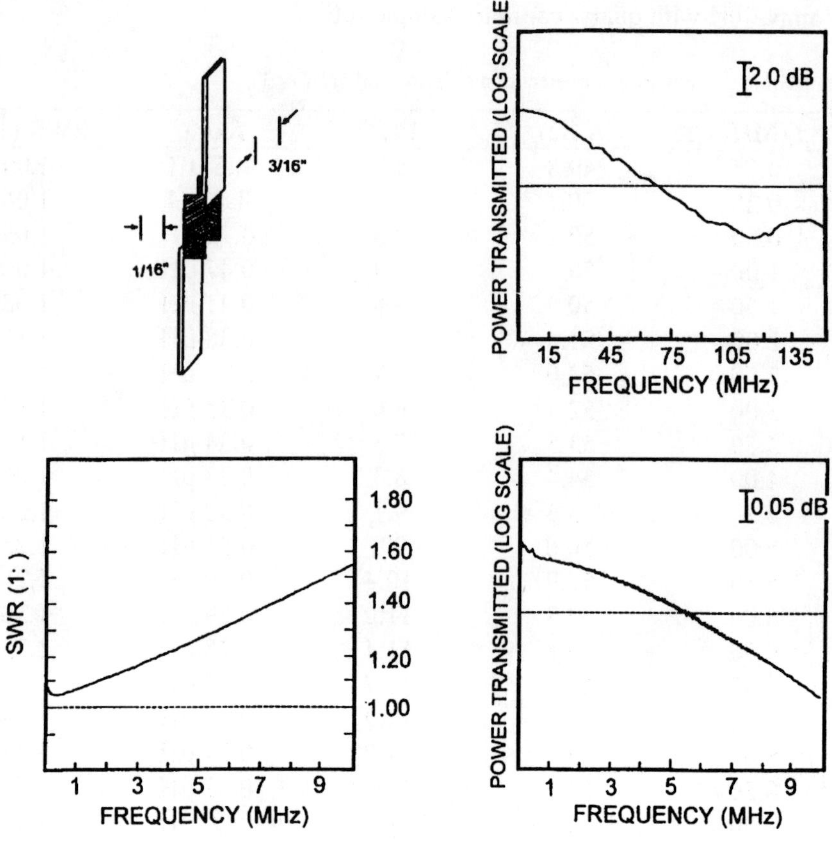

Figure 12. Representative network properties of a smaller Colligani stripline ENDOR coil. Corresponding numerical data are listed in Table 8.

Table 8. Network parameters of a stripline coil 1/16″ × 1″.

f (MHz)	*R* (Ω)	*X* (Ω)	*L* or *C*	SWR (1:)
0.15	49.8	4.3	4.5 μH	1.09
0.25	50.1	2.8	1.8 μH	1.06
0.50	50.3	2.3	0.75 μH	1.05
1.00	50.5	3.0	0.47 μH	1.06
1.50	50.9	3.9	0.41 μH	1.08
2.00	51.4	4.8	0.38 μH	1.10
2.50	52.1	5.8	0.37 μH	1.13
3.00	52.7	6.7	0.35 μH	1.15
3.50	53.6	7.6	0.34 μH	1.18
4.00	54.5	8.4	0.33 μH	1.20
4.50	55.5	9.2	0.33 μH	1.23
5.00	56.7	9.9	0.32 μH	1.25
5.50	58.0	10.5	0.31 μH	1.28
6.00	59.4	11.1	0.29 μH	1.30
6.50	60.9	11.5	0.28 μH	1.33
7.00	62.3	11.8	0.27 μH	1.36
7.50	64.4	12.0	0.25 μH	1.39
8.00	65.4	11.9	0.24 μH	1.42
8.50	68.4	11.7	0.22 μH	1.45
9.00	70.5	11.2	0.20 μH	1.48
9.50	72.7	10.5	0.18 μH	1.51
10.00	74.9	9.5	0.15 μH	1.54

4. IMPEDANCE MATCHING OF COILS

4.1 Matching Devices

The ENDOR coils that have been described in the preceding section possess characteristic impedances that deviate from the 50 Ω standard of transmission line cables and components. For example, the 20 turn wire helix coil (Section 3.2) typically has an impedance of 12 Ω at 15 MHz. For many experiments, such as ^{1}H-ENDOR in the 5–25 MHz range, the mismatch is acceptable because the resultant power loss is small (*cf.* Figure 5), and the power requirements for observing the ENDOR effect with protons are low to moderate. ENDOR studies of other nuclei for which either the nuclear gyromagnetic ratio is small or there exits a nuclear quadrupole moment require higher radiofrequency power so that the radiation-induced spin transitions are temporally competitive with spin relaxation. Under these latter conditions power losses due to the ENDOR coil mismatch may be unacceptable, and one must introduce a matching device in the form of a reactive element or transformer so that the coil has an effective impedance of 50 Ω.

The data compiled in Section 3 show that each coil undergoes a variation of its impedance as the carrier frequency is swept, and therefore no characteristic impedance exists for a given ENDOR coil. At best, one can match a coil to 50 Ω at some prescribed frequency within the sweep range, usually the center, and allow for small deviations of the impedance at frequencies above and below the point at which the coil is matched. The slope of the impedance deviation will be proportional to the coil's inductance, and one may either sweep over a frequency range that is determined by the acceptable fluctuation of power (*e.g.* ± 1.5 dB) or use a low-inductance coil so that the slope of the deviation is smaller. For example, the 12 Ω helix is matched to 50 Ω at 15 MHz, and sweeps ranging from 5–25 MHz are acceptable since the power fluctuation across the sweep range is less than 3 dB, which is the power deviation used to define the pass band of a filter. A different sweep range might be used at another center frequency (*e.g.* 45 MHz) where the slope $\Delta Z / \Delta f$ is different.

One may use broadband transformers or modified networks as a means to match an ENDOR coil. Transformers are the most convenient and intuitive means of matching because they are easily inserted into the transmission line circuit, and their design can be understood in terms of a turns ratio that interrelates the coil impedance Z and the line to be matched Z_0. In the case of the 20 turn helix, the impedance at 15 MHz is matched with a 1:4 transformer; at 5 MHz, where the impedance is approximately 25 Ω, one would use a 1:2 transformer, *etc*.

A very simple transformer that may be understood in terms of a turns ratio is illustrated in Figure 13 and may be found in many handbooks on radio engineering. Two brass tubes (~¼" dia.) are soldered onto a copper plated square of circuit board (two holes drilled to accept the tubes), yielding a horseshoe shaped conductor. Ferrite toroids are stacked on each brass tube, and a second circuit board square is soldered onto the other end of the brass tubes. Unlike the first square, however, the copper plate is etched away between the two solder points so that there is no closed loop of conductor material. Wire is then looped through the brass tubes as many times is necessary in order to provide the desired turns ratio.

An alternative form of transformer that is somewhat less intuitive with respect to the turns ratio has been described by Ruthroff (1959) and Guanella (1944). The advantage of these transformer types, as opposed to the transformer that was described in the preceding paragraph, is that they are very easily rendered compatible with standard transmission lines in a drop-in configuration. These transformers operate on the principle of summing

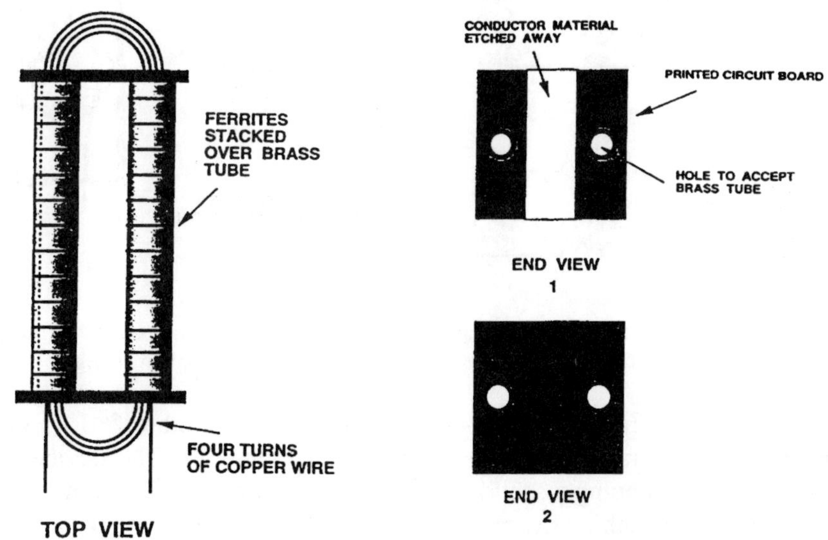

Figure 13. An impedance matching transformer that is based upon a turns ration of conductor loops. A single loop is fashioned from a pair of brass tubes through which are passed several turns of copper wire.

voltage with a delayed voltage at specific terminals, and depending on how one sums the voltages (usually via tapping or multiple stages), one can control the turns ratio (Ruthroff, 1959; Guanella, 1944; Sevick, 1990).

Radio engineers use transformers to match antennas to transmitters, and the transformers of Guanella and Ruthroff are designated according to the symmetry of the antenna. A balanced circuit element is symmetric with respect to a feed point, whereas an unbalanced element is defined as one that is not symmetric with respect to a feed point. For example, a section of coaxial line is unbalanced because the signal radiates down the center conductor only. This is in contrast to dipole antenna, which is balanced because both arms radiate. It therefore follows that the ENDOR coil is an unbalanced device and the type of transformer one wishes to use is the unbalanced-to-unbalanced, or unun, transformer. In a given circuit, the transformer can be imagined as simultaneously being a load to the source and a source to a low impedance load; the ENDOR circuit requires a pair of transformers to separately match the coil/cavity to the power amplifier and the 50 Ω termination. It is also important to recognize that these transmission line transformers operate via reactive coupling and therefore behave as a bandpass filter that should feature a flat response over the desired frequency range.

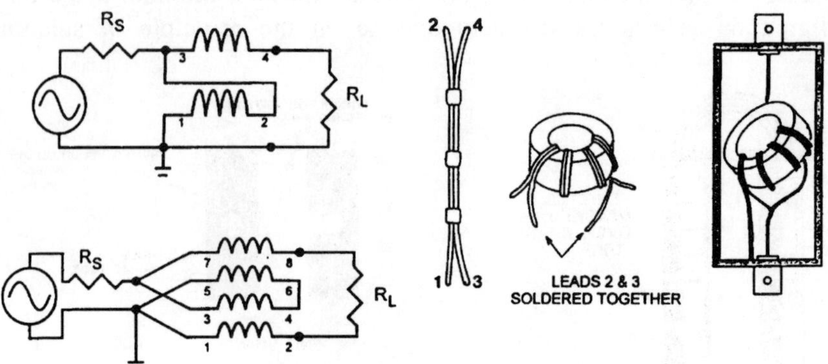

Figure 14. Transmission line transformers of the Ruthroff (top, left) and Guanella (bottom, left) type. Construction details for the Ruthroff transformer are illustrated on the right.

A schematic diagram of both Guanella and Ruthroff 1:4 unun transmission line transformers is illustrated in Figure 14. The devices are constructed by wrapping a section of a conductor pair around a ferrite rod or toroid, and then attaching the exposed leads to the remainder of the circuit as indicated. The wire pair can be fabricated by taping together two segments of enameled magnet wire (side-by-side); alternatively, one may obtain a specialized two-conductor wire (MWS Wire Industries, Westlake Village, CA) or a narrow coaxial cable as the conductor pair. The wire pair may then be tightly wound, twisted, or simply paired. The principal design considerations when constructing a transformer are the matching ratio, the power handling capability, and the characteristic impedance (of the transformer itself). Ruthroff (1959) demonstrated that the optimal performance of a transformer is obtained when its characteristic impedance is the geometric mean of the operating input and output impedances, plus or minus 10–20% (*e.g.* 25 Ω for the 50:12 matching transformer). The characteristic impedance is adjusted by trial and error and based upon experimental trends that follow from the physical dimensions of the wire, the ferrite core, etc. (Demaw, 1980; Sevick, 1990).

The power handling capability of the transformer depends upon the size (diameter and cross-sectional area) of the ferrite and its permeability, both of which determine ability of the coil to dissipate heat. Sevick (1990) reports data that are compiled from many trial and error studies, and, in general, the trends show that lower permeability (less than 300) ferrites better handle high power. Low permeability coils, however, require more turns. According to Sevick (1990), a 1 inch diameter ferrite of permeability less than 300 is adequate for handling power up to 200 W; a 1 kW transformer requires a toroid with a minimum diameter of 1.5 inches. The power handling capability of a transformer is tested by the so-called 'soak test', which entails the application of 1 kW radiofrequency power for a period of time and determining whether the transformer coil becomes overheated.

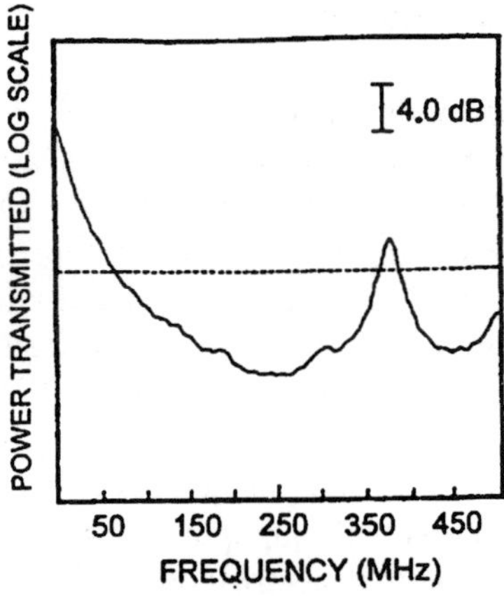

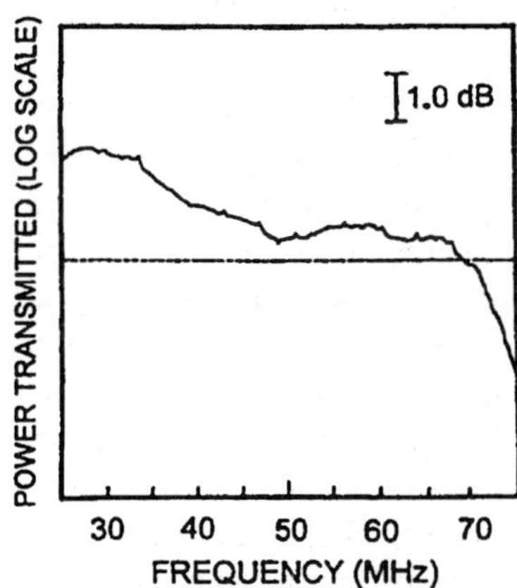

Figure 15. Power transmission plots (log P) for a 20-turn copper helix without (left) and with (right) a 1:4 Ruthroff impedance matching transformer. Since the transformers themselves have only a limited operating bandwidth, the optimum range only is illustrated. Note that the power transmission plot is much flatter in the 25-70 MHz range.

The optimum frequency range of operation can be controlled by altering the characteristic impedance of the transformer with, for example, low-impedance windings. DeMaw (1980) describes transmission line transformers that are optimized for low frequency operation and wound with 25 Ω coax instead of 14–18 gauge wire; stripline also works well (Sevick, 1990). Improved high frequency performance is obtained with a trifilar version of the Ruthroff transformer (Sevick, 1990). Twisting the wire pair apparently has little effect on the performance of large toroidal ($\phi_o \geq 1$ in.) transformers. The rationale behind twisting the wire is that one obtains improved coupling between the lines, but the benefits of this technique seem to be significant only for very thin wire (Sevick, 1990).

Practical impedance matching transformers that have been used for cw-ENDOR were constructed on the Ruthroff design by using 4–5 turns of a twisted pair of 14 gauge wire. Toroids of 0.7" diameter and low permeability (Micrometals, Anaheim CA) work well below 70 MHz and at powers as high as 200 W. Figure 15 illustrates the modified transmission profile of a 20 turn helical ENDOR coil after being matched by a 1:4 Ruthroff transformer. The noteworthy aspect of this network analysis is the flattening of the response, bearing in mind that the ideal 'match' (*i.e.* 1:4, *etc.*) applies at only one frequency in the sweep range. In other words, if one had a purely resistive load of 12.5 Ω, then one would observe a fairly flat response over a bandwidth that is determined by the transformer. One the other hand, if the load is reactive, then the transformer is not going to compensate for changes in Z, and the power will still fluctuate as the frequency is swept.

4.2 Impedance Match & Network Analysis

The ENDOR coil types that were demonstrated in Section 3 were analyzed by using a commercial network analyzer that features microprocessor controlled menus for each type of measurement. These instruments, however, cost tens of thousands of dollars and are not a justified expense for many spectroscopy laboratories. Unless one is affiliated with a school of electrical engineering, the best option is short-term rental from an instrument company during 'production runs', but trial and error ENDOR coil design will often suffice because the effect on cavity Q is the most important factor in determining signal-to-noise of the planned spectroscopic experiment; the impedance match and power throughput at the given frequency of operation are not important in routine spectroscopy.

The impedance can be matched at a fixed frequency by using the properties of a 180° hybrid (the magic tee). The method is described by Fukushima & Roeder (1981), and the procedural details as applied to matching an ENDOR probehead are outlined by Bender & Aisen (1993). In summary, one balances the impedance of two hybrid ports by canceling two reflected waveforms when the corresponding ports achieve match (one port is set at the desired 50 Ω, and the procedure can be performed with an oscilloscope of adequate bandwidth (approximately 4 times the carrier frequency).

SWR measurements may likewise be made by using inexpensive equipment. Commercial bridges such as the Wiltron 60N50 or Eagle RLB150B1 provide balanced test ports for power measurement using a network analyzer or vector voltmeter. The SWR test bridge is simply a bidirectional coupler, and one may measure the waveform amplitude of the forward and reflected signal by using an oscilloscope (matching by inspection of waveform amplitude). The reflection coefficient Γ is defined as V_r/V_f, and the SWR is defined as $(1 + \Gamma) / (1 - \Gamma)$. The rms voltage of an AC waveform is readily measured as 0.707 $V_{p\text{-}p}$. Alternatively, a planar doped barrier diode detector (Hewlett-Packard 8471D, response range 0.1 MHz–2 GHz) will serve as a useful video detector of the radiofrequency circuit performance, and the rectified signal voltages can be compared directly. Finally, an inexpensive SWR meter that operates over a 1–37 MHz frequency range is available from Autek Research (Madiera Beach, FL).

4.3 Radiofrequency Field Measurement

It is sometimes useful to have a measure of the radiofrequency field intensity for an ENDOR coil, for example, when the topic of interest is coherence transfer and requires an accurate measure of linewidth (*cf.* Leniart, 1971). A Bruker Application Note (Biehl, Technical Note GT32-0004) describes a method of measuring the rf field intensity by observing the sideband amplitude modulation splitting of the solution ENDOR line of the *tri-tert*-butyl phenoxyl radical (TTB). Biehl's method may also be applied to solid samples that feature a narrow linewidth, such as a DDPH single crystal, as described by Miyagawa *et al.* (1977). The primary drawback of the Miyagawa method is that it requires a fairly high magnetic field modulation frequency (16 MHz, see Figure 1 of the paper cited), which is problematic for the commercial TM_{110} cavity because of its field modulation coil design.

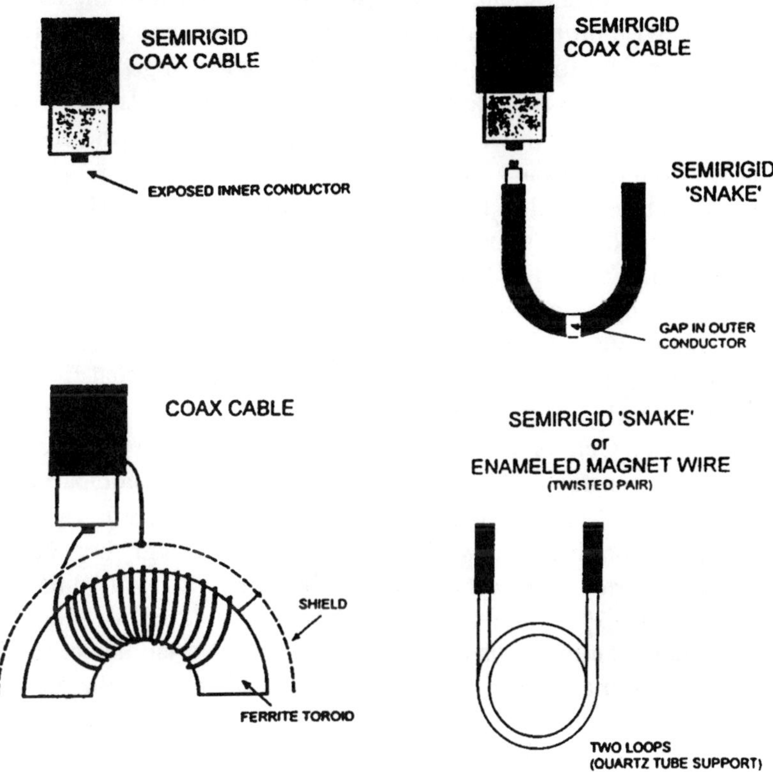

Figure 16. Probes that are suitable for measuring the radiofrequency field intensity within an ENDOR coil. Details of the probe type and references are located in Section 4.3.

Direct field measurements can be made by using a vector voltmeter or oscilloscope that is outfitted with the appropriate probe. Electric field probes may be fabricated from coaxial cable whose inner conductor has been exposed. The magnetic field probe is likewise constructed from coaxial cable, but is shaped into a loop for inductive pick-up. In both cases (Figure 16), very small semi-rigid coaxial cable (MIDISCO 'snake', $\phi_o = 0.034"$, part number MDC5034) is used. A second type of magnetic field probe, which has been suggested as an improvement over the simple loop (Prutchi, 1996), picks up the H-field component via a ferrite bead and offers a high sensitivity and resolution (it is actually intended as a probe for stray fields in communication instruments).

Leniart (1971) and Connor (1972) describe a means of measuring B_{Rot} by using an N-turn wire loop. The probe is constructed as a two-turn wire loop with a 2 foot long twisted wire pair lead (that should be shielded from stray rf), and the loop is situated in the region of maximum field intensity, as measured from the waveform recorded on an oscilloscope (or the reading on a vector voltmeter). The detected oscillatory voltage V is equal to the temporal derivative of the field or flux lines that pass through the loop, and the field is likewise proportional to the number of turns and the area of the loop according to the formula: $\Phi = N \cdot A \cdot B \cdot \sin \omega t$ (using the notation of Leniart, 1971; p64). The time derivative of this formula yields the voltage $V = N \cdot A \cdot B \cdot \omega \cos \omega t$, of which $|V|$ is the measured quantity, whereas N and A are known from the probe's construction.

The probe is terminated at the oscilloscope with a 50 Ω resistor, and the oscilloscope waveform is a measure of the voltage drop across the resistor that is part of a *RL* circuit. The measured voltage, V_{meas}, is scaled to the actual voltage V by an amount that is determined by the impedance of the *RL* circuit, that is, $V_{meas} = 50 \cdot |V| / (50^2 + \chi_L^2)^{1/2}$. The measured voltage $|V_{meas}|$ is thereby used to determine the corrected field voltage V that, in turn, is used to compute B. Determining the loop inductance χ_L constitutes the probe calibration. Connor (1972; p24) calibrates the correction factor (a proportionality constant relating $|V_{meas}|$ to $|V|$) by comparing the measurements of a 50 Ω terminated probe with an unterminated probe (analogous to calibrating a network analyzer with open and short circuits). Finally, B is converted to the rotating frame field strength by dividing by 4.

Neither Leniart (1971) nor Connor (1972) supply details of the probe's construction; apparently it was or was based upon a Varian device. Connor (1972) cites a loop diameter of 5.3 mm, and it is probable that the probe was wrapped around a standard EMR sample tube. A simple homebuilt probe may therefore be fashioned by winding and cementing two turns of 34 gauge enameled wire around the sample region of a sample tube.[9] The leads are

[9] Sample lengths of ~1 cm are used by the author in flowing gas dewars, and therefore the rf probe is wound 0.5–0.7 cm from the tube end. The probe tube is filled with glycerol to a depth of 1 cm. Solution ENDOR rf field probes are positioned in the manner described by Leniart (1971), i.e. at the signal maximum.

then twisted and shielded with copper braid as necessary; a bulkhead connector provides the oscilloscope interface.

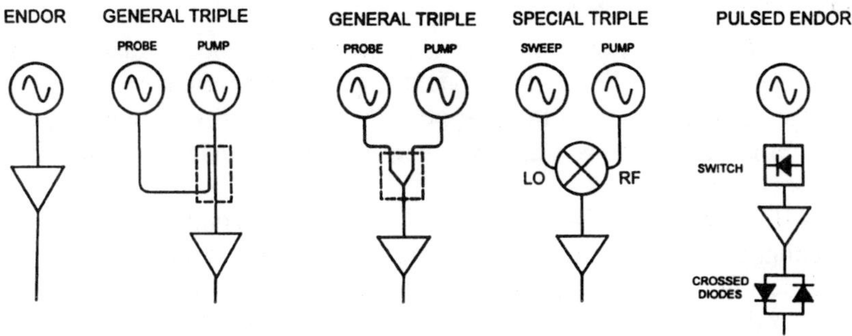

Figure 17. Instrumental configurations for advanced ENDOR methods. Each circuit depicted would terminate with the coil and load (cf. Figure 2).

5. RADIOFREQUENCY SIGNALING CONDITIONING

5.1 Triple Resonance: Waveform Combination

At the risk of oversimplifying the predicted outcome of an ENDOR experiment, ENDOR spectra appear as paired transitions that are symmetrically distributed about some point. For example, the $S = 1/2$, $I = 1/2$ ENDOR effect yields spectral lines that are proscribed by the formulae $\nu_\pm = \nu_n \pm \frac{1}{2} a$ or $\nu_\pm = \frac{1}{2} a \pm \nu_n$, where ν_n is the Larmor frequency of the nucleus in question and a is the nuclear hyperfine coupling constant. Whichever of these two energy terms of the spin Hamiltonian is larger determines whether that term occupies the center point about which the two transitions are observed (*cf.* Kevan & Kispert, 1976). An ENDOR spectrum represents the successive rf-induced population changes of sublevels separated by $h\nu_-$ and $h\nu_+$, and it follows that secondary manipulation of the nuclear sublevel populations by applying a saturating radiofrequency field can be used to obtain additional information about the hyperfine coupling constants (*cf.* Möbius *et al.*, 1982; Kurreck *et al.*, 1988; Spaeth *et al.*, 1992).

The double ENDOR, or Triple, experiment entails the simultaneous saturation of an electron and nuclear magnetic resonance transition. In its 'General' form, Triple is a swept-frequency ENDOR experiment that is conducted with one of the participating NMR transitions saturated. The saturated or 'pumped' NMR transition is therefore transparent to the swept probe frequency and is (ideally) observed as a missing peak from the ordinary ENDOR spectrum. The saturation of the nuclear sublevels corresponding to one NMR transition, however, influences the spin populations of other nuclear hyperfine levels, which increases or decreases the ENDOR enhancement factor of these other nuclei according to certain rules Möbius *et al.*, 1982):

- **a.** The ENDOR transition that is pumped becomes saturated, and the intensity of that line in the ENDOR spectrum is collapsed.
- **b.** The counterpart(s) to the pumped ENDOR transition, that is, ν_+ if ν_- is pumped and *vice versa*, have their intensity enhanced in the ENDOR spectrum.
- **c.** All other ENDOR transitions mimic the pattern of the Triple enhancement as follows: if the sign of the spin Hamiltonian terms that describe the nuclear hyperfine interaction are the same sign as the spin Hamiltonian terms of the nucleus in the pumped transition, then the correlated ENDOR transition pairs intensify and collapse in the same manner; if the signs differ, then the patterns of intensity are mirror images.

For example, suppose there are three different protons in a paramagnetic species, having Fermi Contact interaction a_1, a_2, and a_3. The corresponding ENDOR transitions would be $\nu_{i,\pm} = \nu_H \pm \frac{1}{2} a_i$, according to the spin Hamiltonian description for interacting electron-proton, giving six lines in the ENDOR spectrum. And suppose the signs of the Fermi Contact terms are a_1, $a_2 < 0$ and $a_3 > 0$. If one now saturates the ENDOR line corresponding to $\nu_{1,+}$, then the following intensity pattern is observed:

Proton	Intensity	
	$\nu+$	$\nu-$
1	Decreases	Increases
2	Decreases	Increases
3	Increases	Decreases

The intensity pattern of the Triple Resonance spectrum allows one to determine the sign of the hyperfine interaction parameters. Several examples appear in the review of Möbius *et al.* (1982).

The instrumental aim of General Triple is the combination of two rf frequencies of different power. In general, one uses two rf signal generators, such as the Programmed Test Source synthesized sources, to supply the pump and probe rf frequencies. These two signals are then combined prior to the amplifier. The protocol in this case entails replicate records of the cw-ENDOR spectrum with increasing power (gain) set at the amplifier until one saturates the desired ENDOR line. The frequency of one rf generator is then fixed at this saturating frequency while the second generator is now turned on and used as the swept probe. The power of the swept probe frequency is set at the source, which has a output level control that peaks at ~10 dBm, prior to the amplifier.

The two low power signals may be combined by using either a power divider/combiner or a directional coupler. The former is simply a branched transmission line device in which a single feed point distributes power evenly among two or more branches; it may be used in the forward or reverse direction as a divider or combiner, respectively. A directional coupler may alternatively be used as a means of combining two different radiofrequency signals. The advantage of using the directional coupler instead of the simple power combiner is that the two transmission lines supplying the rf power remain isolated from each other.

A second variant of the Triple experiment is based on the simultaneous irradiation of the ν_+ and ν_- ENDOR transitions. If one refers to the simple formulae that define the transition frequencies of an ENDOR spectrum (in the four-level diagram, Figure 1), it follows that the so-called Special Triple experiment involves the simultaneous sweep of two frequencies, f_+ and f_-. At the start of the experiment, both f_+ and f_- are equal to a symmetry point in the ordinary ENDOR spectrum that is defined by $\frac{1}{2}(\nu_+ + \nu_-)$, which equals the nuclear Larmor frequency ν_n or $\frac{1}{2}a$, depending on the relative magnitude of these two spin Hamiltonian terms (*cf.* Kevan & Kispert, 1976). From this common starting point f_+ and f_- are swept up and down the frequency spectrum while maintaining their symmetric distribution about the starting frequency; in this manner they will simultaneously pass through paired ENDOR transitions ν_+ and ν_-.

The Special Triple experiment is used to enhance the intensity of ENDOR spectra by virtue of modified spin dynamics and relaxation pathway (Möbius *et al.*, 1982). When the nuclear spin-lattice relaxation rate becomes rate determining, one gets a bottleneck in the relaxation pathway that limits the ENDOR enhancement or signal intensity. In a four-level hyperfine state diagram one might envision a state-to-state sequence 4→1→2→3→4, where transitions 4→1 and 2→3 are electron spin flips, and the 1→2 and 3→4 transitions are nuclear spin flips. If the nuclear spin relaxations are slow, then the spin population at state (3) will build up when 4→1→2 are driven in a combined microwave and radiofrequency field, so the intensity of ENDOR transition 1→2 will be low because of the bottleneck at state (3). This bottleneck is removed if both ENDOR transitions 1→2 and 3→4 are simultaneously driven.

The mixer is the device of choice for this simultaneous generation of two swept carrier frequencies that are symmetric about a common point. A mixer combines two input frequencies, designated as f_{RF} and f_{LO}, to yield $f_{RF} \pm f_{LO}$.[10] If one frequency corresponds to the symmetry point, that is $f_{RF} = \frac{1}{2}(\nu_- + \nu_+)$,

[10] The subscripts RF and LO designate specific ports on the mixer; the combined frequencies appear at the IF port.

it follows that sweeping f_{LO} from zero upwards will yield an output frequency that sweeps outward from the specified symmetry point.

Specifying a mixer for Special Triple requires that one examine specifications such as conversion loss (the comparative power level between the IF and RF ports; a measure of the power conversion efficiency), the RF and LO frequency range of operation, and the IF frequency range of operation. Double balanced mixers provide the best isolation of the three ports and suppression of the undesired fundamental frequencies f_{RF} and f_{LO}, plus their harmonics. The lower operating limit of mixer LO ports and many rf generators is 100–200 kHz, and therefore a secondary factor in choosing a mixer is the point at which one wants to begin the sweep, that is, how close to zero one intends to begin the sweep at f_{LO}. Much of the Special Triple literature (Möbius *et al.*, 1982) deals with proton transitions that need to be resolved and identified, and with the exception of matrix lines (defined in detail by Kevan & Kispert, 1976), few proton lines requiring Triple analysis appear within 100 kHz of the proton Larmor frequency. Similarly, when the $\frac{1}{2}a$ spin Hamiltonian term defines the center point of the ENDOR spectrum, the magnitude of the proton Larmor frequency ensures that the ENDOR transitions are sufficiently offset from the center point.

Special Triple is problematic when applied to matrix lines from weakly coupled nuclei or nuclei with very small Larmor frequencies (*e.g.* metals). One problem is that frequency modulation will negate any chance of correlation between the swept carrier and the transition frequency; small signal modulation must be carefully assessed for the effect it might have on the spectral resolution. There are also factors associated with the normal operating band of radiofrequency devices. The metal transitions dispersed as $\frac{1}{2}a \pm \nu_n$ can be separated as necessary by moving to a higher static magnetic field, but nuclei having large Larmor frequencies can only be examined, in principle, by mixing radio and audio frequency signals, and one has to carefully examine the response of the mixer (the problem becomes analogous to frequency modulation and useful circuits can be found in radio engineering texts).

5.2 Triple Resonance of Powder Samples

Continuous wave ENDOR experiments that rely on the modulation of the radiofrequency signal tend to not require any further circuit elements than those described in the preceding sections, that is, a coil network with matching transformers (when necessary) and, for Triple, a power combiner, directional coupler, or mixer. One may find, however, that a Triple resonance experiment does not perform as expected when performed on broadened ENDOR spectra, such as those obtained with powder sample materials. A Triple resonance spectrum that is conducted on a powder

ENDOR spectrum will often result in a hole being burnt in the line selected for pumping, and the enhancement that would be expected in the counterpart line appears as a spike in an otherwise unchanged line. Figure 18 illustrates ENDOR and general Triple spectra recorded from a frozen solution sample containing the benzoquinone anion radical; the lines shown in this 10 MHz frequency sweep correspond to the solvent protons of the ethanolic glass that are coupled to the semiquinone via hydrogen bonding interaction (O'Malley & Babcock, 1986).

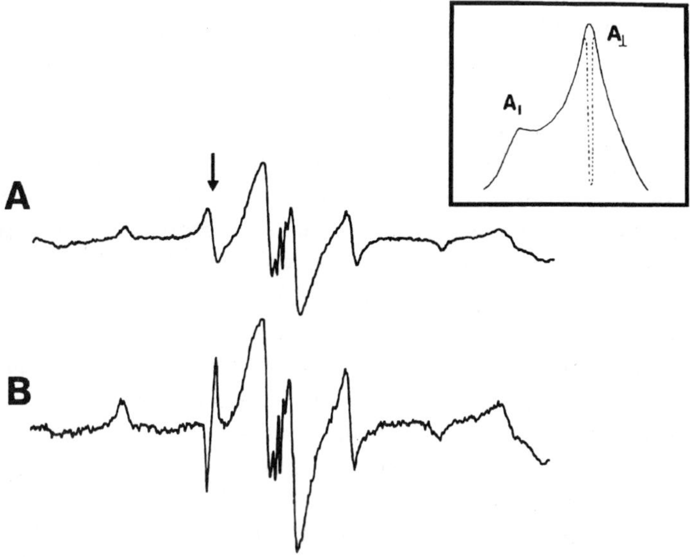

Figure 18. ENDOR and Triple resonance of *p*-benzosemiquinone and a lineshape anomaly attributable to inadequate bandwidth of the rf pump. A conventional Triple pump-probe experiment will merely burn a hole in the powder ENDOR spectrum, giving rise to a derivative 'spike' when using frequency modulation of the rf. One must substantially increase the bandwidth of the pump in order to observe the type of phenomena observed in solution or single crystal spectra.

This behavior, which at first seems to be an artifact, arises from the fact that the powder ENDOR spectrum, like the EMR spectrum from which it comes, is inhomogeneously broadened. When the excitation frequency spectrum of the pump is small in comparison to the width of the ENDOR spectrum line that is to be pumped, a hole is burned in the line rather than total saturation of the pumped transition. This is in contrast to the solution or single crystal ENDOR spectrum for which it is relatively easy to observe textbook examples of the Triple phenomenon because the ENDOR lines that appear in these spectra are narrow. In the illustrated spectra, the dipolar coupled hydrogen bonds are manifest as axial lines with resolved turning points $A_\perp$ and $A_\parallel$; the original intent of this experiment had been to collapse, for example, $A_\perp$, and observe the effect on both $A_\parallel$ and some other proton class for which the sign of the hyperfine coupling is known (*e.g.* α−protons) so that the absolute signs of $A_\perp$ and $A_\parallel$ could be determined.

The lineshape distortion that occurs in the frequency-modulated Triple resonance spectrum illustrated in Figure 18 may be corrected by using dispersion detection and Zeeman field modulation (Allendoerfer, 1973). One also needs to ensure that the pump frequency bandwidth is larger than the width of the ENDOR line that is to be saturated, and there are a variety of methods that may be used to broaden the excitation spectrum of a given signal. For example, one can alter the spectrum of the pump signal via the characteristic sidebands that are introduced by modulation, but this approach has the drawback that the sideband intensity comes at the expense of the center carrier (*cf.* Strauss, 1960). One may also broaden the spectral profile of a signal by altering the shape of the waveform, and this approach is effective at broadening the pump frequency spectrum in a Triple experiment without affecting the detection scheme. A waveform is 'pure' when it is represented by a sine wave of a definite period τ_0; the frequency spectrum of this ideal signal would consist of a vanishingly narrow line of frequency $f_0 = \tau_0^{-1}$. In reality, there is a finite dispersion of frequencies about f_0 because of phase noise and the properties of resonators that are used to generate signals. Fourier's Theorem transforms a signal from the time to frequency domain and *vice versa*, and, accordingly, one may describe a waveform of any shape by a sum of sine and cosine waveforms. The pure sine wave of the idealized signal would, as expected, be defined by a single term. It therefore follows from Fourier's Theorem that one may broaden the excitation spectrum of a signal by distorting its shape; rather than use the pure sinusoidal waveform as a source for the pump frequency in a Triple resonance experiment, one may use a square waveform (50 % duty cycle) at the desired frequency. The square wave is 'synthesized' by many sinusoidal waves, and these many Fourier components of the square wave signal, in effect, broaden the spectrum (see Strauss, 1960; Reich, 1961).

Finally, one may use a white noise source as a third means of broadening the spectrum of the pump frequency signal. The pure carrier frequency of a waveform generator may be mixed with a source of white noise in the hundreds of kilohertz range. The mixer will combine the pure carrier (f_{RF} with the varying f_{LO} to yield a time-variant $f_{IF} = f_{RF} \pm f_{LO}$.

5.3 Pulsed ENDOR

Pulsed ENDOR appears in the literature as a solution to one of several problems: Hyde (1965) recommends pulse modulation as a means of driving a spin system at high powers without heating problems; pulse–modulated cw–ENDOR is used as a means to assess cross-relaxation phenomena in a technique that closely resembles dynamic nuclear polarization (*cf.* Hoogstraate *et al.* 1975a,b; Poot *et al.*, 1979, 1980a,b, 1981; Shcheikin *et al.*, 1974); but in the current jargon the term pulsed ENDOR connotes a detection method that records the ENDOR spectrum as the modulated amplitude of the electron spin echo (Mims, 1965; Davies, 1974). Regardless of whether the experiment is cw– or pulsed, the matching criteria of the rf coil are identical. Rather, the coil design factor that most affects the spectrometer performance in a pulsed ENDOR experiment occurs in the ESE–detected mode of ENDOR: when the transient rf field overlaps with the transient microwave field (of the two- or three-pulse sequence that generates the electron spin echo) there appears to occur a scrambling of the states and/or their phase coherence, with a resultant collapse of the echo amplitude. The first indication of there being interference between the rf and microwave field is a collapse of the electron spin echo when the amplifier is turned on, but without any rf signal being passed to the amplifier input. This interference can be eliminated by using crossed-diodes (see Figure 17, and circuit description below). A second indication of interference may be identified during the adjustment of the rf pulse width when the carrier frequency does not correspond to any NMR transition; if the pulsed rf field persists within the probehead after the amplifier is gated 'off' and completes the pulse (for example, due to ringing), then that field may overlap with the next microwave pulse. This latter interference is remedied by using a shorter rf pulse to compensate the ringdown in the probehead or by modifying the rf network properties of the probehead. This interference has been especially problematic with the Mims transmission probehead, which is very compact.

Crossed diodes are commonly used in the transmitter and receiver circuits of NMR spectrometers, in which a similar problem is encountered because the receiver must be able to detect small signals uncorrupted by the radiofrequency amplifier noise leakage (Fukushima & Roeder, 1981). Power amplifiers are inherently noisy devices, and it is common to encounter both electron (shot) noise and harmonic generation (Bava & Bava, 1969; TechPress, 1969). The directionality imposed by a diode on the flow of current is therefore a very simple remediation for the taming of background noise; the desired amplified radiofrequency signal (*i.e.* the pulse) exceeds the noise background and pass the diode block because the amplitude of the signal exceeds the voltage threshold level.

In a sense, silicon diodes act as switches because they conduct only when the bias voltage exceeds a threshold value. For example, a 1N914 signal diode features a V_F of 0.7 V, and therefore noise of about 0.7 V or less will be blocked. If the amplifier noise is greater than 0.7 V, additional noise must be added in series until the sum of their V_F's exceeds the anticipated noise level. The noise level will scale with the gain setting on the amplifier, however, and therefore one should anticipate this need for scaled noise attenuation by building modular diode arrays. An alternative to serial diode arrays, however, is a backward-oriented Zener diode that is used in combination with a single silicon diode (Fukushima & Roeder, 1981).

6. CONCLUSION

6.1 Summary of Design Principles

An ENDOR coil is any conductor that can be used to generate an AC magnetic field. Uniformity of the magnetic field over the sample volume is required, however, and therefore the coil should have a geometry that is symmetric about the sample holder. The three principal ENDOR coil types are the wire helix, parallel post, and parallel strip conductors. The magnitude of the AC magnetic field is proportional to the AC current that passes through the conductor, and one therefore wishes to reduce the reactance of the ENDOR coil, which means that low-inductance devices are preferred.

The change in the microwave power at the detector is proportional to the microwave resonator *Q*–factor, and therefore the most important consideration for ENDOR coil design is whether the geometry of the coil is compatible with the microwave resonator. The *Q*–factor is a measure of power losses from an otherwise confined standing wave (resonant mode). Any dielectric material will alter *Q* because the E-component of the microwave field will accelerate charged particles (electrons), and this interaction with the charges dissipates power from the field. The electron motion in a thin wire is parallel to the wire axis, and if the E-component of the AC field is perpendicular to this direction of motion there will be minimal coupling between the field and electron motion, hence minimal dissipation of the field energy and spoiling of resonator *Q*. It follows therefore that the conductive material of an ENDOR coil should be positioned away from the E-field lines of the microwave resonator's operating mode, but at the very least small diameter linear conductors may pass through the E-field perpendicular to the field direction. For example, parallel posts run off center and parallel to the main axis of a TE mode resonator, and a wire helix is wound around the center axis of the TM mode.

For the simple identification of the ENDOR transition frequencies in order to record the hyperfine spectrum, one need not be concerned with the power or variations of power during the frequency sweep, so long as adequate rf power can be supplied to the sample and that the ENDOR transitions may be observed. Matching a given ENDOR coil to 50 Ω is necessary only if the coil's impedance is so badly mismatched that the ENDOR transitions cannot be driven; for many of the devices studied in this review, matching became desirable above 30 MHz. Bear in mind that transformers, like filters, tend to have an operating bandwidth and that the impedance of the ENDOR coil network will still fluctuate as the frequency is swept. The impedance fluctuation is commensurate with the magnitude of the inductance.

Coil performance is critical only when one intends to use ENDOR as a means to provide more information than spectral line position. Any measurement that depends upon the linewidth or power dependence of lineshape (*e.g.* coherence transfer effects, exchange rates, *etc.*, see Freed, 1979; Kirste *et al.*, 1982; Mehring, *et al.*, 1986) absolutely requires that the performance of the coil over the range of frequencies swept be known. Matching of the coil becomes more critical and, unless the broadband transformers offer a very flat response, servo-tuned PI-networks (Leniart, 1971; Forrer *et al.*, 1977) are preferable.

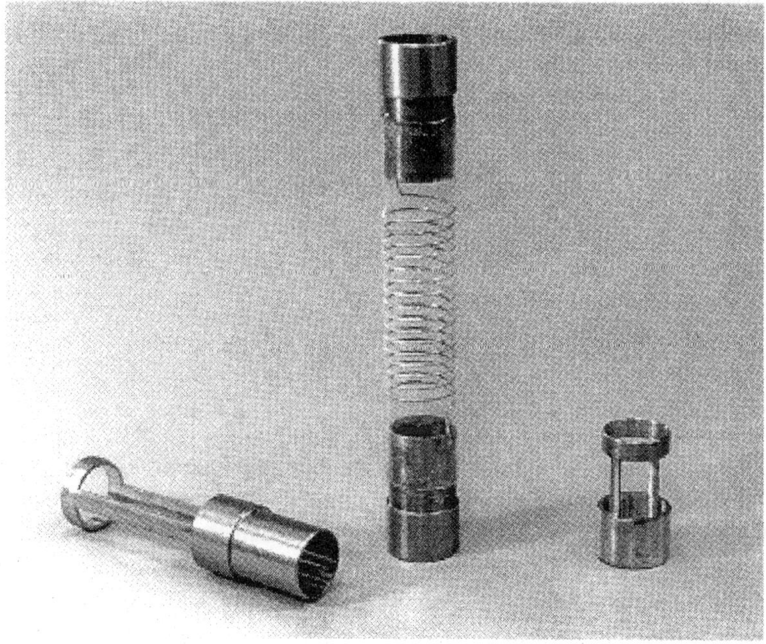

Figure 19. Representative (free-standing) ENDOR coil and some modified end pieces. On the left is an end piece with a handle that extends above cavity stack for *in situ* adjustments. On the right is an end piece with non-essential metal cut away in order to reduce capacitance with respect to cavity stack. Coil end piece prototypes were made at the suggestion of Bob Kreilick.

6.2 Cavity & Coil Prototyping

It is recommended that wire loops and helices be constructed and tested on 10 cm sections of pyrex or quartz tubing for prototype work, depending on what property of the coil is to be tested. Quartz tubing sections are used only for in-cavity testing of coil designs that are to be mounted on a dewar insert. A dewar mock-up consists of two quartz tubes, representing the dewar inner and outer wall, that are mounted concentrically by using teflon spacers; with this arrangement one can optimize coil and dewar geometry so that the rf field is maximized and the perturbation to the cavity Q is minimized.

Free-standing wire helix coils of the type described by Hurst *et al.* (1982) are first prototyped as above, since the end pieces that make electrical contact to the cable connectors in the commercial cavity are outside the cavity. After the dimensions of the coil have been optimized, the actual coil to be used in the probehead during an experiment is wound on a teflon form as described by Hurst *et al.* (1982). In practice, however, the construction requires the application of heat-shrink tubing, and the teflon form may begin to soften and lose its shape with repeated use. A more stable coil form is constructed by replacing the teflon rod with a teflon tube through which a brass rod is pushed. The brass core prevents the coil form from bending as the heat shrink tubing is applied; it also enables one to mount the coil form on a metal worker's lathe so that it may be held rigid as the wire is wound (by hand, *cf.* Bender & Aisen, 1993).

Coils may be tested in the cavity resonator that one intends to use, or one may use brass cavity mock-ups. I use several cylindrical cavity bodies that vary with respect to length and diameter. A circular iris port is positioned at the midsection so that the cavity may be coupled to a rectangular waveguide section. TE and TM modes are excited within the cavity mock-up by mounting the cavity with the broad wall of the rectangular guide parallel (TE) or perpendicular (TM) to the cylindrical axis. These cavities likewise feature interchangeable end plates, some of which are drilled breadboard fashion so that parallel post configurations may be tested.

Finally, standard low power resistive loads are a convenient aid in preparing the transformers and matching networks. These may be constructed by arranging 6 or more carbon composition resistors (½ W) in parallel and in a pattern that is symmetric as possible. The load is housed in a can and connected via a UHF or TNC connector. The same technique may be used to fabricate a reactive load by installing a network (combinations of inductors and/or capacitors) instead of resistors. I might also mention that when making impedance matching transformers, be aware that connectors affect the impedance properties of the device. For example, one can put a 50 Ω BNC connector on to the transformer port that will mate with the 50 Ω cable. But in seeking to make transformers drop-in devices, one often finds that the other port likewise has a standard 50 Ω BNC connector; this should not be the case. The transformer should be connected directly to the mismatched coil/network, or alternatively, the interface between coil and transformer should use UHF connectors without an impedance specification. The introduced mismatch may be trivial in light of the added convenience of having drop-in transformers. One may examine this by connecting an rf network to a 12.5 Ω resistive load via connectorized and non-connectorized 1:4 Ruthroff or Guanella transformers (this experiment was described in the ESR Society Newsletter, *7(1)*, **1995**).

6.3 Basic Concepts of Network Theory

The ENDOR coil is a two-port junction in an otherwise uniform transmission line (*i.e.* the 50 Ω cables). A port is simply a reference point in an electrical network, and in the case of an ENDOR coil spanning two 50 Ω lines, ports 1 and 2 correspond to the coil ends. A wave propagating along this line encounters the junction (*i.e.* the coil), and will be altered by reflection and losses. This alteration of the traveling wave is mathematically described by transformation and scattering matrices. Figure 20 illustrates a two-port device and directionality of the propagating wave. Let **A** designate a 1×2 matrix whose elements a_1 and a_2 are the incident wave amplitudes, and let **B** represent a similar matrix whose elements correspond to the reflected wave amplitudes. These two matrices enable one to describe the properties of the junction in terms of coefficients.

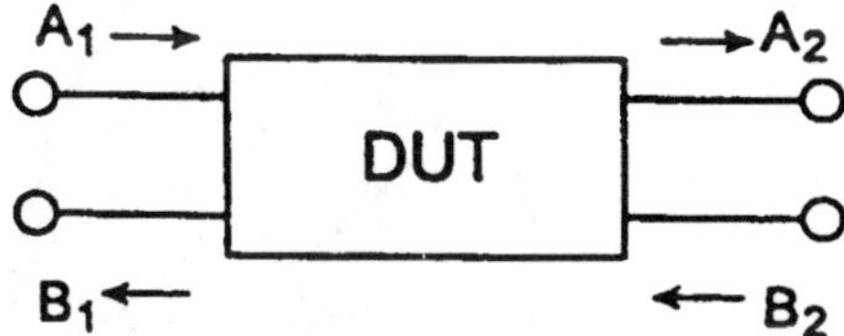

Figure 20. S-matrix convention for a two-port device.

The incident and reflected wave amplitudes are related to one another via the reflection coefficients, and this is conveniently expressed as $\mathbf{B} = \mathbf{S} \cdot \mathbf{A}$. The terms **B** and **A** are 1×2 matrices, and **S** is a 2×2 transformation matrix whose elements s_{ij} are the coefficients that are solutions to the linear equations:

$$b_1 = s_{11}a_1 + s_{12}a_2$$
$$b_2 = s_{21}a_1 + s_{22}a_2$$

The elements s_{ij} are known as the S-parameters and can be obtained by measuring the reflection and transmission coefficients under controlled conditions. For example, if $a_2 = 0$, then $s_{11} = b_2 / a_1$, and $s_{12} = b_2 / a_1$. In short, S-parameters are a convenient and alternative method of describing the total impedance parameters of a device.

A scalar reflection coefficient is defined as the ratio of the traveling wave amplitudes moving in the negative sense (direction) and the positive sense; the positive sense is defined as the direction of propagation that is towards the load. In the context of the previous paragraph, the reflection coefficient is defined as $\Gamma = |B| / |A|$, for which there are three defining cases: $\Gamma = 0$ (match, no reflected wave), $\Gamma = -1$ (open circuit), and $\Gamma = +1$ (short circuit). The inter-relationship between the SWR and Γ is SWR = $(1 + |\Gamma|) / + (1 - |\Gamma|)$, which in turn yields the familiar SWR = 1 (match) and SWR = ∞ (full reflection). In the case of a two-port junction (Figure 20) one defines two reflection coefficients b_1 / a_1 and b_2 / a_2, depending on how the direction of the traveling wave is defined. Similarly, there are transmission coefficients a_2 / a_1 and a_1 / a_2 that define the fractional wave amplitude that is transmitted by the junction.

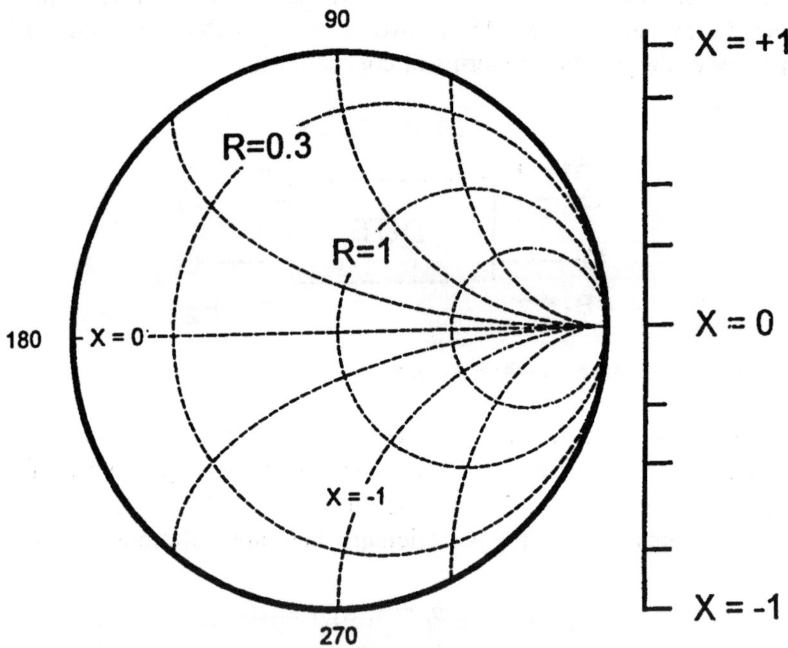

Figure 21. The Smith Chart. Circular patterns represent constant R contours. Divergent arcs represent constant X contours. Impedance is represented by an R, X pair that is specified by the contours. Quadrants are specified by the polar angles indicated.

Still another convention for describing the impedance properties of an electrical network is the Smith Chart (Figure 21), which is a graphical method of representing the impedance Z in the complex plane (Smith, 1983; Caron, 1989). The perimeter of the chart is circular, and polar angles (0–360° are delineated in a counterclockwise direction around this perimeter. This is a familiar convention in the designation of complex numbers *cf.* Rubenfeld, 1985; Churchill, 1984).

The chart itself is drawn as a series of tangential nested circles, each of which represents some resistance, or the real component of the complex valued impedance. The circles are bisected by a horizontal line that represents the condition at which the imaginary component, reactance, is zero. The rightmost intercept, where all the nested circles touch, is the polar origin ($\theta = 0°$), and arcs are drawn that emanate from this point and diverge away from the center bisecting line. Each of these arcs represents a constant reactance contour.

The chart is normalized so that a perfect match $Z_0 = R + iX$ is defined as $Z_0 = 1\,\Omega + i\,0$, and this is the center point of the chart. The reader may have noted that Smith Chart data reported in Section 3 appear to originate at the center of the chart and spiral outward as the impedance and SWR increase (Figures 6 & 7). For the most part $Z_0 = 50\,\Omega$, which is then used as the normalization factor. The polar angles that designate quadrants are drawn, and the center point represents a point of symmetry. Impedance points that are displaced an equal distance from the north-south center line (not drawn in the figure) are therefore reciprocals, and this is a means of computing the admittance from an impedance, and *vice versa*.

The ENDOR apparatus is comprised of a shielded coil that is terminated by a 50 Ω resistive load, and the ideal operating condition would be at a 'match' between the output impedance of the power amplifier (*i.e.* $Z_0 = 50\,\Omega$) and the coil/load. In order to match a coil network to the power amplifier, the coil impedance is $Z_{COIL} = R_{COIL} + i \cdot X_{COIL}$, and one would plot this point on the Smith Chart. A circle is then drawn that is centered at Z_0 and whose radius is the distance to Z_{COIL}. One now proceeds clockwise on the circle's perimeter from the point Z_{COIL} to the intersection with the horizontal line where $Z' = R' + i \cdot 0$. The ratio R': Z_0 is equal to the SWR.

One then continues clockwise along the circle's perimeter until the R = 1 contour is crossed, and at this intercept $Z'' = 1 - i \cdot X''$ represents the series reactance that will normalize (*i.e.* match) the starting impedance Z_{COIL}. The point on the circle directly opposite Z'' through the center point yields the parallel susceptance that will likewise match Z_{COIL}. Details of the procedure and the theory of the Smith Chart are described in the texts by Smith (1983) and Caron (1989).

7. ACKNOWLEDGEMENTS

This chapter is based on cumulative studies of ENDOR coil design that began during a postdoctoral appointment at the Chemistry Department of Michigan State University and completed at the NIH Biotechnology Research Resource for Pulsed EPR at the Albert Einstein College of Medicine. I thank Kermit Johnson (MSU), Hans van Willigen (University of Massachusetts), Theodore Camenisch (Medical College of Wisconsin), and Martin Baumgarten (Freie Universität Berlin) for advice and suggestions during my early foray into rf engineering, and Dr. Giorgio Romanenghi who kindly supplied a copy of his doctoral thesis, upon which Section 3.5 is based.

8. REFERENCES

Abragam, A., 1961, *The Principles of Nuclear Magnetism*, Clarendon Press, Oxford.

Abragam, A., 1962, In *Topics in Radiofrequency Spectroscopy,Proceeedings of the International School of Physics <<Enrico Fermi>>,XVII Course* (A. Gozzini, ed.), Academic Press, New York. p281.

Abragam, A., and Bleaney, B., 1970, *Electron Paramagnetic Resonance of Transition Ions*, Clarendon Press, Oxford.

Abrie, P.L.D., 1985, *The Design of Impedance-Matching Networks for Radio-frequency and Microwave Amplifiers*, Artech House, Norwood.

Alger, R.S., 1968, *Electron Paramagnetic Resonance: Techniques and Applications*, Interscience, New York.

Allendoerfer, R.D., and Maki, A.H., 1969, *J. Amer. Chem. Soc.*, **91**: 1088.

Allendoerfer, R.D., and Maki, A.H., 1970, *J. Magn. Resonance*, **3**: 396.

Allendoerfer, R.D., and Eustace, D.J., 1971, *J. Phys. Chem.*, **75**: 2765.

Allendoerfer, R.D., 1973, *J. Magn. Res.*, **9**: 226.

Andrew, E.R., 1958, *Nuclear Magnetic Resonance*, Cambridge University Press, Cambridge.

Bava, E., and Bava, G.P., 1969, *Intermodulation Noise in Wideband Amplifiers with Multicarrier Input*, International Institute of Communication, Genoa.

Bender, C.J., Sahlin, M., Babcock, G.T., Barry, B.A., Chandrashekar, T.K., Salowe, S.P., Stubbe, J., Lindström, B., Petersson, L., Ehrenberg, A., and Sjöberg, B.-M., 1989, *J. Amer. Chem. Soc.*, **111**: 8076.

Bender, C.J., and Babcock, G.T., 1992, *Rev. Sci. Instrum.*, **63**: 3523.

Bender, C.J., and Aisen, P., 1993, Continuous Wave ENDOR, in *Methods of Enzymology: Metallobiochemistry, Part D.* Vol 227 (J.F. Riordan and B.L. Vallee, eds.), Academic Press, New York. p190.

Biehl, R., Lubitz, W., Möbius, K. and Plato, M., 1977, *J. Chem. Phys.*, **66**: 2074.

Böttcher, R., and Heinhold, D., 1986, *Exp. Techn. Physics*, **66**: 2074.

Böttcher, R., Heinhold, D., and Windsch, W., 1990, *Z. Naturforch.*, **45a**: 570.

Bowick, C., 1982, *RF Circuit Design*, Howard Sams, Indianapolis.

Box, H.C., 1977, *Radiation Effects: ESR and ENDOR Analysis*, Academic Press, New York.

Breed, G.A., 1993, *Power Amplifier Handbook*, Cardiff Publications, Englewood.

van Camp, H.L., Scholes, C.P., and Isaacson, R.A., 1976, *Rev. Sci. Instrum.*, **47**: 516.

Caron, W.N., 1989, *Antenna Impedance Matching*, Amateur Radio and Relay Leagure Press, Newtington.

Castner, T.G., and Doyle, A.M., 1968, *Rev. Sci. Instrum.*, **39**: 1090.

Chacko, V.P., 1978, *Rev. Sci. Instrum.*, **49**: 1012.

Champion, K.S.W., 1950, *Proc. Phys. Soc.*, **B63**: 795.

Chen, W.-K., 1976, *Theory and Design of Broadband Matching Networks*, Pergamon Press, Oxford.

Chodorow, M., and Chu, E.L., 1955, *J. Appl. Phys.*, **26**: 33.

Christidis, T.C., and Heineken, F.W., 1973, *J. Phys. E*, **6**: 432.

Churchill, R.V., 1984, *Complex Variables and Applications*, 4th ed., McGraw-Hill, New York.

Colligiani, A., Pinzino, L., and Bertolini, M., 1979, *J. Magn. Res.*, **33**: 511.

Colligiani, A., and Romanenghi, G., 1991, *Rev. Sci. Instrum.*, **62**: 81.

Connor, H.D., 1972, *Electron Nuclear Double Resonance (ENDOR) and Electron Spin Resonance (ESR) Studies of Spin Relaxation in Solutions of Semiquiones*, Thesis, Cornell University, Ithaca.

Coremans, J.W.A., Polucktov, O.G., Groenen, E.J.J., Canters, G.W., Nar, H., and Messerschmidt, A., 1998, *J. Amer. Chem. Soc.*, **119**: 4726.

Dalton, L.A., and Dalton, L.R., 1979, In *Multiple Electron Resonance Spectroscopy* (M.M. Dorio & J.H. Freed, eds.), Plenum Press, New York. p169.

Davies, E.R., 1974, *Phys. Lett. A*, **47**: 1.

DeMaw, M.F., 1980, *Ferromagnetic Core Design and Application Handbook*, Prentice-Hall, Engelwood Cliffs.

DeMaw, M.F., 1982, *Practical RF Design Manual*, Prentice-Hall, Engelwood Cliffs.

Dwek, R.A., Richards, R.E., and Taylor, D., 1968, In *Annual Review of NMR Spectroscopy* (E.F. Mooney, ed.), Academic Press, London. p293.

Dye, N., 1993, *Radiofrequency Transistors: Principles and Practical Applications*, Butterworth-Heinemann, Boston.

Elleman, D.D., Manatt, S.I., and Pearce, C.D., 1965, *J. Chem. Phys.*, **42**: 650.

Engelson, M., 1984, *Modern Spectrum Analyzer Theory and Applications*, Artech House, Norwood.

Feher, G., 1956, *Phys. Rev.*, **103**: 500.

Feher, G., and Gere, E.A., 1956, *Phys. Rev.*, **103**: 501.

Feher, G., and Gere, E.A., 1959, *Phys. Rev.*, **114**: 1245.

Forrer, J., Schweiger, A., and Günthard, Hs.H., 1977, *J. Phys. E*, **10**: 470.

Forrer, J., Schweiger, A., and Günthard, Hs.H., 1981, *J. Phys. E*, **14**: 565.

Forrer, J., Pfenninger, S., Eisenegger, J., and Schweiger, A., 1990, *Rev. Sci. Instrum.*, **61**: 3360.

Freed, J.H., 1972, In *Electron Spin Resonance in Liquids* (L.T. Muus and P.W. Atkins, eds.), Plenum Press, New York. p387.

Freed, J.H., 1979, In *Multiple Electron Resonance Spectroscopy* (M.M. Dorio & J.H. Freed, eds.), Plenum Press, New York. p73.

Fukushima, E., and Roeder, S.B.W., 1981, *Experimental Pulse NMR: A Nuts and Bolts Approach*, Addison-Wesley, Reading.

Goslar, J., Piekara-Sady, L., and Kispert, L.D., 1994, In *Handbook of Electron Spin Resonance. Data Sources, Computer Technology, Relaxation, and ENDOR* (C.P. Poole and H.A. Farach, eds.), American Institute of Physics Press, New York. p360.

Gothe, K.H., 1970, *Z. tech. Hochsch. Ilmenau*, **16**: 67.

Grant, W.J.C., 1964a, *Phys. Rev. A*, **134**: 1554.

Grant, W.J.C., 1964b, *Phys. Rev. A*, **134**: 1565.

Grover, F.W., 1946, *Inductance Calculations: Working Formulas and Tables*, van Nostrand, New York.

Gruber, K., Forrer, J., Schweiger, A., and Günthard, Hs.H., 1974, *J. Phys. E*, **7**: 569.

Guanella, G., 1944, *Brown-Boveri Rev.*, **31**: 327.

Hoogstraate, H., Poot, J., Wenckebach, W.Th., and Poulis, N.J., 1975, *Physica B + C*, **79**: 499.

Hoogstraate, H., Wenckebach, W.Th., and Poulis, N.J., 1975, *Magnetic Resonance and Related Phenomena, Proc. Congr. AMPERE 18th* (P.S. Allen, E.R. Andrew and C.A. Bates, eds.),,North-Holland, Amsterdam. p427.

Hurst, G., Kraft, K., Schultz, R., and Kreilick, R., 1982, *J. Magn. Res.*, **49**: 159.

Hyde, J.S., 1965, *J. Chem. Phys.*, **43**: 1806.

Hyde, J.S., Rist, G.H., and Eriksson, I.R., 1968, *J. Phys. Chem.*, **72**: 4269.

Jasik, H., 1961, *Antenna Engineering Handbook*, McGraw-Hill, New York.

Jordan, E.C., and Balman, K.G., 1968, *Electromagnetic Waves and Radiating Systems*, 2nd ed., Prentice-Hall, Engelwood Cliffs.

Kevan, L., and Kispert, L.D., 1976, *Electron Spin Double Resonance Spectroscopy*, Wiley, New York.

Kilgus, C.C., 1969, *IEEE Trans. Ants. Prop.*, **AP-17**, 349.

Kirste, B., Krüger, A., and Kurreck, H., 1982, *J. Amer. Chem. Soc.*, **104**: 3850.

Kraus, J.D., 1988, *Antennas*, 2nd ed., McGraw-Hill, New York.

Kraz, V., 1995, *Compliance Eng.*,May/June, 43.

Kurreck, H., Kirste, B., and Lubitz, W., 1988, *Electron Nuclear Double Resonance Spectroscopy of Radicals in Solution: Applications to Organic and Biological Chemistry*, VCH Publishers, New York.

Kwiram, A.L., 1971, *Ann. Rev. Phys. Chem.*, **26**: 133.

Leniart, D.S., 1971, *Electron Nuclear Double Resonance of Organic Free Radicals in Solution*, Thesis, Cornell University, Ithaca.

Leniart, D.S., 1979, In *Multiple Electron Resonance Spectroscopy* (M.M. Dorio & J.H. Freed, eds.), Plenum Press, New York. p5.

Lenk, J.D., 1992, *Lenk's RF Handbook*, McGraw-Hill, New York.

Medhurst, R.G., 1947, *Wireless Eng*, February, 35.

Medley, M.W., 1993, *Microwave and RF Circuits: Analysis, Synthesis, and Design*, Artech House, Boston.

Mehring, M., Höfer, P., Grupp, A., 1986, *Phys. Rev. A*, **33**: 3523.

Mims, W.B., 1965, *Proc. Roy. Soc. (London)*, **283**: 452.

Mims, W.B., and Peisach, J., 1978, *J. Chem. Phys.*, **69**: 4921.

Miyagawa, I., Hayashi, Y., and Kotaki, Y., 1977, *J. Magn. Res.*, **25**: 183.

Möbius, K., and Biehl, R., 1979, In *Multiple Electron Resonance Spectroscopy* (M.M. Dorio & J.H. Freed, eds.), Plenum Press, New York. p475.

Möbius, K., Plato, M., and Lubitz, W., 1982, *Phys. Rep.*, **87**: 171.

Murdoch, J.B., 1970, *Network Theory*, McGraw-Hill, New York.

Newton, M.E., and Hyde, J.S., 1991, *J. Magn. Res.*, **95**: 80.

Nilson, J.W., 1983, *Electric Circuits*, Addison-Wesley, Reading.

O'Malley, P., and Babcock, G.T., 1984, *J. Chem. Phys.*, **80**: 3912.

Overhauser, A., 1953, *Phys. Rev.*, **92**: 411.

Piekara-Sady, L., and Kispert, L.D., 1994, In *Handbook of Electron Spin Resonance. Data Sources, Computer Technology, Relaxation, and ENDOR* (C.P. Poole & H.A. Farach, eds.), American Institute of Physics Press, New York. p312.

Poot, J., Wenckebach, W.Th., and Poulis, N.J., 1979, *Magn. Reson. Related Phenom., Proc. Congr. AMPERE 20th* (E. Kundla, E. Lippmaa and T. Saluvere, eds.), Springer-Verlag, Berlin. p284.

Poot, J., Wenckebach, W.Th., and Poulis, N.J., 1980, *Physica B + C*, **101**: 329.

Poot, J., Wenckenbach, W.Th., and Poulis, N.J., 1980, *Physica B+C*, **101**: 354.

Poot, J., Wenckenbach, W.Th., and Poulis, N.J., 1981, *Physica B+C*, **106**: 368.

Prutchi, D., 1996, *Circuit Cellar INK*, #71 June, 30.

Reich, H.J., 1961, *Functional Circuits and Oscillators*, van Nostrand, Princeton.

Rhea, W., 1994, *HF Filters and Computer Simulation*, Noble Publishing, Atlanta.

Romanenghi, G., 1985, *Progettazione e Realizzazione di un Apparato ENDOR ad elevata Potenz e a larga Banda*, Thesis, University of Pisa.

Rubenfeld, L., 1985, *A First Course in Applied Complex Variables*, Wiley, New York.

Rudin, M., 1992, *In vivo Magnetic Resonance Spectroscopy I: Probeheads and Radiofrequency Pulses, Spectrum Analysis*, Springer-Verlag, Berlin.

Ruthroff, C.L., 1959, *Proc. IRE*, August: 1337.

Schelkunoff, S.A., and Friis, H.T., 1952, *Antennas. Theory and Practice*, Wiley, New York.

Schweiger, A., Rudin, M., and Günthard, Hs.H., 1979, *Mol. Phys.*, **37**: 1573.

Schweiger, A., and Günthard, Hs.H., 1981, *Mol. Phys.*, **42**: 283.

Schweiger, A., Rudin, M., Forrer, J., and Günthard, Hs.H., 1982, *J. Magn. Res.*, **50**: 86.

Schweiger, A., 1987, *Electron Spin Resonance: A Specialist Periodical Report, Vol. 10B*, Royal Society of Chemistry, London. p138.

Schweiger, A., and Jeschke, G., 2001, *Principles of Pulsed Magnetic Resonance*, Oxford University Press, Oxford. Chapters 12 & 13.

Sevick, J., 1990, *Transmission Line Transformers*, 2nd ed., Amateur Radio and Relay League, Newtington.

Shcheikin, V.D., Vainshtein, D.I., Saikin, K.S., Vinokurov, V.M., and Dantov, R.A., 1974, *Fiz. Tverd. Tela.*, **16**: 3469.

Simpson, A.I.F., 1947, *Electronic Eng.*, November: 353.

Smith, P.H., 1983, *Electronic Applications of the Smith Chart*, Krieger, Malabar.

Solomon, I., 1955, *Phys. Rev.*, **99**: 559.

Spaeth, J.-M., Niklas, T.R., and Bartram, R.H., 1992, *Structural Analysis of Point Defects in Solids: An Introduction to Multiple Magnetic Resonance*, Springer-Verlag, Berlin.

Strauss, L., 1960, *Wave Generation and Shaping*, McGraw-Hill, New York.

Talpe, J., 1971, *Theory of Experiments in Paramagnetic Resonance*, Pergamon Press, Oxford.

TechPress, Inc., 1966, *Principles of RF Power Amplifiers*, TechPress, Brownsburg.

Terman, F., 1955, *Electronic and Radio Engineering*, 4th ed., McGraw-Hill, New York.

Thal, H.L., 1979, *IEEE Trans. Theory Tech.*, **MTT-27**: 982.

Thomann, H., and Bernardo, M., 1993, *Methods of Enzymology: Metallobiochemistry, Part D* (J.F. Riordan and B.L. Vallee, eds.), Vol. 227, Academic Press, New York. p118.

Tien, P.J., 1953, *Proc. IRE*, **41**: 1617.

Tsvetkov, Yu.D., and Dikanov, S.A., 1987, In *Metal Ions in Biological Systems* (H. Sigel, ed.) Vol. 22, Dekker, New York. p207.

Vizmuller, P., 1987, *Filters with Helical and Folded Helical Resonators*, Artech House, Norwood.

Wang, W., and Chasteen, N.D., 1995, *J. Magn. Res.*, **116A**: 237.

Welsby, V.G., 1960, *The Theory and Design of Induction Coils*, 2nd ed., MacDonald, London.

Witte, R.A., 1991, *Spectrum and Network Measurements*, Prentice-Hall, Englewood Cliffs.

Yagi, H., Inoue, M., Tatukawa, T., Yaeguchi, O., and Kato, S., 1970, *Jap. J. Appl. Phys.*, **9**: 1534.

Zweyart, W., Thanner, R., and Lubitz, W., 1994, *J. Magn. Res.*, **109A**: 172.

Chapter 5

The Generation and Detection of Electron Spin Echoes

Christopher J. Bender
Department of Chemistry, Fordham University, 441 E. Fordham Road, Bronx, New York 10458 USA

1. SCOPE & RATIONALE

Time domain and pulsed electron magnetic resonance (EMR) have become important and experimentally facile techniques in biology and chemistry because the technology of microwave circuits has advanced to the state where pulse sequences can be synthesized on a time scale that is commensurate to the state-to-state dynamics of the electron spin. For this reason it is now feasible to conceive of experiments that are analogous to those of modern pulsed NMR, a field of study that enjoyed a surge of popularity as both a fundamental and analytical research technique when the technology of radiofrequency circuits similarly evolved.

This chapter is intended as a sequel to Chapter 1 of this volume and serves as a review of the high frequency technology and techniques that are salient to pulsed EMR. Related articles and books provide the motivation and underlying theory of the technique (Bagguley, 1992; Kevan & Schwartz, 1976; Kevan & Bowman, 1990; Schweiger, 1990; Gemperle & Schweiger, 1991; Höfer, 1994; Schweiger, 1995; Freeman, 1997; Schweiger & Jescke, 2001), and two specialized articles on specific instrument design treat the digital logic underlying the generation of the pulse sequence and data acquisition (Thomann *et al.* 1984; Quine *et al.* 1987). The topics covered in this chapter will therefore be limited to those microwave methods that are relevant to the generation and detection of the electron spin echo.

2. INSTRUMENTAL ORIGINS OF TIME-DOMAIN SPIN RESONANCE

2.1 Measurement of Relaxation Times: FID and Echo Methods

Spin resonance was initially of interest because it was regarded as the ideal laboratory for examining the nature of dynamics among quantum states during radiation-induced transitions (*cf.* Macomber, 1976; 1995). The sources of radiation that correspond to spin transitions are very nearly coherent and monochromatic, and the transitions could be detected on a time scale that was compatible with routine electronic detection instruments.

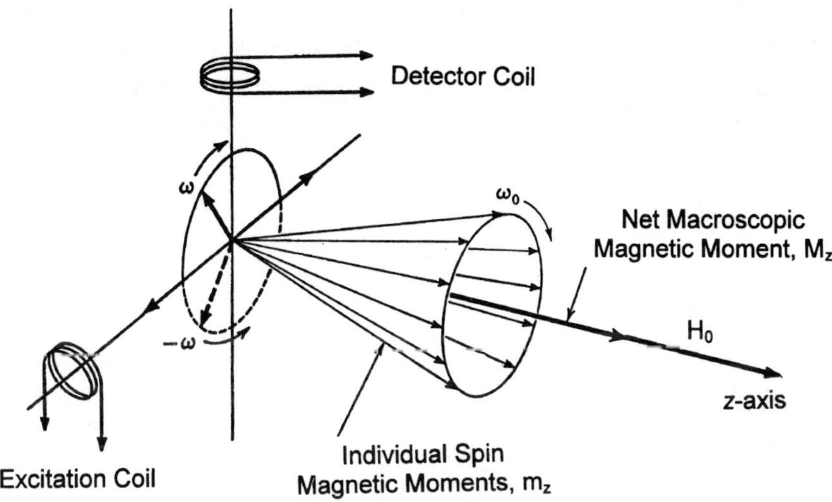

Figure 1. A schematic representation of spin dynamics and detection as bulk magnetization. The spin state condition depicted is prior to excitation by the resonant field, and no net magnetic moment is detectable in the *x-y* plane.

A magnetic resonance spectrometer may be regarded simply as a pair of magnetic pick-up coils that are orthogonally oriented (Figure 1). A sample material may therefore be placed within these two coils so that a magnetic field generated by one coil is detected by the second coil only if it is transmitted via a perturbation in the sample material; the sample must act as a transducer in this idealized scenario. For a sample material consisting of atoms or molecules that each possess a magnetic moment, the coils are used to effect or detect changes in the orientation of the sample's magnetic moment, depending on whether the coil acts as a AC field transmitter or receiver, respectively.

The magnetic moment or magnetization that one observes in a sample material is the bulk magnetization M, which is the vector summation of the magnetic moments of the individual atoms or molecules m_i:

$$\mathbf{M} = \sum_i m_i \tag{1}$$

In the absence of any polarizing field, the individual atomic magnetic moments may be described as vectors of no preferred orientation, and therefore $\mathbf{M} = 0$.

At the microscopic level, the magnetic moment of an atom (or molecule) arises from the quantized orbital and spin angular momentum of the atomic nucleus or its associated electrons. In a uniform polarizing field, the atomic magnetic moment becomes quantized along an axis whose direction is defined by the polarizing field and designated as the *z*-axis. One may now conceptually decompose the individual magnetic moment vectors m_i as two projections: one onto the *z*-axis, and a second onto the *x-y* plane. The bulk magnetization is therefore decomposed in a likewise manner as

$$\mathbf{M}_{\text{z}} = \sum_i m_{z,i} \tag{2a}$$

$$\mathbf{M}_{\text{xy}} = \sum_i m_{xy,i} \tag{2b}$$

The individual magnetic moments $m_{z,i}$ are directed along the *z*-axis at proscribed orientations that are specified by the quantum mechanics, and assume a value among the set $\left(-m_S, -m_{S+1}, -m_{S+2}, \cdots, m_{S-1}, m_S\right)$, where m_S is the spin quantum number (*e.g.* 1/2). The number of individual magnetic moments having a specific value of m_S (*e.g.* +1/2 or −1/2 for the case cited) is determined by the Boltzmann statistics because each orientation of the spin magnetic moment corresponds to some energy of interaction between the microscopic magnetic moment and the external field. The bulk component $\mathbf{M}_{\text{z}}$ is therefore nonzero and aligned along the *z*-axis.

In classical electromagnetic theory, a magnetic dipole that is aligned by an external field will oscillate with a characteristic frequency about the axis defined by the external field direction (Becker, 1982). This oscillation frequency is determined by the magnetic dipole moment and the external field strength. The analogous model at the microscopic level is that the magnetic moment of atoms or molecules is said to precess about the axis defined by the external polarizing field with a characteristic frequency defined as $\omega = g{\cdot}\beta{\cdot}H_z/2\pi h$.[1] There is no quantization in either the *x*- or *y*-direction, and therefore the precession about the z-axis is randomized so that the bulk magnetization in the *x-y* plane is still zero.

Given the coil configuration of Figure 1, therefore, a sample material that is subjected to a uniform magnetic field (applied as a ideal step function) would be polarized, and the transient magnetization then be detected in the orthogonally located coil. Suppose now that the sample material is polarized by a DC magnetic field so that there exists a net bulk magnetization in the *z*-direction $\mathbf{M}_z$, but none in the *x-y* direction. The coil designated as the transmitter in Figure 1 is now turned on with an AC signal whose frequency corresponds to the precession frequency of the magnetic moment (as defined in the preceding paragraph). A resonance condition is thus established between the source of radiation and the sample, and the magnetic moments are 'redirected' to one of the other spatial quantization states allowed by the selection rules. In effect, the resonant radiation changes the direction of each microscopic magnetic moment with respect to the *z*-axis, and the sum total of these changes in the microscopic magnetic moments is reflected in $\mathbf{M}_z$.

For example, suppose a sample material consists of 1000 microscopic spin magnetic moments with allowed states of $\pm 1/2$ and that the applicable statistics dictates that 900 of the moments assume the state specified as $+1/2$ and 100 assume the state $-1/2$ in some arbitrary DC field prior to the imposition of the AC resonant field. The bulk magnetization will therefore be $\mathbf{M}_z = (900 - 100)\, m_z$. If one now turns on the resonant AC field as described in the preceding paragraph, each of the microscopic magnetic moments undergoes a transition (assuming 100 % efficiency) to its counterpart state (*e.g.* $+1/2 \rightarrow -1/2$, and *vice versa*). After the transitions among all the microscopic magnetic moments the populations of the two states in our scenario are reversed, $\mathbf{M}_z = (100 - 900)\, m_z$, and the bulk magnetization is effectively, reversed with respect to the z-axis. In other words, we have rotated $\mathbf{M}_z$ by 180°.

There is a finite time associated with the radiation-induced transition that, in principle, could be observed by the inductive coupling of the reversing magnetic field $\mathbf{M}_z$ to the receiver coil in Figure 1. This transition time is very fast, however, and rather one tends to record the return of the system to thermodynamic equilibrium after the radiation-induced transition, the so-called relaxation from the excited to ground state. This relaxation transition, and its associated time T_1 dispose of the excess energy via non-radiative coupling to the lattice (*cf.* Standley & Vaughan, 1969), and the accompanying change in the magnetization may be observed in the receiver

[1] The terms *g*, β, and H_z correspond to the g-factor, the Bohr magneton, and the applied DC magnetic field, respectively. *g* is a measure of the deviation between the electron's local and applied field, where $g_e = 2.0023 = \beta\, H_z / h\nu$ (no local field effect). $\beta = eh / 2m_e = 9.274 \times 10^{-24}\ J{\cdot}T^{-1}$ (J = Joules; T = Tesla).

coil as a transient signal of exponential decay according to a first order law (Bloch, 1946)

$$dM_z/dt = \frac{1}{T_1}\left(M_0 - M_z\right) \tag{3a}$$

which yields the decay relation for the time-dependent magnitude of $\mathbf{M}_z$

$$M_z = M_0 \exp\left(-\frac{t}{T_1}\right) \tag{3b}$$

The description of the radiation-induced transitions and accompanying relaxation from the excited state in the preceding paragraphs dealt only with changes in the vector $\mathbf{M}_z$, but there is a second phenomenon that accompanies the resonant interaction of radiation with the precessing magnetic moments and affects the projection of the magnetic moment onto the *x-y* plane. In the resonant AC field, the precessing magnetic moments that are otherwise randomly oriented in the *x-y* plane become coherent, and therefore $\mathbf{M}_{xy}$ is rendered non-zero. This vector $\mathbf{M}_{xy}$ then rotates about the *z*-axis at the characteristic frequency dictated by the DC field strength and the microscopic particle's magnetic moment. And like $\mathbf{M}_z$, once the resonant AC field is turned off, the system returns to its initial state with a characteristic time T_2 as the magnetic moments dephase and render $M_{xy} = 0$.

The redirection of the $\mathbf{M}_z$ vector and its consequential effect on the measured magnitude M_z is described in terms of a 'turning angle' θ. The degree to which the precessing spin ensemble can be made coherent and hence the magnitude of the turning angle θ depends upon the driving conditions and the power and coherence of the radiation. A so-called 90° pulse corresponds to the situation in which the spin ensemble is not only made coherent, but the populations along $+z$ and $-z$ are equalized, thereby rendering $\mathbf{M}_z = 0$. Consequently, there are two relaxation responses at work when the radiation is removed: one is the dephasing of the precessing spin (*i.e.* transverse relaxation), and the second is the return to thermal population (*i.e.* longitudinal relaxation). In general, the transverse relaxation rate, T_2^{-1}, is very much faster than the longitudinal rate, T_1^{-1}.

The free induction decay is formally defined as a transient magnetization signal that is measured after a saturating resonant pulse (*cf.* Hahn, 1950a), and it is a measure of excited state dynamics as the system 'relaxes' to the ground state. Blume (1958) examined the transverse relaxation rate T_2^{-1} of solvated electrons (sodium-ammonia solutions) by measuring the free induction decay after a 300 ns 90° rf pulse. These measurements were made in precisely the manner outlined above by using an NMR spectrometer whose excitation and detection circuits (coil) axes were fixed at a relative angle of 90°. The spectrometer operated at a very low frequency (17.4

MHz), and the FID was therefore measured directly from the decay of the resonance signal by using an oscilloscope (Blume, 1958).

Measurements of relaxation rates via the free induction decay are problematic for samples for which the spectral line is inhomogeneously broadened and T_2 is very short. These measurements are difficult is because the magnetization is a very weak signal relative to the excitation pulses that must be used to drive the spin system; the sensitive circuits of the receiver must therefore be protected while the transmitter is on and the interval between transmitter shut-down and receiver turn-on (the so-called 'deadtime') may be unacceptably long on the timescale of the FID. For such systems the spin echo technique was introduced (Hahn, 1950a,b). This multi-pulse technique overcomes the timing limitations of the single-pulse FID measurement by, in effect, imposing a time-reversal process onto the much slower longitudinal relaxation process. The existence of multiple rate processes is predicted in Bloch's equations of transient phenomena, and the phenomenon is experimentally verified when rates are measured as a function of the Zeeman field. Recognizing that two effects, dephasing and spin-lattice relaxation, are occurring, one saturates the spin system to maximize M_{xy} and render the spins coherent (*i.e.* 90° pulse). The spins now dephase within the receiver deadtime, but may be temporally 'reversed' by a second pulse that does not affect M_{xy} (*i.e.* 180° pulse). The spins again become coherent with a resurgence of the measured magnetization (the 'echo') and then dephase again. Given that the temporal stage upon which these dynamics occur is now determined by the longitudinal rate T_1, the echo amplitude traces an envelope that behaves according to the longitudinal spin relaxation rate $\exp(-t / T_1)$ (Hahn, 1950a,b; Carr & Purcell, 1954).

Analyses of the FID and spin echo phenomena were published by Jaynes (1955) and Bloom (1955). Both are noteworthy because they express the spin dynamical response in a density matrix form, but Bloom's paper (1955) is useful for its depiction of experimental observation and comparative analysis of the FID and echo. In particular, the analysis demonstrates how the shape of the FID is not determined by the field (or line) inhomogeneity, but rather the rf level. For a given $\gamma H_1 \Delta t_p$, a long weak rf pulse will yield a long FID, and a short intense rf pulse will yield a short FID. In other words, according to Bloom (1955), as one shortens Δt_p so that $T_2 >> \Delta t_p$, one destroys ones chance to measure an FID; the spin echo rescues the measurement because the echo is shaped like two FIDs place back to back (Hahn, 1950a).

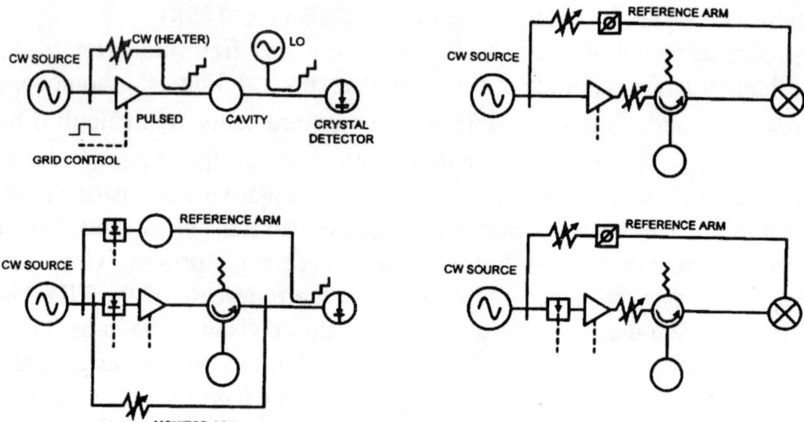

Figure 2. A catalog of pulsed EMR spectrometer designs, indicating various excitation and detection modes. Pulse-perturbation spectrometer, in which a cw signal is monitored following a high-power pulse (now called coherent raman beats) are illustrated on the left hand side of the figure (Top: Bowers & Mims, 1959; Bottom: Brown & Sloop, 1970). Homodyne spin echo spectrometers are illustrated on the right hand side (Top: Taylor et al., 1969; Bottom: Modern variation with PIN diode modulator generating low-power pulses for amplification). Dashed lines indicate a pulse-modulated component; open circles depict sample resonators. Routine protection devices are not indicated in the figure. For spectrometers employing dual pulsed sources (magnetron), see Kaplan et al. (1961) and Rowan et al. (1965).

The electron spin echo was likewise used to measure spin lattice relaxation times. For example, Blume (1958) artificially created the condition of spectral line inhomogeneity by using a set of coils to create inhomogeneity in H_0 and was thereby able to detect an echo from a sample of sodium in ammonia. The echo was otherwise detected by the same manner as the free induction decay using the NMR spectrometer. Gordon & Bowers (1958) reported the detection of microwave spin echoes from silicon crystals, and in this case the spectrometer that was used resembled the autodyne cw-EMR spectrometers of the time (Figure 2; described at length in Chapter 1). Bowers and Mims (1959) measured the electron spin relaxation rate of magnetically dilute $NiSiF_6 \cdot H_2O$ by a technique that bears a resemblance to the coherent raman beats experiment (Brewer & Hahn, 1973). A cw signal is applied to the spin system and acts as a monitor of the spin system. Relatively low power (ca. 1 Watt) pulses are superimposed upon this signal to 'heat' the spin system (*i.e.* put the spin population into a metastable, for example, saturated, state), and the spin dynamics after the pulse is turned off is observed as a 'beat' on the cw-signal.

High power, short excitation pulse spin echo measurements of T_1 and T_2 at microwave frequencies are reported in Na/NH_3 solutions by Cutler &

Powles (1962), who modified a commercial radar transmitter that delivered 7 kW 50 ns pulses. The sample resided in a low-Q (ca. 300) cylindrical TE_{011} cavity, and the receiver seems to have been that of the radar system and featured a deadtime of approximately 750 ns.

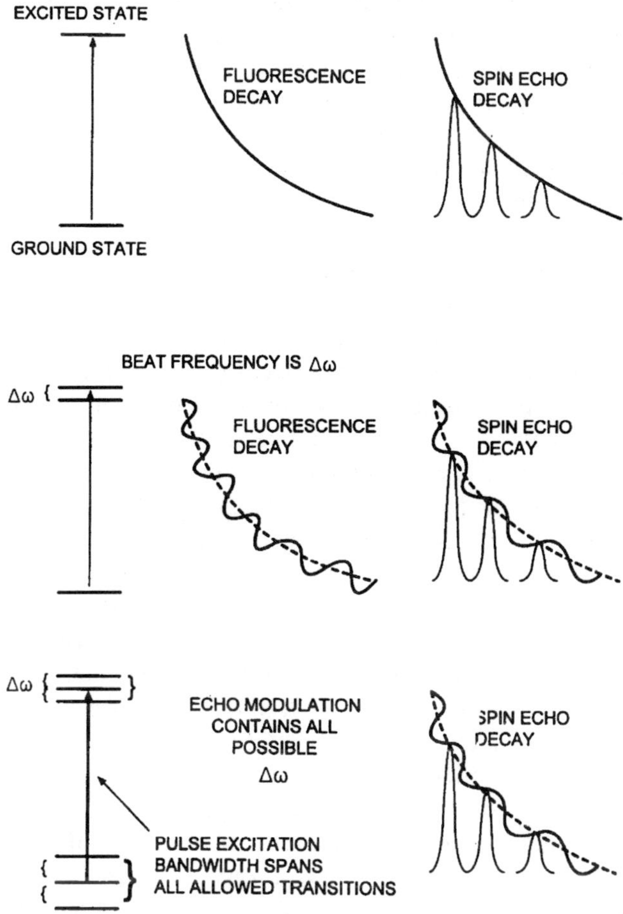

Figure 3. The analogy between the quantum beats phenomenon of fluorescence spectroscopy and electron spin echo modulation. When two spectroscopic transitions are indistinguishable in a given experiment, for example, when the excited states are energetically closer than the spectral bandwidth of the exciting radiation source, the observed de-excitation process (decay) exhibits oscillations or 'beats' that arise from the interference between what would otherwise be distinct state-to-state decays. In the pulse EMR scenario, the pulse bandwidth is $f_c \pm 2\Delta t_p^{-1}$, where Δt_p is the pulse width, and the ground hyperfine states usually simultaneously excited into states that may be brought into a very nearly crossing condition. The echo, whose amplitude reflects the de-excitation rate, is therefore modulated.

2.2 Echo Modulation

With the discovery of the spin echo it was reported that the amplitude of the spin echo varied as a function of the interpulse spacing (Hahn, 1950a,b; Hahn & Maxwell, 1952), and the authors recognized that this so-called modulation was an interferogram generated by the beats of spin precession frequencies in the sample material. In a classical mechanical sense, the spin echo modulation experiment is analogous to observing the behavior of a gyroscope that is coupled to one or more other gyroscopes precessing with at different frequencies. The quantum mechanical description of the beating phenomenon is schematically illustrated in Figure 3 and occurs when two or more states cannot be distinguished in a given experiment. In the case depicted, which corresponds to the classic resonance fluorescence experiment, two excited states are energetically close to the extent that the radiation-induced transition does not select one excited state in preference over the other. Both excited states are therefore populated and the fluorescence emission (the exponential decay) features an oscillatory 'beat' that corresponds to the interference between the two emitted light quanta. The actual frequency of this beat is the separation between the two near-degenerate levels. The fluorescence emission behaves as a linear mixer of the frequencies extant in the level (anti-) crossing. The corresponding spin echo modulation state diagram is illustrated in the lower panel of Figure 3, and as may be seen from the figure, the echo modulation beats correspond to the frequencies of spin-spin interaction: chemical shifts, in the case of nuclear resonance (Hahn & Maxwell, 1952), and hyperfine spectra for electron resonance. In at least one paper (Newman & Rowan, 1972), electron spin echo modulation is termed FT-ENDOR because the modulation spectrum is analogous to the high-resolution nuclear hyperfine spectra normally obtained by electron-nuclear double resonance.

The early echo modulation data were recorded by an oscillographic technique that entailed a time-exposed photograph of the triggered oscilloscope display as the pulses (and resultant echo) are temporally stepped (Hahn, 1950b). The frequency spectrum of the echo modulation was then determined by simulation and fitting a cosine series to the experimentally recorded time series data (Newman & Rowan, 1972). One of the earliest examples of electron spin echo ENDOR was performed at least in part in order to determine the transition frequencies that comprised the echo modulation and so avoid the trial and error fitting of modulation patterns (Liao & Hartmann, 1972; S. Hartmann, personal communication). Mims (1965b) described a spectrometer that used a boxcar averager to collect modulation data as an integrated amplitude of some sampled portion of the echo (see also Brown & Sloop, 1970; Huisjen & Hyde, 1974), and current methods of echo modulation analysis use discrete Fourier transform techniques (see Mims, 1984) or other spectral estimation method (De Beer *et al*., 1988a,b; Jones & Hore, 1991) to convert a time series array of echo amplitudes into a frequency spectrum.

3. PULSE MODULATION

3.1 Timing Factors & Excitation Bandwidth

There are two fundamental differences between pulsed NMR and EMR. The first of these is the relative widths of the magnetic resonance (sample) and pulse excitation (stimulating radiation) spectra. The second difference concerns the time scales upon which the spin dynamics occurs. It therefore follows that the timing factors in a pulsed magnetic resonance experiment will vary according to the spin system and type of information one wishes to acquire.

The interrelationship between a waveform and its frequency spectrum is familiar from Fourier transform analysis and spectral theory. For example, speaking of a sine or cosine wave connotes a smooth oscillatory function that corresponds to a single frequency. The frequency spectrum of a general sine wave function of the form $\sin(\omega t + \varphi)$ would therefore resemble a δ-function when plotted as signal intensity *vs*. frequency.

In practice, however, we recognize that the term φ may be time dependent (see Chapter 1), and that a time-varying phase factor $\varphi(t)$ distorts the pure wave by, in effect, introducing new frequencies into the spectrum, which may be represented by replacing the single sine term with a sum of the original plus several new sine terms. For such a simple representation the spectral domain may be generated by plotting the individual frequencies. In general, it is possible to represent any waveform by a series of sine and cosine functions (Brown & Churchill, 1993), and the Fourier transform is one mechanism of interconverting the time and frequency domains. Two texts that specifically deal with the Fourier transform in the context of spectroscopy are those by Hoch & Stern (1996) and Williams (1996).

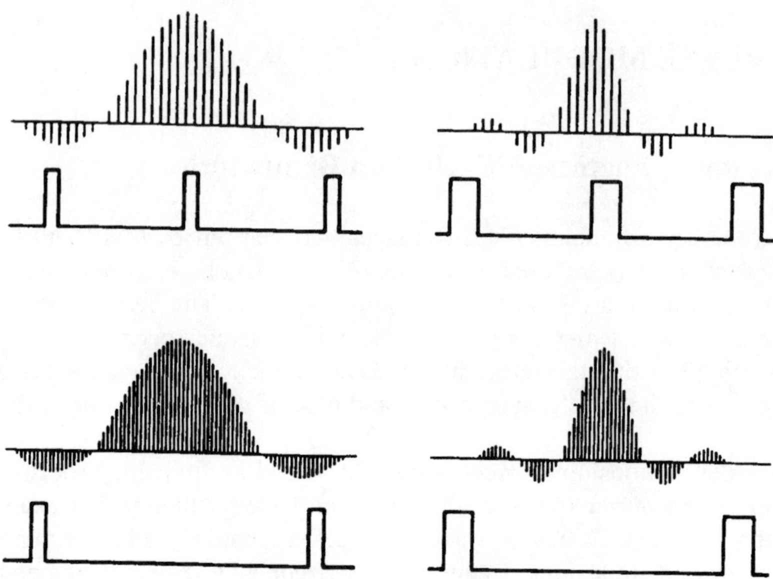

Figure 4. The spectral dispersion of a pulsed frequency source as a function of pulse width and repetition rate (After Schwartz, 1959).

For the purpose of this review, it will be assumed that the pulsed EMR spectrometer in question generates rectangular pulses by modulating a continuous wave signal of frequency f_c. It is possible to reproduce a rectangular waveform, such as a pulse, with a superposition of sine and cosine functions, and the requisite number of sine functions of frequency ω_i is inversely proportional to the time interval between the leading and falling edges of the 'pulse' (see Brown & Churchill, 1993). It will be stated here without proof (see Schwartz, 1959; Brown & Churchill, 1993) that a rectangular waveform superimposed onto a (pure) oscillatory signal of frequency f_c transforms into the frequency domain as $f_c \pm 1/2\Delta t_p^{-1}$ where Δt_p^{-1} is the pulse width. Practical application of this concept to pulsed EMR leads to recognition that a 20 ns rectangular pulse of 9.0 GHz carrier would therefore have an excitation spectrum that is centered at 9.0 GHz and span 8.975 to 9.025 GHz; the shape of the spectrum, that is, the power distribution, is theoretically sin(x) / x, but will vary according to the actual shape of the pulse, which is affected by the network properties of the modulator that generates the pulse and the transmission line along which the pulse propagates (see Schwartz, 1959).

It follows from the preceding paragraphs and section that for many hyperfine spectra it will not be practical to create the conditions of a

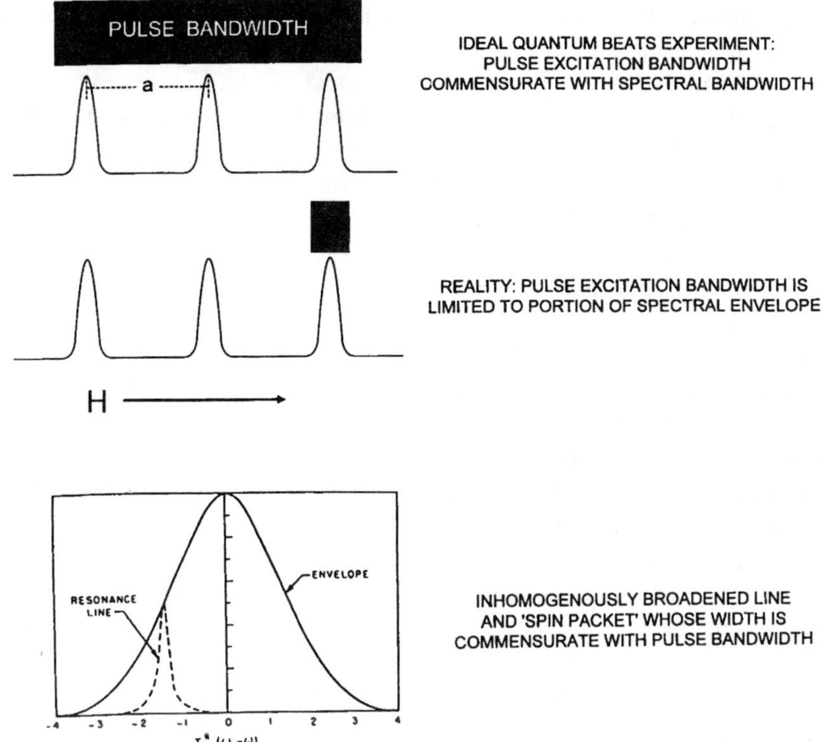

Figure 5. The relationship between the cw-EMR spectrum and the pulse excitation spectral bandwidth. The top trace is an idealized situation (analogous to modern NMR) where the pulse's frequency spectrum spans the entire EMR spectrum (here represented as an S = 1⁄2, I = 1 hyperfine spectrum) that is applicable to weakly coupled ^{14}N. The middle trace is a more realistic representation of spectral coverage by a pulsed carrier for a strongly coupled S = 1⁄2, I = 1 hyperfine spectrum. Finally, the bottom trace depicts an inhomogeneously broadened (i.e. powder) spectrum as a superposition of narrow homogeneous lines that may be individually excited (spectral diffusion studies).

quantum beats experiment and achieve an interferogram superimposed on an FID or echo decay profile. To take the simple example of Fremy's salt (^{14}N hyperfine pattern, S = 1/2, I = 1), one might record a three-line cw-EMR (solution) spectrum similar in lineshape to the profile depicted in Figure 5 (*cf.* Windle & Wiersema, 1963). The splitting between the three lines is dominated by the Fermi Contact interaction *a*, which may typically be 13 Gauss (*i.e.* ~36 MHz). In order to span the entire three-line spectrum one therefore needs a pulse excitation bandwidth that is greater than 72 MHz, which corresponds to a rectangular pulse width of less than 14 ns. This switching speed is approximately the limit of conventional low power switches (PIN or GaAs FET), and as the Fermi Contact energy increases it is very difficult to achieve the requisite pulse widths for spanning a set of $\Delta m_S = 1$ transitions. This limitation can be played to advantage, however, for studies of relaxation processes.

Pulsed EMR is therefore an experiment in which Δt_p^{-1} is tied to the width of the spectrum one wishes to excite. For some free radicals it is possible to span the entire hyperfine spectrum and achieve true FID derived spectra in analogy to modern NMR, but for the most part pulsed EMR is more accurately described as a spectral hole-burning experiment. In the case of solution or single crystal samples for which individual $\Delta m_S = 1$ transitions may be resolved and driven by pulsed excitation one can easily rationalize some of the general rules of pulse width, spectral dispersion, and the transverse relaxation time T_2. The width of the cw-EMR line may be regarded as a dispersion of frequencies, which is inversely proportional to the transverse relaxation time (see Poole & Farach, 1971). Given the relationship between pulse width and excitation bandwidth, the rule $\Delta t_p \ll T_2$ makes intuitive sense because the rf pulses must force profound changes in the spin populations of a sample on a timescale that is shorter than the natural tendency of the system to restore thermodynamic equilibrium.

The transverse relaxation time varies with the sample type, and therefore the requirements placed upon the pulse modulator vary with the type of system. For example, transition metal ions are typically studied by using rf pulses on the order of ten nanoseconds, whereas Δt_p on the order of a hundred nanoseconds is adequate for organic radicals. A useful compilation of electron spin relaxation rates may be found in several reviews (Banci *et al.*, 1986; Bertini *et al.*, 1994).

3.2 Instrumental Techniques for Manipulating Excitation Bandwidth

The shape of pulses that are used to excite a spin resonance spectrum are not limited to the rectangle; the pulse spectral envelope may be 'tuned' by deviating from the rectangular pulse shape (Schwartz, 1959). Shaped pulses have been the staple of NMR, and the pulses that are shaped for NMR spectroscopy tend to be created by imposing a DC waveform on mixers (Warren & Silver, 1988; McDonald & Warren, 1991). The imposition of a DC waveform upon a mixer is not practical for EMR because the pulse widths are still necessarily short and pulse-shaping circuits are not amenable to complex forms on the tens or hundreds of nanoseconds time scale. Digital arbitrary waveform generators (*e.g.* LeCroy, New York; WaveStation LW410) are a second option, but these too depend on the bit rate at which the D/A converter can put out the data (typically 400 MHz, see LeCroy 1996 catalog specifications). A feasible alternative to DC waveform shaping of rf pulses is the delay line, in which two (or, in principle, more) rf signals are combined in such a manner that their phase incoherence causes their sum to assume some shape that may be controlled by a suitable attenuator and/or phase shifter (Crepaeu *et al.*, 1989; Gorcester & Freed, 1990). One point worth noting that in the circuit diagram depicted in Figure 31 by Gorcester & Freed (1990), the reflective switch and associated delay line will pass a continuous wave signal; one needs to implement a second modulator or use the traveling wave tube amplifier grid as a modulator in order to generate discrete pulses.

An alternative approach to increasing the excitation bandwidth may lie with the carrier frequency rather than its modulation. Doppler radar methods, like FT-EMR, require narrow excitation pulses in order to increase the spectral bandwidth of the pulses and thereby obtain more information from the returning echo. There are techniques routinely used in radar that are not as yet applied to EMR: these are (1) the use of non-sinusoidal carrier waves (Harmuth, 1981) and (2) pulse compression (Barton, 1975). A third method, (3) monopulse or monocycle pulse (Rhodes, 1980; Sherman, 1984), is actually a DC method that in some respects resembles the pulse compression technique.

The application of non-sinusoidal waves is straightforward; one takes advantage of the frequency dispersion of an arbitrary waveform in order to increase the spectral bandwidth of the carrier prior to pulse modulation. If a square wave is substituted for an otherwise sinusoidal carrier of the same carrier frequency (*i.e.* a 9 GHz square wave *vs.* 9 GHz sine wave), the frequency dispersion is, for reasons described in preceding paragraphs, increased, and a broader excitation spectrum is thus obtained. A square wave generator operating at microwave frequencies is impractical, however there exists devices known as comb generators that deliver short DC pulses at a rapid repetition rate sufficient to generate a microwave spectrum.

Chirped pulses and pulse compression are related to the use of non-sinusoidal signals. The former technique refers to a procedure in which the carrier waveform is frequency modulated during the course of the pulse. For example, pulse chirping could be performed by applying a sawtooth waveform on the reflector voltage of a reflex klystron in order to ramp the electrical tuning of the electron tube. Klauder *et al.* (1960) introduced the technique for radar applications, and in 1963 Mims introduced a method of chirped signal detection via the electron spin echo.[2] Jeschke and Schweiger (1995; see also Forrer *et al.*, 1996) adopted the method in electron spin echo ENDOR experiments in which the radiofrequency carrier (nuclear transitions) was 'chirped', but the technique's application to the electron spin transitions is problematic because one must chirp the microwave source on a

[2] The spin dynamics described by Mims in this technique correspond to an ELDOR process.

very short time scale. The procedure might be feasible when performed on an electron tube, but, as described by Klauder *et al.* (1960), specific requirements are imposed upon the corresponding receiver.

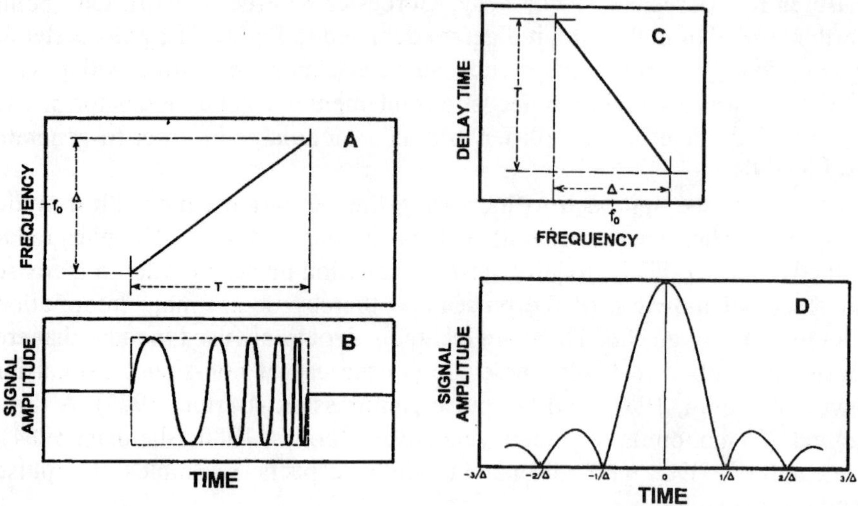

Figure 6. The principles of pulse chirping. A: A frequency sweep about some central carrier f_0 during some time interval τ, for example, sweeping a reflex klystron reflector voltage; B: the resultant pulse waveform in the time domain; C: Imposition of a delay function over the frequency sweep; and D: The resultant pulse, whose width is determined by the frequency sweep, or chirp. Figure is adapted from Klauder *et al.*, 1960; Figures 2 & 3.

Pulse compression is a generic term that is used to describe a waveshaping process that is produced as a propagating waveform is modified by the electrical network properties of the transmission line. Pulse compression originated with the desire to amplify the transmitted impulse (peak) power by temporal compression (see Cook, 1960). Linear frequency modulation (pulse chirping) plus network filtering are used together to yield a very narrow pulse of a characteristic carrier frequency; group delay factors of the filter overlay portions of the linear FM pulse in order to produce the ultra-short pulse.

Optical pulses are often compressed via a combination of frequency dispersion and interference between wave groups, which is conceptually identical to the technique that is described in the preceding paragraph. Rather than chirping the frequency of a pulse, however, a quasi-optical approach would entail the manipulation of a short pulse rather than the carrier frequency source. The frequency dispersion that is otherwise created at the source is created by a self-phase modulation process by a delay line. The pulse that exits the delay line is effectively 'chirped' in a linear fashion and subsequently recombined by a filter as described above; for optical pulses, this filter is a pair of gratings.[3] This technique of pulse compression yields a narrowing factor on the order of 100 and is described at length in a text by Robertson (1995).

One final broadband technique of radar origin is the monopulse technique. The monocycle method relies on the inherent bandwidth of narrow pulses that are carrier-free. Solid-state technology has made it possible to deliver a high voltage impulse on the sub-nanosecond time scale. Again, the formula describing the bandwidth of such a pulse can be used to describe its spectrum, but in this case it is a DC pulse (*i.e.* $f_c = 0$) rather than a modulated rf signal, and the spectral distribution of the pulse therefore extends from DC to >2 GHz. With the proper antenna system (Mushiake, 1995) it is possible to imagine that a remarkably broad excitation spectrum might be achieved.

3.3 Modulation Circuits and Devices

One may pulse modulate a carrier frequency by one of two ways. The pulse element may be a control device that affects the driving conditions of an otherwise continuous wave source, or it may be an independent 'gate' of the frequency source (see Tsypkin, 1964 for theory). The tuning modes of a reflex klystron and their dependence upon the reflector voltage provide a means of control over the tube's output, and this represents an example of the former case. The latter scenario in which a cw-source is gated by an independent element is more convenient and may simply entail the imposition of a PIN diode switch on the output port of the source. The technique of choice, however, will depend on factors such as the desired pulse repetition rate, pulse duration, pulse power, and pulse shape.

3.3.1 Solid State Diode Switches

The most convenient method of generating low- to moderate- (*i.e.* less than 1 Watt) power pulses is via a gate that is external to the continuous wave frequency source because very fast solid-state switches can be purchased as drop-in circuit components that are already compatible with TTL or ECL logic. The operating principles of diode or transistor switches at microwave frequencies is the same as low frequency devices and behave as voltage-controlled resistors. In other words, a PIN diode or GaAs FET (field-effect transistor) operates as a fast switch when the semiconductor interface of the device is subjected to a DC bias level. Gated operation for pulse generation is accomplished by driving the diode or FET with a pulsed DC bias.

[3] Demonstrable at microwave frequencies by using rexolite optics

PIN diode switches have been traditionally used to gate microwave sources in pulsed EMR. Older devices were waveguide-based and entailed a so-called 'stick' diode mount that was connected across the gap between waveguide walls. This configuration, however, introduces stray inductance (the diode stick resembles an inductive post), and improved performance is obtained when ridged waveguide is used.

Presently, stripline PIN devices with coaxial drop-in packages tend to be used for switching and phase-shifting applications. The bandwidth in such stripline configurations is limited only by the bias circuit and any resonance that may arise in the diode circuit. These stripline devices therefore feature a multi-octave operation bandwidth (*e.g.* 2–18 GHz for General Microwave or similar devices, see manufacturer's catalog) and are usually preconfigured with a TTL interface. Design for broadband operation is only limited by an accompanying increase of insertion loss and poor flatness of power throughput across the operating band. With the introduction of MMIC technology PIN switches are being supplanted by GaAs FET switches as the latter become available as drop-in devices, and these GaAs switches tend to be faster because of the higher carrier mobility through the semiconductor (InP based devices would presumably be yet faster for the same reason). Yet both PIN and GaAs FET devices are subject to power limitations by their respective low breakdown voltages and poor thermal dissipation.

3.3.2 MOSFET and Tube Switches

The principal advantage of diode switches is their speed of operation, and low power pulses of width 15–20 ns can be generated with a single-stage unit. The disadvantage of such switches is their low power rating (ca. 1 W), which is due to the damage imposed on the semiconductors, and therefore the high power pulses that are necessary for spectral hole-burning must be created either by amplifying the low-power pulses or modulating a high power frequency source. Since amplification of the low power pulses is best achieved by electron tubes, it follows that at least a portion of any pulsed EMR spectrometer will contain high power switches. It is possible to modulate the electron gun directly, but it is better to modify the gun by inserting a gridded electrode between the cathode and anode, which is then used as a switching device (see Section 4). In either case, the TTL pulse logic is amplified to power levels that are much higher than those used by a PIN diode driver, and specialized circuits are used.

It is possible to modulate an electron beam by varying the voltage on the focusing electrodes, but this requires very large voltage swings and the accompanying difficulty of achieving sufficiently fast rise/fall times. It is therefore preferable to use a control grid that functions in the same manner as focusing electrode modulation, yet requires only a more manageable high voltage pulse on the order of 100V and whose amplitude decreases as the grid is placed closer to the cathode.

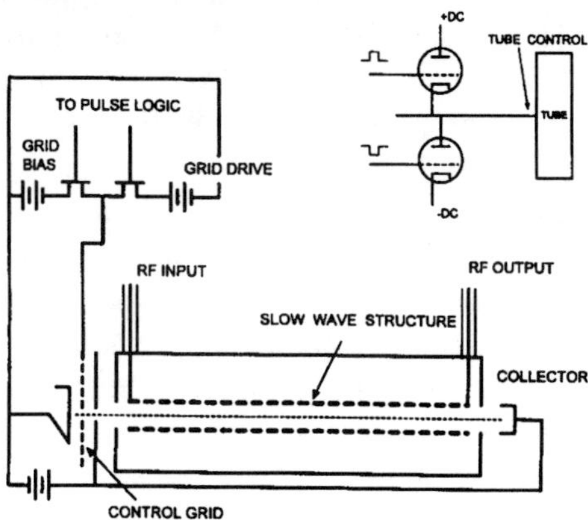

Figure 7. Grid modulation of an electron tube by a floating deck MOSFET or tube circuit. A complete MOSFET schematic is illustrated connected to a TWT grid (after Sivan, 1994), and a rough block diagram is depicted for the analogous tube circuit (after Ewell, 1981).

The introduction of a grid between an anode and cathode, in effect, alters the bias voltage specifications because of the close proximity of the grid to the cathode (*cf.* Gewartowski & Watson, 1965). The cathode-anode bias voltage can therefore be maintained at some optimum value, but the grid voltage (relative to the cathode) can be set to some value that modifies the electrostatic field between cathode and grid with the result that the electrons are not impelled across the space charge region. The principal advantage of this architecture is the lower grid voltage that facilitates electron beam modulation. The grid has two drawbacks, however. The first is an adverse effect on the beam quality because even when 'on', the tube's architecture causes imperfect focusing of the beam. Secondly, the grid is subject to physical damage from the electron bombardment, which contributes to tube failure.

Control grids are generally driven by MOSFET or vacuum tube modulator circuits whose function is to convert a low voltage control pulse to a pulse of sufficiently high voltage and current so that the electron beam is deflected. The modulator is often operated in a 'floating' configuration at the cathode potential (*i.e.* −10kV), and therefore the modulator circuit design must incorporate isolation between the high voltage power supply and the low voltage pulse circuits. A second design constraint put upon the modulator is high pulse fidelity; droop and ripple in the 'beam on' pulse will introduce phase noise in the rf pulse because grid voltage fluctuations will modulate the electron beam.

Figure 7 illustrates a representative example of a floating deck modulator intended as a means of grid control in a traveling wave tube. Two DC power supplies provide the negative grid bias (beam off) and positive grid drive (beam on) voltages. The switches (MOSFET or tube) must be isolated from the low voltage (<10V) pulse circuits, and this is accomplished via conventional transformers or opto-couplers. Typical grid bias and grid drive voltages are –150V and 250V, respectively.

The driver switches typically generate 250 V pulses with rise and fall times of approximately 50–100 ns and a minimum pulse width of 100 ns. Pulse jitter is likewise typically rated at 5 ns. It follows therefore that short pulses for electron magnetic resonance are generated in the following manner: the TWTA is switched to a 'beam on' status by a 150–200 ns grid pulse, and a short (*e.g.* 20 ns) low power microwave pulse is passed to the tube's rf input during this 'beam on' time window. The location of the low power pulse within this window is optimized in order to minimize spectrometer deadtime (*i.e.* situate pulse close to grid driver's falling edge) while at the same time avoiding a period in the grid pulse during which there is any voltage transient (*i.e.* the edges or any associated ripple). This optimization is a trial and error procedure whose outcome depends on the quality of the grid driver pulse, particularly the fall time. So-called 'tail biter' circuits may be used to sharpen the grid driver pulse's falling edge.

4. HIGH POWER PULSES: SOURCES & AMPLIFIERS

Electron tubes (Kleen, 1958; Hutter, 1960; Gilmour, 1986; Liao, 1988) are used as a means to generate very high power microwave signals on the principle of energy transfer from electron beams, a DC signal, to an initially small AC signal. The tubes operate at high voltage, and they may be modulated in order to provide the requisite rf pulse by gating the electron beam (*i.e.* the high energy DC source). The two common methods of gating the electron beam are: (1) modulation of the cathode voltage, which has the disadvantage of requiring very large voltage swings (*e.g.* > 10 kV) on a very short time scale; and (2) beam deflection and interception by a grid, which may be achieved with much lower voltage transients. This section contains descriptions of three electron tube devices that are used in electron magnetic resonance; they represent but a few of the available options.

4.1 The Pulsed Klystron

The klystron mode diagram was introduced in Chapter 1 in order to demonstrate the interplay among the various electrical and geometric parameters that must assume optimal conditions in order for the tube to operate. Amplitude/pulse modulation of the signal output may therefore be effected by varying the tube's electrical drive conditions so that it may be driven between operable and inoperable states. A reflex klystron, for example, may be modulated by varying the reflector, grid, or beam voltage, and each manner of control has its own relative merits.

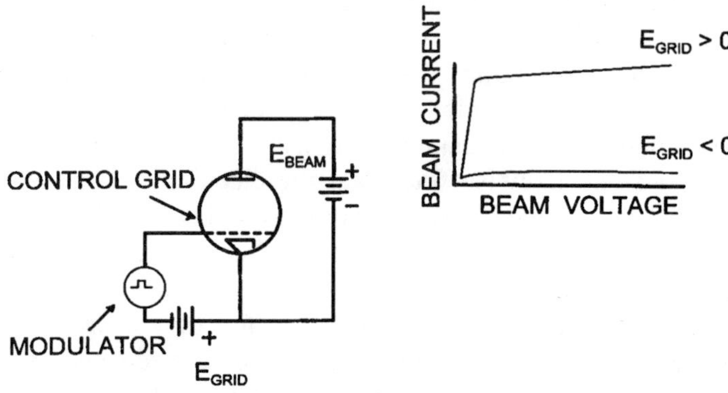

Figure 8. Grid controlled modulation of an electron tube. The electron beam is turned off by a negative bias that deflects the electron beam; this is represented as the beam *vs*. beam voltage plot (inset). The grid bias must become increasingly negative as the beam voltage is increased, and therefore the optimum bias and 'pulse on' condition varies with the beam voltage.

Reflex klystrons are preferentially modulated at the reflector electrode, which draws very little current and is therefore compatible with fast (hard-tube) modulator circuits. The klystron, however, is an *RLC* circuit device and subject to a 'build-up' time that affects the tube's suitability for pulsed operation. Once put into an operating mode, the time interval that ensues before useful power is generated will depend on the physical dimensions of the transit region and electrical parameters such as coupling between beam and cavity. Reflex klystrons, in general, have a short build-up time because of their compact geometry, high load and unitary coupling (Harrison, 1947), but build-up times increase as one increases the number of cavities in the klystron.

The klystron is a variant of the triode or tetrode type of vacuum tube and, in principle, may be grid modulated. In triodes, the control grid is a wire screen (or parallel posts) that is situated close to the cathode in order to control the electron flow between cathode and anode, but without drawing current. If the control grid is placed close to the cathode and biased so that it is negative relative to the cathode, then space charge effects influence the electron beam trajectory and affect the beam current. Since the grid controls the electron beam trajectory and current via space charge effects, it follows that an interplay exists between the grid bias voltage, grid proximity to the cathode, and the beam voltage. The grid bias voltage (to shut off the tube) must therefore be increased if either the beam voltage or distance between grid and cathode electrodes is increased. A representative circuit depicting the power supply connections to a gridded triode and the effect of bias voltage on the beam current are illustrated in Figure 8.

The advantage of the control grid (triode) or control grid with screen (tetrode) is that the configuration affords control over the tube's power output with very low voltages (tens of volts) and current. The primary disadvantage, and presumably the reason that one does not find many gridded klystrons, is the interplay between geometry, grid voltage, and beam voltage. At microwave frequencies the geometries of the tube become constrictive, and one likewise finds that the electrical parameters become problematic. The so-called klystrode is one example of a gridded transmitter tube that is capable of pulse mode operation, but it tends to be used below 500 MHz. One should bear in mind however that factors causing a tube design to be rejected by radio engineers may not be a cause to reject that design for spectroscopy. Many tube configurations are rejected because they offer poor performance as radar or radio transmitters, which demand enormous power output; spectroscopists do not require such high power and can work with a tube design or circuit configuration that might not be optimal from the standpoint of communications (so long as other factors such as pulse edge times frequency/phase coherence *etc.* are compatible with our experimental needs).

As an alternative to pulse generation by 100% amplitude modulation, the reflector voltage may be varied only to the extent that the carrier frequency of the reflex klystron is shifted and brought into resonance with the sample resonator for short interval. Gordon & Bowers (1958) used this technique in their spin echo study of donor impurities in silicon. The klystron was ordinarily tuned and locked to a frequency approximately 14 MHz above the sample resonator's frequency, and the response time of the synchronizer circuit was sufficiently slow that it could not respond to the short duration shift.

Pulse modulation of a reflex or two-cavity klystron may be used for low power applications, such as saturation recovery and transfer, but the power levels attained with these tubes are, by themselves, inadequate for pulsed EMR experiments that require very short high power pulses. For the latter, allow power klystron may be used with a second klystron amplifier, as found in the description of spectrometer by Lick *et al.* (1973). A low power klystron oscillator was pulse modulated as described above, and the low power pulses were amplified by a hard-tube modulated second klystron to yield 0.2–1 μs pulses rated at 7 kW peak; shorter pulse widths were achieved by modulating the low power klystron with an external PIN diode switch (Lick *et al.*, 1973).

Klystron tubes with two or more cavities may be used in a configuration in which a small signal is imposed on the electron beam, which is then velocity modulated and transfers its energy to a second or higher stage cavity (Figure 9, top left). Depending on the number of stages, it is possible to achieve gains of 45 dB upwards, although bandwidths are inherently narrower than those of traveling wave tubes that use slow-wave structures. Despite what would seem to be a limitation in pulsed EMR, however, the cavity resonators of klystron amplifiers may be mechanically tuned in order to provide the requisite frequency coverage, or, if desired, the cavity resonators may be stagger-tuned (see Figure 9, inset) in order to make a given amplifier operable over a wide band width, albeit at the 'cost' of lower gain. This gain loss is not problematic, however, because many klystron amplifiers tend to yield power outputs that greatly exceed amounts required to drive the spin system.

Finally, klystron amplifiers may be advantageous at millimeter frequencies, although a direct scaling of a cavity klystron is fraught with problems. Hybrid tubes that are known as Extended Interaction devices possess features of both klystrons and traveling wave tubes. A slow wave structure surrounds the electron beam, as in a TWT, while the rf signal is amplified in successive cavity resonators (Figure 9). Typical devices (Varian Associates, VK series devices) are available as cw or pulsed sources with kilowatt power outputs. Extended Interaction Amplifiers are specified for operation at 95 GHz and ~30 dB gain. One should be aware, however, that the voltages that are used with these tubes are extremely large; the cathode voltage is typically 21 kV for a 1 kW tube, and the grid bias is on the order of 1 kV. Modulator circuits for such tubes will therefore be specialized.

4.2 The Pulsed Traveling Wave Tube Amplifier

The most common microwave amplifier that is used in pulsed EMR (*cf.* Taylor *et al.*, 1969; Tan *et al.*, 1984) is based on the traveling wave tube, or TWT. The traveling wave tube amplifier, or TWTA, is an integrated unit that consists of a high voltage (~10 kV) power supply, a high voltage (~300 V) pulse generator, and a traveling wave tube. Microwave amplification is performed by the tube, which, like the klystron described in Chapter 1, transfers kinetic energy from an electron beam to the high frequency signal. The advantage of using an integrated amplifier, as opposed to linking a tube with off the shelf power supplies and pulse generators, is that the former is better engineered to protect the tube from damage incurred from improper operation. Because they are not required to generate pulses on the order of 10ns, however, the modulators of commercial TWTAs do not have very short pulses edges, but they have been greatly improved with the introduction of electro-optic devices. Although they do not publish the circuit, Taylor *et al.* (1969) report the construction of a home-built modulator with 6 ns pulse edges that, together with the Litton L-5022 TWT (discontinued), yielded kilowatt pulses with 2 ns edges and on/off ratio in excess of 120 dB.

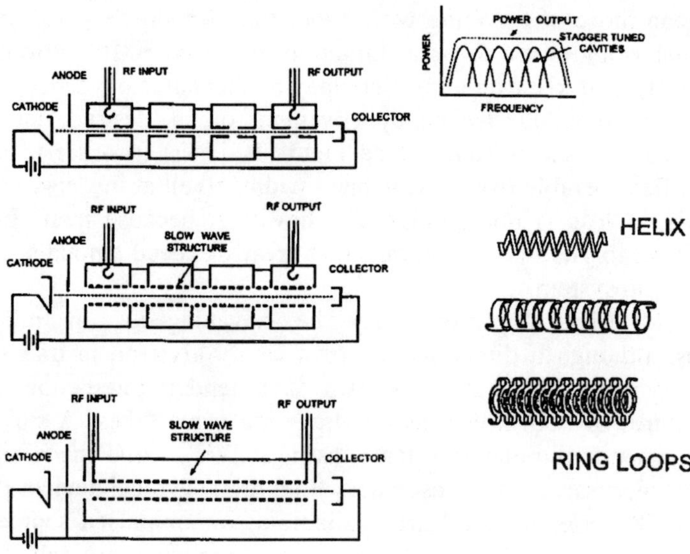

Figure 9. Electron tube amplifiers. Top: the multi-stage klystron, which may be rendered broadband by staggered tuning of the cavities. Middle: an Extended Interaction Amplifier schematic diagram, illustrating features of both a multi-stage klystron amplifier and traveling wave device. Bottom: A traveling wave tube amplifier. Representative examples of slow-wave structures are illustrated to the right.

The traveling wave tube (TWT; Figure 9, bottom illustration) is an electron tube that consists of a slow wave structure and an electron gun. The slow wave structure and electron beam are geometrically arranged so that the beam and radiofrequency field may interact; for example a helix through which the electron beam passes along the cylindrical axis. As it enters the slow wave structure, the group velocity of the radiofrequency wave is altered so that it is comparable to the velocity of the electron beam that passes through it. As a result, the radiofrequency field causes bunching of the electron beam along the propagation length, and the beam transfers kinetic energy to the traveling wave. The gain of the tube is therefore proportional to the length of the slow wave structure; typical tube lengths are 30–40 cm.

At X-band and lower frequencies, the slow wave structure of choice is the helix or its variant, the ring bar (or ring loop). The advantage of these structures is that they are inherently broadband (octave or more) and their geometry is well suited for accommodating the electron beam. Some aspects of helix design are described in Chapter 1, but in general, their diameter is very small, and they become impractical at frequencies above 18 GHz. TWTs that operate at high frequencies are therefore designed by using a coupled cavity architecture, which has an inherently narrower operating bandwidth. The functional attributes of both the helix and coupled cavity architectures may be understood as filters, as described in Chapter 1.

The electron gun operates in the same fashion as in the klystron tube. Electrons are emitted thermionically from a cathode that is in close proximity to an anode. The beam passes the anode and travels through the slow wave structure before being collected at the tube's end. Because the continuous electron bombardment is detrimental to metal, the beam must be focused to minimize collisional encounters with the helix and anode. The coupled cavity structure tends to be more robust and is the structure of choice for very high power tubes. Focusing is achieved by an electrode that is held very near the cathode potential and situated between the cathode and anode; its sole purpose is to reduce the cross-sectional area of the beam.

The cathode is very often constructed of a metal oxide material such as calcium, barium, or aluminum oxide that is incorporated into a metal matrix. This design, as opposed to a conventional metal or coated metal, is preferable because it is less prone to poisoning by volatile material in the tube and damage when operated at high current densities. The proper rate thermionic electron emission is facilitated by a heater that is usually tungsten based. The heater is operated at high currents and a DC voltage to ensure minimal interaction between the heater source and electron beam; it is also biased negative (typically −6.3 V) relative to the cathode in order to ensure that the heater current does not induce electrolysis of the cathode. The cathode bias, designated E_c, is negative relative to ground (typically −10 kV), and the beam is collected at the opposite side of the tube by a multistage collector that is likewise biased negative relative to ground (albeit only some small fraction of E_c).

One or more anodes are used to generate the requisite electron velocity. These anodes are fixed at a voltage close to ground and may be connected to the tube body, but some designs call for a resistor to be inserted between the anode and ground as a protective measure against overcurrent if the cathode arcs to the anode.

All components of the TWT are held under a very high vacuum, which is necessary to ensure that the cathode does not degrade because of surface 'poisoning'. Cathode poisoning is defined as the reaction of and/or deposition of volatile materials on the cathode surface, which lowers the beam current by introducing an effective surface resistance. Furthermore, if the surface 'corrosion' of the cathode surface is not uniform, then electron emission will likewise be non-uniform and the beam will be defocused. The source of the volatile material is all exposed surfaces, including the electrodes, slow wave structure, and ceramic casing. Despite degassing at the factory, volatiles may still be released by the tube components post-production if the tube develops abnormal hot spots or is overdriven (excessive high voltage or rf input). Although counterintuitive, a tube that is suspected of being gassy may be prepared for operation by turning on the heater and/or high voltage for a short period prior to passing an rf signal to it. The heating of the cathode and associated components of the electron gun volatilizes material that then condenses on the cooler walls of the tube.

At very high gains, it is impractical to operate a traveling wave tube continuously because of the damage the electrodes and helix incur from the electron beam, and the beam is gated for operation as necessary. The TWT grid structure resembles a nested pair of sieves that are used in kitchen sinks, and the grid closest to (and often mounted on) the cathode is called a shadow grid. The shadow grid takes the brunt of the damaging electron bombardment while the second control grid is actually used to modulate the beam. The grid is constructed of very fine wire and is very sensitive to burnout. At 'beam on' the grid is positive relative to the cathode, whereas the 'beam off' state puts the grid negative relative to the cathode. It follows therefore that the standard set up and operation procedure entails the application of a negative DC bias relative to the cathode and then pulsing the grid with a positively-going DC voltage whose amplitude is about twice the grid bias voltage. The voltage swing for 'beam on' is typically 250 V for a grid bias of −110 V, and the grid is a capacitive load on the order of 100 pF (direct measure of grid lead to tube body). A MOSFET transistor is the preferred method of pulse generation (Section 3.3.2).

Finally, high frequency (*e.g.* V-band) amplification can be achieved with a device known as a peniotron (see Liao, 1988), which resembles a TWT in layout, but the electron beam follows a helical trajectory through the rf structure. The peniotron rf structure is a section of rectangular waveguide along whose broad walls run a pair of double ridges that support the TE_{10} mode and concentrate the DC electric field.

4.3 The Pulsed Magnetron

The performance of klystrons and traveling wave tubes depends upon the quality of the electron beam, which must remain focused over the cathode-anode distance and in spite of such necessary appurtenances as the control grid and the resonant (or slow wave) structure. The end face of the cathode must therefore be (cylindrically) symmetric both in respect to the surface finish and material quality. But if the electron beam emanates from the side of a cylinder in a coaxial electrode arrangement, then one obviates the problem of beam focusing and relaxes some of the design constraints. The magnetron is a self-oscillatory device that, like the klystron tube, couples the power of a DC source to an *RLC* circuit in order to generate an AC signal, but unlike the klystron or traveling wave tube, the electron trajectories are cyclotronic because the magnetic field is perpendicular to the electric field within the tube.

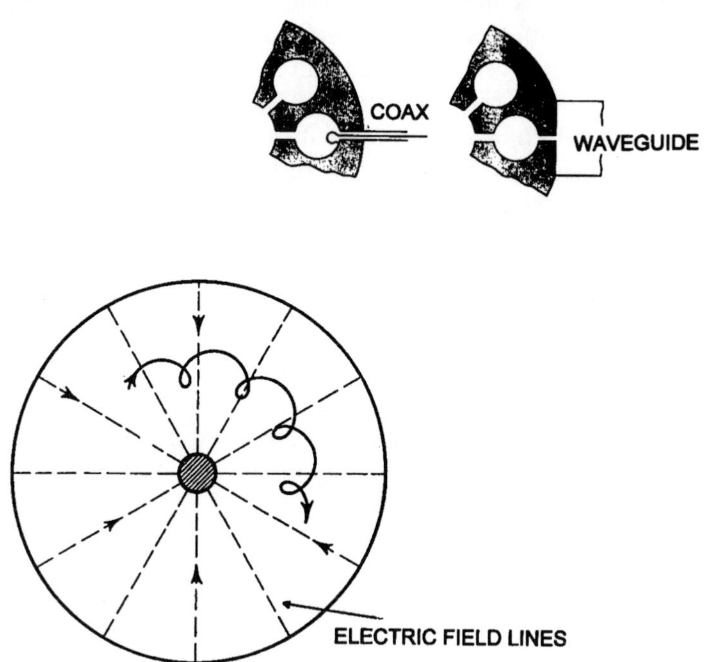

Figure 10. Schematic diagram of the magnetron. The cyclotronic path of the electron within the device is illustrated (the DC magnetic field lines run perpendicular to the page). The cathode corresponds to the center post, whereas the anode corresponds to the circular boundary in the diagram. The anode is actually a ring of resonant structures, hence the coupling of electron beam energy to an oscillatory signal (cf. Chapter 1). Two representative means of coupling the magnetron to an rf circuit are shown.

As an rf power source alternative to the klystron, the magnetron provides the advantage of operating at significantly higher efficiency (40–70 %), which enables one to achieve high power yields with a compact device. Like the klystron, the magnetron may be mechanically or electrically tuned (<10 % bandwidth) and features a low pulse-pulse frequency jitter (10–100 kHz, depending on the design). In the free-running mode of operation the magnetron tends to be noisier than the klystron, but it may be stabilized (like the klystron) with a low-level reference signal and phase locking. Its principal drawbacks include a strong sensitivity to loading and a problem known as 'moding', which refers to a generation of spurious and harmonic frequencies that apparently arise from the anode, which is a multi-resonator structure. The latter problem is remedied, however, by antiphasing the modes of the adjacent anode resonators, and this may be accomplished together with enhanced frequency stability by inserting the anode within a cylindrical TE_{011} cavity (*i.e.* coaxial magnetron).

Figure 10 illustrates the components of the magnetron. The device is axially symmetric, with a centrally-located cathode and surrounding annular anode. The DC (electric) field lines and direction of electron travel therefore extend radially from the cathode to anode. The AC signal is obtained by two modifications of the axial electrode system. The first modification entails the segmentation of the anode by machining resonant structures into the annulus so that they are electrically coupled to the electron beam and magnetically coupled to each other. The second modification affects the trajectory of the electrons. A magnetic field is applied in the direction perpendicular to the DC electric field, with the result that the electrons move with a cycloid trajectory. As the magnetic field is increased from zero, the radius of the electron's trajectory decreases, but the anode current remains constant until the magnetic field reaches some critical value H_c This critical value of the magnetic field occurs when the radius of the electron's trajectory brings its path tangential to the anode surface; beyond the radius, the electron returns to the cathode.

The resonators that are located in the annular anode are oriented so that their capacitive element is directed inward. The AC field E-field lines of the resonators project into the gap between the anode and cathode, and coupling to the electron's cycloid trajectory takes it through the resonator's E-field lines, and the electrons bunch as they pass through the field projections. A retarding field captures the energy of the electron and transfers it to the rf field, whereas an accelerating field causes back bombardment of the electron back into the cathode. Antiphasing and imposing electron beam conditions so that the electron takes a half period to move between adjacent resonators ensures that the electrons encounter a retarding field and enhance the rf power.

Magnetrons have been used in the design of pulsed EMR spectrometers (Cowen *et al.*, 1962; Greenslade, 1991; 1993), but one should be aware of device characteristics that must be corrected. Because magnetrons are self-oscillatory, pulsed operation is accompanied by a period of instability as the electrical parameters assume values that allow for stable power output. The moding problem is particularly acute because of the various voltages (and corresponding electron velocities) that may coincide with a resonant frequency. As a rule, magnetrons are contraindicated for operation with short pulse widths (<100 ns) and for coherent systems. Moding in conventional magnetrons is also attributed to spurious or missing pulses in radar transmitters, but this is apparently ameliorated by the coaxial design. These design limitations may, however, be corrected (see Silvan, 1994) and in fact make the magnetron well suited to pulsed ELDOR experiments because they underlie the device's high frequency agility.

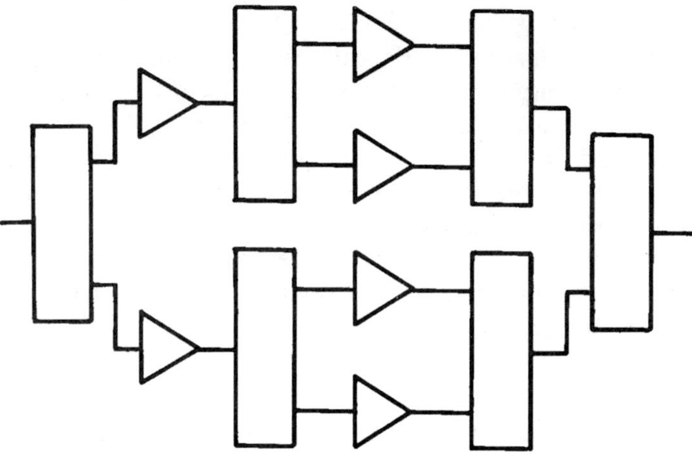

Figure 11. Functional schematic diagram of a solid-state power amplifier. Low power transistors are combined in series and in parallel via power combiner/dividers (rectangular symbols) to provide the requisite high powers.

4.4 Tube vs. Solid-State Amplifiers

Solid-state devices (Liao, 1985; Ostroff & Borkowski, 1985) are gradually replacing low power rf sources and are increasingly competitive as amplifiers below 1 GHz. Among the cited advantages of using solid-state devices, as opposed to tubes, are (1) high bandwidth, and (2) no need for a pulse modulator when operating as a Class-C device because the rf drive itself provides the necessary bias for amplifier turn-on. GaAs FET amplifiers are particularly well suited to applications in the 1–20 GHz frequency range because the transistors are easily integrated into circuits containing passive devices and stripline design, and IMPATTS or GaAs Gunn devices may be used to obtain cw or pulsed power above 30 GHz (Kramer, 1981).

The self-biasing of Class-C solid state amplifiers characteristically leads to pulse sharpening, which can complicate applications requiring shaped pulses, but the principle limitations of solid state amplifiers is the breakdown voltage at the collector and heat disposal limitations of semiconductors that together limit the power handling capabilities of a transistor. Amplifiers therefore tend to be constructed as distributed arrays of transistors that are arranged in parallel and series stages (Figure 11), which requires careful design of the power combiners and splitters so that there is adequate isolation in parallel and series. The DC power supply is also stringently designed so that power is supplied uniformly over the transistor array throughout the pulse-up period because each transistor must be driven identically in order to ensure proper phasing of the pulses as they are combined.

Solid-state amplifiers characteristically feature a low ratio of peak pulse power to cw power output, in contrast to tubes. Radar transmitters therefore tend to be constructed of 102 to 103 amplifier modules that each consist of many parallel and series arrays of microwave transistors. Typical peak power output per module is 300 W (L- or S-band units) and drives one member of a phased array antenna. Unlike a radar transmitter, pulsed EMR spectrometers require less than a half dozen such modules whose output may, in principle, be combined for delivery to a sample resonator by using a multiport waveguide section.

5. TRANSMISSION LINES & RESONATORS IN PULSED SYSTEMS

5.1 General Principles

The design considerations for transmission lines in pulse-modulated systems are similar to that of continuous wave systems, which can be described in terms of standing wave interactions (*cf.* Altschuler, 1963), but in the latter case the pulse response may vary because of the *RLC* properties of the transmission line. As a familiar example, a rectangular DC pulse can be 'shaped' by passing it through a network that consists of reactive circuit elements (capacitors and inductors). High frequency transmission lines possess an inherent *RLC* network characteristic that may be optimally matched to some characteristic impedance as described in Chapter 1. An important difference between cw- and pulsed circuits, however, is that although the circuit match may be optimized for some fixed frequency f_c and perform well under cw operation, the pulsed circuit must pass a band of frequencies that depends on the pulse width (Section 3). In other words, the pulsed circuit must be designed as a bandpass filter.

Two-port junctions were described in Chapter 1 in terms of a standing voltage wave ratio (VSWR) that is related to impedance, and three conditions define the limiting cases that describe the response of a traveling (cw) wave at a junction: (1) open circuit ($Z_x = \infty$), in which the wave amplitude is identical at the junction as it was at the origin; (2) ideal 'match' ($Z_x = Z_0$), in which the wave amplitude is zero at the boundary; and (3) short ($Z_x = 0$), in which case the incident wave is reflected. Figure 12 depicts these conditions in the case where the incident energy is delivered as a pulsed waveform, with the resultant observation that the pulse envelope (pulse/amplitude) responds in a manner that is similar to that of a standing wave. It follows that two-port junctions such as terminations and joints need to be considered carefully in pulsed systems if one wishes to avoid spurious pulses that arise from reflections; rather than simply affecting the transmitted power, a poor match can introduce reflected pulses that may affect the measurement.

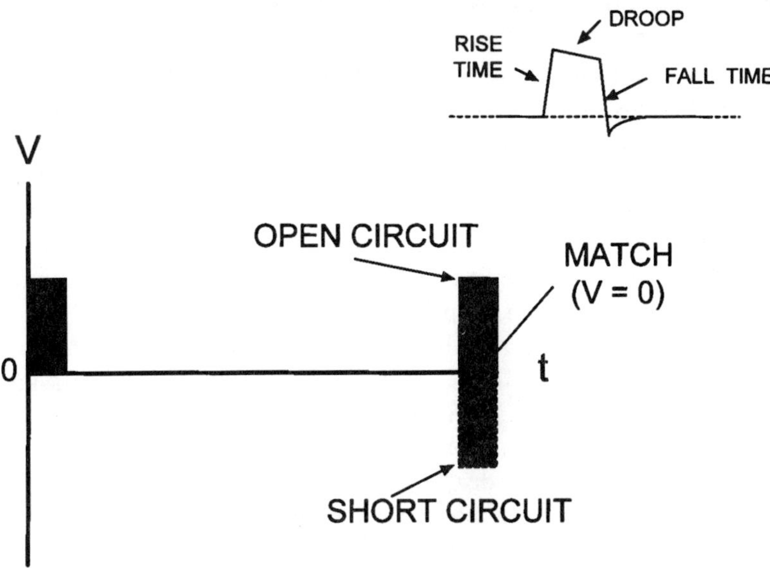

Figure 12. Pulse modification along a transmission line. An initially square pulse at some origin (t = 0) will be reflected with a phase shift that is dependent upon the impedance of the junction (represented as a maximum along the time axis). A perfect impedance match across the junction yields no reflection. The inset pulse profile indicates the typical distortion of an ideal rectangular pulse that will occur with a real transmission line because of phase shifts and losses associated with the finite capacitance and inductance along the line.

In general, radar system design guides recommend that the transmission lines be short and the number of joints and discontinuities be minimized. It follows, therefore, that there are potential benefits to applying MMIC technology to the design and construction of pulsed EMR spectrometers rather than assembling a system tinker-toy fashion using discrete connectorized components. Until such time that a MMIC EMR spectrometer is designed and tested, however, a judicious use of isolators may reduce unwanted reflections. Figure 12 (inset) illustrates a representative example of pulse shape distortion that can result from the discrete components approach. For an otherwise uniform transmission line, however, the pulse distortion is only manifest as a variation of rise time that is primarily due to skin depth and radiative losses (Cazenave, 1951).

The manner by which a reflected (spurious) pulse may enter the experimental pulse sequence is outlined by describing a simple experimental scenario. In the practice of pulsed EMR, the setting of pulse timing and power parameters so that π and $\pi/2$ pulses are created differs from the procedures that are described in NMR texts (Fukushima & Roeder, 1981). For example, one sets up a three-pulse stimulated echo pulse sequence by setting one's logic generator so that three pulses of equal length are issued per experimental 'frame'. One would then pass these pulses to the TWTA and adjust their power by using an attenuator (located between the amplifier and the sample resonator) to maximize the echo amplitude, using the assumption that the maximized echo amplitude indicates that the pulse timing-power characteristics now correspond to the requisite $\pi/2$ per pulse condition. If there are losses at junctions due to mismatches, however, one has been compelled to actually compensate for losses by producing pulses that exceed the $\pi/2$ per pulse condition. In other words, what one monitors on the oscilloscope during set up is the pulse power-timing combination that provides a $\pi/2$ pulse sequence in spite of losses due to reflection.

A reflected pulse will travel back in the direction of the transmitter (TWTA) where it may be eliminated from the circuit by isolators and/or terminated 3-port circulators (see Section 6). But such a spurious pulse could remain in the circuit if a second reflection turned it back towards the sample. One might, for example, consider the following scenario in which there are situated two junctions between the transmitter and the sample resonator. One of these junctions is simply the coupling iris between the transmitter and the resonator, and the other can be any other junction, such as the attenuator that is used to set the pulse power level. It stands to reason that as the junction mismatches increase the proportion of the pulse power that is reflected (and thereby compensated by admitting more power from the transmitter), the reflected pulse is going to have a power that represents a significant turning angle and effect on the spin dynamics. Reflections also occur on the receiver side of the sample resonator because there is a gating switch that is closed during delivery of the high power pulses so that the receiver is not damaged. If this gating switch is reflective, and there are no devices interposed to prevent the reflected pulse from returning to the sample resonator, then these reflected pulses will spuriously enter into the driving of the spin dynamics.

The handling of spurious reflections is exemplified in the spectrometer circuit and sample resonator design of Mims (1965b). In this design there are separate transmitter and receiver branches of the circuit, which are spanned by a $\lambda/2$ transmission line resonator (*cf.* Section 5.3.3). The sample resonator spans (and, in effect, couples) two tapered waveguide sections, but the taper on the receiver branch is steeper than the one on the transmitter side of the resonator. The steep taper offers a high VSWR to waves/pulses that might be reflected back from the receiver gate, while the gradual taper on the transmitter side presumably improves the match of the resonator to transmitter and optimizes delivery of the incident pulses (*i.e.* preventing reflections back towards the transmitter).[4]

5.2 Tapered Lines & Terminations

The so-called Gordon Coupler is a convenient means to achieve a match between the spectrometer bridge and sample resonator. As described in Chapter 1, a section of transmission line possesses *RLC* characteristics that vary according to its geometric parameters. In the case of a coaxial line, *RLC* will vary as the thickness of the inner conductor is varied, and the *RLC* characteristics of a hollow waveguide will likewise vary as the cross sectional dimensions change. It follows therefore that two waveguide sections of impedance Z_a and Z_b can be matched by an intervening tapered section whose impedance varies continuously along its length. The taper is analogous to an impedance matching transformer and has a 'turns ratio' similar to a conventional power transformer. Unlike a conventional transformer, however, the impedance ratio that may be matched by a single taper is small. Frank (1943) demonstrated that a tapered dielectric plug of high permitivity increases the impedance ratio, and this is the essence of what EMR spectroscopists call a Gordon Coupler.

The coupler consists of a section of hollow waveguide whose dimensions are smaller than the primary waveguide transmission line, and therefore does not pass the carrier frequency.[5] This hollow waveguide section abuts the sample resonator whose match to the characteristic impedance is desired, and the waveguide section is filled with a dielectric plug that lowers the cutoff frequency of the otherwise hollow guide. A tapered waveguide section mates the standard hollow waveguide and the filled section, thus creating a 'match'. The dielectric section is adjustable along the propagation direction in order to facilitate coupling between the guide and the resonator.

For a given tapered waveguide section, the impedance varies along the propagation direction x according to the formula:

$$Z_x = (L_x / C_x)^{1/2} \tag{5}$$

in which L_x and C_x may be expressed as linear, exponential, or Gaussian functions. The most convenient taper, with respect to fabrication, is the linear taper, although the latter may outperform the linear taper in certain

[4] Some years ago a colleague suggested to me that the taper disparity in the Mims resonator was intended as a means to bias the direction of the echo power. A test of this suggestion was made and is described in Section 5.3.3 together with details of the Mims resonator design.

[5] For example, for an X-band spectrometer using WR-90 rectangular guide, the coupler might be fashioned of a short section of WR-62 guide.

situations. Kaufman (1955) lists a bibliography of analyses and design principles of various tapered line transitions, and Frank (1943) describes the analysis of the linear taper. Regardless of the style of the taper, however, the dielectric plug must likewise be tapered (usually as a mirror image of the opposing tapered waveguide wall) so as to increase the effective turns ratio for matching and avoid the worst case scenario of causing reflections that would occur if the incident pulse encountered a flat face of a dielectric (an abrupt change in a waveguide property, including dielectric load, corresponds to an abrupt change in the impedance of the line).

In order to avoid pulse shape distortion, the frequency 'band' of the pulse must pass the transition with the same delay time. This property of the device under test (here the transition, but generally applicable to filters, *etc.*) is called the 'group' delay of the pulsed signal and should ideally be uniform across the frequency spectrum of the signal. The time delay of a transmission line is equal to the square root of the product of L_x and C_x and therefore varies with x for all cases except the exponential taper. Analysis of the commonly used linear taper (Frank, 1943) reveals several optimal design principles.

The optimal taper angle is frequency-dependent, which follows from the analogy to the abrupt step in waveguide dimensions. Clearly, as the wavelength increases the length of the taper may, by comparison, appear to be a steep step rather than a gradual taper. At low frequencies, therefore, a tapered line matching transformer becomes impractical. In general, the minimal distortion of pulse shape is achieved when the fractional change of C or L per wave length is less than 6, and for a given pulsed carrier the wavelength selected for the analysis corresponds to the lowest frequency component. For nanosecond pulses, however, the rate of line taper relative to the pulse length (and the dispersion of frequencies) is unimportant.

5.3 Sample Resonators

EMR sample resonators are analogous to transmission line bandpass filters and may be designed along similar principles (see Malherbe, 1979). A general description of filters appears in Chapter 1, but some of the salient features of bandpass filters will be repeated here in the context of pulsed EMR. A sample resonator, like a bandpass filter, responds to a select band of incident frequencies in the electromagnetic energy spectrum, which is determined by the *RLC* properties of the resonator. For example, Figure 13 depicts a network analyzer trace for a filter/resonator; the plot represents the power transmitted (in dB) as a function of frequency. The difference between the peak power transmitted and the incident power level is defined as the inherent loss of the filter, and the pass band is defined by the range of frequencies spanned by the upper and lower frequency limits at which point the transmitted power is attenuated relative to the center frequency power by 3 dB.

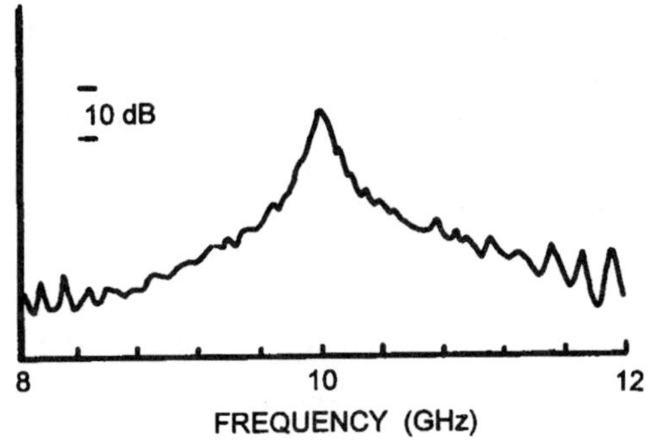

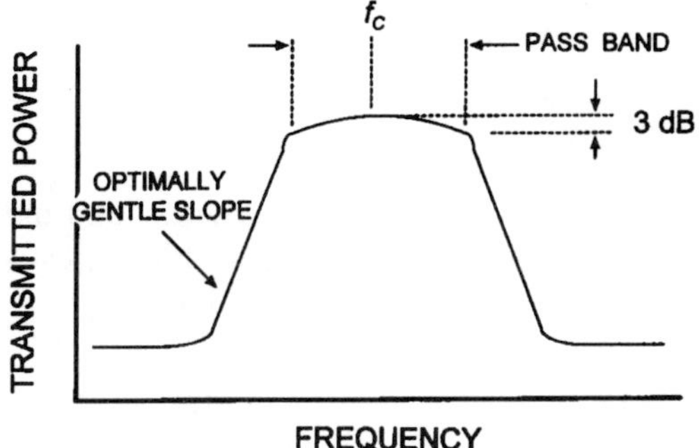

Figure 13. Power transmission of a pass band filter or sample resonator. The lower portion of the figure illustrates the optimal parameters for a given resonator; the specified 'gentle slope' minimizes group delay (see text). The upper portion of the figure is an actual pulsed network analyzer trace (recorded using a Hewlett-Packard Transition Analyzer – essentially a very fast oscilloscope with FFT capability) recorded with a Mims style resonator. Specific details of the Mims resonator analysis are given in Section 5.3.3.

A microwave pulse of carrier frequency f_c and width Δt_p will possess a power spectrum spanning $f_c \pm \Delta t_p^{-1}$. One can define a set of frequencies $\{f_i \mid f_c, f_1, f_2, f_3, \cdots\}$ that encompasses all of the δ-function representations of the energy spectrum defined by the band $f_c \pm \Delta t_p^{-1}$, and this set will be called a group, following the parlance of wave physics. A traveling wave, or pulse, of frequency f_i will be transmitted along a constant impedance line with some phase shift $\phi(x)$, and an important question in the design of filters, in general, is the group or set ϕ_i variation among its members, the so-called 'group delay'. A bandpass filter or resonator may therefore propagate a given element of the energy spectrum f_i differently than another element of the group f_j. In other words, a given filter is defined by both its frequency and time domain response.

The group delay, or time variation, among the pass band spectrum is greater for filters that feature rapid cut-off. The sharper the profile of transmitted frequencies, as depicted in Figure 13, the greater the variation among the set f_i and the greater propensity towards ringing (*cf.* Williams & Taylor, 1988; Rhea, 1994). An important attribute of the Mims resonator, therefore, is the gentle slope of the network trace above and below resonance; an overcoupled cavity resonator with comparable *Q* and bandwidth does not have cut-off regions as gently sloping as the Mims resonator (or some variants of the loop-gap) and therefore tends to ring longer following an impulse (Bender, unpublished data). The ideal qualities of a sample resonator for pulsed EMR are: (1) a 3 dB pass band that exceeds the excitation spectrum of the pulses used; (2) minimal loss at the pass band center (the Mims cavity losses are typically near 1.5 dB, Bender, unpublished data); (3) a broad and gently sloping cut-off.

5.3.1 Cavity Resonators

Because resonators/filters are *RLC* circuit devices, and the pass band of a filter is determined by a mathematical relationship between *R*, *L*, and *C*, which is designated by a quality factor *Q* (see Chapter 1 and Poole, 1967; 1983; Rinard *et al.*, 1994), one typically aims for low-*Q* sample resonator designs. Cavity resonators constructed of metal boxes or cylinders are inherently high *Q* devices, although the actual *Q* achieved depends on the operating mode (see Poole, 1967, 1983). The cylindrical TE_{111} is perhaps the lowest *Q* cavity resonator type, but still possesses *Q*-factors in the range of 1000–2000 (loaded unfinished copper or brass cavity operating at 9.5 GHz; Bender, unpublished). For pulsed EMR applications, therefore, cavity resonator *Q* may be lowered by 'spoiling'. Surface finish affects the current-carrying properties of the cavity walls, and *Q* may be lowered by plating the cavity with metals of low conductivity (*e.g.* tin or platinum, as opposed to copper, silver, or gold). Another means for lowering *Q* of a standard cavity entails sandblasting the cavity interior with a fine grit so that the surface finish is uniformly dull; aluminum cylindrical TE_{111} cavities that were sandblasted featured *Q*-factors approximately 10% lower than the polished aluminum counterpart and approximately 30–40% lower than the polished copper counterpart (S. Gedam & C. Bender, unpublished).

Dielectric materials in the cavity space or on the wall surface will lower *Q*, and the sandblasting results described in the preceding paragraph are to some extent attributable to increasing the oxide layer surface area on the metal surface. Metal-plated dielectric (*e.g.* quartz cylinders) cavities may be rendered low-*Q* if the metal layer is much less than the skin depth (so that the microwave field extends partially into the dielectric wall). These approaches to *Q*-spoiling, however, are not entirely satisfactory because the degree to *Q* is lowered is not large enough to significantly affect ringdown the spectrometer deadtime.

An *RLC* circuit and its *Q*-factor is affected by external loading, and therefore the coupler may be used to affect *Q* and pulsed performance. A cavity may be coupled to a transmission line either inductively or capacitively, but in general, EMR systems use inductive coupling between the sample resonator and transmission line. Two popular methods of coupling are the wire loop, which is used with coaxial transmission lines, and the circular aperture and adjustable post, which is used with waveguide. The mode that is excited in the cavity is determined by the orientation of the junction. The H-field of the loop should coincide with the H-field of the desired mode of the cavity, whereas the orientation of the waveguide determines the excited mode in a hole-coupled cavity.

The diameter of the loop or circular aperture affects the degree of coupling, and therefore the dimensions and geometry of the loop or hole affect the cavity *Q* because the *L* and *C* of the coupling element are part of the external load that must be incorporated into the equivalent circuit of the cavity resonator. A common practice with resonant cavities that are used with pulsed EMR spectrometers is overcoupling, that is, couple the cavity by using an oversized loop or aperture (Tan *et al.*, 1984). The added inductance and capacitance that is introduced by the larger size of the coupling element lowers the *Q* of the loaded cavity.

There was at one time a popular adoption of bimodal sample resonators that enabled the user to isolate the transmitter and receiver arms of the spectrometer in a manner that resembles the practice of NMR with its separate transmit and receiver coils. Often this entailed the operation of a resonant cavity in two degenerate modes (*e.g.* TM and TE modes in cylindrical cavities; see Barendswaard *et al.*, 1984) or symmetric magnetron-like structures (Conciauro *et al.*, 1973, Figures 7-9). The rationale behind using these structures is that the transmitter and receiver are isolated (often by as much as 80 dB, Barendswaard *et al.*, 1984), and therefore the noise of the traveling wave tube pulse is isolated and the echo can be detected at a shorter time interval past the last pulse.

In principle, bimodal structures offer substantial advantages over conventional resonators, but their practical application is problematic. There is, for example, a temptation to increase sensitivity by keeping the quality factor high in the resonator. Although the modes may still be decoupled, the bimodal operation does not affect the bandwidth of the 'filter' or ringdown of the incident microwave pulse. In other words, the 'transmit' mode of the structure may ring long after the incident pulse and, although this does not affect the receiver, it does affect the spin system. One might therefore be recording a response on the 'receiver' mode on a time scale that does not accurately reflect the spin dynamics. Unfortunately, increasing the bandwidth by lowering *Q* makes it more difficult to decouple modes (see, however, Wallace & Silsbee, 1991, on bimodal response of microstrip resonators, and Piasecki *et al.*, 1996, on bimodal loop gap resonators).

5.3.2 Low Q Structures: Wire, Stripline and Loop-Gap Resonators

A catalog of probehead designs that are based on the coupling of a standing wave to a low-*Q* structure is illustrated in Figure 14. In a sense, the designs are depicted as an evolutionary trend on the theme of bandpass filters. A short section of wire whose length corresponds to a fraction of the incident wavelength, a so-called linear resonator, will electrically couple to an incident standing wave created in a shorted waveguide section. The efficiency of coupling is controlled by pivoting the wire in the plane perpendicular to the guide's broad wall, and the quality factor is determined by the composition of the wire (Pifer & Magno, 1971).

The linear resonator 'evolves' to the rectangular strip. As illustrated in the figure, the strip is positioned near the shorting plane of a transmission line. In this case, however, coupling via adjustments of the strip's orientation within the EM field is not compatible with sample placement. One solution to this problem was described by Lin *et al.* (1985), who placed a folded strip in a shorted waveguide section and adjusted the coupling to the structure by using an iris. In the illustration, the concept is represented in the same context as the linear resonator; the resonator is placed in a shorted waveguide section whose dimensions were below cut-off for the incident carrier frequency (or in a cylindrical 'shield' at the end of a waveguide, see Lin *et al.*, 1985; Dirksen *et al.*,1990). A Gordon Coupler is therefore used to critically couple the resonator to the incident wave.

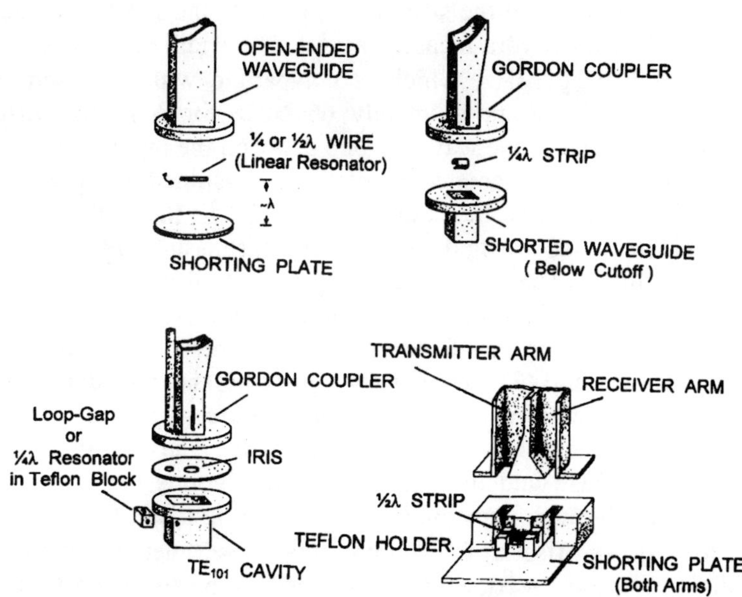

Figure 14. A catalog of some low-Q sample resonator arrangements. The low-Q structure (a strip or folded strip) is coupled to a standing wave (imposed by a short or a resonant structure). The Mims cavity (bottom right) is shown in cut-away form. See text for citations.

Low-Q resonators such as rectangular strips, loop-gaps, and dielectric structures may be coupled to a resonant field in high-Q structures such as cavity resonators. The design constraint is simply a matter of choosing a resonant mode in a rectangular cavity, and putting the low-Q structure with identical resonant frequency into the proper field lines within the cavity. For example, a rectangular cavity operating in the TE_{101} mode is characterized by H-field lines that form a loop along the narrow walls that delineate the cavity (see Ishii, 1966). A loop gap or similar structure whose H-field lines pass along the cylindrical axis may be arranged along any of the narrow walls so that the H-field lines run parallel (Figure 14), and the structures will couple (Britt & Klein, 1987).

Finally, the so-called Mims cavity, shown as a cutaway diagram in Figure 14 (bottom, right), couples two transmission lines via a rectangular strip and in this regard resembles the textbook example of $\lambda/2$ resonant filter. For this reason its description will be given in a separate section.

5.3.3 The Mims Sample Resonator

The Mims transmission resonator is a bandpass filter that is based upon the use of a $\lambda/2$ resonant structure (Malherbe, 1979). Losses at the bandpass midpoint are typically less than 2 dB and may be optimized to 1.5 dB or less by adjusting the distance of the strip's center line to the shorting plane (C.J. Bender, unpublished network measurements). The width of the pass band is approximately 70 ± 10 MHz, which is compatible with a pulsed carrier frequency bandwidth of approximately 66 MHz for a 15 ns pulse and therefore imposes little in the way of group delay on the input signal.

The resonator and its associated couplers are illustrated in Figure 14. The body of the device is a brass block that is bolted onto a flange through which two parallel rectangular waveguides open for transmission of the microwave pulses into the block. The probehead is typically used with a liquid helium immersion dewar and therefore constructed in three sections: (1) a lower copper waveguide section that features coupling tapers; (2) a stainless steel heat barrier section; and (3) a copper waveguide bulkhead feedthrough. The joints are bolted together using half flanges and are staggered because the flange lacks the capacity to suppress signal leakage (the adjoining surfaces are not perfectly flush and in electrical contact, nor is there a means to 'choke' the field, although one could wrap the joint in metal foil to attenuate radiative losses). An offset of 0.5" (approximately λ/2) between the joints in the parallel waveguide sections reduces crosstalk between the transmitter and receiver arms.

The brass block that constitutes the sample resonator serves as a shunt for the two parallel waveguide transmission lines. A rectangular strip is located at approximately λ/4 above the 'floor' of the cavity (*i.e.* the waveguide short), and without the strip the two parallel sections are isolated (~40 dB). The strip therefore couples the two parallel guides via the electric field lines of what would otherwise be a standing wave (VSWR ∞), and the resonator is therefore analogous to the common resonant bandpass filter. Although the strip is modeled as λ/2, the resonant frequency is offset from the ideal because the rectangular strip is held in place by a teflon fork (see figure), and the dielectric loading renders the relationship between the physical length of the strip and the actual resonant wavelength to be approximately λ/3.

The waveguide tapers on the transmitter and receiver arms of the probehead manipulate the directionality of the excitation pulses. The taper on the receiver side is steeper than the taper on the transmitter side (Figure 14), which affects VSWR of a wave that is incident on the resonator. The gradual slope on the transmitter side is meant to offer a low VSWR to the pulses incident on the sample resonator from the transmitter, whereas the steep slope offers a high VSWR to the pulses that are then reflected back to the cavity by the receiver's gate (see Section 6).

An alternative interpretation of the tapers' function, however, might come from the standpoint of the echo, which might be viewed as a shock wave originating from the sample resonator. For example, ultrasonic shock waves are subject to scattering phenomena in ways that are similar to pulsed electromagnetic radiation and are prone to media effects similar to preferential VSWR (see Ben-dor, 1992). In this case, it might then be plausible that the unequal tapers direct the echo's energy preferentially towards the receiver.

The effect of unequal tapers on the signal amplitude distribution between two transmission lines was tested by measuring the vector voltage of a signal propagating from a Tee junction spanning two tapers of different angles of inclination (C.J. Bender, unpublished). A standard WR-90 waveguide opened into a modified Tee junction for which the E-plane dimension was reduced from 0.500" to 0.300". The two arms of the reduced Tee opened up to standard guide dimensions via the tapers. The total length of each taper section was 12", but the length of the tapers varied from 0.5" to 3". No significant difference in the (continuous) wave amplitude was detected regardless of the taper slope combination, and one can draw the conclusion that the waveguide taper disparity has no significant effect on the distribution of continuous wave energy.

The Mims sample resonator's single practical flaw is that the device is not compatible with any commercial helium cryostat and must be used in a liquid helium immersion dewar. In its original form, the Mims probe is open via the flanges to the liquid helium, which tends to fill the waveguide and thereby introduce a phase shift because of the now (dielectric) loaded transmission line. As the helium level varies in the course of the experiment, so too does the phase of the electromagnetic wave traveling through the waveguide, which can alter the signal intensity over long periods of data acquisition (homodyne detection, *cf.* Chapter 1).

The Mims cavity may be modified to exclude helium with the added benefit of making possible operation as a variable temperature cryostat in the manner of Swenson & Stahl (1954). A malleable metal such as indium is generally used to make vacuum seals in cryogenic systems, and the half-flanges that are used to prevent cross-talk between the transmitter and receiver waveguide sections can be sealed if one makes a small modification to the waveguide wall. The connection to the cavity is left unaltered, but a vacuum seal is achieved via an outer can that surrounds the cavity. A narrow gauge stainless steel tube then passes through the bottom flange and serves as the only free passage between the liquid helium chamber and the cavity. Since the cavity is open to the central waveguide of the probehead, a vacuum port through the waveguide at the bulkhead can be used to create a negative pressure differential between the cavity and helium storage chambers. Cold helium gas can then be drawn in through the tube from the liquid helium boil-off, or the tube can be connected to an external helium gas manifold. A prototypical design (Bender, unpublished) is illustrated in Figure 15.

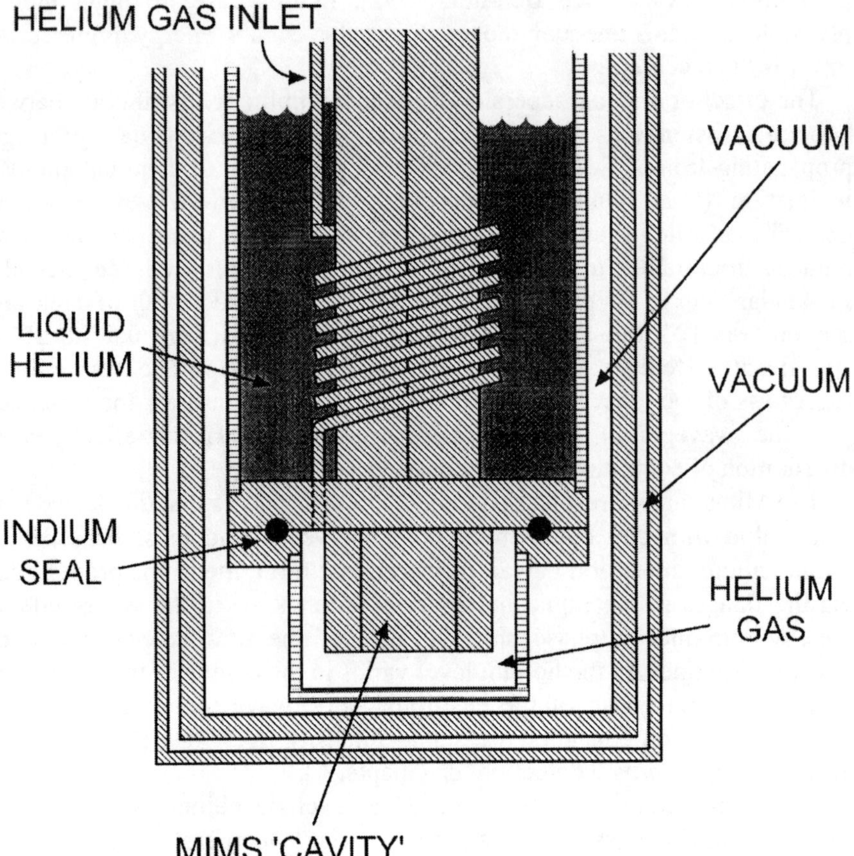

Figure 15. Svenson-style cryostat for variable temperature operation of the Mims sample resonator. Dry helium gas is flowed through a liquid helium reservoir and into a chamber containing the 'cavity'. Return flow is via the waveguide sections that mate with the cavity via a flat flange. The thick barrier separating the two vacuum chambers is copper and in contact with a liquid nitrogen reservoir (not shown). A second design that facilitates sample loading features a removable bottom flange and modified dewar cross-section.

5.3.4 Slow Wave Structures

Wire helices are used in traveling wave tubes for amplification because the structure is inherently broadband (Section 4). Wire helices radiate at high frequency and may be therefore also used as antenna (Kraus, 1988; Chatterjee, 1988). There are some examples of helical resonators being applied to EMR (Poole, 1967; p562ff, 752ff; Webb, 1962), but these applications are few. In general, helical resonators are most attractive for pulsed EMR applications that require broadband response; as appertains pulsed EMR, this broadband response is a manifestation of a very low *Q*-factor and therefore one expects a short ringdown time.

The open architecture of the wire helix also lends itself well to certain experiments that are of interest to the spectroscopist. For example, the helix only minimally blocks light from reaching the sample material within, which facilitates photochemical studies. The helix may also be counter-wound so that a second coil carrying an rf signal enables one to perform electron spin echo detected ENDOR. And finally, the helix may be used in a such as fashion as to mimic the probehead architecture of NMR spectrometers: if a helix is wrapped around a sample material and placed so that its cylindrical axis is parallel to the propagation direction in a waveguide, the waveguide and helix are decoupled. The waveguide acts as a pump arm, whereas the helix serves as a low power probe of spin dynamics in the sample material it surrounds. This is the basis of a maser design of Geusic *et al.* (1959) in which the helix was wrapped around a ruby rod and inserted into a waveguide section. The rod was pumped via the waveguide, and the stimulated emission was obtained from the helix; the experimental arrangement of the device is illustrated in Figure 16.

Helical resonators for microwave frequencies tend to be difficult to design properly for EMR applications, which may explain their infrequent use. The difficulty lies with dependence of the radiating mode shape on the helix shape and size (Kraus, 1988; Chatterjee, 1988). The axial mode of radiation is achieved with a large helix, whereas a small helix radiates in the normal mode. This size/mode interdependence is true of many radiating and resonant structures types, such as solid conductors and loops (see Kraus, 1988; Figure 7-1).

Small helices can be designed by using a simple pair of criteria. The diameter of the helix should be approximately 0.1λ, and the interwinding spacing should be 0.05λ (Kraus, 1988; Vizmuller, 1987). A large helix that radiates in the axial mode is typically 0.32λ in diameter and has an interwinding spacing of 0.25λ (Kraus, 1988). When the circumference of the helix c, is less than $2\lambda/3$ (small helix), the interior current distribution becomes sinusoidal, but it becomes uniform as $c \approx \lambda$ (Marsh, 1951). In the axial mode, the diameter of the conductor has little effect on the helix characteristics (Tice & Kraus, 1949), but this is not the case for small antennas that radiate in the normal mode. These design parameters have been analyzed by Pearlman & Webb (1967) for EPR applications.

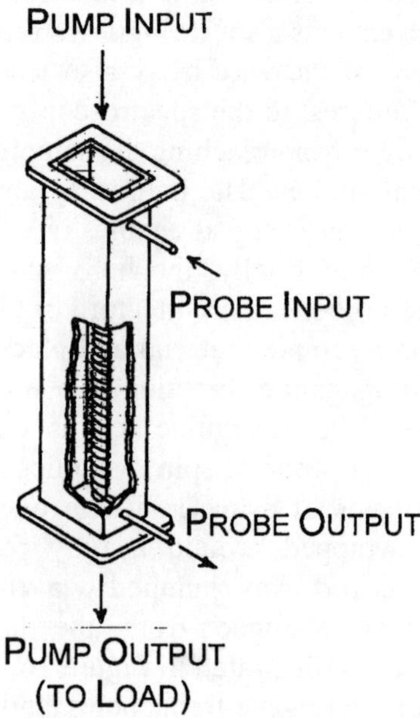

Figure 16. A schematic diagram of an optical maser system that may potentially be adapted for pulsed EMR. A waveguide section houses a ruby rod (or sample material) around which is wrapped a wire helix. The helical slow-wave probe is decoupled from the high power pumping signal that is transmitted via the waveguide circuit and therefore may be used to detect a weak echo signal generated by the sample without elaborate protective devices on the receiver. Adapted from Guesic *et al.*, 1959.

Helices can be free-standing or fixed, but the latter is preferred because it is less prone to the possibility of microphonic noise (particularly relevant with high frequency helices). Prototypes may be wound onto a quartz capillary and permanently affixed by using low temperature varnish such as GE 7031. Larger helices for low frequency can be plated onto a dielectric tube, for example Polyflon, which is a plated teflon product.

6. THE RECEIVER

6.1 Protection

The pulsed-EMR spectrometer receiver circuits are subject to damage by the high power microwave pulses and must be protected just as with radar and NMR receivers, but they are otherwise similar to the receiver that is used in a cw-EMR spectrometer (Chapter 1). In each case one is faced with the problem of coupling a high sensitivity rf detection circuit to the same probe/antenna that is used to carry the high power excitation signal. The solution entails any combination of switches and decoupled devices. Side-by-side schematic diagrams of radar and NMR/EMR systems are illustrated in Figures 17 and 18. High power rf/microwave pulses are generated and passed to an antenna/sample probe that is also used to receive the information-containing signal. In practice, the system behaves as two sub-systems that operate independently in transmit and receive modes. One can therefore imagine that a switch alternately changes the connection of the antenna/probehead during the different stages of operation.

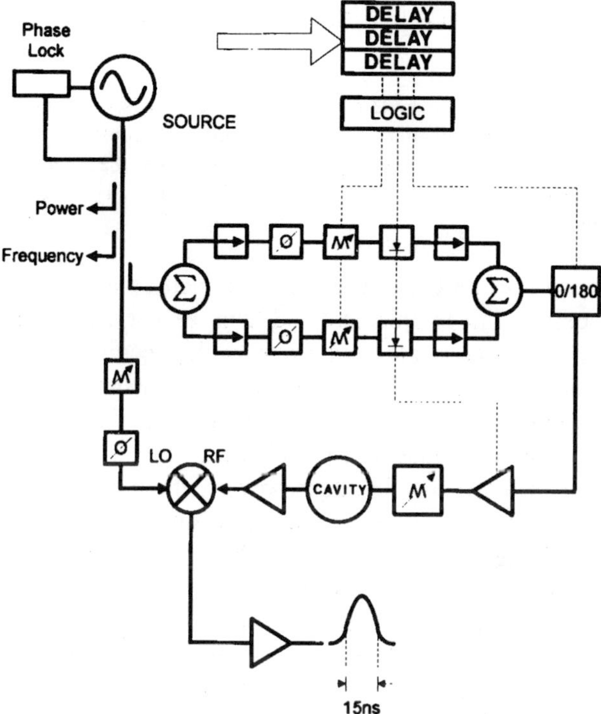

Figure 17. A schematic diagram of a two-channel pulsed electron magnetic resonance spectrometer. Low power nanosecond microwave pulses are generated on one of two channels that permit independent control of pulse amplitude and/or width. Phase modulation permits phase cycling for the elimination of unwanted echoes. A 1 kW pulsed traveling wave tube (tied to the pulse logic controller) amplifies the nanosecond pulses to powers adequate for driving the spin 'acrobatics'. Homodyne detection of the spin echo yields a video signal that is recorded by using a gated integrator (i.e. boxcar averager). The layout depicted corresponds to operation with a transmission (e.g. Mims) cavity/resonator; for operation with a reflection cavity one would replace the cavity symbol with a circulator (or hybrid Tee) and the reflection cavity (see Figure 2).

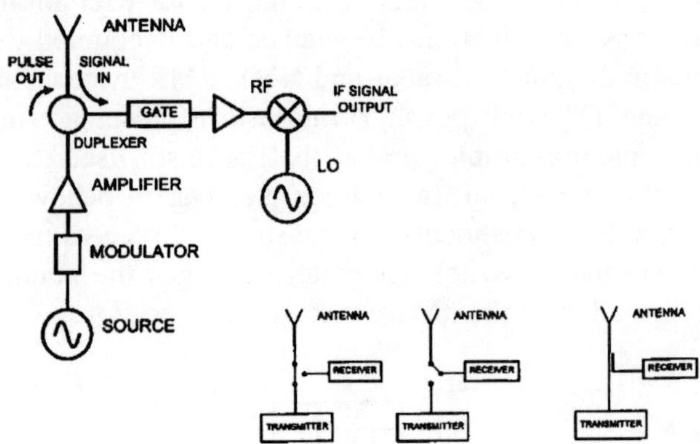

Figure 18. Functional diagram of a radar system. The transmitter consists of a high power source or an amplified (e.g. TWTA) modulated low power pulse. A duplexer, which is a generic term for a device that directs the high power transmitter pulses to the antenna and the weak returning signal to the receiver, acts as a SPDT switch (lower left schematic) and may be a circulator or directional coupler.

In general, the transmitter and receiver are combined as a single circuit that is called a transceiver, but remain isolated from each other by devices that block the high power transmitter pulses from the receiver. A simple version (Figure 18) consists of a directional coupler or circulator and a type of spark gap that is called a T-R tube. The T-R tube is a gas-filled section of waveguide with a conical (bandpass) filter element that is tuned to the desired operating frequency. At low incident powers, the device acts as an ordinary filter, but at high power, the close proximity of the cone tips supports the generation of electric fields that sufficiently strong ionize the gas and thus cause a spark and blocks transmission of the microwave power.

A T-R tube is one of many devices that are classified as limiters. A typical commercial device (Thomson Tubes Electronique, Product Guide TTE 165-2) will possess a 500 MHz pass band and a 1.2 dB insertion loss. T-R tubes, like gas tubes in general, are being superceded by solid-state devices. Diode shunts, for example, accomplish the function as the T-R tube (*cf.* Fukushima & Roeder, 1981) and are available as coaxial octave devices that are capable of blocking pulse power up to 1 kW peak, but with a higher insertion loss (2.8 dB; Herley-MDI, Woburn, MA MD-030 series).

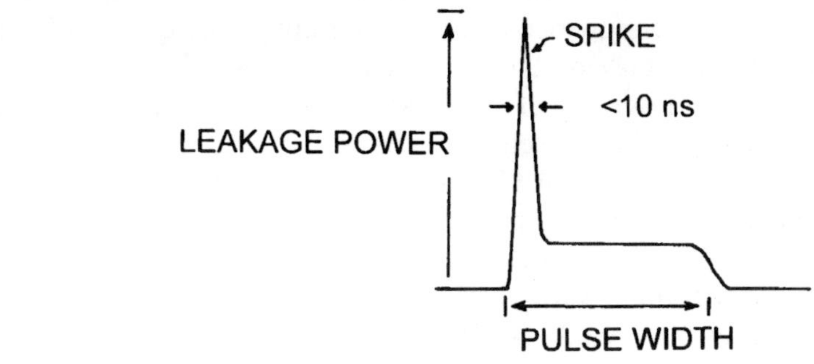

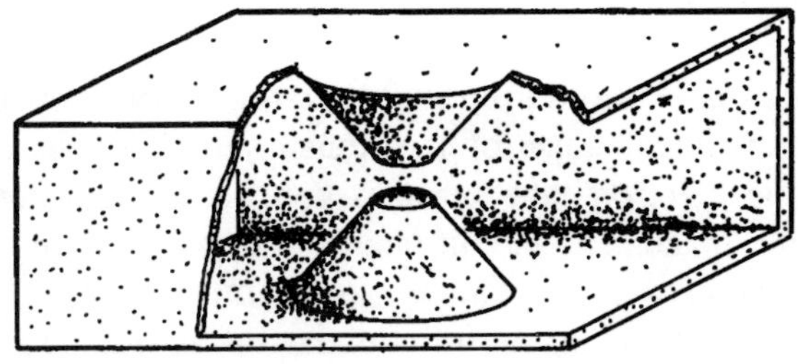

Figure 19. A T-R tube (cutaway view) and its typical response, which is similar to that of many diode limiters. Many of these devices suffer from a lag period on the order of 10 ns, which renders them superfluous when one operates the spectrometer using pulses with nearly the same temporal period.

Both T-R tubes and diode shunts protect the circuit components due to a catastrophic event, namely the breakdown of a gaseous or dielectric material. As such, these field-induced electrical breakdowns introduce design constraints by the very nature of the breakdown process. For example, spike leakage denotes the leading edge power that passes a limiter device prior to breakdown (gas or dielectric) and attenuation. The inset of Figure 19 is a typical profile of pulse attenuation by a T-R tube (*cf.* Ridenour, 1964), and shows that a spike on the order of ten nanoseconds passes the limiter. Such a response is therefore useful only for protecting the receiver from long (*i.e.* >1 μs) pulses, but useless for EMR experiments in which 10–20 ns pulses are used. A second design constraint is the recovery time, which is the time interval that must pass before the limiter will conduct a low level signal after attenuating the high power pulse. For example, the recover time of a T-R tube corresponds to the time required for the gap to be cleared of ions that provide the electrical short circuit.

Finally, by protecting the receiver one may reflect high power pulses back towards the transmitter and therefore one must also find some means of disposing that energy. In particular, one must avoid the return of the excitation pulses to the sample resonator, which may, in effect, change the nature of the pulse sequence and give spurious echoes. One solution to this potential problem may be the use of an absorptive SPST switch as a front end gate, or the insertion of an isolator between the sample resonator and receiver front end if a reflective switch is used, but in both cases one must address the power that must be dissipated. For example, a 15 ns pulse at approximately 60 dBm maximum delivers about 25 W of power. A typical ferrite isolator attenuates the reverse signal by 20 dB, which implies that some power will find its way back to the sample. A higher isolation solution is depicted in Figure 20, in which a circulator is used to isolate the transmitter and sample from reflected pulses.

6.2 Amplification & Detection of the Echo

The PIN diode switch is closed during the excitation pulse sequence and blocks the energy and residual noise of the transmitter. It opens at some point beyond which the electron tube noise is considered acceptable and echo detection may take place. The circuit elements beyond the limiter and gating switch are selected to optimize spectrometer sensitivity by maximizing gain while minimizing the introduction of additional noise. A typical receiver front end (Section 6.3) therefore features a low noise solid-state amplifier and DC block prior to the RF port on the mixer. The amplifier is selected for a low noise figure and a modest gain of approximately 25–30 dB. The best performance is obtained from GaAs MMIC units that span a single octave, although multi-octave devices can be obtained. In spite of the overwhelming use of solid-state amplifiers, however, a cw-TWT has the unusual advantage as a front end because it is robust and tends to block the transmitter pulses while passing only the low power echo signal (Taylor *et al.*, 1969).

6.3 A Representative Receiver Front End

A typical receiver front end is illustrated in Figure 20. The desired signal reaches the RF port of a double balanced mixer via an isolator (or terminated circulator), a 0–20 dB attenuator, a limiter, a gating switch, a bandpass filter, a low noise amplifier, and a DC block. For reasons cited in the preceding paragraphs, one may choose to forgo the limiter and the bandpass filter. The attenuator is used to ensure that the echo amplitude is not so great as to saturate the amplifier; alternatively, in such circumstances a fixed value drop-in attenuator may be used. The IF port of the receiver (*i.e.* reference arm) consists of a variable attenuator, phase shifter, and isolator.

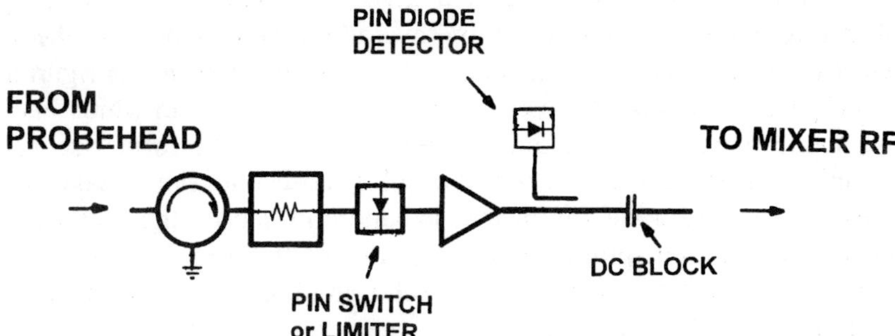

Figure 20. Representative receiver front end. An absorptive PIN switch and/or diode limiter protects the amplifier and mixer from the TWTA pulses. The circulator and attenuator are optional, but have been used by the author in order to minimize reflections and ensure that the echo amplitude is at all times in the linear response region of the low-noise GaAs FET amplifier.

The receiver is conveniently constructed by using discrete coax devices that are inherently broadband and readily operable up to approximately 20 GHz. Above 20 GHz, however, waveguide-based devices are advantageous (see Johnston, 1980; Bhartia & Bahl, 1984), and the receiver may be waveguide-based up to and including the double balanced mixer.

6.4 Recording the Echo Modulation

The resultant video signal obtained at the IF port of the mixer is the spin echo envelope, which is a bell-shaped pulse of width on the order of ten nanoseconds. As cited in the introductory sections, early spectrometers recorded the echo modulation as a timed exposure of the triggered oscilloscope display as the echo was recorded while the pulse timing parameters were stepped incrementally (Hahn, 1950b). Most spectrometers now use an instrument known as a boxcar averager, which is a device that samples a signal for a specified time interval, but modern oscilloscopes now work on a very similar principle and have adequate bandwidth to sample the echo waveform so that it (the echo waveform) may be integrated numerically rather than via analog circuits.

6.4.1 The Boxcar Averager

The so-called boxcar averager or gated integrator is a device that is used to sample repetitive waveforms (*e.g.* a spin echo) over a very short time interval. In such a manner it is therefore possible to record a waveform piece-wise by stepping the time interval between the trigger input (start of the waveform) and the point at which the boxcar samples the waveform. The fact that the measurements are made on repetitive waveforms means that one may make repetitive samples at a given time relative to the trigger and therefore reduce signal noise by averaging procedures. In the case of electron spin echo experiments, one must distinguish configurations in which one records the echo envelope (FID) from those in which one simply wishes to record the echo amplitude (ESEEM). In the former scenario, the sampling interval must be on the sub-nanosecond time scale since one wishes to record a waveform that is superimposed on the bell-shaped spin echo. The electron spin echo envelope modulation experiment, on the other hand, requires that one record the echo amplitude (or the integrated echo amplitude) as the time interval between the excitation pulses is varied. Because one is only interested in recording this pulse (*i.e.* the echo) amplitude, the sampling interval is commensurate to the echo width, which is taken as the half-peak-half-height. The advantage of using the latter criterion for selecting the sampling interval is that it is insensitive to phase-induced distortions of the echo envelope associated with homodyne detection.[6]

The boxcar device itself is very simply an *RC* integrator circuit coupled with an operational amplifier and coupled with a diode or transistor 'gate' that allows the signal to reach the integrator/amplifier for only a short sampling period (called the aperture time). The gating may be controlled in the time domain and it is therefore possible to sample a waveform as a sequence of windowed events; many of the modern digital oscilloscopes operate on the principle of sampling the waveform in a time-swept manner. Transistors tend to be used as gates for high-speed applications, and JFETs are preferred at aperture widths below 100 ns.

[6] At the start of an experiment one optimizes the echo amplitude at the receiver by adjusting the phase of the reference arm. Due to drift or changes in the electrical length of the transmission lines (as cryogen levels change), the phase match between RF and LO mixer ports may alter the echo shape, often skewing its symmetry. One therefore wishes to avoid using a sample interval that would base its measurement on the 'wings' near the baseline, which may dip below baseline (W.B. Mims, personal communication).

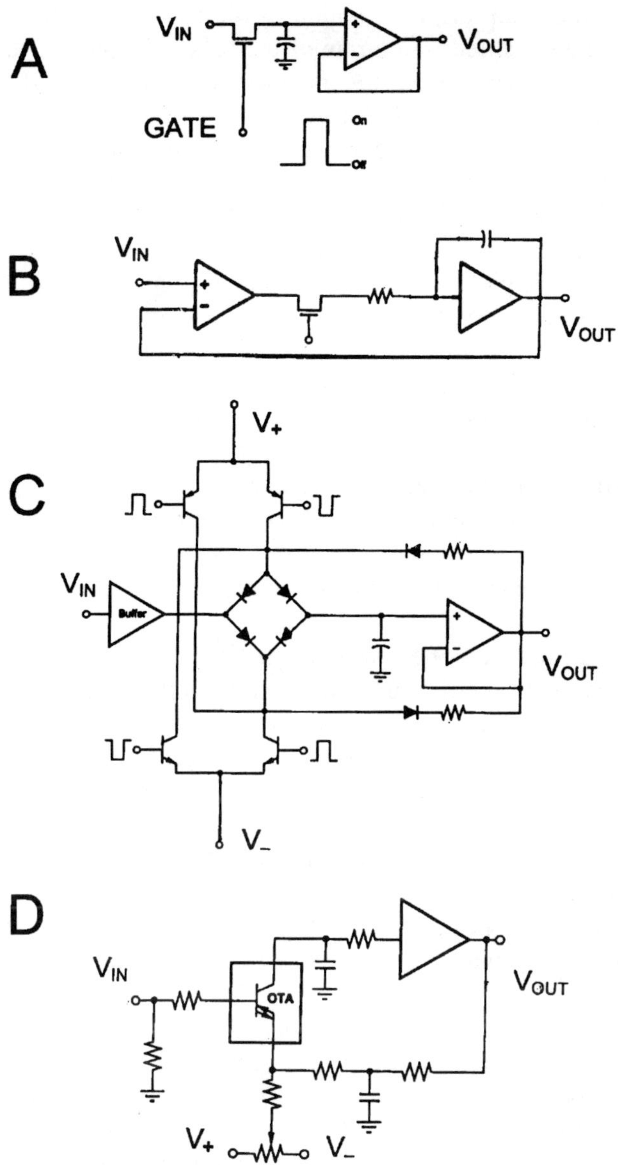

Figure 21. Representative circuits for 'boxcar' echo detection. A – Basic sample/hold circuit. B – Arrangement for integrating the sampled signal. C – Very fast sample/hold circuit using a diode bridge as a gate instead of JFET. D – Nanosecond pulse integrator (no gate) using an operational transconductance amplifier (OTA) and buffer (monolithic). Circuits shown are adapted from Burr-Brown Application Bulletin 27A, Linear Technologies Application Note 47, and Princeton Applied Research boxcar service manuals.

Sampled data acquisition introduces noise whose frequency spectrum is determined by the sampling rate, which in the case of pulsed EMR would correspond to the so-called frame rate of the echo sequence. The frame rate of spin echo modulation experiments is typically less that 100 Hz, and therefore one's data will be influenced by very low frequency noise. One particular problem that has been encountered is a slow baseline variance that appears to be arising from the electron tube (Bender, unpublished). Electron beam fluctuations are manifest as phase shifts, and these are detected as amplitude variations in the homodyne receiver. On very slow data acquisitions these are picked up as a low-frequency modulation.

The addition of a second boxcar channel that samples the baseline on each frame improves the fidelity of echo (or FID) sampling (Fessenden, 1973; Percival & Hyde, 1975). The second channel samples the baseline after some delay (following the echo or after the FID has fully decayed) and processes the signal under the same conditions used to sample the echo signal. This baseline signal is then subtracted from the echo signal on a per frame basis and therefore provides a running correction of the baseline modulation.

6.4.2 Oscillographic Integration of the Echo

Modern digital oscilloscopes, such as the Tektronics 11400A (bandwidth 3 GHz) or Hewlett-Packard 54616 (bandwidth 500 MHz; 2 GSamples per second), and transient event digitizers, such as the Perkin-Elmer/EG&G Eclipse (262144 points at 500 ps per point) and Tektronics SCD5000 (1024 pts at 5 ps intervals; 11-bit resolution), have very high sampling rates that enable one to capture the echo envelope as a series of discrete points. It is therefore possible to replace the gated integrator by a digitizer or storage oscilloscope by passing the video IF signal directly from the mixer to the oscilloscope, and then perform the integration of the area under the echo waveform by using some discrete numerical algorithm (*e.g.* Simpson's Rule).

As with any instrumental configuration, there are both pros and cons to replacing the stand-alone boxcar averager with a digital waveform analyzer. Historically, boxcar averagers were used to record waveforms where conventional oscilloscopes lacked adequate bandwidth, and in fact, sampling oscilloscopes are scanning boxcar devices. One may appreciate this fact by comparing the Tek 7104 oscilloscope and the Princeton Applied Research 162 boxcar integrator, which both used the same sampling heads to set the gate aperture. Both the boxcar and the waveform analyzer are sampling devices that enable one to construct a repetitive waveform by gating a port and sampling the voltage at this port for a finite, albeit short, time interval. The relative merits of either device are a matter of assessing whether it is better to (1) sample and integrate over a 10-15 ns window over the echo, and use this integrated amplitude as the temporal datum, or (2) for each temporal point in the stepped pulse sequence, collect the echo as a set of discrete points and then either use the oscilloscope to measure the pulse amplitude directly or integrate this echo waveform by using some discrete numerical algorithm.

Echo modulation studies are better served by the first option because the second technique is needlessly cumbersome. If one chooses, for example, a 10 ns window of the echo as the measurement criterion, one first has to decide how many sample points will be required so that the numerical integration is sufficiently sensitive to accurately reproduce the true echo amplitude changes (also true of the *RC* elements comprising the analog integrator). The numerical integration will also require a baseline correction step if phase changes distort the echo's shape (*cf.* footnote 6). Finally, one also need be concerned with the so-called quantization error of the digitizer, which arises from the inherent limitation of the instrument's A/D conversion resolving power. A typical high bandwidth digitizer features an 8- to 10-bit (full-scale) amplitude resolution. This latter constraint means that the displayed waveform is superposed on a vertical scale of 8 or 10 bits. As the echo amplitude decays (due to relaxation effects) and the echo modulation becomes shallow one is actually dynamic range limited by the digitizer's A/D converter. Quantization error is the voltage resolution 'lost' when an otherwise continuous voltage range is represented bitwise, and one must be concerned about the relative magnitude of the modulation depth and this quantization error.

7. SPECIAL TECHNIQUES

7.1 ESE-Detected EMR and ENDOR

The electron spin echo may be used as a general detection scheme for magnetic resonance experiments. One can take advantage of the spectral hole-burning aspect of the pulse-echo experiment and 'tune' the pulse excitiation bandwidth to excite spin packets in an inhomogeneously broadened line (Moerner, 1988; Sild *et al*., 1988). One can exploit the timing parameters of the pulse sequence to detect varying relaxation rates in the inhomogeneously broadened line spectral envelope (Mims *et al*., 1961; Mims, 1968). Aspects of this detection scheme are discussed in detail by Clark *et al*. (1994).

The EMR spectrum may be recorded as the DC magnetic field dependent echo amplitude, and in so doing may eliminate modulation effects, or for that matter, the need for a cw-EMR spectrometer. This is essentially a fixed timing pulse-echo experiment in which the magnetic field is stepped on successive echo generation collection frames. The pulse sequence, however, can be broken up into preparation and detection time frames, and one may manipulate the spin dynamics at certain times during the multi-pulse (≥ 3 pulses) sequence (Schweiger, 1990; 1995). Electron spin echo detected ENDOR (ESE-ENDOR) is one such technique: an rf pulse is interspersed between the microwave pulses, and the carrier frequency of the rf pulse is swept in order to record a spectrum as echo intensity *vs*. NMR transition frequency (Mims, 1965a).

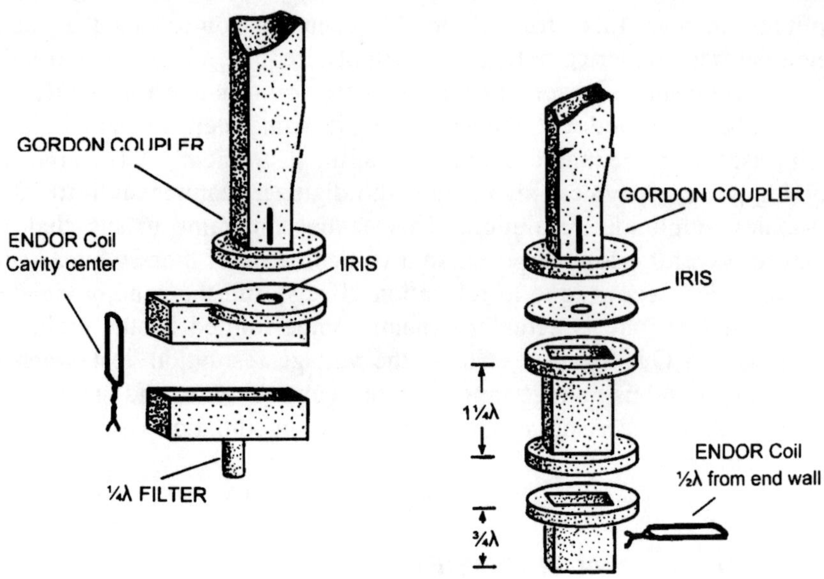

Figure 22. Rectangular TE_{10n} cavities for use as echo-detected EMR or ENDOR sample resonators. Adapted from Danilov & Manoogian (1972, right) and Abragam & Borghini (1964, left). Operation with an ENDOR coil *in situ* requires that the iris and sliding dielectric plug of the Gordon Coupler be supplemented by a tuning post (Bender, unpublished results). The teflon sample holder around which the coil is wrapped is not shown.

Deadtime is less important during the practice of echo-detected magnetic resonance (EDMR) than when collecting modulation time series because no Fourier transform of a time series is involved and no need to record an echo amplitude in the time limit $t \rightarrow 0$. This lifts some of the constraints on cavity design that otherwise compel one to design a cavity that features a long ringdown and broad pass band.

Figure 22 illustrates two sample resonator designs that are compatible with the probehead that was described in Section 5. On the left it a rectangular TE_{102} cavity resonator into which is inserted a wire loop that is wound around a plastic form (Abragam & Borghini, 1968). A $\lambda/4$ choke is attached in order to decouple the microwave and rf curcuit. A rectangular

TE_{104} cavity is depicted on the right hand side of Figure 22 (Danilov & Manoogian, 1972). Again, the coil is wound onto a plastic form and inserted into the cavity. The quality factor of the TE_{10n} cavities is affected by loading (*i.e.* the dielectric coil form and the sample holder), but, more, importantly, it is controlled by the size and shape of the iris. Besides large circular irides, low-Q coupling and resultant short deadtimes can be achieved by using crossed slots as irides (S. Gedam & C. Bender, unpublished results). The TE_{104} ENDOR cavity, operated with a 1/8" thick delrin sample holder and 1/4" circular iris, yielded a room temperature quality factor of 500, as measured by the $f_c/\Delta f$ procedure. Four turns of 28 Ga. magnet wire were used. It was found that in some cases that the dielectric plug of the Gordon Coupler alone did not allow adequate coupling range; in such cases an adjustable post mounted in the flange as in conventional cw-EMR iris coupler was used to successfully critically or over- couple the cavity.

Other sample resonator configurations that have been tested by the author include an axially-shortened (lowers Q) cylindrical TM_{110} cavity, dielectric-filled wire-walled cylindrical TE_{011} cavities, and the Mims transmission resonator. One noteworthy comment on the Mims resonator is in order: because of its compact size, the rf fields produced for a given power level tend to be larger than with cavity resonators, and care must be exercised to distinguish artifacts from ENDOR transitions. On the positive side, the Mims resonator yields very good ENDOR spectra with low power input; excellent ^{1}H-ENDOR spectra have been obtained with a 25 W system amplifier.

7.2 EMR Hole Burning Experiments - The Linear Electric Field Effect

Inhomogeneously broadened EMR spectral lines may be described as a contiguous set of overlapping homogeneously broadened lines of some natural width. Modern optical spectroscopies likewise encounter this problem and recover lost information by techniques that are collectively known as 'spectral hole burning' (Moerner, 1988; Sild *et al.*, 1988), and these techniques afford an increase in sensitivity over direct methods because the spectroscopic detection and measurement is limited to those lines on the edge of the 'hole', which are manipulated by some external perturbation.

At a fixed microwave carrier frequency and DC magnetic field, a hole may be burned in the EMR spectrum by saturating a packet that represents those paramagnets whose orientations are optimal for resonance. Any perturbation that shifts adjoining, non-saturated packets into the saturating resonant field can be used to perform hole burning high resolution spectroscopy on an individual packet. In the case of magnetic resonance, the ideal perturbative force is an electric field that perturbs the paramagnet via the Stark Effect.

The linear electric field effect is observable via continuous wave and pulsed EMR methods and has been described in detail by Mims for both cases (Mims, 1976). The Stark perturbation, or electric field shift $\Delta\varepsilon$, is written as

$$\Delta\varepsilon = \frac{\sum_{i} \langle \Psi_i H_e \Psi_m \rangle \langle \Psi_i H_0 \Psi_m \rangle}{(E_m - E_i)},$$

which is linear for the first order term in the perturbation expansion. The electric field Hamiltonian H_e mixes the ground and excited states, and its odd parity dictates that the linear electric field effect is measurable only when the ground and excited states are of mixed parity (*i.e.* noncentrosymmetric system), and in this regard there is a connection between LEFE and the optical effect known as second harmonic generation in crystals. It follows that the measurement of LEFE is accurate only at low electric field strengths because it is possible to introduce higher order nonlinear terms of the perturbation expansion (including even parity). This nonlinearity is evident in Figure 23, which is a plot of packet shift as a function of applied electric field strength; the packet shift is measured as the suppression of the two-pulse electron spin echo by the electric field. The implication of this plot in the practice of LEFE measurement will be elaborated below.

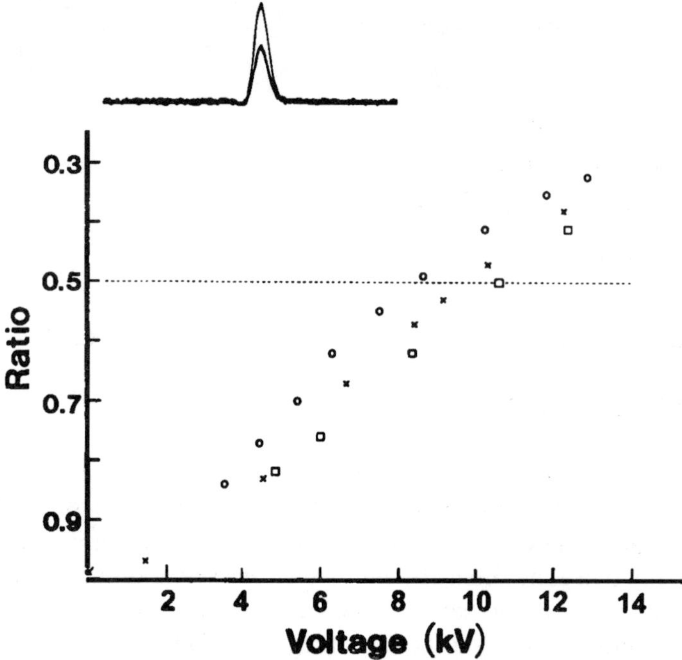

Figure 23. The measure of echo amplitude reduction as a function of an applied electric field. The ratio of the y-axis corresponds to the ratio of the integrated echo amplitudes (field on: field off, note oscilloscope traces at top of figure). The plots depict the ratio variation as the electric field strength (as applied voltage) increases at various points along the EMR spectral envelope, and these illustrate the deviation from linearity at certain *g* values (data recorded using rusticyanin, a Type I copper protein).

The high voltage pulse generator that is used for measurement of the linear electric field effect via spectral hole burning is illustrated in Figure 24 (after Mims, 1965b). It consists of an *RC* network that is discharged by a thyratron; the cavity electrode is the $\lambda/2$ rectangular strip of the Mims resonator described in Section 5 (Mims, 1974), which is tapped for connection to the high voltage pulse generator at the center of one edge (24 Ga. lead). The deuterium thyratron switch is one of several possible methods for rapidly discharging the capacitor, but it is the most convenient means for producing accurately timed pulses of medium to high power. It is a gridded tube containing mercury vapor and hydrogen (or deuterium, see Gewartowski & Watson, 1965), and therefore has the advantageous property of operating as a switch of very low inductance. Like thermionic tubes described in Chapter 1, a heater filament aids in the emission of electrons from the cathode.

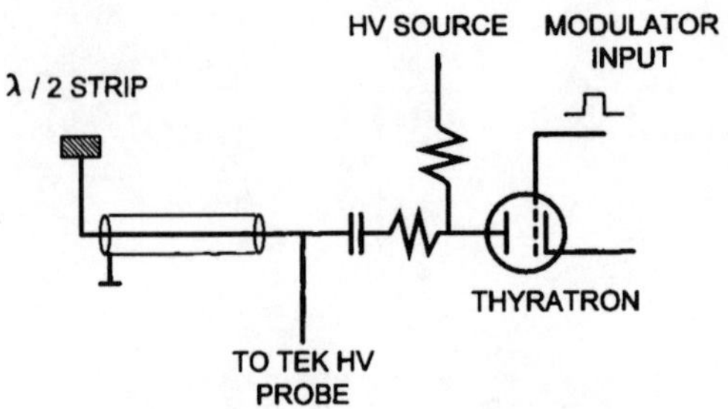

Figure 24. High voltage pulse circuit for LEFE, as implemented by Mims and McCracken.

7.2.1 HV Pulse Width Requirements and the Generator Circuit

The pulse generator that was described in the preceding paragraphs is suitable for the basic LEFE experiment provided that one can work with sufficiently long tau values (*i.e.* ≥ 400 ns). The LEFE measurement is derived from the relative echo intensity with the electric field on *vs.* off, and therefore it is necessary that one have a large signal because of quantization errors in the A/D converters of the boxcar integrator. Samples that can be prepared in high concentration and whose echo has good phase memory are therefore best suited to LEFE measurements. In those situations where the echo amplitude drops off rapidly as tau increases (either due to poor spin phase memory or profoundly deep modulation), it is necessary to make the LEFE measurement with narrow high voltage pulses that feature a rapid rise time. In such cases more sophisticated HV pulse generators that feature pulse-shaping networks are specified. Radar pulse generator circuits of the type that formerly delivered high voltage pulses to spark gaps (see Ewell, 1981) can be used on an LEFE apparatus to obtain 20 kV pulses of 10-100 ns duration. The sole difficulty of using such complicated networks is that their performance greatly depends on the network properties, and care must be taken when adding the three foot coaxial line that is terminated by the cavity and its 200 pF capacitive load (see below). Many of the pulse generator designs described by Ewell (1981) and the MIT Radiation Laboratory Handbooks are rated for 600 Ω resistive loads, but a resistive shunt may be used to assist the match for LEFE (Bender, unpublished).

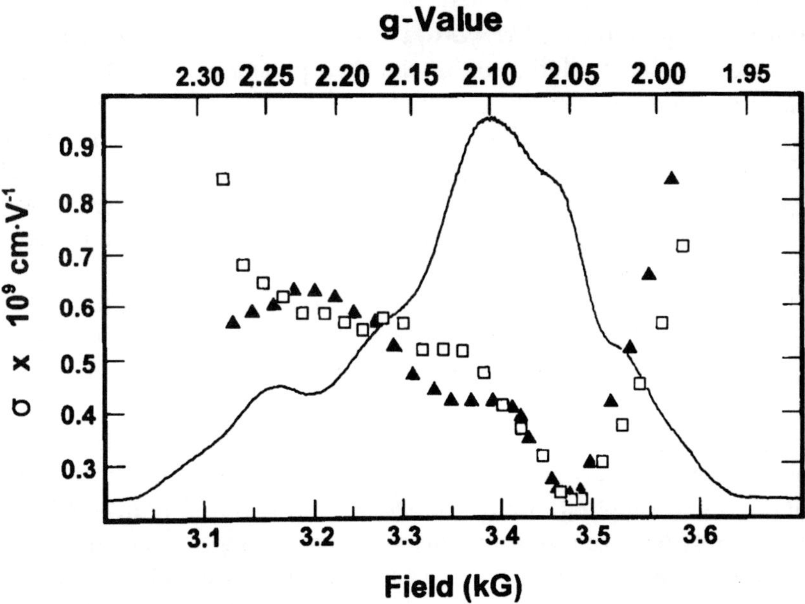

Figure 25. Echo-detected EMR spectrum (solid line) of rusticyanin, and the LEFE recorded over the spectral envelope. Filled triangles correspond to σ obtained with E ⊥ H vs. E ∥ H (open squares).

The cavity acts as a terminus for the pulse-forming network. In the original Mims (1974) configuration, the cavity/electrode are located at the end of a ~1 meter stainless steel coaxial line (C_s ~ 100 pF at 120 Hz). With the sample compartment unloaded and at room temperature, the capacitance between the strip and the cavity wall is 225 pF at 120 Hz, and the impedance is approximately 7.5 MΩ (Bender, unpublished measurements). A loaded cavity containing 1:1 ethylene glycol/water at liquid nitrogen temperature (cavity contacted with cryogen but not immersed) yielded a capacitance of 3 nF and an impedance of 2 MΩ. One simple impedance matching solution entailed the placement of a 550 Ω resistor parallel to the cavity, which altered the impedance of the circuit for an adequate match to a 600 Ω radar pulse unit while retaining the field-generating capacitance of the cavity.

7.2.2 Graphical Approach to LEFE Measurement

The linear electric field effect measurement is made in a two-pulse electron spin echo experiment (Mims, 1974). A ½π–π pulse sequence burns a hole in the EMR spectral envelope, whereas the echo represents a recovery. If an electric field is applied to the sample, a Stark shift is introduced to that an adjacent spin packet (line) is shifted into the region left transparent after the hole-burning. This Stark shift is equivalent to a *g*-value shift of line, and the experiment entails the measure of a so-called shift parameter, represented as $\sigma = d(6f\tau V_{1/2})^{-1}$, as a function of the magnetic field strength. In the definition of the shift parameter, the term $V_{1/2}$ is the voltage that reduces the echo amplitude by a factor of two, d is the sample thickness, f is the spectrometer operating frequency, and τ is the timing parameter of the two pulse echo experiment (Mims, 1974). Parallel and perpendicular orientations of **E** relative to **H** are measured, and for convenience the plots can be superimposed over an absorption mode (echo-detected) EMR spectrum (Figure 25).

The inset of Figure 23 depicts the oscilloscope trace of the Hahn echo on alternating sample frames. The larger of the two spin echoes is unperturbed by an applied electric field, whereas the smaller echo trace is recorded when the sample is subjected to a pulsed DC electric field on the order of 10 $kV\cdot cm^{-1}$. The experimental criterion of measuring the g-shift is made on the reduction of the echo intensity by 0.5, and the plot of the echo ratio as a function of the applied voltage illustrated in Figure 23 indicates that some caution must be taken with the measurement. In short, there are regions of the EMR spectral envelope where echo reduction by a factor of 0.5 can only be achieved by using high voltages/fields, but as the data show, these high fields introduce non-linear field effects.

The phenomenon of interest is linear in the applied voltage, but second order terms in the perturbation expansion are nonlinear and become increasingly significant as the magnitude of the applied electric field becomes large. In a given LEFE experiment, the voltage required to reduce the echo amplitude by 0.5 is plotted as a function of magnetic field, or g-value (Figure 25). Regions of the spectrum where the g-shift, σ, is small (*i.e.* near $g = 2.05$ in Figure 25) will require a very large electric field be applied to the sample in order to suppress the echo by 0.5, and one may introduce nonlinearities that affect the accuracy of the measurement (*e.g.* comparing shifts at one point of the spectrum where $\mathbf{E} = 70\ kV\cdot cm^{-1}$ *vs.* another point where $\mathbf{E} = 35\ kV\cdot cm^{-1}$.

Besides nonlinearity, Figure 23 also indicates that the slope of the echo suppression varies among regions of the EMR spectrum. A graphical approach to measuring the linear electric field effect is therefore recommended in order to ensure linearity and, when necessary, to find the 0.5 suppression point by extrapolation when the g-shift is small.

8. SUMMARY & CONCLUSION

The generation and detection of spin echoes can be achieved through a variety of instrumental options, which allow for flexibility in accommodating experimental design and the material available for the spectrometer's construction. The basic needs for the transmitter are a gated low power frequency source and high power amplifier, whereas the receiver requirements include a robust switch, a low-noise amplifier, and a mixer (double-balanced or quad). The video signal (echo amplitude) is recorded via a sample-and-hold device that may be a simple home-built boxcar averager or commercial transient digitizer.

Digital logic and associated computer hardware have been omitted from this review because such instrumentation tends to be a matter of personal preference. Home-built systems are described in the NMR and EMR literature, and two representative systems have been described in sufficient detail that they might be duplicated (Thomann *et al.*, 1984; Quine *et al.*, 1987). Echo modulation experiments are not complicated, and a workable system can be assembled by using nanosecond delay generators under the control of a slow microchip (such as 6505) because one is often frame rate limited to a sub-100 Hz data acquisition sequence by the spin system. As with the microwave components, however, there are many new options, and one may use programmable bit-word timing generators (*cf.* Interface Technology) to create custom multi-channel TTL or ECL waveforms for experimental design. Therefore, as regards the design of the digital systems for pulsed EMR, one should consult the description of novel systems in the original papers for the philosophy and specifications.

9. ACKNOWLEDGEMENTS

The author is grateful to John McCracken, Bill Mims, Sven Hartmann, and David Greenslade for many helpful discussions regarding instrument design and their rationale.

10. REFERENCES

Abragam, A., and Borghini, M., 1964, In: *Progress in Low Temperature Physics* (C.J. Gorter, ed.), Volume IV, North-Holland, Amsterdam. p384.

Adkins, L.R., and Nolle, A.W., 1966, *Rev. Sci. Instrum.*, **37**: 1404.

Ager, R., Cole, T., and Lambe, J., 1963, *Rev. Sci. Instrum.*, **34**: 308.

Alexandrov, E.B., 1964, *Opt. Spectroscopy*, **17**: 522.

Altschuler, H.M., 1963, *Rev. Sci. Instrum.*, **34**: 1441.

Bagguley, D.M.S., ed., 1992, *Pulsed Magnetic Resonance: NMR, ESR, and Optics*, Oxford University Press, Oxford.

Banci, L., Bertini, I, and Luchinat, C., 1986, *Magn. Resonance Rev.*, **11**: 1.

Barendswaard, W., Disselhorst, J.A.J.M., and Schmidt, J., 1984, *J. Magn. Res.*, **58**: 477.

Barton, D.K., 1975, *Pulse Compression*, Artech House, Dedham.

Beauchamp, K.G., 1987, *Transforms for Engineers. A Guide for Signal Processing*, Clarendon Press, Oxford.

Becker, R., 1982, *Electromagnetic Fields and Interactions*, Dover Publications, New York.

Ben-dor, G., 1992, *Shock Wave Reflection Phenomena*, Springer-Verlag, Berlin.

Bertini, I., Martini, G., and Luchinat, C., 1994, In: *Handbook of Electron Spin Resonance* (C.P. Poole and H.A. Farach, eds.), American Institute of Physics, New York. p80.

Bhartia, P., and Bahl, I., 1984, *Millimeter Wave Engineering and Applications*, Wiley-Interscience, New York.

Bhartia, P., Rao, K.V.S., and Tomar, R.S., 1991, *Millimeter-Wave and Microstrip and Printed Circuit Antennas*, Artech House, Boston.

Bloch, F., 1946, *Phys. Rev.*, **70**: 460.

Bloom, A.L., 1955, *Phys. Rev.*, **98**: 1105.

Blume, R.J., 1958, *Phys. Rev.*, **109**: 1867.

Bowers, K.D., and Mims, W.B., 1959, *Phys. Rev.*, **115**: 285.

Breene, R.G., 1964, In: *Handbuch der Physik* (S. Flügge, ed.) Band XXVII: Spektroskopie I., Springer-Verlag, Berlin. p1.

Breene, R.G., 1981, *Theories of Spectral Line Shapes*, Wiley, New York.

Brewer, R.G., and Hahn, E.L., 1973, *Phys. Rev. A*, **8**: 464.

Britt, R.D., and Klein, M.P., 1987, *J. Magn. Res.*, **74**: 535.

Brown, J.M., and Sloop, D.J., 1970, *Rev. Sci. Instrum.*, **41**: 1774.

Brown, J.W., and Churchill, R.V., 1993, *Fourier Series and Boundary Value Problems*, 5th edition, McGraw-Hill, New York.

Brunner, W., and Gentsch, F., 1971, *Ept. Tech. Phys.*, **19**: 39.

Carr, H.Y., and Purcell, E.M., 1954, Phys. Rev., **94**: 630.

Cazenave, R., 1951, *Cables et Transm.*, **5**: 271.

Chatterjee, R., 1988, *Advanced Microwave Engineering: Special Advanced Topics*, Ellis Horwood Publishers, Chichester.

Cherry, C., 1950, *Pulses and Transients in Communications Circuits*, Dover Publications, New York.

Chiba, M., and Hirai, A., 1969, *Jap. J. Appl. Phys.*, **8**: 1523.

Clark, W.G., Hanson, M.E., Lefloch, F., and Ségransan, P., 1994, *Rev. Sci. Instrum.*, **66**: 2453.

Collins, F.G., and Katchinowski, R., 1973, *Rev. Sci. Instrum.*, **44**: 1178.

Conciauro, G., Puglisi, H., Franconi, C., Galuppi, P., and Randazzo, E., 1973, *J. Magn. Res.*, **9**: 363.

Cook, C.E., 1960, *Proc. IRE*, **48**: 310.

Cowan, J.A., and Kaplan, D.A., 1961, *Phys. Rev.*, **124**: 1098.

Cowan, J.A., Kaplan, D.E., and Browne, M.E., 1962, *J. Phys. Soc. Jpn.*, **17**(B-1 Suppl.): 465.

Creed, F.C., 1989, *The Generation and Measurement of High Voltage Impulses*, Center Book Publishers, Princeton.

Crepeau, R.H., Dulcic, A., Gorchester, J., Saarinen, T.R., and Freed, J., 1989, *J. Magn. Res.*, **84**: 184.

Cutler, D., and Powles, J.G., 1962, *Proc. Phys. Soc. (London)*, **80**: 130.

Danilov, A.G., and Manoogian, A., 1972, *Phys. Rev. B*, **6**: 4103.

de Beer, R.D., van Ormondt, D., Pijnappel, W.W.F., and van der Veen, J.W.C., 1988, *Israel J. Chem.*, **28**: 249.

de Beer, R.D., van Ormondt, D., Pijnappel, W.W.F., Bosman, Th.J.L., and Hilbers, C.W., 1988, *Signal Processing*, **15**: 293.

Dirksen, P., Disselhorst, J.A.J.M., van der Meer, H., Wenckenbach, W.Th., 1990, *J. Magn. Res.*, **87**: 516.

Dodd, J.N., Kaul, R.D., and Warrington, D.M., 1964, *Proc. Roy. Soc. (London)*, **84**: 176.

Ernst, R.R., 1965a, *Rev. Sci. Instrum.*, **36**: 1689.

Ernst, R.R., 1965b, *Rev. Sci. Instrum.*, **36**: 1696.

Ewell, G.W., 1981, *Radar Transmitters*, McGraw-Hill, New York.

Feher, G., Gordon, J.P., Buehler, E., Gere, E.A., and Thurmond, C.D., 1958, *Phys. Rev.*, **109**: 221.

Fernbach, S., and Proctor, W.G., 1955, *J. Appl. Phys.*, **26**: 170.

Fessenden, R.W., 1973, *J. Chem. Phys.*, **58**: 2489.

Forrer, J., Pfenninger, S., Sierra, G., Jeschke, G., Schweiger, A., Wagner, B., and Weilnad, Th., 1996, *Appl. Magn. Res.*, **10**: 263.

Frank, N.H., 1943, *Tapered Transmission Lines and Wave Guides*, Radiation Laboratory Report 189. Massachusetts Institute of Technology, Cambridge.

Freeman, R., 1997, *Spin Choreography: Basic Steps in High Resolution NMR*, University Science Books, Oxford.

Fukushima, E., and Roeder, S.B.W., 1981, *Experimental Pulsed NMR: A Nuts and Bolts Approach*, Addison-Wesley, Reading.

Gemperle, G., and Schweiger, A., 1991, *Chem. Rev.*, **91**: 1481.

Gent, A.W., and Wallis, P.J., 1946, *Proc. IRE*, **93(3a)**: 559.

Geusic, J.E., Schultz-DuBois, E.O., deGrasse, R.W., and Scovil, H.E.D., 1959, *J. Appl. Phys.*, **30**: 1113.

Gewartowski, J.W., and Watson, H.A., 1965, *Principles of Electron Tubes. Including Grid-Controlled Tubes, Microwave Tubes, and Gas Tubes*, van Nostrand, Princeton.

Gilmour, A.S., 1994, *Traveling wave Tubes and Their Theory*, Artech House, Norwood.

Gorchester, J., Milhauser, G.L., and Freed, J., 1990, In: *Modern Pulsed and Continuous-Wave Electron Spin Resonance* (L. Kevan and M.K. Bowman, eds.), Wiley-Interscience, New York.

Gordon, J.P., and Bowers, K.D., 1958, *Phys. Rev. Lett.*, **1**: 368.

Gordon, J.P., 1961, *Rev. Sci. Instrum.*, **32**: 658.

Greenslade, D.J., and Higgs, J., 1993, *J. Chem. Soc., Faraday Trans.*, **89**: 3733.

Greenslade, D.J., 1991, *Spec. Publ.--R. Soc. Chem.*, **90**: 35.

Grivet, P., 1970, *The Physics of Transmission Lines at High and Very High Frequencies*, Academic Press, London.

Hahn, E.L., 1950a, *Phys. Rev.*, **77**: 297.

Hahn, E.L., 1950b, *Phys. Rev.*, **80**: 580.

Hahn, E.L., 1950c, *Nutation of the Nuclear Magnetic Moment and Related Effects*, University of Illinois, Urbana.

Hahn, E.L., and Maxwell, D.E., 1952, *Phys. Rev.*, **88**: 1070.

Hardy, W.N., and Whitehead, L.A., 1981, *Rev. Sci. Instrum.*, **52**: 213.

Harmuth, H.F., 1981, *Nonsinusoidal Waves for Radar and Radio Communications*, Academic Press, New York.

Harrison, A.E., 1947, *Klystron Tubes*, McGraw-Hill, New York.

Hoch, J.C., and Stern, A.S., 1996, *NMR Data Processing*, Wiley, New York.

Höfer, P., 1994, *J. Magn. Res., Ser. A*, **111**: 77.

Huisjen, M., and Hyde, J.S., 1974, *Rev. Sci. Instrum.*, **45**: 669.

Hutter, R., 1960, *Beam and Wave Electronics for Microwave Tubes*, van Nostrand, Princeton.

Hyde, J.S., Froncisz, W., and Oles, T., 1989, *J. Magn. Res.*, **82**: 223.

Ishii, T.K., 1966, *Microwave Engineering*, Ronald Press, New York.

Jaynes, E.T., 1955, *Phys. Rev.*, **98**: 1099.

Jeschke, G., and Schweiger, A., 1995, *J. Chem. Phys.*, **103**: 8329.

Johnston, S.L., 1980, *Millimeter Wave Radar*, Artech House, Dedham.

Jones, J.A., and Hore, P.J., 1991, *J. Magn. Res.*, **92**: 276.

Kaplan, D.E., Browne, M.E., and Cowan, J.A., 1961, *Rev. Sci. Instru.*, **32**: 1182.

Kaufman, J, 1955, *IRE Trans. Antenna Prop.*, **AP-3**: 218.
Kettlewell, E., 1971, *The Magnetron Oscillator*, Mills & Boon, London.
Kevan, L., and Schwartz, R.N., eds, 1979, *Time Domain Electron Spin Resonance*, Wiley, New York.
Kevan, L., and Bowman, M.K., eds., 1990, *Modern Pulsed and Continuous-Wave Electron Spin Resonance*, Wiley-Interscience, New York.
Klauder, J.R., Price, A.C., Darlington, S., and Albersheim, W.J., 1960, *Bell Syst. Tech. J.*, **39**: 745.
Kleen, W.J., 1958, *Electronics of Microwave Tubes*, Academic Press, New York.
Kohl, W.H., 1960, *Materials and Techniques for Electron Tubes*, Rheinhold, New York.
Kramer, N.B., 1981, In: *Infrared and Millimeter Waves*, Volume 4 (K.J. Button and J.C. Wiltse, eds.), Academic Press, New York. p151.
Kraus, J.D., 1988, *Antennas*, 2nd ed., McGraw-Hill, New York.
Kreuchen, K., 1957, *J. Electronics*, 2(May):
Lawson, J.L., and Uhlenbeck, G.E., 1950, *Threshold Signals*, McGraw-Hill, New York.
Liao, P.F., and Hartmann, S.R., 1972, *Phys. Rev. B*, **8**: 69.
Liao, S.Y., 1985, *Microwave Solid-State Devices*, Prentice-Hall, Engelwood Cliffs.
Liao, S.Y., 1988, *Microwave Electron-Tube Devices*, Prentice-Hall, Engelwood Cliffs.
Lick, T.A., Unruh, W.P., and Culvahouse, J.W., 1973, *Phys. Rev. B*, **8**: 4941.
Lin, C.P., Bowman, M.K., and Norris, J.R., 1985, *J. Magn. Res.*, **65**: 369.
Macomber, J.D., 1976, *The Dynamics of Spectroscopic Transitions*, Wiley, New York.
Macomber, J.D., 1995, *Dynamics During Spectroscopic Transitions: Basic Concepts*, Springer-Verlag, Berlin.
McDonald, S., and Warren, W.S., 1991, *Concepts Magn. Res.*, **3**: 55.
Malherbe, J.A.G., 1976, *Microwave Transmission Line Filters*, Artech House, Norwood.
Marsh, J.A., 1951, *Proc. IRE*, **39**: 668.
Metzger, G., 1969, *Transmission Lines with Pulsed Excitation*, Academic Press, New York.
Mims, W.B., Nassau, K., and McGee, J.D., 1961, *Phys. Rev.*, **123**: 2059.
Mims, W.B., 1963, *Proc. IEEE*, **51**: 1127.
Mims, W.B., 1965a, *Proc. Roy. Soc. (London)*, **283**: 452.
Mims, W.B., 1965b, *Rev. Sci. Instrum.*, **36**: 1472.
Mims, W.B., 1968, *Phys. Rev.*, **168**: 370.
Mims, W.B., 1974, *Rev. Sci. Instrum.*, **45**: 1583.
Mims, W.B., 1976, *The Linear Electric Field Effect in Paramagnetic Resonance*, Clarendon Press, Oxford.
Mims, W.B., 1984, *J. Magn. Res.*, **59**: 291.
Moerner, W.E., 1988, *Persistent Spectral Hole-Burning*, Springer-Verlag, Berlin.
Moss, H., 1968, *Narrow Angle Electron Guns and Cathode Ray Tubes*, Academic Press, New York.
Mushiake, Y., 1996, *Self-Complementary Antennas. Principle of Self-Complementarity for Constant Impedance*, Springer-Verlag, Berlin.
Newman, F.C., and Rowan, L.G., 1972, *Phys. Rev. B*, **5**: 4231.
Okamura, S., 1993, *History of Electron Tubes*, Institute of Physics, Burke.

Ostroff, E.D., and Borkowski, M., 1985, *Solid-State Radar Transmitters*, Artech House, Dedham.
Page, F.M., and Goode, G.C., 1969, *Negative Ions and the Magnetron*, Interscience, London.
Percival, P.W., and Hyde, J.S., 1975, *Rev. Sci. Instrum.*, **46**: 1522.
Perkins, R.D., 1977, *Practical Theory and Operation of Traveling Wave Tubes, and Microwave Glossary*, Perkins, San Diego.
Pearlman, M.R., and Webb, R.H., 1967, *Rev. Sci. Instrum.*, **38**: 1264.
Pfenninger, S., Forrer, J., and Schweiger, A., 1988, *Rev. Sci. Instrum.*, **59**: 752.
Piasecki, W., Froncisz, W., and Hyde, J.S., 1996, *Rev. Sci. Instrum.*, **67**: 1896.
Pifer, J.H., and Magno, R., 1971, *Phys. Rev. B*, **3**: 663.
Pollak, V.L., 1961, *J. Chem. Phys.*, **34**: 864.
Poole, C.P., and Farach, H.A., 1971, *Relaxation in Magnetic Resonance, Dielectric, and Mössbauer Applications*, Wiley, New York.
Poole, C.P., 1967, *Electron Spin Resonance. A Comprehensive Treatise on Experimental Techniques*, Wiley, New York.
Poole, C.P., 1983, *Electron Spin Resonance. A Comprehensive Treatise on Experimental Techniques*, 2nd ed., Wiley, New York.
Portis, A.S., and Gossard, *J. Appl. Phys.*, **31**: 205S
Quine, R.W., Eaton, G.R., and Eaton, S.S., 1987, *Rev. Sci. Instrum.*, **58**: 1709.
Quine. R.W., Rinard, G.A., Eaton, G.R., and Eaton, S.S., 1992, *J. Magn. Res.*, **99**: 571.
Rabek, J.F., 1982, *Experimental Methods in Photochemistry and Photophysics, Part 1*, Wiley, Chichester.
Rhea, R.W., 1994, *HF Filter Design and Computer Simulation*, Noble Publishers, Atlanta.
Rhodes, D.R., 1980, *Introduction to Monopulse*, Artech House, Dedham.
Ridenour, L.N., 1964, *Radar System Engineering*, McGraw-Hill, New York.
Rigden, J.S., 1986, *Rev. Mod. Phys.*, **58**: 433.
Rinard, G.A., Quine, R.W., Eaton, S.S., Eaton, G.R., and Froncisz, W., 1994, *J. Magn. Res.*, **108A**: 71.
Robertson, W.M., 1995, *Opto-Electronic Techniques for Microwave and Milli-meter Wave Engineering*, Artech House, Norwood.
Rosebury, F., 1993, *Handbook of Electron Tube and Vacuum Technique*, American Institute of Physics, New York.
Rowan, I.G., Hahn, E.L., and Mims, W.B., 1965, *Phys. Rev. A*, **137**: 61.
Rumsey, V.H., 1966, *Frequency Independent Antennas*, Academic Press, New York.
Schmalbein, D., Witte, A., Röder, R., and Laukien, G., 1972, *Rev. Sci. Instrum.*, **43**: 1664.
Schwartz, M., 1959, *Information, Transmission, Modulation, and Noise. A Unified Approach to Communications Systems Analysis*, McGraw-Hill, New York.
Schweiger, A., 1990, In: *Modern Pulsed and Continuous-Wave Electron Spin Resonance* (L. Kevan and M.K. Bowman, eds.), Wiley, New York. p43.
Schweiger, A., 1995, *J. Chem. Soc., Faraday Trans.*, **91**: 177.
Schweiger, A., and Jeschke, G., 2001, *Principles of Pulse Electron Magnetic Resonance*, Oxford University Press, Oxford.
Sherman, S.M., 1984, *Monopulse Principles and Technique*, Artech House, Dedham.
Sild, O., and Haller, K., 1988, *Zero-Phonon Lines and Spectral Hole Burning in Spectroscopy and Photochemistry*, Springer-Verlag, Berlin.
Sivan, L., 1994, *Microwave Tube Transmitters*, Chapman & Hall, London.
Slater, J.C., 1942, *Microwave Transmission*, McGraw-Hill, New York.
Standley, K.J., and Storey, B.E., 1962, *Rev. Sci. Instrum.*, **33**: 394.
Standley, K.J., and Vaughan, R.A., 1969, *Electron Spin Relaxation Phenomena in Solids*, Hilger, London.

Stillman, A.E., and Schwartz, R.N., 1976, *Mol. Phys.*, **32**: 1045.
Suter, D., 1997, *The Physics of Laser-Atom Interactions*, Cambridge University Press, Cambridge.
Sutphin, H.D., 1972, *Rev. Sci. Instrum.*, **43**: 1535.
Swain, D.W., 1970, *Rev. Sci. Instrum.*, **41**: 545.
Swenson, C.A., and Stahl, R.H., 1954, *Rev. Sci. Instrum.*, **25**: 608.
Tan, S.L., Waugh, J.S., and Orme-Johnson, W.H., 1984, *J. Chem. Phys.*, **81**: 576.
Taylor, D.R., Gillen, R.P., and Schmidt, P.H., 1969, *Phys. Rev.*, **180**: 427.
Thomann, H., Dalton, L.R., and Pancake, C., 1984, *Rev. Sci. Instrum.*, **55**: 389.
Tice, T.E., and Kraus, J.D., 1949, *Proc. IRE*, **37**: 1296.
Tsypkin, Ya.Z., 1964, *Sampling Systems Theory and Its Applications, Volume 1*, MacMillan, New York.
Vizmuller, P., 1987, *Filters with Helical and Folded Helical Resonators*, Artech House, Norwood.
Volino, F., Csakvary, F., and Servoz-Gavin, P., 1968, *Rev. Sci. Instrum.*, **39**: 1660.
Wallace, W.J., and Silsbee, R.H., 1991, Rev. Sci. Instrum., **62**: 1754.
Walter, C.H., 1990, *Traveling Wave Antennas*, Peninsula Publications, Los Altos.
Wanlass, L.K., and Wakabayashi, J., 1961, *Phys. Rev. Lett.*, **6**: 271.
Ware, D., and Mansfield, P., 1966, *Rev. Sci. Instrum.*, **37**: 1167.
Warren, W.S., and Silver, M.S., 1988, *Adv. Magn. Res.*, **12**: 247.
Webb, R.H., 1962, *Rev. Sci. Instrum.*, **33**: 732.
Williams, A.B., and Taylor, F.J., 1988, *Electronic Filter Design Handbook: LC, Active, and Digital Filters*, McGraw-Hill, New York.
Williams, R., 1996, *Spectroscopy and the Fourier Transform: An Interactive Tutorial*, VCH Publishers, New York.
Windle, J.J., and Wiersema, A.K., 1967, *J. Chem. Phys.*, **39**: 1139.

Chapter 6

Convolution-Based Algorithm: from Analysis of Rotational Dynamics to EPR Oximetry and Protein Distance Measurements

Convolution-Based Fitting of EPR Spectra

ALEX I. SMIRNOV AND TATYANA I. SMIRNOVA
Department of Chemistry, North Carolina State University, Raleigh, NC 27695, USA

Key words: electron paramagnetic resonance, least-squares fitting, Levenberg-Marquardt optimisation, convolution, nitroxide, spin-labeling, site-directed spin-labeling, high field EPR, high frequency EPR, rotational dynamics, oximetry, nitroxide bioreduction, dipolar interaction, distance measurements.

1. INTRODUCTION

Spectral simulations and optimization of simulation parameters to fit experimental data are very powerful tools for data analysis in EPR spectroscopy. Data modeling serves two purposes, namely, (i) an aid to the interpretation of raw spectra, and (ii) reducuction of the data file, which usually contains from a few hundred to thousands of data points, to a much smaller set of meaningful physical parameters. It has been shown (*e.g.*, Halpern *et al.*, 1993) that the use of the whole EPR spectrum instead of a few characteristic data points can significantly, sometimes by an order of magnitude, increase the accuracy of the extracted spectral parameters. If the theoretical model is suitable for the experiment, and the experimental noise has a Gaussian or similar distribution, then the least square's criterion (χ^2) can be chosen for parameter optimization. Since the EPR spectrum usually depends on several (≥3) parameters, a multi-dimensional minimization algorithm is required to optimize parameters based on the value of χ^2.

In this chapter we wish to describe how the spectral parameters can be extracted accurately and quickly for one class of EPR spectra, that is, those which are uniformly broadened along the magnetic field coordinate and can be described by a convolution integral:

$$I(B) = \int p(B - B')f(B')dB' = p(B) \otimes f(B) \tag{1}$$

where $\otimes$ is the convolution symbol, $p(B)$ is the envelope function (or an unbroadened spectrum), and $f(B)$ is the line shape of an individual spin packet (or broadening function).

The convolution integral (1) is well suited to modelling isotropic inhomogeneously broadened EPR spectra. A spectrum is represented by electron spin packets that are distributed under an envelope whose shape is determined according to a mechanism of inhomogeneous broadening. The model describing the inhomogeneous line shape requires that the broadening must come from interactions outside the spin system and must be adiabatic during the time of the spin transition (Portis, 1953). Sources of inhomogeneous broadening include anisotropy broadening, hyperfine interaction, inhomogeneities in the magnetic field, etc., and at least to some extent, all EPR spectra are inhomogeneously broadened. But Equation (1) works well only for the spectra that are broadened uniformly, such as, for example, individual nitrogen hyperfine components of spin label EPR spectra in the fast motional limit. In this case the envelope $p(B)$ can be assigned to the proton hyperfine structure. This envelope can be depicted by a stick diagram and directly used in the simulations or approximated by a Gaussian envelope (Bales, 1989). The resonance line profile of an individual spin packet is given by a Lorentzian function with its center at B_0 and a width $\Delta B_{1/2}$ at half of the line height:

$$f(B) = \frac{2A}{\pi} \frac{\Delta B_{1/2}^2}{\Delta B_{1/2}^2 + 4(B_o - B)^2} \tag{2}$$

where A is the value of the double integral that is proportional to the number of spins, and $\Delta B_{1/2}$ is the line width that is determined by the rotational relaxation and spin-spin interactions. For spin labels in fast motional limit the individual line widths, $\Delta B_{1/2}(m_I)$ should be assigned to each of the m_I nitrogen hyperfine transitions.

Slow-motion EPR spectra require a more sophisticated simulation analysis, which generally involves a greater number of parameters than the fast limit spectra (Budil *et al.*, 1996). But even these spectra may contain isotropically broadened components. For example, if some hyperfine interaction (*e.g.*, with protons of methyl groups) can be considered as "pre-averaged" by a fast (on the EPR time scale) intramolecular rotation, then those interactions will contribute to an anisotropic slow-motional EPR spectra as a uniform broadening. This kind of broadening can be again modeled by Equation (1) in which the slow-motional spectrum is treated as an envelope *p(B)* and *f(B)* is assigned to the additional uniform broadening.

A second and, from the standpoint of spin labeling experimentation, very important source of uniform broadening are magnetic interactions between similar or dissimilar electron spins. These include Heisenberg spin exchange (*e.g.,* broadening by molecular oxygen) and isotropic dynamic dipolar interaction. The latter dominate in interactions of spin-labels with some paramagnetic metal ions such as Gd^{3+} in non-viscous liquids (Hyde *et al.*, 1979). Spin exchange between the nitroxides in solutions at concentrations corresponding to slow exchange also leads to line broadening, but the shape remains Lorentzian only to a first approximation (Molin *et al.*, 1980). Another case of uniform spectral broadening is caused magnetic interactions between two nitroxide labels introduced into a protein by site-directed spin-labeling methods. Under certain conditions, for example, if the spins in such a pair are separated by >10-12Å and the rotational correlation time of the interspin vector are short enough, the static dipolar interaction is averaged out and the broadening function is again represented by a Lorentzian and can be considered to be uniform across the nitroxide spectrum (*e.g.,* Mchaourab *et al.*, 1996). In all these cases, the "envelope" spectrum can be assigned to that taken in absence of magnetic interactions. For example, in EPR oximetry experiments, the envelope is measured in absence of oxygen or at a lower oxygen concentration while in spin-label pair distance measurements the envelope is the sum of double-integral-normalized EPR spectra from single-labeled protein mutants.

Rabenstein and Shin (1995) suggested that even when the static part of dipolar interaction is not averaged out, the dipolar broadening effects on the EPR spectra can, under certain conditions, be approximated by a convolution integral (although more rigorous treatment shows that, generally, these effects are not uniform across the nitroxide spectra). In their experiments, Rabenstein and Shin measured EPR spectra of single- and double-labeled mutants at low temperatures (*ca.*77 K) to eliminate dynamic averaging of dipolar interactions. Then these authors followed the original paper by Pake (1948) and treated polarization of dipoles as static, so only shifts in the resonance frequency are produced. The EPR spectrum in presence of the dipolar interaction was approximated by a convolution of an unbroadened EPR spectrum (which we can consider as an envelope in Equation (1)) with the Pake's dipolar pattern (Pake, 1948). A somewhat similar approach to that of Rabenstein and Shin was independently developed by Steinhoff *et al.* (1991, 1997). In studies of frozen spin-label samples, Steinhoff also treated the dipolar interaction as static and approximated the EPR spectrum from spin-label pairs as a convolution integral of an unbroadened spectrum and the dipolar broadening function; however, the broadening function was constructed digitally under the assumption of a Gaussian distribution for the interspin distances.

A third source inhomogeneous broadening is local heterogeneity of biological samples. In some cases this heterogeneity could be manifested in a strain broadening - a distribution in hyperfine coupling constants and *g*-factors due to local variations in electrostatic and hydrogen bonding environment of spin labels. Heterogeneity of biological membranes could also result in a distribution of line widths due to different mobilities of spin probes located in different membrane domains. Superposition of EPR spectra with different widths could be treated as an additional inhomogeneous broadening (Sankaraman and Marsch, 1999). Characteristic parameters of such a spectrum such as, for example, an intensity ratio (see *ibid.*), can be used to deduce statistical distribution of the spin labels.

Overall, the examples described above demonstrate that in many cases EPR spin-label spectra can either be fully described by a convolution integral (1) or have a spectral contribution which is given by Equation (1). This chapter is intended to (*i*) provide a practical guide to least-squares simulations of such spin-label spectra, and (*ii*) discuss the range of problems to which this convolution-based simulation approach can be applied. We will show that our approach is very accurate and flexible enough to accommodate a variety of experiments (*e.g.*, Gaussian-Lorentzian convolution often being calculated with accuracy better than 10^{-6}). The main advantage of the method is that it allows one to extract the amount of the uniform broadening from the spectra directly and with fewer parameters than the simulations of the full spectra. The algorithm is combined with an efficient Levenberg-Marquart optimization algorithm, which converges rapidly and allows for estimates of parameter uncertainties. Applications of the technique are demonstrated with selected examples ranging from studies of rotational diffusion to EPR oximetry in membranes and structural protein studies.

2. CONVOLUTION ALGORITHM WITH LEVENBERG-MARQUARDT OPTIMIZATION FOR FITTING INHOMOGENEOUS EPR SPECTRA

2.1 Simulations of Inhomogeneous EPR Spectra with Fourier Transform

While many EPR spectra can be presented in a form of a convolution integral (Equation (1)), direct digital calculation of this integral is computationally inefficient because it involves integration over N data points and therefore requires $O(N^2)$ operations. A more efficient way of evaluating Equation (1) is based on the properties of Fourier transform:

$$F(\nu) = \Im(f(B)) = \int_{-\infty}^{+\infty} f(B)\exp(2\pi i B\nu)dB \quad (3)$$

where $2\pi\nu=\omega$ and the inverse transformation is given by:

$$f(B) = \Im^{-1}(F(\nu)) = \int_{-\infty}^{+\infty} F(\nu)\exp(-2\pi i B\nu)d\nu \quad (4)$$

While the function $p(B)\theta\ f(B)$ is determined in the field domain by Equation (1), in the frequency domain it corresponds to a much simpler transformation pair which is also known as a convolution theorem:

$$p(B)\otimes f(B) = P(\nu)F(\nu) \quad (5)$$

Thus, in the frequency domain the multiple convolutions can be carried out as a series of simple multiplications.

For a discrete set of data, Fourier transforms can be calculated by a computationally efficient Fast Fourier Transform (FFT) algorithm, which requires only *O(Nlog2N)* operations. Then the digital convolution, carried out by transforming each function into frequency domain using FFT, multiplying Fourier images, and transforming the product back to the field domain, also takes *O(Nlog2N)* operations.

This efficiency of calculating convolution integrals by FFT in computer simulations of EPR spectra was recognized a long time ago (*cf.* Evans *et al.*, 1978). These authors made use of the FFT convolution method to simulate isotropic EPR spectra and found the method to be particularly efficient when a large number of hyperfine transitions are present. If the second- and higher order effects can be neglected and the EPR spectrum is isotropic, then the hyperfine structure observed in the field domain corresponds to simple harmonic functions in the frequency domain (*cf.* Evans *et al.*, 1978; Dunham *et al.*, 1980). For example, the hyperfine term for an $I=1/2$ nuclear spin is:

$$H_i(\nu) = \cos(\pi\nu a_i/\Delta) \quad (6)$$

where a_i is the hyperfine coupling constant and Δ is the scan range. Expressions for other nuclear spin values were reported by Evans *et al.* (1978) and Duling (1994). Thus, in the frequency domain an isotropic EPR spectrum can be simulated by multiplying simple harmonic terms corresponding to hyperfine couplings with different magnetic nuclei. This makes computation time of the FFT simulations method virtually independent upon the number of hyperfine transitions. Accounting for isotope abundances is also significantly simplified in this method, as is modeling of instrumental distortions caused by magnetic field modulation and spectrometer time constant (Evans *et al.*, 1978). (For most accurate calculations of modulation broadening effects, derived from the first principles, see Robinson *et al.*, 1999.) Gaussian and Lorentzian line shapes are easy to introduce by multiplication of the corresponding Fourier images in the frequency domain. The resulting theoretical spectrum can be least-squares fitted to the experimental data in either field or frequency domain (Dunham *et al.*, 1980; Duling, 1994). Even earlier, before least-squares methods became widely used in EPR, analysis of EPR spectra in the frequency domain was exploited by Silsbee (1966) to examine EPR spectra for partly resolved hyperfine transitions.

2.2 Gaussian-Lorentzian Convolution and Gaussian-Lorentzian Sum Approximation

While applying FFT-methods to the simulation and optimization of simple isotropic EPR spectra with just a few hyperfine transitions does not much improve the computation time compared with direct simulations in the field domain, the efficiency of computations does become important when the number of transitions is large and when some or all of those multiple transitions are approximated by a Gaussian envelope. The Gaussian envelope approximation is, for example, especially useful for correcting for inhomogeneous broadening of spin label spectra whose proton superhyperfine structure is not precisely known (for a review see Bales, 1989). It has been shown that in many practical cases an unresolved proton hyperfine structure can be replaced by a Gaussian, thus eliminating the need for detailed hyperfine analysis (Bales, 1989).

Because the Gaussian-Lorentzian convolution, also called a Voigt function,[1] is a rather simple line shape, it can be characterized in terms of peak-to-peak width and amplitude parameters, which then can be used for correcting for inhomogeneous broadening without computer-assisted fitting (*ibid.*). However, for least-squares simulations of the EPR spectra the whole spectral shape should be calculated. Simulations of the Voigt line shape directly in the field domain are computationally inefficient because this function can only be expressed in a form of an integral thus requiring $O(N^2)$ operations. The calculation time can be significantly improved by approximating the Voigt profile as a GL-sum, that is, a sum of Gaussian and Lorentzian functions of a common line width (Wertheim *et al.*, 1974). This allows one to approximate the integral (1) with good accuracy (0.5% for the first derivative and 0.7% for the first integral forms) in only $O(N)$ operations. Using this approximation, Halpern and coworkers (1993) developed an efficient simulation method in conjunction with the Levenberg-Marquardt least-squares optimization for fitting inhomogeneously broadened EPR spectra. In their method the use of the GL-sum approximation simplifies the

[1] Editor's Note: Numerical algoriths in Fortran 90 and *Mathematica* for this and ther functions in this chapter may be found in W.J. Thompson, *Atlas for Computing Mathematical* Functions, Wiley-Interscience, New York 1997.

calculations of parameter partial derivatives required by the Levenberg-Marquardt algorithm and directly provides the overall line width and its position. Also, it has been shown that if the modulation amplitude is less than the EPR peak-to-peak line width, then the modulation broadening contributes substantially only to the Gaussian component of the derivative Voigt line (Peric and Halpern, 1994). The GL-sum approximation can be also used for modeling the Voigt line shape with a dispersive component (Halpern *et al.*, 1993), which is usually present in a continuous-wave EPR spectrum obtained with single-channel detection.[2] It is often the case that the problem of a signal phase shift becomes much more severe in high frequency/high field (95-250 GHz) EPR (Earle *et al.*, 1993). It is also sometimes encountered in low-frequency EPR (0.25-1.5 GHz; Nilges *et al.*, 1989). However, we find no literature describing accuracy of the GL-sum approximation in modeling dispersive Voigt line shapes.

In spite of its clear computational advantages in simulation of EPR spectra with a small number of hyperfine transitions, the GL-sum method neither yields Gaussian and Lorentzian components of the line width directly and conveniently nor insures the flexibility to accommodate deviations from the Gaussian-Lorentzian convolution model. Many practical applications necessitate accounting for other parameters in the mathematical model of EPR spectra such as instrumental distortions by time constant effects or a modulation amplitude that does exceed the peak-to-peak line width. Other non-Gaussian distribution functions for the spin packets are very common (*e.g.*, Pake and powder patterns). Finally, *anisotropic* EPR spectra that are *isotropically* broadened by magnetic interactions cannot be simulated with the GL-sum method. We shall illustrate these and other applications of the convolution-based simulations and fitting later in this Chapter.

In this section we will give a general description of the convolution-based fitting algorithm. We will also show that by properly using the properties of the convolution integral, partial derivatives of the line shape required for thc Levenberg-Marquardt optimization algorithm can be calculated effectively and accurately. In brief, each function is expressed in a form of an analytic function, and only the Fourier transform is calculated digitally. Thus, there is no digital differentiation involved. All fitting parameters, including Lorentzian and Gaussian line width contributions, are extracted directly. The dispersion component of the EPR signal can be introduced easily into the fitting algorithm as desired. A hyperfine and/or superhyperfine structure pattern (or envelope function) may be included in the fitting procedure without any significant increase of computational time.

[2] A more expensive quadrature detection scheme can eliminate this problem.

2.3 The Use of Fast Fourier Transform Algorithm for Calculating Gaussian-Lorentzian Convolution

The FFT algorithm for fast calculation of the convolution integral is based on a discrete convolution theorem which assumes that one function, $p(B)$, has a finite duration on a interval included in the spectral window $[B1, B2]$ (*i.e.*, $p(B)\equiv 0$ for $\forall B \notin [B_a, B_b]$, $[B_a, B_b] \in [B_1, B_2]$), and the other function, $f(B)$, is periodic with a period equal to the spectral window. A Gaussian envelope function satisfies the first assumption rather well because it rapidly vanishes to zero; its value becomes negligible over the interval of just a few line widths. A Lorentzian function is not periodic and it vanishes less quickly. Because the Lorentzian shape has a non-zero intensity even far from its center (for example, $F(B)/F(B0)\approx 10^{-4}$ at $(B-B_0)\approx 50\Delta B_{1/2}$), this function will be truncated by any reasonable scan width. When such a truncated Lorentzian is used in FFT convolution, the result $F(B)$ will include wrap-around data over the interval B_b-B_a equal to the duration of the function $p(B)$ (*cf.* Press *et al.*, 1986).

Figure 1 illustrates the wrap-around effect on an example of a digital convolution of Gaussian and Lorentzian functions of equal line widths at the half-heights ($\Delta B_{1/2}$). These functions were initially simulated in the field domain, then digitally transformed into the frequency domain, multiplied, and consequently converted back to the field domain using FFT. If the spectral window is equal to $5\Delta B_{1/2}$, then the wings of the function computed with the FFT are visibly distorted compared to an essentially exact Voigt function (Figure 1A and B). When the spectral window was set to $10\Delta B_{1/2}$ (Figure 1C), the residual, defined as the difference between these two function, was improved by an order of magnitude (Figure 1D). The maximum of the residual, about 0.3% of the Voigt function amplitude, was better than approximation of the Voigt function by the GL-sum. Analysis shows that in the case of Figure 1C, the spectrum is approximated to better than 0.001% over the central 80% spectral window. This accuracy is sufficient for most applications.

Generally, it is desirable that the accuracy of the simulated approximation be better than experimental noise. Even better accuracy in computing the Gaussian-Lorentzian convolution is achieved by using larger spectral windows or by padding the line shape function with zeros according to the methods discussed by Press *et al.* (1986). At times experimental data are recorded over a field interval, which is less than $10\Delta B_{1/2}$.Then, in order to avoid the wrap-around effects in FFT convolution, one should carry out the simulations on a larger field interval (the width of the interval is determined by the accuracy required) and then compare only the central portion of the simulated spectrum with the experiment. Therefore, implementation of the

FFT convolution algorithm generally does not require large field intervals for data acquisition.

Calculations of the Voigt line shape can be simplified if one starts with simulations in the frequency domain. Expressions for Fourier images of Lorentzian and Gaussian functions are simple and can be found elsewhere (*e.g.*, Duling, 1994). However, simulations in the frequency domain do not eliminate the wraparound problem when the Voigt function is transformed back into the field domain by use of FFT.

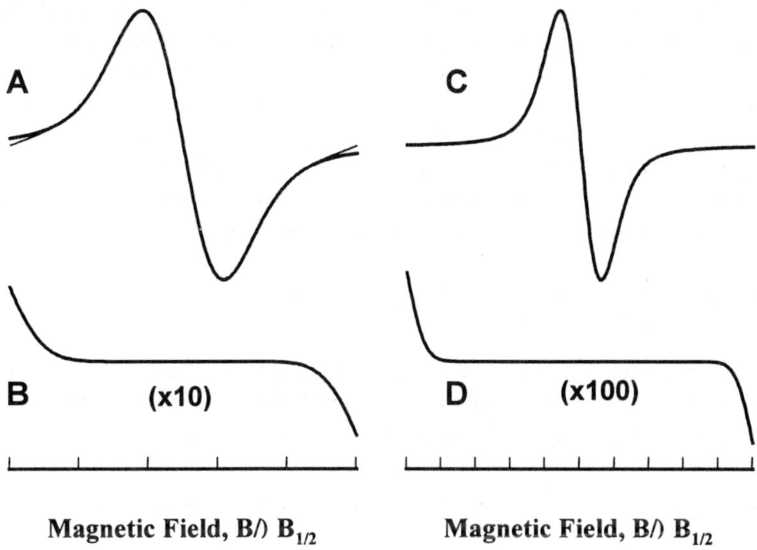

*Figure 1.*The wrap-around effect in the digital convolution of Gaussian and Lorentzian functions. (A) Result of digital convolution using the FFT algorithm (thick) line is superimposed on essentially exact Voigt line shape (thin line). The difference is noticeable only in the wings region and is clearly seen in the residual (B), which is shown in tenfold amplification. Result of a convolution on a twice as large spectral window and essentially exact Voigt function are indistinguishable on the plot (C). Residual (D), shown in a hundred-fold amplification, was improved by an order of magnitude. Reproduced with permission from Smirnov and Belford, 1995.

Another issue to consider while using FFT methods is the digital resolution and the choice of the sampling interval ΔB. In principle, the experimental spectra are always bandwidth limited because the EPR signal passes through amplifiers with limited frequency response and/or additional filters with known time constants are employed. In those cases, the signal is typically recorded at the rate Δt equal twice the maximum frequency passed through the amplifier or filter. This rule originates from the Nyquist theorem, which relates the Nyquist critical frequency f_c to the sampling rate Δt:

$$f_c = \frac{1}{2\Delta t} \tag{7}$$

The theorem states that if a continuous function is sampled at a rate Δt and is bandwidth limited to frequencies smaller in magnitude than $/f_c/$, then that function is completely determined by its samples. In other words, its information content is fully preserved in the digital representation.

If an EPR spectrum is digitized with a sampling interval ΔB, then the Fourier transform is defined only between plus and minus the Nyquist critical frequency $1/(2\Delta B)$. When a spectrum contains any higher frequencies, the power outside the Nyquist frequency range is folded over or "aliased" into that range (see Press *et al.*, 1992).This property of digital Fourier transform could cause some problems if simulations are carried out starting with the frequency domain. Indeed, let us consider a Lorentzian function $f(B)$ which has a simple Fourier transform pair:

$$f(B) \propto \frac{1}{A^2 + B^2} \Leftrightarrow F(\nu) \propto \frac{1}{A}\pi \exp(-|A\nu|) \tag{8}$$

In order to center this function at B_0, one needs to convolve $f(B)$ with a delta function $\delta(B\text{-}B_0)$. In the frequency domain this is equivalent to multiplying $F(\upsilon)$ by a Fourier image of $\delta(B\text{-}B_0)$, namely $exp(-2\pi i\upsilon B_0)$:

$$f(B - B_0) \propto \frac{1}{A^2 + (B - B_0)^2} \Leftrightarrow \frac{1}{A}\pi \exp(-2\pi i\nu B_0)\exp(-|A\nu|) \tag{9}$$

For cases in which B_0 is not precisely at the digitizing point $n\Delta B$ but rather at $B_0=n\Delta B+B_1$, where $B_1<\Delta B$, then

$$\exp(-2\pi i\nu(n\Delta B + B_1)) = \exp(-2\pi i\nu n\Delta B)\exp(-2\pi i\nu B_1) \tag{10}$$

The first exponential term in the right hand part of Equation (10) is periodic on the interval $\pm 1/(2\Delta B)$ while the second is not. Moreover, because $B_1<\Delta B$, less than a single period of the second function would fit inside the $\pm 1/(2\Delta B)$ interval. This would result in small but visible oscillations after the function (9) is digitally transformed into the field domain. These oscillations arise from improper sampling in the frequency domain.

The example above demonstrates that if the simulations are started in the frequency domain, the positions of the lines cannot be simulated any better than the digitizing step in the corresponding field domain. Generally, this is not of concern because if the sampling intervals are sufficiently small then the line positions could be chosen exactly at the sampling points. However, in simulations of spectra with very good signal-to-noise ratios, this would cause an unnecessary round off of the simulation parameters. Figure 2 illustrates this on an example of two least-squares simulations of a model (Lorentzian) spectrum with a high signal-to-noise (=100). The first simulation was carried out with a "continuous' adjustment of the line center (residual is shown in Figure 2 B in five fold amplification), while in the second simulation the line center was adjusted with discrete steps coinciding with the field digitizing. The latter simulation yields a discernible residual (Figure 2C).

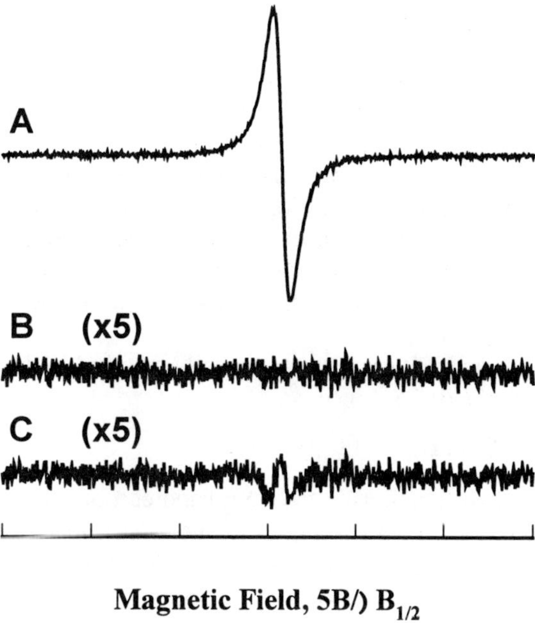

Figure 2. Effect of rounding-off the position of the Lorentzian first-derivative line to the nearest sampling point.(A) The line with $\Delta B_{1/2}$ peak-to-peak width was simulated over a spectral interval of $30\Delta B_{1/2}$.The line was properly digitized with 20 data points per peak-to-peak width and positioned so its center was almost exactly between the digitizing points. A Gaussian noise was also added to the spectrum (signal-to-noise ratio =100). The spectrum (A) was least-squares fitted to a Lorentzian shape. During the first fit the line center (and other parameters) were allowed to adjust continuously while during the second fit the center was adjusted in discrete steps coinciding with the digitizing. The corresponding residuals are shown in the fivefold amplification (B and C respectively).

Although this problem with digital resolution could be somewhat overcome by an additional interpolation in the field domain, we prefer to start our simulations of the line shape in the field domain but carry out convolutions by the FFT algorithm. This way we avoid all the artifacts except some rather minor wraparound effects we discussed earlier.

2.4 Convolution Algorithm with Levenberg-Marquardt Optimization for Fitting Inhomogeneous EPR Spectra

We write the general function for fitting inhomogeneously broadened EPR lines based on Equation (1) with addition of a polynomial baseline term of the order N (typically, N=1 or =2 is sufficient):

$$I(B) = f(B) \otimes p_1(B) \otimes p_2(B) \otimes \ldots \otimes p_k(B) + \sum_{i=0}^{N} l_i B^i \qquad (11)$$

The line shape is given by $f(B)$, and the $p_j(B)$'s are optional envelope functions. Typically, one of the envelope functions is a Gaussian, which is a very common model of inhomogeneous broadening. Other envelopes are assigned to isotropic hyperfine interactions with magnetic nuclei and also may include a non-adjustable envelope function, which is read as a file. The latter could be, for example, an experimental spin-label EPR spectrum in absence of oxygen and/or some other broadening function due to isotropic dipolar interactions. Modulation and time constant broadening can be also included in these envelopes.

Continuous-wave EPR spectra are usually detected in the form of a first derivative, $I'(B)=\partial I(B)/\partial B$. Since

$$I'(B) = p'(B) \otimes f(B) = p(B) \otimes f'(B) \qquad (12)$$

only one of the two functions in the right-hand part of Equation (12) should be differentiated in order to describe the experimental first-derivative spectrum. This property of the convolution integral is also very useful for efficient calculations of partial derivatives required for the Levenberg-Marquardt optimization.

The model of inhomogeneous line shape we describe here is general and allows for different line shapes f(B). One practically important example is a dispersion shape or a mixture of dispersion and absorption. Dispersion and absorption signals are related by Kramers-Kronig equations. If the contribution from dispersion is measured as a phase shift $\Delta\varphi$ ($\Delta\varphi$=0 corresponds to absorption line), then the first-derivative Lorentzian EPR line is given by:

$$f(B) = \frac{2A}{\pi} \frac{8\Delta B_{1/2}^{L}(B_0 - B)\cos\Delta\varphi + \left[4(B_0 - B)^2 - (B_{1/2}^{L})^2\right]\sin\Delta\varphi}{\left[(\Delta B_{1/2}^{L})^2 + 4(B_0 - B)^2\right]^2} \qquad (13)$$

where $\Delta B^{L}{}_{1/2}$ is the Lorentzian width measured at half height of an absorption signal, A is the area under the resonance absorption curve or (double-integral) intensity of the signal, B_0 is the resonance field. Then the inhomogeneous EPR spectrum for an arbitrary $\Delta\varphi$ is given by a convolution of $f(B, B_0, \Delta B^L_{1/2}, \Delta\varphi)$ with an inhomogeneous envelope. Thus, only one function in the convolution integral has to be corrected for the dispersive component. Although the dispersion line shape can be computed for both Gaussian and Lorentzian functions, expression for the dispersive Lorentzian shape is much simpler while calculations of the dispersive Gaussian shape require evaluations of the error function integral.

Accounting for the dispersive contribution in least-squares fitting of spin-label EPR spectra is especially important because those spectra are often collected from aqueous samples, which have high dielectric constants. For these samples the EPR signal always has some dispersive component because the microwave phase varies along the sample. However, even in this case the observed EPR signal $I(B)$ appears as a mixture of absorption, $A(B)$, and dispersion, $D(B)$, signals with an effective phase shift $\Delta\varphi_e$. Indeed, if $\Delta\varphi(\vec{r})$ is the microwave phase expressed as a function of a spatial vector $\vec{r}$, then:

$$\begin{aligned} I(B) &= \int_V \left[A(B)\cos(\Delta\varphi(\vec{r})) + D(B)\sin(\Delta\varphi(\vec{r}))\right] d\vec{r} = \\ &= A(B)\langle\cos(\Delta\varphi(\vec{r}))\rangle + D(B)\langle\sin(\Delta\varphi(\vec{r}))\rangle = \\ &= R\left[A(B)\cos(\Delta\varphi_e) + D(B)\sin(\Delta\varphi_e)\right] \end{aligned} \tag{14}$$

where

$$R^2 = \langle\cos(\Delta\varphi(\vec{r}))\rangle^2 + \langle\sin(\Delta\varphi(\vec{r}))\rangle^2$$

and

$$\Delta\varphi_e = \tan^{-1}\frac{\langle\sin(\Delta\varphi(\vec{r}))\rangle}{\langle\cos(\Delta\varphi(\vec{r}))\rangle} \tag{15}$$

Thus, the shift observed in such an experiment cannot be always experimentally compensated by adjusting the spectrometer reference arm. The phase shift problem becomes much more severe in high frequency (95-250 GHz) EPR spectrometers and also can be a problem in certain low-frequency (0.5-2.0 GHz) EPR experiments when a single channel is used for the signal detection. These phase shifts can significantly distort line shapes and lead to data misinterpretation if they are not accounted for. This can be overcome by using Equation (13) in the simulations or by correcting experimental spectra by mixing in an out-of-phase component calculated using Kramers-Kronig equations.

For simulations, initial values of the parameters in function Equation (11) are determined by a search for crude local extrema in the experimental spectrum $E(B)$. The initial microwave phase $\Delta\varphi$ is usually set to 0 or π. Both values correspond to pure absorption spectra at different modulation phases. Baseline coefficients are set to zero. A multiparameter line shape function, $f(B, \alpha_i)$, is derived analytically and then, if desired, is digitally convoluted with optional envelope functions $p_i(B)$'s using the FFT convolution algorithm. Envelope functions are derived analytically or read as a file. Some of the envelopes, such as Gaussian and modulation and/or time constant broadening should be centered over the spectral window and are easily simulated in the frequency domain. The use of FFT requires that the simulated spectrum $I(B_i)$ has n data points, where n is a power of 2. If the experimental spectrum was collected with a different number of data points, it can be interpolated to the required number of data points or else the theoretical spectrum can be calculated over a wider spectral window.

The simulated function $I(B_i, \vec{a})$ is compared with the experimentally measured spectrum $E(B_i)$ by a square norm χ^2 test:

$$\chi^2(\vec{a}) = \sum_{i=1}^{N} \left[E(B_i) - I(B_i, \vec{a})\right]^2 \tag{16}$$

where $\vec{a}$ is a parameter vector and the standard deviations at each experimental data points are assumed to be the same. The parameter vector $\vec{a}$ is then adjusted to minimize the value of $\chi^2(\vec{a})$.

For the minimization, we have utilized the Levenberg-Marquardt algorithm (*cf.* Press *et al.*, 1986, Ball *et al.*, 1973). Many modifications of this algorithm require a user-supplied subroutine which, for the input value B_i, returns the value of the model function $I_i=I(B_i, a_j)$ and the vector of the derivatives $\partial I(B_i, a_j)/\partial a_j$ (*cf.* Press *et al.*, 1986). In order to take the advantage of the FFT convolution algorithm, we modified this procedure as follows. Instead of calculating one value of I_i and the vector $\partial I(B_i, a_j)/\partial a_j$ at each value of B_i at a time, we calculate these functions for the whole set of B_i; $i=1,\ldots,n$. Then the calculation procedure is organized as follows:

(1) The functions $p_k(B_i)$, $f(B_i, a_j)$, and $\partial f(B_i, a_j)/\partial a_j$ are generated from analytically derived expressions for all values of B_i; $i=1,\ldots,n$. Alternatively, a non-adjustable envelope function is read as a file.

(2) The convolution $p(B)$ of functions $p_1(B)$, $\ldots p_k(B)$ ($p(B)=p_1(B)\otimes p_2(B)\otimes\ldots\otimes p_k(B)$) is computed digitally by the FFT convolution algorithm and all intermediate Fourier images of envelopes and their products are stored in the computer memory.

(3) All functions generated in step (1) are convoluted with $p(B)$; $f(\omega, a)$ is stored. Since $p(\omega)$ was previously stored during step (2), each convolution requires only two Fourier transforms instead of three. Convolution of $p(B)$ with $\partial f(B_i, a_j)/\partial a$ gives a matrix of partial derivatives $\partial I(B_i, a_j)/\partial a_j$ for all values of B_i, since:

$$\frac{\partial I(B)}{\partial a_j} = p_1(B) \otimes p_2(B) \otimes \ldots \otimes p_k(B) \otimes \frac{\partial f(B, a_j)}{\partial a_j} \tag{17}$$

The vector of the simulated function $I(B_i)$ is computed by adding a baseline term (Equation 11).

(4) If envelopes contain adjustable parameters σ_j, the vectors $\partial p_k(B_i, \sigma_j)/\partial\sigma_j$ are computed from analytical expressions and then are convoluted digitally with other envelopes $p_m(B)$ and $f(B, \vec{a})$. The number of required Fourier transforms can be decreased by using vectors $p_m(\omega_i)$ and $f(\omega_i, \vec{a})$ stored during steps (2) and (3).

(5) Partial derivatives with respect to the baseline parameters and with respect to the amplitude parameter are computed without additional convolutions. A complete matrix of partial derivatives $\partial I(B_i, a_j)/\partial a_j$ is formed and then supplied for Levenberg-Marquardt optimization. The iterations are stopped after the value of $\chi^2(\vec{a})$ is improved insignificantly (*i.e.*, by 10^{-3} of the previous value) for a second time. The estimates of parameter uncertainties are calculated in a standard manner (Press *et al.*, 1986).

The advantages of the described algorithm are such that it does not require digital differentiation of simulated functions with respect to the simulation parameters. Most functions required for calculations are derived in analytical forms. The final step involves digital convolution of the generated functions by FFT. Saving the intermediate results can be used to reduce the number of Fourier transforms. For convolutions with Gaussian and/or modulation and time constant broadening functions, simulating these functions directly in the frequency domain from the analytical expressions, and then digitally transforming the product back to the field domain can further decrease the computational time. Depending on experimental needs, the fitting function (11) can be extended by summation over non-equivalent line shape functions $f_j(B, a_i)$ and/or non-equivalent paramagnetic centers.

Simulation/optimization time with this algorithm depends upon complexity of the model, number of adjustable parameters, number of data points, and computer hardware. Typically, with the use of modern Pentium-based PC, each iteration for a typical nitroxide in the fast motion limit collected with the digital resolution of 1024-2048 data points is completed a fraction of a second, with 4-6 iterations required to find the optimal fit. During the iterations, more than 95% of the time is spent by the FFT subroutine, making an efficient FFT algorithm essential for the program performance.

2.5 Fitting of Model Spectra

The algorithm was tested on sets of computer-simulated spectra. In the absence of explicitly introduced noise, the accuracy of the extracted parameters is determined by the computer round-off errors. To prevent wrap-around effects, the spectral windows used for intermediate FFT convolutions were chosen to be twice as wide as those for the synthetic spectra. We have also tested the algorithm in the presence of a noise pattern having a Gaussian (normal) distribution, which is a good approximation of the noise in EPR experiments (*cf.* Duling 1994). In this case, the covariance matrix provided by the Levenberg-Marquardt method can be used to estimate the standard deviations of fitting parameters (Press *et al.*, 1986).

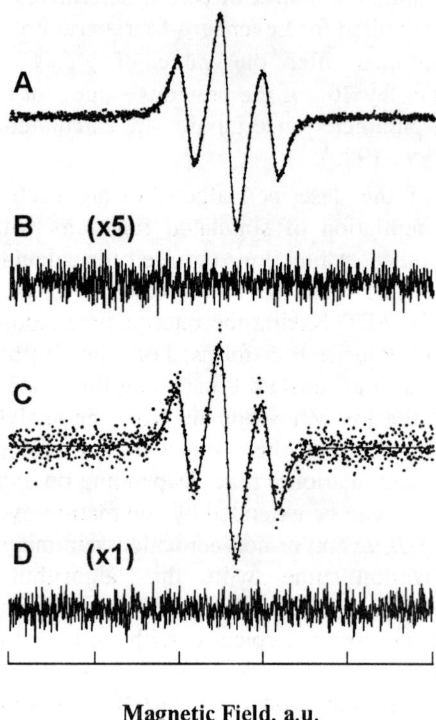

Figure 3. Fitting results for a synthetic spectrum with various levels of computer-generated Gaussian noise. Details of simulations are given in the text. Noise standard deviations for the spectra (A) and (B) are respectively 1.0 % and 5% of the spectrum amplitude. Residuals (B, shown in fivefold amplification, and D) reveal no systematic differences between the spectra and the fits. Comparison of differences between exact and extracted parameters and estimated parameter discrepancies are summarized in the Table.

Table 1. Percentage Deviation of Line Width Parameters Extracted from a Three-Line Spectrum Containing Gaussian Noise of Various Amplitude

Fit	Noise σ=%I_{p-p}	$\delta(\Delta B^G{}_{p-p})$, %	Estimate of $\delta(\Delta B^G{}_{p-p})$, %	$\delta(\Delta B^L{}_{p-p})$, %	Estimate of $\delta(\Delta B^L{}_{p-p})$, %
1	0.1	0.083	0.12	0.16	0.14
2	1.0	0.58	1.2	0.059	1.5
3	5.0	4.8	6.2	8.7	7.1
				0.14	1.9
4	1.0	0.45	1.3	0.11	1.6
				0.84	1.9
				4.5	2.0 (4.0)
5	1.0	0.14	1.4	0.85	1.7 (3.4)
				6.2	2.0 (4.0)

Note. Parameters of the three-line spectrum simulated with a dispersion contribution and the fitting functions used in different fits are given in the text. Noise standard deviations are given in % of the signal amplitude I_{p-p}.$\delta(\Delta B^G{}_{p-p})$ is the difference between extracted Gaussian contribution to the line width and that used in the simulation and is expressed in % of the exact Gaussian line width.$\delta(B^L{}_{p-p})$ values correspond to Lorentzian component or components depending on the fitting function. Estimates of $\delta(\Delta B^G{}_{p-p})$ and $\delta(B^L{}_{p-p})$ were calculated using a covariance matrix and correspond to 68% confidence intervals. Values in brackets correspond to the estimated 94.5% confidence intervals.

Table 1 and Figure 3 summarize fitting results for a model synthetic spectrum. The spectrum was constructed as a convolution of Gaussian and Lorentzian functions with equal line widths at half height $\Delta B^G{}_{1/2}=\Delta B^{L1}{}_{/2}$ and a 5% contribution of the dispersion signal. A hyperfine splitting of the spectrum corresponds to two equivalent S=1/2 nuclei with $a_{iso}=2.5\Delta B^G{}_{1/2}$. First, the fitting function was constructed with a fixed hyperfine constant a_{iso}. Lorentzian and Gaussian line widths, phase of the signal (or dispersion weight parameter), spectrum center, amplitude, and baseline were the adjustable parameters. Line width parameters were assumed to be the same for each of the hyperfine components. Before the fitting, a computer-generated Gaussian noise with a characterized standard deviation, σ_{noise}, was added to the synthetic spectrum. Discrepancies between the best-fit parameters and the exact parameters used to generate the synthetic spectrum were compared with the estimated parameter discrepancies (68% confidence intervals) derived from a covariance matrix for three standard deviations of the Gaussian noise (Table 1, fits 1-3, σ_{noise}=0.1%, 1%, and 5% respectively; σ_{noise} is given in % of the signal peak-to-peak amplitude Ip-p). Figure 3 shows the simulated "noisy" spectra (σ_{noise} =1% of I_{p-p} for A, σ_{noise} =5% of I_{p-p} for C), which are superimposed with best fits and the residuals (B and D respectively). The residuals reveal no systematic deviation between synthetic and best-fit spectra. The estimated parameter uncertainties are proportional to the standard deviation of the noise. Estimated parameter discrepancies, δ, and the discrepancies between extracted and exact parameters used for simulations are in reasonable agreement.

When the number of fitting parameters is increased, one would expect a decrease in accuracy of extracted parameters. As an example, we have fitted the spectrum A (Figure 3, σ_{noise} =1% of $I_{p\text{-}p}$) as a superposition of three overlapping spectra with the same Gaussian envelope function. (Lorentzian line width, position, and intensity of each spectrum were adjustable.) The estimates of parameter uncertainties increased only slightly (Table 1, fit 4).

Simulated spectra (Figure 3) have a small (5%) dispersion contribution that is hardly noticeable from visual analysis. To check how neglecting such a small dispersion contribution would affect the goodness of the fit, we fitted the spectrum A (Figure 3) with a microwave phase shift set to zero (Table 1, fit 5). The rest of the parameters were varied as in fit 4. We found that the data can be fitted with this "incorrect" model quite well: the residual norm was increased by only 13% compared to fit 4. However, the extracted Lorentzian line widths are out of the predicted 95.4% confidence intervals (Table 1, fit 5, 95.4% confidence intervals are shown in brackets). Several other Gaussian noise patterns were added to the simulated noise-free spectrum and the fitting was repeated with the microwave phase shift set to zero. In all cases some extracted parameters were out of the estimated discrepancy range, as we had expected for an incorrect fitting model.

3. APPLICATIONS

In this section we consider practical applications of the least-squares convolution-based fitting in spin labeling experiments. We start with EPR experiments carried out with lossy samples at low and high frequencies, which often produce dispersive line shapes. Then we will discuss how the convolution-based fitting improves the accuracy in CW EPR oximetry experiments, and extends this technique to spin labels in intermediate to slow motional regimes. We will also discuss the extension of the EPR oximetry method to multi-site experiments in which the spatial resolution is achieved by applications of magnetic field gradients. We also show that least-squares fitting is essential for correct data interpretation when nitroxide bioreduction is followed in course of time. Finally, we consider applications of convolution-based fitting for determining distances between dipolar-coupled nitroxides attached to proteins and discuss the limitations.

3.1 Phase Shift in Low Frequency EPR Experiments with Lossy Samples

In examples of fitting model spectra (section 2.5), we have demonstrated that a line shape model that properly accounts for a dispersion contribution is essential for accurate parameter determination. One type of EPR experiment in which this is often the case is low-frequency EPR carried out with small animals or animal organs (*e.g.*, Smirnov *et al.*, 1993). For example, when a low-frequency surface resonator probe is used for EPR detection, the microwave field B_1 not only decreases with penetration through a biological object, but also changes in phase (Nilges *et al.*,1989).These phase changes can significantly distort line shapes and lead to errors in, for example, Lorentzian width. The latter is very undesirable because this width serves as a sensitive indicator of biologically important parameters such as oxygen permeability and microviscosity.

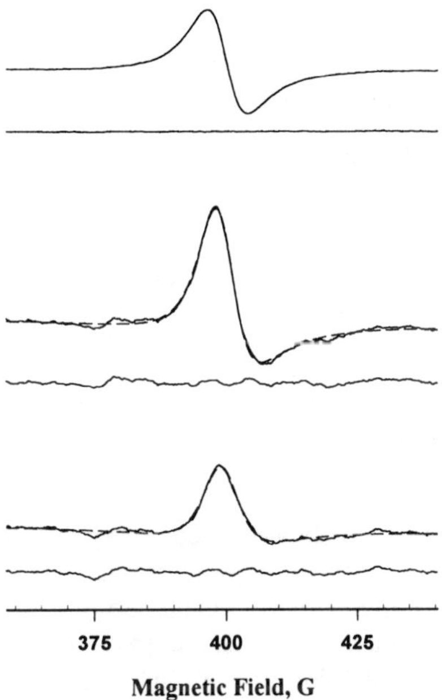

Figure 4. Experimental 1.13 GHz (L-band) EPR spectra from a small solid Tempol speck at different distances from the end of the surface probe resonator: A- 1.8 mm, C- 6.6 mm, and E- 9 mm. The fits, are shown in dashed lines and are superimposed with the experimental spectra. Microwave phase shift, line width, resonance field, double integrated intensity, and baseline coefficients were optimized. B, D, and F are corresponding residuals. C-F are shown in tenfold amplification. Reproduced with permission from Smirnov and Belford, 1995.

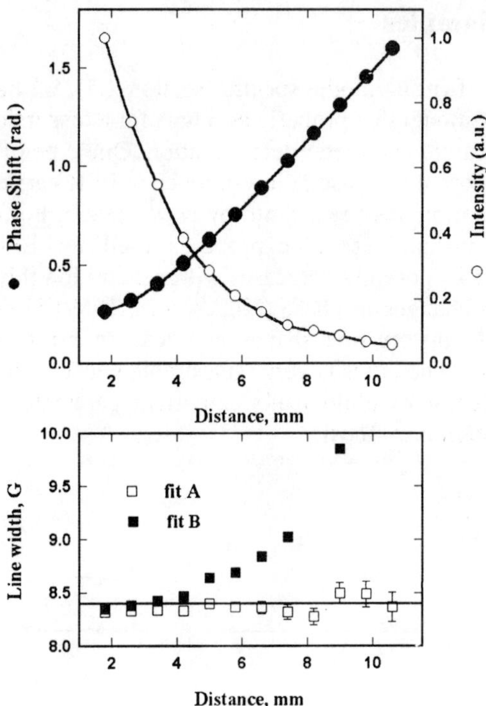

Figure 5. Top: Microwave phase shift (filled circles) and intensity (open circles) of the EPR signal from a Tempol speck as functions of distance from the end of the surface probe L-band resonator. Experimental details are given in the text. Estimated parameter errors are less than size of the plot points. Bottom: Extracted Lorentzian line widths (peak-to-peak) plotted for various probe positions: (open square) - fit with adjustable microwave phase shift, (filled square) - fit with microwave phase shift set to zero. Error bars correspond to 95.4% confidence intervals. Reproduced with permission from Smirnov and Belford, 1995.

We have tested our convolution-based fitting algorithm on EPR spectra collected at 1.13 GHz (L-band) from a "point" test sample surrounded by phosphate buffered saline (PBS) which served as a model of a lossy biological medium (for more details see Smirnov and Belford, 1995). A thin-wall quartz capillary containing a small speck of a solid nitroxide Tempol (4-hydroxy-2,2,6,6-tetramethylpiperidinyloxy) was moved incrementally from the resonator end. The EPR spectrum of solid Tempol is exchange narrowed and is quite well described by a Lorentzian function.Typical experimental spectra, the fits, and the residuals are shown in Figure 4. The spectra were fitted with a Lorentzian function and adjustable microwave phase. When the test sample was close to the resonator surface (Figure 4A; distance=1.8 mm), the dispersion contribution was small and resulted in only a moderate spectral asymmetry. Fitting results for this spectrum were virtually identical with experiment (residual is shown in Figure 4B), giving the phase shift of $\Delta\varphi$=0.2505±0.0006 rad. Upon increase of the distance from the resonator surface, the signal decreased in amplitude and both phase shift and dispersion contributions increased (Figure 4C and 4E, 6.6 and 9.0 mm from the probe surface respectively; spectra shown with tenfold amplification). Residuals (Figure 4 D, 4 F) show some systematic deviations related to so-called "microphonic" effects - interactions of eddy currents with the modulation field.

The results of all L-band measurements are summarized in Figure 5. With increasing sample-to-resonator distance, the phase shift approached a linear asymptote, as we had expected from the modeling of the penetration of the microwave field into the sample. A pure dispersion signal was observed with the Tempol speck positioned 11 mm away from the resonator. In spite of a significant signal decrease, the peak-to-peak line width could be extracted quite well when the microwave phase shift was adjusted during the fitting procedure (Figure 5, fit A). The presence of a microphonic effect can explain why the estimates of parameter uncertainties are somewhat less than the observed variations in the extracted line widths. When the dispersion component was neglected during least-squares simulations, the line width error was raised significantly with increasing the sample-to-resonator distance. In fit B (Figure 5), we fixed the phase shift at $\Delta\varphi$=0 and adjusted all other parameters. The results show a noticeable discrepancy in the extracted line width when dispersion contribution increases; the error estimates (95.4% confidence intervals) become meaningless.

3.2 High Field EPR: Rotational Diffusion in Fast Motional Limit

Another use of the convolution-based fitting algorithm is in accurate processing of high field/ high frequency EPR spectra from spin-labels in the fast motional limit. At EPR frequencies of 95 GHz and higher the anisotropy of the Zeeman term in the spin Hamiltonian dominates over the nitrogen hyperfine interaction. Modulation of the g-factor alone by rotational diffusion contributes to the homogeneous line width $\Delta B^L_{p\text{-}p}$ as $(\Delta g\nu)^2$, where ν is the resonant frequency. Therefore at high magnetic fields the proton hyperfine typically become unresolved, even for the nitroxides with well resolved features at conventional X-band (9.5 GHz). In addition, an admixture of a dispersion signal, which results in asymmetric line shapes, often complicates the analysis.

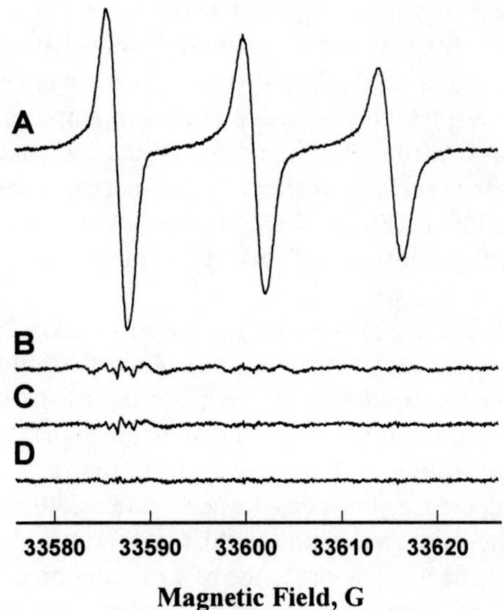

Figure 6. A: Experimental W-band (94.3 GHz) spectrum from deoxygenated 0.5 mM solution of 3-doxyl-17β-hydroxy-5α-androstane in *o*-xylene (T= 9 °C) fitted to different line shape models as described in the text. Corresponding residuals are shown as B, C, and D. Reproduced with permission from Smirnova *et al.*, 1995.

Figure 6 shows a typical high field (W-band, 94.3 GHz) EPR spectrum from 0.5 mM solution of 3-doxyl-17β-hydroxy-5α-androstane (androstane spin probe) in *o*-xylene (T=9.6 °C). Although the rotational motion of the probe still falls into the fast motional regime, the contribution to the homogeneous line width $\Delta B^{L}_{p\text{-}p}$ arising from the rotational motion is larger than that at X-band, and the proton superhyperfine structure is not resolved. (Compare to Figure 7 showing an X-band spectrum at the same temperature.) The spectra collected at 94.3 GHz (Figure 6) are asymmetric because of an admixture of the dispersion signal.

In the motional narrowing regime, the homogeneous line width ($1/T_2$ or $\Delta B^{L}_{p\text{-}p}$) is a function of the nitrogen nuclear spin number (m_I):

$$\frac{1}{T_2} = A + Bm_i + Cm_i^2 \tag{18}$$

where A, B, and C are the line width parameters, which can be expressed in terms of spectral density functions and also magnetic parameters of the radical and the rotational diffusion tensor (Freed, 1964, Budil *et al.*, 1993).

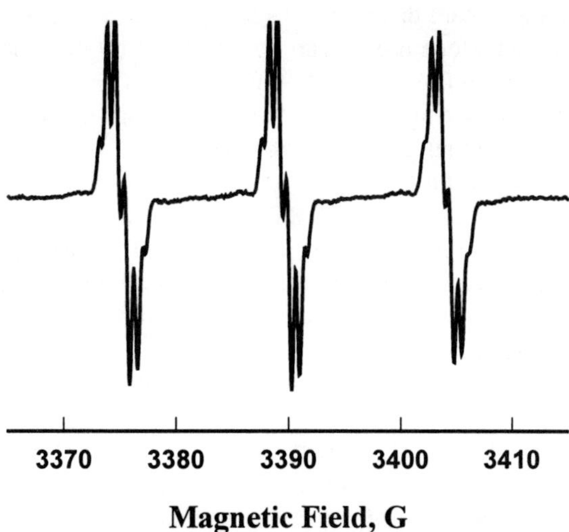

Figure 7. X-band (9.05 GHz) experimental spectrum from deoxygenated 0.5 mM solution of 3-doxyl-17β-hydroxy-5α-androstane in *o*-xylene (T= 9 °C) corresponding to fast motional regime. Reproduced with permission from Smirnova *et al.*, 1995.

While the proton hyperfine structure is well resolved in X-band EPR spectra (Figure 7), the contributions from $\Delta B^L_{p\text{-}p}$ are small when compared with the overall inhomogeneous width and, thus, are difficult to measure even with least-squares fitting. At high magnetic fields, the motional contribution to the overall line width is increased; however, inhomogeneous broadening still contributes somewhat significantly to the spectra, making direct estimations of $1/T_2$ inaccurate. Correction for inhomogeneous broadening, especially for poorly resolved proton hyperfine structure (or superhyperfine - shf), is a challenging task in EPR (Bales, 1989). One approach is to measure proton hyperfine-coupling constants by NMR or ENDOR (Ottaviani, 1987; Bales, 1989). When the coupling constants are much smaller than the homogeneous line width, the correction for inhomogeneous broadening can be simplified significantly by approximating the proton hyperfine pattern by a Gaussian envelope (Bales, 1989). We explored this and other models (all with the help of the convolution-based algorithm!) to obtain the best possible fit of our W-band spectra.

First, we have modeled the inhomogeneous broadening by a Gaussian envelope (model (1), corresponding best fit shown in Figure 6). Variable Lorentzian line width $\Delta B^L_{p\text{-}p}(m_I)$ was assigned to each of the nitrogen hyperfine components m_I and the phase of the signal was adjusted to account for a visible asymmetry due to dispersion. Although the ratio of the merit function χ^2 at the minimum to the square norm of the experimental signal was small ($X^2=\chi^2/(\sum Y^2(B_j))=1.57\cdot10^{-3}$), the residual shown in Figure 6 B revealed some significant discrepancies between the experiment and the fit.

The Gaussian envelope model can be improved by including ^{13}C satellite lines arising from two sets of non-equivalent ^{13}C nuclei (corresponding hyperfine constants being estimated from fitting of low-noise X-band spectra) as an additional non-adjustable envelope function $p_{13C}(B)$ in our fitting function (Equation (11)). The norm of the residual for this model (2) was improved significantly (X^2=0.86·10-3; or by 45% compared to the model (1)), although the deviations were still noticeable, especially in the vicinity of the sharpest m_I=1 transition (residual is shown in Figure 6C). This served as an indication that the Gaussian approximation for the proton shf structure was somewhat inaccurate.

In model (3), the fitting function was constructed by use of the temperature-dependent proton hyperfine constants, $a^j_{iso}(T)$, j=1,2, measured from least-squares fitting of X-band data. The non-adjustable ^{13}C satellite lines were included as in the model (2). During the fitting procedure, the following parameters were adjusted: Lorentzian line width, $\Delta B^L_{p\text{-}p}(m_I)$ for each of the nitrogen component m_I; Gaussian envelope function, $\Delta B^G_{p\text{-}p}$ (to account for an additional broadening, *e.g.*, inhomogeneities of magnetic field); microwave phase shift, $\Delta\varphi$; signal intensity, A; and the coefficients for the linear baseline. The best fit is superimposed on the experimental spectra in Figure 6A. The residual norm (Figure 6D) was further significantly improved (X^2=0.41·10-3). The ratio of the merit function (16) to the estimated standard deviations of the amplitude noise $\sigma^2(B)$ (the same $\sigma^2(B_j)$'s being assumed for each of the spectral data points) at its minimum $\chi^2_{min}/\sigma^2\approx2370$, was close to the number of degrees of freedom (η=2038; 2048 data points collected and 10 parameters in all were adjusted), assuring a "moderately" good fit (Press *et al.*, 1986). Clearly, the model (3) seems to be the most suitable for fitting the W-band spectrum shown in Figure 6

With decreasing temperature, the contribution from the homogeneous line width to the spectrum will further increase and this will gradually eliminate all remaining features of the proton superhyperfine structure. Eventually, the model (2) is expected to fit the experimental spectrum as well as the model (3). However, at temperatures higher than 10°C, the information on proton superhyperfine coupling can be crucial for extracting the correct Lorentzian width parameters. (For more details see Smirnova *et al.* (1995).)

Although model (3) was found to be the most accurate, it does require time consuming measurements of the proton and ^{13}C hyperfine constants and the assignment of all these constants to the nuclei. The measurements of the homogeneous line width, $\Delta B^L_{p\text{-}p}$, can be significantly simplified if the whole inhomogeneous envelope function $P(B)=p_1(B)\otimes p_2(B)\otimes\ldots\otimes p_k(B)$ is known. Then, Equation (11) is simplified:

$$I(B) = f(B) \otimes P(B) + \sum_{i=0}^{N} l_i B^i \quad (19)$$

The envelope function *P(B)* for the androstane spin probe can be well approximated from the X-band measurements. Indeed, at X-band and $T>$ 10 °C, the main contribution to the homogeneous line width of this nitroxide probe is frequency-independent. The middle $m_I=0$ component of the X-band spectrum is the narrowest and can be taken as an approximation of the envelope function. The second-order effects for the proton hyperfine are estimated to be less than 1 mG and can be neglected. Then each of the nitrogen hyperfine components of the W-band spectrum at the same temperature can be approximated by Equation (19), where the frequency-dependent part of the homogeneous broadening is included in $f(B, B_0, \Delta B^L_{p-p}(m_I), \Delta\varphi)$. Frequency-independent homogeneous broadening is already included in the function $F_0(B)$.(Convolution of two Lorentzian functions gives a Lorentzian function with a width equal to the sum of the initial widths.) When this model was applied to the fitting of the W-band spectra, it was found that to account for the effects of small additional inhomogeneites of the magnetic field and/or g-strain, it was necessary to include an additional Gaussian envelope $p_G(B, \Delta B^G_{p-p})$ in the fitting functions as follows (Smirnova *et al.*, 1995):

$$I(B) = f(B) \otimes P(B) \otimes p_G(B) + \sum_{i=1}^{N} l_i B^i \quad (20)$$

The introduction of a Gaussian envelope also can be helpful for modeling the W-band spectra in a broader temperature range. For a nitroxide such as the androstane spin probe, both proton hyperfine constants decrease slightly with temperature (less than 30 mG per 100 °C, Smirnova *et al.*, 1995), and an additional Gaussian width may "compensate" for such temperature variations. Fitting function (20) also accounts for any ^{13}C satellite lines (since they are present in the X-band spectrum *P(B)*) without actual measurements of the corresponding hyperfine constants and their assignments to the nuclei.

When Equation (20) was used to fit the W-band spectrum at 9.6 °C (Figure 6; the $m_I=0$ component of the X-band measured at T=61 °C was taken as *P(B)*), the residual norm (X^2=0.40·10-3) was about the same as that for the best fit under model (3) (X^2=0.41·10-3). That no significant deviations between the fit and the residual were found demonstrates an acceptable fit.

Since *P(B)* in the model (4) is taken from the X-band experiment, the magnetic field sweeps of X- and W-band EPR spectrometers must be accurately calibrated. Well-resolved nitrogen hyperfine splitting, when corrected for higher order effects at X-band, provides ideal means for such a calibration. Additionally, the samples for both frequencies should be equally deoxygenated, and the spin probe concentration should be kept the same since the frequency-independent line width and the proton hyperfine constants vary with the concentration (mainly because of the spin-exchange between identical species; this interaction between identical radicals is frequency-independent). The same requirements should be fulfilled for the model (3) as well.

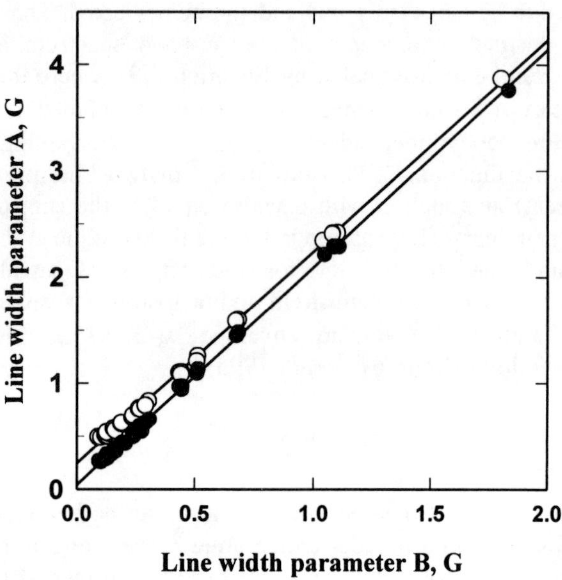

*Figure 8.*A comparison between A and B line width parameters (measured as Lorentzian width, peak-to-peak) for 0.5 mM solution of 3-doxyl-17β-hydroxy-5α-androstane in *o*-xylene.Parameters extracted under the most accurate line shape models (3) (open circles) and (4) (filled circles) from the experimental 94.3 GHz EPR spectra over the temperature range -49 °C < *T* < 63 °C .The fitting models are described in the text. Estimated experimental errors (68% confidence interval) are less than size of the plot symbols. Linear regression of two data sets over all temperatures gave very similar slopes (A/B=2.00 for model (3), A/B=2.05 for model (4)).The intercept for model (4) was close to the zero (A'=0.036 G), as it should be since the frequency independent line width was already subtracted from A during the fitting. Reproduced with permission from Smirnova *et al.*, 1995.

Smirnova *et al.* (1995) further tested the model (4) (Equation 20) by fitting the same set of W-band spectra over the whole temperature range. A comparison between A and B parameters extracted with models (4) and (3) is shown in Figure 8. The data obtained with model (4) can be well approximated by a straight line with a slope of 2.05, which is close to that found with model (3) (2.00). The intercept of 0.036 G is close to zero, as it should be since the frequency independent homogeneous line width *A'* was already accounted for in the function $F_0(B)$. Figure 8 also shows that the *A/B* data obtained with the model (4) are linear over whole temperature range; this is in good agreement with the theory.

The model (4) can be especially useful when detailed information on the proton hyperfine coupling is difficult to obtain because of poor resolution (even at X-band) and/or when assignment of the hyperfine lines is not trivial (Smirnova *et al.*, 1995). At the same time, the use of Lorentzian line shape functions modified for admixture of dispersion (Equation 13) allows one to account for a phase shift observed at W-band. Thus, the convolution-based algorithm we described here simplifies the analysis of HF EPR data from spin-labels in the fast motional limit. Other sample applications of this algorithm in HF EPR studies of rotational dynamics of nitroxides include experiments with model phospholipid bilayers (Smirnov *et al.*, 1995, Smirnova *et al.*, 1997), model lubricants (Britton *et al.*, 1997), and identification of spin adducts (Smirnova *et al.*, 1997). The latter paper also describes accurate least-squares simulations of EPR spectra from various phenyl *ter*-butylnitrone (PBN) spin adducts. These simulations allowed the authors to differentiate various spin adducts by rotational correlation time and rotational anisotropy. This characterization might be particularly useful when the hyperfine parameters of spin adducts are essentially the same but sizes and shapes of adduct molecules are rather different. The same paper also provides useful estimates for the EPR field/frequency limits at which the resolution still can be gained for typical nitroxide-containing probes diffusing freely in organic solvents and water.

3.3 EPR Oximetry

Over the last decade, EPR oximetry has become a valuable tool in biomedical studies (*e.g.*, Liu *et al.*, 1993; Swartz *et al.*, 1994; Kuppusamy *et al.*, 1994; Smirnov *et al.*, 1994). While the nitroxide radicals remain the most widely used molecular probes in EPR oximetry, several new solid-state oxygen probes have been introduced including synthetic carbohydrate chars (Boyer & Clarkson, 1994; Clarkson *et al.*, 1999), fusinite (a fraction of naturally occurring coal, *e.g.*, Vahidi *et al.*, 1994), Indian ink (*e.g.*, Liu *et al.*, 1993), and lithium phthalocyanine (*e.g.*, Swartz *et al.*, 1994). For the nitroxide radicals and for most solid-state probes, the main mechanism of oxygen sensitivity is shortening of the relaxation time by the Heisenberg exchange during bimolecular collisions of probe molecules with oxygen.

For nitroxides, the oxygen effect on CW EPR spectra is rather small - about 0.5 G per 1 mM of oxygen concentration.The oxygen-induced broadening becomes difficult to measure at low, biologically important oxygen concentrations (<15 μM, corresponding to changes in line width <7.5 mG). This problem is somewhat eliminated with particulate probes; however, those cannot replace nitroxides for measurements on a submicroscopic or molecular level - in membranes or at a particular protein residue (accessibity of spin-labeled residues to molecular oxygen in site-directed spin-labeling experiments). In some EPR oximetry experiments, particularly for those carried out *in vivo* at low (0.2-2.0 GHz) EPR frequencies, a poor signal-to-noise ratio often imposes another obstacle for accurate oxygen readings.

In this section we describe how the convolution-based fitting algorithm improves the accuracy of EPR oximetry with nitroxides and solid state probes. It also enables extension of line-width-based EPR oximetry experiments to nitroxide probes exhibiting intermediate and restricted anisotropic motion, such as in biological membranes or attached to proteins. We will also discuss the use of convolution-based fitting method in multi-site EPR oximetry experiments *in vivo*, in which the spatial resolution is achieved by application of magnetic field gradients. Finally, we discuss the issue of accuracy in EPR oximetry experiments and the factors affecting it.

3.3.1 EPR Probes with Complex Hyperfine Structure

One of the first most commonly used probes in EPR nitroxide oximetry is CTPO (3-carbamoyl-2,2,5,5-tetramethyl-3-pyrrodine-1-yloxyl). At room temperature, the EPR spectrum from CTPO in aqueous solution has three nitrogen hyperfine components. Under conditions of a low (<0.5 mM) CTPO concentration, absence of oxygen, and a fast tumbling rate (*e.g.*, aqueous solution at room temperature), additional splittings of the components are observed because of the hyperfine interactions with 12 equivalent and 1 different protons (Hyde and Subczynski, 1984). Figure 9A shows a typical spectrum of the central (m_I =0) CTPO EPR line after purging the 0.2 mM aqueous solution with nitrogen for a half-hour. Upon re-equilibrating this solution with increasing amounts of oxygen, the lines broaden; the resolution between the proton superhyperfine (shf) components deteriorates, and finally disappears. Figure 9C shows an EPR spectrum from the same solution re-equilibrated in air. Commonly, apparent peak-to-peak intensities of individual shf components relative to the absolute peak-to-peak amplitude of the spectrum are calibrated for various oxygen concentrations and this calibration is used for oxygen measurements (*e.g.*, Lai *et al.*, 1982).

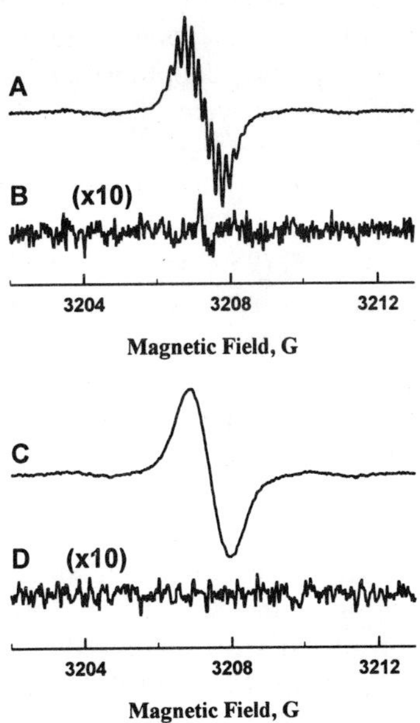

Figure 9. Central nitrogen hyperfine component (m_I=0 transition) of X-band experimental spectrum from 0.2 mM CTPO aqueous solution equilibrated with nitrogen (A) is superimposed with results of the fit assuming two sets of non-equivalent protons. Residual (B) is shown in tenfold amplification. Extracted parameters and details of the fitting functions are given in the text. Central hyperfine component (m_I=0 transition) of an experimental spectrum from air-saturated 0.2 mM CTPO aqueous solution is superimposed with result of the fit using Equation (19) (C).The fitting function was constructed with the use of the experimental spectrum (A). Lorentzian broadening, intensity, resonance field, and coefficients of a linear baseline were used as adjustable parameters. Residual (D) is shown in tenfold amplification. Reproduced with permission from Smirnov and Belford, 1995.

As an example, we have tested the convolution-based fitting algorithm by least-squares simulations of the central line of CTPO aqueous solution equilibrated with nitrogen. Fitting results and experimental spectra are virtually identical; they are superimposed in Figure 9A. Line position, intensity, line widths, and two proton hyperfine constants were used as adjustable parameters. Only the strongest ^{13}C lines were accounted for in the simulations and the corresponding hyperfine splitting was adjusted. Accounting for the ^{13}C lines in the fitting model, did not change the values of the extracted proton hyperfine coupling constants. The residual (Figure 9B; tenfold amplification) shows practically no deviations between the fit and the experimental data. A small peak on the residual at a position close to the center of the CTPO spectrum is likely due to some instability of magnetic field and/or microwave frequency (these deviations varied from experiment to experiment). A better digital resolution of the central part of the spectrum was achieved in a 4 G magnetic field scan per 1000 data points. For room temperature, the hyperfine coupling constants to twelve equivalent and one strongly coupled protons were determined as 0.193±0.002 G and 0.517±0.002 G respectively. These values agree with previously reported data (Hyde and Subczynski, 1984). The error was estimated to be less than half of the digitizing step (4 mG). The estimates of errors provided by the algorithm were better than 2 mG (95% confidence interval).

Once the hyperfine and Gaussian width parameters are measured for CTPO spectra at low oxygen concentration when the superhyperfine (shf) is well resolved, those parameters can be used to simulate CTPO spectra at any other oxygen concentration, even when the proton shf is not resolved at all (air-equilibrated solution, Figure 9C). There is no need to adjust proton hyperfine couplings because oxygen has no effect on the hyperfine splitting (Jones, 1963). However, this kind of fitting calls for an accurate model for the shf structure and involves some extra calculations while there is only one parameter that is changing - the Lorentzian line width. This is fully analogous to the problem of extra broadening that we discussed in section 3.2. Thus, one can use Equation (19) to fit the CTPO (and other nitroxide) spectra at any oxygen concentration. In this case, $P(B)$ is the spectrum in absence of oxygen and $f(B)$ is the Lorentzian line with width proportional to the rate of molecular collisions of oxygen with the probe. Again, the use of Equation (19) simplifies the data fitting in EPR oximetry by using an one line-width-parameter model. Equation (19) automatically accounts for additional lines caused by the presence of ^{13}C and ^{15}N in natural abundance without measuring and assigning the corresponding hyperfine constants.

Figure 9C shows a typical result of the fitting of an experimental EPR spectrum (1000 data points) from a CTPO air-equilibrated aqueous solution with the help of the convolution equation (19) and adjusting the Lorentzian broadening function. The residual shows no systematic deviation of the fit from the experimental data. The standard deviation of the residual corresponds to the level of experimental noise. We have tested this model for various oxygen concentrations and two concentrations of CTPO - 0.2 and 0.4 mM (Smirnov and Belford, 1995). All data fit the expected straight line (Figure 10) with a close to zero intersect (0.5 mG). Line width error bars (68% confidence intervals) are less than the size of the plot points shown on the graph. Accuracy of prepared air/nitrogen mixtures was better than 2%. Oxygen concentrations were calculated from the solubility data summarized by Dean (1985) for $T = 22\ ^{\circ}C$. The slope of the line (Figure 10) was 540±10 $mG{\cdot}mM^{-1}$, which is consistent with the low-frequency data reported by Halpern and Peric (1993).

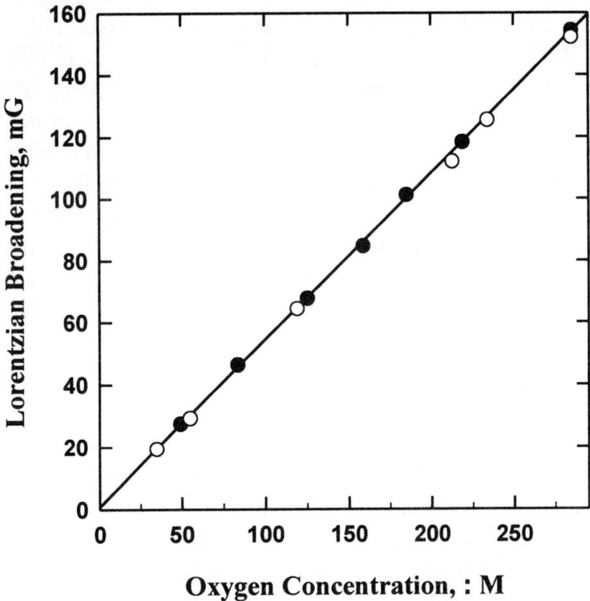

Figure 10. Lorentzian broadening (measured as peak-to-peak line width) of the central nitrogen hyperfine component (m_I=0 transition) of 0.2 mM (filled circles) and 0.4 mM (open circles) CTPO aqueous solution EPR spectra vs. molecular oxygen concentration. The fitting function is given by Equation (19).$P_0(B)$ functions (*i.e.*, EPR spectra for nitrogen-equilibrated samples) for both concentrations were experimentally measured. Estimated errors of line width measurements (68% confidence intervals) are less than size of the plot points. Data were fitted with the hypothesis of identical straight line for both concentrations. The slope of the line is 540±10 mG·mM-1 with 0.5 mG intersect. Reproduced with permission from Smirnov and Belford, 1995.

3.3.2 Spin-labeling CW EPR Molecular Oxygen Accessibility Measurements in viscous and anisotropic environments: Oxygen Permeability and the Bilayer Structure

It is well known that continuous-wave EPR spectra of nitroxide probes (labels) introduced into phospholipid bilayers or covalently attached to proteins are sensitive to molecular oxygen. However, accurate determination of oxygen broadening from these experiments is complicated by the complex shapes of EPR spectra, which are strongly influenced by anisotropic restricted motion of the probe molecules. Only a few attempts, such as measurements of the peak-to-peak line width of the central component of the nitroxide spectrum over a few characteristic data points (Windrem, and Plachy, 1980), observations of oxygen effects on second harmonic out-of-phase absorption spectra (Popp and Hyde, 1981), and theoretical predictions of oxygen effects on EPR spectra from slow-tumbling nitroxides (Hyde and Subczynski, 1989) have been documented.

When a spin probe is incorporated into a membrane or attached to a protein, the oxygen - spin probe interaction is still dominated by the Heisenberg spin exchange; the dipole-dipole interaction in most cases can be neglected (*ibid.*). When the diffusion constant of oxygen, $D(O_2)$, is much larger than that of the spin probe, $D(SP)$, (as is likely the case for spin-labeled phospholipids or doxyl-stearic acid spin probes in the phospholipid bilayer and is certainly true for many spin-labeled proteins), then the frequency ω of bimolecular encounters between oxygen and the spin probe is given by (*ibid.*):

$$\omega = 4\pi RcD(O_2)C(O_2) \tag{21}$$

where R is the collision distance, c is the collision probability, and $C(O_2)$ is the oxygen concentration. If the medium (solution or phospholipid bilayer) is in an equilibrium with the gas phase, then the oxygen concentration in the medium can be expressed as a product of the oxygen partial pressure $p(O_2)$ and the solubility coefficient $S(O_2)$. Then Equation (21) can written as:

$$\omega = 4\pi RcP(O_2)p(O_2) \tag{22}$$

where $P(O_2)=D(O_2)S(O_2)$ is the permeability coefficient of the medium.

Effects of bimolecular collisions of a spin probe with molecular oxygen or other spin-relaxing agents can be observed in pulsed EPR experiments (usually saturation recovery method) by comparison of the relaxation rates, $1/T_1$, measured with and without spin-relaxing agents (*e.g.*, Kusumi *et al.*, 1982; Subczynski *et al.*, 1989). By labeling phospholipid bilayers with nitroxides at a known location, x, Hyde and co-workers studied the oxygen effect on $T_1(x)$ by pulsed saturation recovery EPR (*ibid.*). They defined the oxygen transport parameter $W(x)$ as:

$$W(x) = \frac{1}{T_1(p(O_2) = p_{air}(O_2), x)} - \frac{1}{T_1(p(O_2) = 0, x)} = \\ = 8\pi c r_0 D(O_2, x)C(O_2, x) \tag{23}$$

where the interaction distance between oxygen and spin label is r_0 ($\approx$4.5 Δ), and $D(O_2,x)$ and $C(O_2,x)$ are the local oxygen diffusion coefficient and local concentration, respectively. They measured T_1 for both air-equilibrated, $p(O_2)=p_{air}(O_2)$, and nitrogen-equilibrated, $p(O_2)=0$, samples of phospholipid bilayers with spin-labels at a given location x and the same temperature. From repeated measurements with nitroxide labels at various membrane locations, the variation of $W(x)$ across the membrane was determined.

The effect of bimolecular collisions of molecular oxygen with spin probes can be observed in continuous-wave EPR as additional broadening of the homogeneous contribution to the line width $\delta(\Delta B^L_{p\text{-}p}(x))$ (e.g., Subczynski and Hyde, 1984):

$$\omega(x) = \frac{\sqrt{3}}{2}\gamma\delta(\Delta B^L_{p-p}(x)) = \\ = \frac{\sqrt{3}}{2}\gamma\left[\Delta B^L_{p-p}(p(O_2),x) - \Delta B^L_{p-p}(p(O_2)=0,x)\right] \tag{24}$$

where γ is the magnetogyric ratio of the electron. From Equations (22) and (24), the local oxygen permeability coefficient $P(O_2,x)$ (or the product of the oxygen diffusion and solubility coefficients) is proportional to $\delta(\Delta B^L_{p\text{-}p}(x))$:

$$P(O_2,x) = \frac{\sqrt{3}\gamma\delta(\Delta B^L_{p-p}(x))}{8\pi Rcp(O_2)} \tag{25}$$

The product $(R{\cdot}c)$ for doxyl-stearic acid probes is not accurately known, but it is usually assumed that its value (≈2 Δ) does not vary with position of nitroxide moiety (*e.g.*, Windrem and Plachy, 1980) across the bilayer.

Because oxygen effects on spin label EPR spectra are dominated by the Heisenberg spin exchange, the broadening should be uniform across the spectrum and, thus, can be described by Equation (19). Thus, the oxygen permeability profile can be measured from CW EPR data by simulating the oxygen broadening effects using our least-squares convolution approach. We have tested this algorithm by measuring oxygen permeability profiles of model membranes composed of 1,2-dimyristoyl-sn-glycero-3-phosphocholine (DMPC) above and below the main phase transition (Smirnov *et al.*, 1996).

Typical experimental X-band EPR spectra of a DMPC aqueous dispersion at 36.5 °C (above the main phase transition) labeled with 5-doxyl stearic acid (5-DS) and equilibrated with nitrogen and subsequently with oxygen are shown in Figure 11. Although the oxygen effect on these double-integral-normalized spectra is noticeable, the broadening parameter $\delta(\Delta B^L_{p\text{-}p})$ is difficult to measure directly because of the evident slow motional and ordering effects. Complete simulations of this spectrum require a detailed and accurate motional model for this membrane probe. However, to measure broadening effects arising from spin-exchange with oxygen requires only a much simpler equation (19). The use of this equation is illustrated in Figure 12.

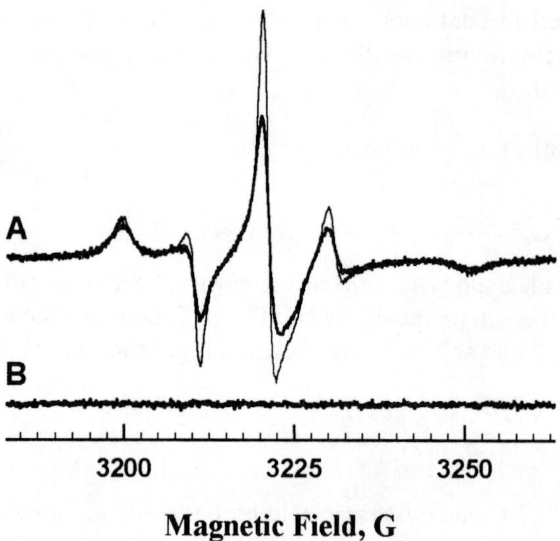

Figure 11.(A) Experimental X-band EPR spectra (first-derivative presentation) of a DMPC aqueous dispersion at 36.5 °C (above the main phase transition, membrane phase is a fluid bilayer structure (L")) labeled with 5-DS and equilibrated with nitrogen (thin line) and consequently with oxygen (thick line).Experimental spectra were normalized by the double integral values. The 5-DS EPR spectrum broadened by oxygen was least-squares-fitted according to Equation (19) under an assumption of a homogeneous (Lorentzian) broadening of the spectrum taken in absence of oxygen. The residual of the fit (*i.e.*, experimental spectrum minus simulated) is shown at the bottom (B).The broadening caused by molecular oxygen extracted from the fit was $\delta(\Delta B^{L}_{p\text{-}p})$ =620±5 mG. Reproduced with permission from Smirnov *et al.*, 1996.

When this model (Equation (19)) was applied to fit 5-DS EPR spectra, we found that the experimental spectrum and the least-squares fit were almost identical. Since the fit and the experimental spectrum are indistinguishable on the plot, we show the residual of the fit (a difference between experimental and simulated spectra; Figure 11 C). Homogeneous broadening caused by molecular oxygen was $\delta(\Delta B^{L}_{p\text{-}p})$=620±5 mG. The estimated error, ±5 mG, corresponds to the 68% confidence interval. The excellent agreement between the experiment and the fitting model given by Equation (19) confirms applicability of this convolution-based model.

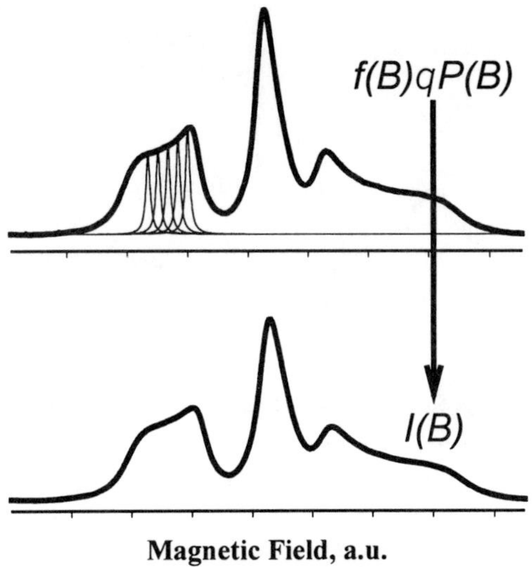

Figure 12. Illustration of the use of Equation (19) to extract Lorentzian broadening from EPR spectra of membrane nitroxide probes.(A) EPR spectrum (thick line, first-integral representation) of nitrogen equilibrated sample, which serves as an envelope to assemble Lorentzian broadening functions (thin lines) of equal widths. (B) The results of digital convolution which is compared with the experimental spectrum.

Below the main phase transition, oxygen permeability decreases and the rotational correlation time of the probes increases; therefore, the oxygen effect is less noticeable in the spectra. Figure 13 shows the oxygen effect on 5-DS/DMPC EPR spectra taken at 18.0 °C. It is apparent that oxygen broadening is even more difficult to measure without computer simulation because the line shape changes caused by oxygen are very moderate and may be obscured the noise present. Another problem in extracting oxygen-induced broadening from such data is that the digitizing steps become comparable with the the magnitude of the broadening. For example, fitting with Equation (19) gave $\delta(\Delta B^{L}_{p\text{-}p})$=126.8±5.6 mG while the digitizing step was 40 mG; therefore, the Lorentzian broadening function might be undersampled.

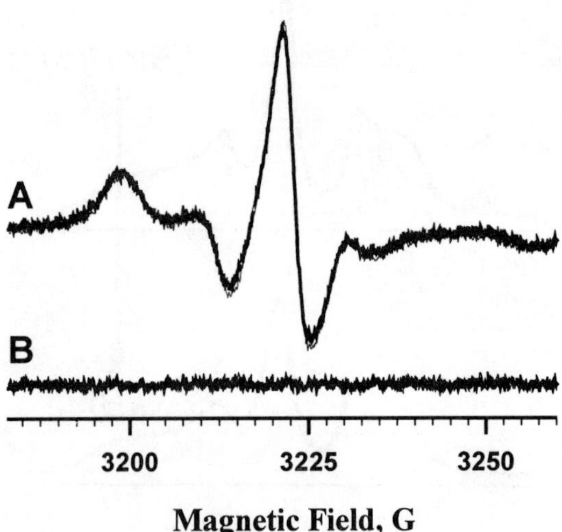

Figure 13 Experimental X-band EPR spectra of a DMPC aqueous dispersion at 18.0 °C (below the main phase transition, membrane phase is a ripple structure (P'_{β})) labeled with 5-DS and equilibrated with nitrogen (thin line) and consequently with oxygen (thick line). Residual of the least-squares simulations according to Equation (19) is (B).Extracted homogeneous broadening was $\delta(\Delta B^{L}_{p\text{-}p})$=126.8±5.6 mG or 3.17±0.14 in data intervals, which were 40 mG each. Reproduced with permission from Smirnov *et al.*, 1996.

Problems with digital resolution can be solved by collecting data over a narrower spectral window such as the most "sensitive" region of the spectrum - its central component (Figure 14). Equation (19) still can be used for the fitting, but the fitting interval should be decreased appropriately to avoid "wrap-around" effects characteristic of the digital convolution algorithm. As shown in Figure 14, the central component can be fitted with a residual that reveals no systematic deviations between the experiment and the fit. The extracted Lorentzian broadening was $\delta(\Delta B^{L}_{p\text{-}p})$=11.52±0.31 (data intervals; 10 mG each), or 115.2±3.1 mG, which is close to the result with a fourfold broader scan (Figure 13). The predicted accuracy for oxygen broadening extracted from data shown in Figure 14 is ≈±2.7%, which is sufficient for many practical applications.

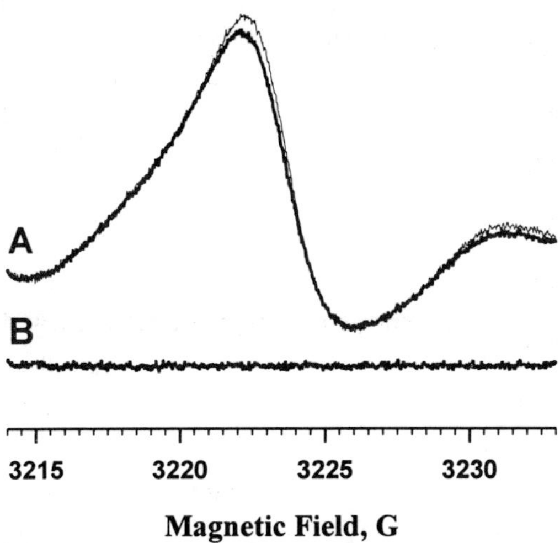

Figure 14. (A) Central component of the experimental X-band EPR spectra of a DMPC aqueous dispersion at 18.0 °C labeled with 5-DS and equilibrated with nitrogen (thin line) and subsequently with oxygen (thick line).Homogeneous broadening caused by molecular oxygen was extracted by fitting only a portion of the spectrum to Equation (19); (B) is the residual of the fit to oxygen-equilibrated spectrum. Compared to the larger spectrum window (see Figure 13), the digital resolution is improved: $\delta(\Delta B^{L}_{p\text{-}p})$=11.52±0.31 (in data intervals; 10 mG each), or 15.2±3.1 mG. Reproduced with permission from Smirnov *et al.*, 1996.

In order for Equation (19) to be valid, the spectra both of a sample equilibrated with nitrogen and of that sample subsequently oxygen-equilibrated must be taken at essentially the same temperature. Small temperature variations are inevitable in EPR experiments even after gas flow stabilization, and this can affect the accuracy of measured $\delta(\Delta B^{L}_{p\text{-}p})$. To check this, we have monitored the kinetics of line width changes after switching the gas bathing the sample from nitrogen to oxygen (Smirnov *et al.*, 1996). Figure 15A shows results for two temperatures (36.5 °C and 18.0 °C) for DMPC membranes labeled with 5-DS. At those two temperatures, spectra of corresponding nitrogen-equilibrated samples were taken as references for fitting the data. Before experiments, the samples were equilibrated with nitrogen for at least 30 min. In both cases, equilibrium was achieved within 6 minutes. Predicted errors for 36.5 °C are within the size of the symbols (hollow dots) of the plot. Changes in the homogeneous spectral width can be fitted by first-order kinetics (Figure 15A, solid lines), which corresponds to oxygen diffusion through the gas-permeable TFE (poly(tetrafluoroethylene)) capillary containing aqueous membrane sample.

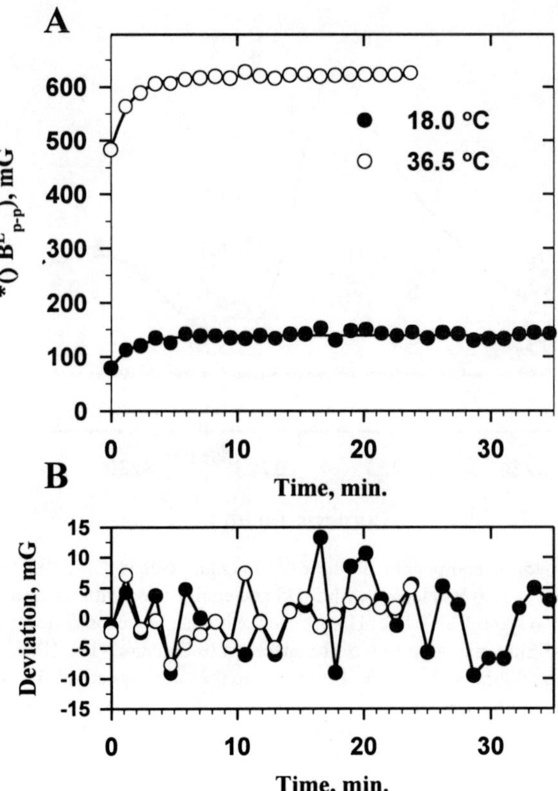

Figure 15.(A) Kinetics of re-equilibration of DMPC sample with molecular oxygen as monitored by Lorentzian broadening of the central component of a 5-DS spectrum at 36.5 °C (open circles) and 18.0 °C (filled circles).Estimated errors are within the size of the symbols. Before the experiments, the samples were equilibrated with nitrogen for at least 30 min. .Kinetics of $\delta(\Delta B^L_{p-p})$ were fitted by first-order kinetics (solid lines), which approximates the process of oxygen diffusion into the sample.(B) Deviations of individual data points from the least-squares kinetic curves. Reproduced with permission from Smirnov *et al.*, 1996.

Figure 15B shows the deviations of individual data points from the least-squares kinetic curves. It is evident that about 68% of all data points taken at 36.5 °C lie within a ±5 mG band. Dispersion of data points taken at 18.0 °C is only slightly larger. From analysis of all 5-DS spectra collected at two temperatures, the following values of oxygen broadening were obtained: for 36.5 °C – $\delta(\Delta B^L_{p-p})$=620±1 mG, for 18.0 °C – $\delta(\Delta B^L_{p-p})$=139±1 mG. The errors reported here were calculated as predicted standard deviation of the equilibrium broadening measured from fitting the entire broadening curves (such as shown in Figure 15A) to a first-order kinetic of final stages of oxygen diffusion into the capillary.

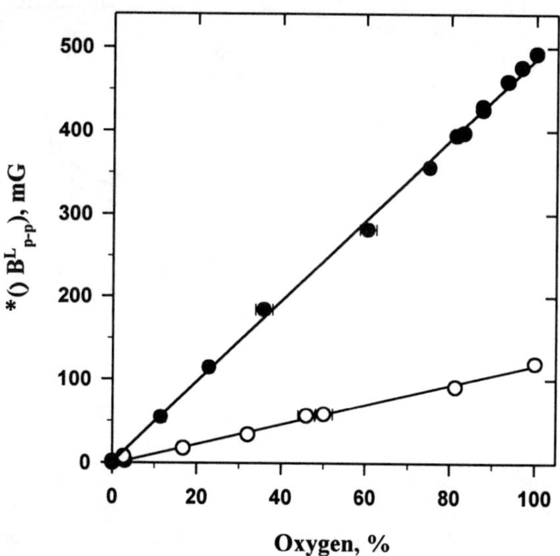

Figure 16. Lorentzian broadening $\delta(\Delta B^{L}_{p\text{-}p})$ *vs.* oxygen concentration for 5-DS/DMPC at two temperatures: filled circles - 27.2 °C, open circles -18.0 °C. Reproduced with permission from Smirnov *et al.*, 1996.

Let us now estimate how small temperature variations would effect the measurements. For example, for DMPC labeled with 5-DS, the temperature dependence of oxygen broadening as estimated from data given reported by Smirnov *et al.* (1996) is ≈10 mG/°C at 36.5 °C. Therefore, ±0.1 °C corresponds to only additional ±1 mG (±0.16%) error in the line width values, which is smaller than the other errors. At 18.0 °C, temperature dependence of oxygen broadening as measured by 5-DS is about 20 mG/°C and the same temperature variations would result in ±2 mG error, which is comparable with the variance for individual data points (±3 mG). This might explain why the dispersion for the 18.0 °C data set is larger than predicted.

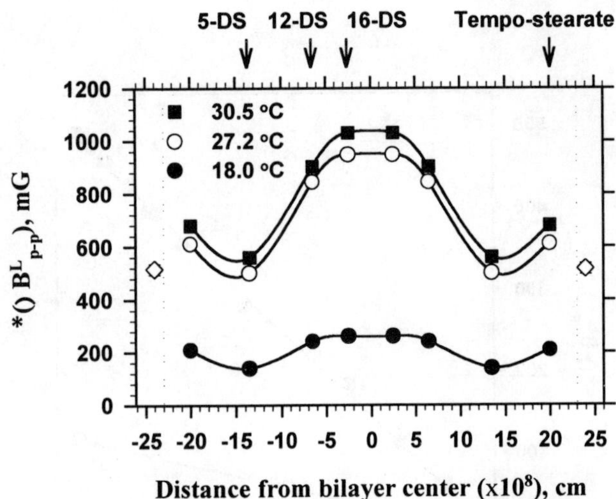

Figure 17. Profiles of homogeneous broadening parameter $\delta(\Delta B^L_{p\text{-}p})$ across DMPC membranes for several temperatures. Arrows show approximate locations of nitroxide probes in the bilayer.Parameter $\delta(\Delta B^L_{p\text{-}p})$ for aqueous phase (open diamond), was measured with the nitroxide tempo-choline. Dotted lines indicate approximate location of the aqueous phase. Reproduced with permission from Smirnov *et al.*, 1996.

Figure 16 shows the dependence of $\delta(\Delta B^L_{p\text{-}p})$ for 5-DS/DMPC upon the concentration of molecular oxygen in the gas mixture bathing the sample at 18.0 and 27.2 °C (*i.e.*, below and above the phase transition temperatures). For each flowmeter/mixer setting, the flow rate for each case was kept constant (±0.5%); therefore, the accuracy of oxygen concentration for all points (except a few values at the middle of the scale when higher-than-standard gas pressures were employed) was estimated to be better than ±1.0%. For both temperatures (Figure 16), we found the dependence to be linear; this agrees with previous assumptions that the Heisenberg exchange is the dominant mechanism for intermolecular interactions between molecular oxygen and doxyl-stearic acid spin labels (Hyde and Subczynski, 1989).

Smirnov *et al.* (1996) have measured parameter $\delta(\Delta B^L_{p\text{-}p})$ for various membrane probes and temperatures below and above the main phase transition of DMPC. Profiles of $\delta(\Delta B^L_{p\text{-}p})$ across the DMPC membrane for several temperatures are shown in Figure 17. Parameter $\delta(\Delta B^L_{p\text{-}p})$ for the was measured with a tempo-choline nitroxide and is also shown for comparison. Since this parameter changes insignificantly with the temperature ($\delta(\Delta B^L_{p\text{-}p})$ =500±1 mG at 18 °C and 530±2 mG at 30.5 °C), an average value is shown. Approximate location of the nitroxide probes in the DMPC membrane is taken from the literature (Schreier-Muccillo *et al.*, 1976; Windrem and

Plachy, 1980; and the papers cited therein). These profiles are consistent with literature data on the oxygen permeability profile measured by T_1-sensitive techniques (*e.g.*, Subczynski *et al.*, 1989).

Smirnov and coauthors (1996) also made comparisons with saturation-recovery results (*i.e.*, T_1-data reported by Subczynski *et al.* (1989)). For example, at 29 °C the oxygen transport parameter W recalculated for oxygen partial pressure 1 atm was reported as ≈16 μs^{-1} for 16-DS and ≈10 μs^{-1} for 5-DS (*ibid.*). Corresponding values calculated from $\delta(\Delta B^L_{p\text{-}p}$ are 15 μs^{-1} (16-DS) and 8 μs^{-1} (5-DS). As was discussed by Hyde and Subczynski (1989), it is possible that, for spin exchange between spin probe and oxygen, the collision rate observed in the T_1 experiment is not quite equal to that observed in the T_2 experiment. It was experimentally shown (*ibid.*) that the ratio of event probabilities $p(T_1)/P(T_2)$ from saturation-recovery and line width data varies between 2/3 and 2 over a wide range of conditions where the Heisenberg exchange can be expected to be the dominant intermolecular mechanism. The T_2 data derived from the convolution-based fitting are within 20% of those measured from saturation-recovery experiments. This serves as an additional proof that the model of Heisenberg exchange is appropriate for spin-spin interactions between doxyl-stearic acid probes and molecular oxygen, and confirms the applicability of our convolution-based fitting method. The accuracy of our method appears to be sufficient for many practical applications, such as measurements of oxygen permeability profile through phospholipid bilayers, determination of membrane location of spin-labeled molecules including membrane proteins through measurements of molecular oxygen accessibility in site-directed spin-labeling experiments, and effects of various biochemical compounds on membrane structure and oxygen permeability. Some other examples of the use of the convolution-based oxygen accessibility method include studies of ethanol effect on oxygen permeability of model DMPC bilayers (Smirnov *et al.*, 1996) and partitioning and interactions with membranes of several lipophilic Gd^{3+} complexes used as contrast agents in MRI (Smirnova *et al.*, 1998).

Finally but not lastly, measurements of molecular oxygen accessibility of spin-labeled sites by our convolution-based approach appear to be more convenient and may be more accurate than widely used continuous wave saturation experiments that involve fitting of the "rollover" intensity plot versus incident microwave power. The CW EPR rollover saturation method requires measurements of multiple EPR spectra upon stepping the microwave attenuator. These changes in incident microwave power may also require spectrometer re-tuning. Another problem of EPR saturation experiments with lossy aqueous samples is that the sample temperature is not controlled and this could be a source of errors. For example, depending on the sample configuration, Barnes *et al.* (1996) reported sample heating from ca.3 to 15 °C at 128 mW of incident power versus the 2 mW power level. By using the convolution-based fitting described here this undesirable heating can be avoided by keeping the microwave power acceptably low in course of measurements.

3.3.3 The use of convolution-based fitting in multi-site EPR oximetry

Many applications of EPR oximetry *in vivo* involve repetitive line width measurements of particulate oxygen sensitive probes, which, once placed in the tissue of interest, would stay in place for the entire span of the measurement series (Swartz and Clarkson, 1998). While many of these measurements are carried out at a particular site with a single spec of oxygen-sensitive material, one could probe oxygen at multiple sites simultaneously by measuring EPR spectra in presence of magnetic field gradients (Smirnov *et al.*, 1993). We called such experiments a "multi-site EPR oximetry" (*ibid.*).Generally, if the size of the implanted oxygen-sensitive probe particles is sufficiently small compared with the distances between the implants, then the magnitude of the magnetic field gradient could be chosen such so the individual EPR spectra from the implants become resolved and not distorted in any significant extend by magnetic field gradients (*ibid*). If this could be accomplished, then the spectra from the individual sites can be processed separately to extract oxygen-induced broadening.

While useful for variety of applications, the multi-site EPR oximetry as described above has several limitations. Most importantly, the method lacks corrections of the measured line width for distortions caused by magnetic gradients of arbitrary magnitudes and for arbitrary shape/density function of solid-state paramagnetic materials implanted in the tissue. Typically, the shape of paramagnetic particulate probe is very irregular and several specs of the material can be present in a close proximity to each other. Moreover, because the oxygen tension might not be uniform throughout the tissue (*e.g.*, one could expect quite steep gradients of pO_2 in tissue) the local oxygen concentration could vary significantly even over the length of the implanted particle (site).

In principle, corrections for the line shape distortions caused by magnetic field gradients could be easily carried out once the distribution of a paramagnetic material at the site under study is determined. Then one should convert this distribution into the magnetic field coordinate and use the obtained function as a non-adjustable envelope $P(B)$ in the Equation (19) to extract the exact amount of oxygen broadening. Alternatively, if one could take the EPR spectra from the same system with and without the oxygen but at exactly the same magnetic field gradient, then the Equation (19) can be used in the same way as described in the Section 3.3.2. Unfortunately neither of these two simple approaches are practically feasible because once injected into an animal the distribution of the paramagnetic particulate material is largely unknown, the magnitude of the gradient is limited, and the spectrum without the oxygen and exactly the same orientation of the gradient with respect to the animal cannot be easily obtained.

Recently, Grinberg, Smirnov, and Swartz (2001) proposed to overcome some of the previous limitations of the multi-site EPR oximetry by a consequent application of magnetic field gradient with the same direction but with different magnitude and processing the results by the least- squares convolution-based fitting. We shall describe this method below.

Generally, EPR spectrum *J(B)* is given by an integral

$$J(B) = \int_V f(B - B_0(\vec{r}))\mu(\vec{r})d\vec{r} \tag{26}$$

where $\mu(\vec{r})$ is the distribution (density) function of paramagnetic centers and $f(B(\vec{r}))$ is the line shape function. When the external magnetic field is changing linearly, *i.e.*, $B(r)=B_0+r(\partial B/\partial r)= B_0+r\cdot gradB$, Equation (26) is simplified to a convolution integral which can be written in either spectral (Equation 27) or spatial (Equation 28) coordinates:

$$J(B, gradB) = \int p(B')f(B - B', \Delta B_{p-p})dB' = p(B) \otimes f(B) \tag{27}$$

$$J(r, gradB) = \int p(r')f(r - r', \Delta r_{p-p})dB' = p(r) \otimes f(r) \tag{28}$$

In these equations *p* is the projection of the density of paramagnetic centers on the direction of the gradient, and *f* is the line shape of EPR spectrum in absence of the gradient. We will use spatial coordinates (*i.e.*, Equation (28)) for the sake of convenience. It is worth nothing to say that in this scale the projection function *p(r)* is independent of the magnitude of magnetic field gradient *gradB* but the line shape function $f(r,\Delta r_{p\text{-}p})$ changes with the magnitude of the magnetic field gradient because $(r-r_0)=(B-B_0)/gradB$. The EPR line shape for many oxygen-sensitive particulate probes is exchange-narrowed and is well approximated by a Lorentzian function. When transformed from the spectral coordinate scale *B* to the spatial coordinates *r* the width of the line shape function scales with the gradient:

$$\Delta r_{p-p} = \frac{\Delta B_{p-p}}{gradB} \tag{29}$$

We record two EPR spectra $J_i(r,(gradB)_i)$; i=1, 2 at two different magnetic field gradients $(gradB)_2 > (gradB)_1$ applied in the same direction. Thus

$$\begin{aligned} J_1(r,(gradB)_1 &= p(r) \otimes f(r,(\Delta r_{p-p})_1) = \\ &= p(r) \otimes \left[f(r,(\Delta r_{p-p})_2 \otimes f(r,(\Delta r_{p-p}) \right] = \\ &= J_2(r,(gradB)_2) \otimes f_1(r,\Delta r_{p-p}) \end{aligned} \tag{30}$$

where

$$\begin{aligned} \Delta r_{p-p} &= (\Delta r_{p-p})_1 - (\Delta r_{p-p})_2 = \\ &= \Delta B_{p-p} \left[\frac{1}{(gradB)_1} - \frac{1}{(gradB)_2} \right] \end{aligned} \tag{31}$$

Equation (30) could be derived because the convolution of two Lorentzian functions gives a Lorentzian function with a width equal to the sum of the initial widths. Equation (30) shows that when expressed in spatial coordinates, the EPR spectrum $J_1(r)$ taken at a smaller gradient $(gradB)_1$ can be presented as a convolution of a spectrum $J_2(r)$ taken with a larger gradient $(gradB)_2$ and a Lorentzian function $f(r,\Delta r_{p-p})$ with a line width ,Δr_{p-p} that depends on the undistorted line width ΔB_{p-p} and the magnitudes of the two gradients as given by Equation (31). When the second gradient is much larger than the ratio of EPR line width to the characteristic size of the solid state probe that gives rise to the EPR signal, the projection function $p(r)$ is well approximated by $J_2(r)$ and Equation (30) reduces to Equation (28). This means that if the projection function $p(r)$ is known from another experiment, then, as we discussed above, the Equation (27) or (28) could be applied to derive the width of EPR spectrum and no correction given by Equation (31) is necessarily. In either case, the problem of extracting the width of Lorentzian width is reduced to least-squares fitting of two experimental spectra to Equation (19) with a Lorentzian broadening function of a variable width. This one-line-width-parameter fitting results in a significant decrease of the number of parameters that are needed to recover the oxygen from quite noisy EPR spectra which are typical for low frequency *in vivo* experiments.

Additional details of application of convolution-based fitting for high spatial resolution multi-site EPR oximetry are discussed by Grinberg and coworkers (2001). These authors also discussed the problem of partially overlapping spectra, the proper choice of the magnitudes for the gradients and fitting intervals, and also demonstrated this approach in model and *in vivo* experiments.

3.3.4 Accuracy of Oxygen Measurements by EPR

Increase in the accuracy of spectral parameters is one of the main reasons for carrying out spectral fitting along with developing models to understand the spectra. One of the advantages of using the Levenberg-Marquardt algorithms for minimization, is that it also provide a convenient way to evaluate parameter uncertainties and correlations between the parameters using the covariance matrix (*e.g.*, Press *et al.*, 1986). The magnitude of parameter uncertainties depends on several factors. We shall discuss this in application to convolution-based fitting on example of EPR oximetry measurements.

In EPR oximetry the accuracy of oxygen measurements is an important parameter, which determines suitability of a particular for experiments. We have already discussed that EPR experiments *in vivo* are very demanding because of a low signal-to-noise ratio and only moderate line width changes caused by low oxygen concentrations typically found in tissues For some time researchers focused on development of the probes with a larger line width response to oxygen concentration. Clearly, another spectral parameter, signal-to-noise, also contributes to experimental error and is as important as the line width response to oxygen.

Typically, the amplitude noise in EPR experiment has a normal (Gaussian) distribution (Duling, 1994). Then, if the fitting model is suitable for the experiment, the variance estimates in the extracted parameters (estimated parameter errors) are given by a product of χ^2 per degree of freedom and the square root of the corresponding diagonal elements of the covariance matrix *[C]* (Press *et al.*, 1984):

$$[C] = [\alpha]^{-1} \tag{32}$$

where

$$\alpha_{kl} = \frac{1}{2}\frac{\partial^2 \chi^2}{\partial \alpha_k \partial \alpha_l} \tag{33}$$

The covariance matrix is generated after an acceptable minimum is found by the Levenberg-Marquardt algorithm.

In order to check how suitable these estimates (and thus the model given by Equation (19)) are for CW EPR oximetry experiments, Smirnova *et al.* (1995) have measured and processed sets of EPR spectra (100 spectra each) for several EPR oxygen probes recorded at identical experimental conditions — ambient oxygen concentration and temperature. The spectra were fitted through Equation (19) by use of an experimentally measured "envelope" - a spectrum taken from a nitrogen equilibrated sample.The histograms of the extracted line width broadening (distribution of extracted broadening vs.counts of measurements) for two common EPR oxygen probes, CTPO and ^{15}N PDT (^{15}N-substituted 4-oxo-2,2,6,6-tetramethylpiperidine-d$_{16}$-1-oxyl), are shown in Figure 18. For each nitroxide the histograms can be approximated by Gaussian distributions. The half widths of the Gaussian distributions (given in the Figure 18 caption) agreed well with the variances estimated according Equations (32, 33). This shows that the estimates provided by the Levenberg-Marquardt algorithm are in fact correct and can be used to analyze the accuracy of such measurements.

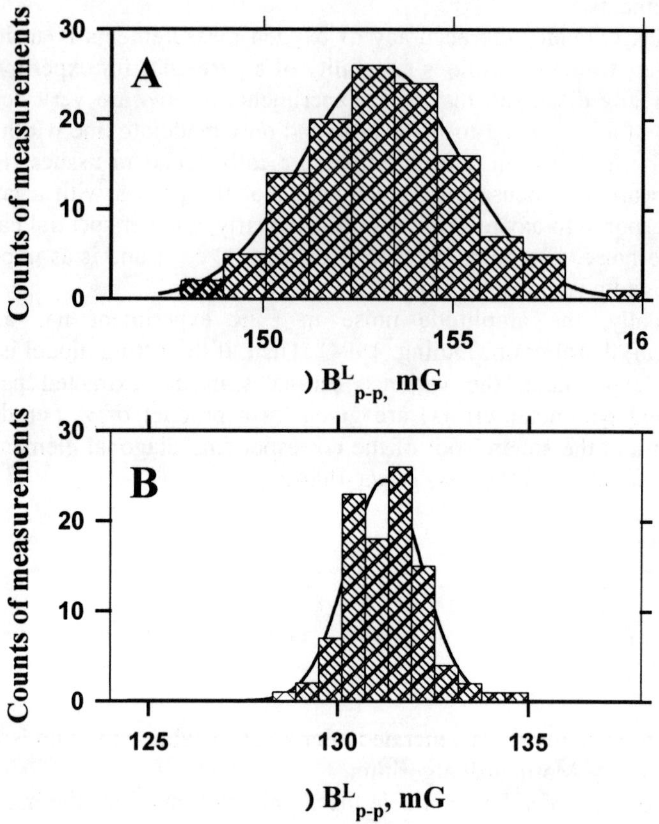

Figure 18. Distributions of $\Delta B^L_{p\text{-}p}$ defined by results of least-squares fitting for a series of 100 X-band EPR experimental spectra. Gaussian distributions characterized by variances σ^2 are shown in solid lines. (A): CTPO in air-saturated water, σ^2 =2.07 mG, predicted variance according to Equation (32) σ^2 =2.00 mG. (B): ^{15}NPDT in air-saturated water σ^2 =0.99 mG, predicted variance σ^2 =1.4 mG. Reproduced with permission from Smirnova *et al.*, 1995.

If the measurement errors are normally distributed and the fitting model can be replaced by a linear model within the region of parameter uncertainties (Press *et al.*, 1986), the parameter uncertainties $\sigma(a_i)$ can be estimated as:

$$\sigma^2(a_i) = \frac{C_{ii}}{(N-M)^2} \sum_{j=1}^{N} \left[E(B_j) - I(B_j, \vec{a})\right]^2 \tag{34}$$

where N is the number of experimental data points and M is the number of adjustable parameters. Typically $N>>M$ and N-$M \approx N$. Then $\sigma(a_1) \approx \sigma\ (C_{11})^{1/2}$, where σ is the standard deviation of the amplitude noise of the EPR spectrometer. Then the accuracy of the extracted oxygen concentration $\sigma([O_2])$ can be approximated as follows:

$$\sigma([O_2]) \approx \sqrt{C_{11}} \frac{\sigma}{\beta} \tag{36}$$

where C_{11} is the covariance matrix diagonal element related to the line width parameter $\Delta B^L_{p\text{-}p}$ and $\beta = \partial(\Delta B^L_{p\text{-}p})/\partial([O_2])$. For the same line shape, both functions $E(B)$ and $I(B, \vec{a})$ can be amplitude-normalized (*e.g.*, by the peak-to-peak amplitude, A):

$$E(B) = A \cdot e(B); \quad I(B) = A \cdot i(B) \tag{36}$$

Then

$$C_{11} = \frac{1}{A^2} c_{11} \tag{37}$$

where c_{11} is the diagonal element of the covariance matrix calculated for normalized functions $e(B)$ and $i(B)$. Then expression (35) can be rewritten as:

$$\sigma([O_2]) \approx \sqrt{c_{11}} \frac{1}{\beta} \frac{\sigma}{A} = \frac{\sqrt{c_{11}}}{2\beta R} \tag{38}$$

where $R = A/(2\sigma)$ is the signal-to-noise ratio.

Expression (38) shows that the accuracy of oxygen measurements depends not only on the sensitivity coefficient β, but also on signal-to-noise ratio R and the coefficient c_{11}. Inverse proportionality of the standard deviation of the extracted oxygen broadening to the signal-to-noise ratio R was also confirmed experimentally by processing sets of data from the same sample recorded with different R (Smirnova *et al.*, 1995). Those results showed that at a signal-to-noise ratio $R \approx 1000$, the error in line width determination could be as low as 1 mG, which corresponds to an average error in oxygen measurement of 0.1 μM.

The signal-to-noise consideration is very important for comparing various oxygen probes. For example, most biomedical applications set a limit for nitroxide concentrations because of their toxicity at concentrations greater than 1 mM (*e.g.*, Swartz *et al.*, 1992) or some concerns about perturbing the membrane system. (Typically, membrane spin probe-to-phospholipid ratio is less than 1:100.) Therefore, in practice, EPR spectra from nitroxides and, particularly membranes and proteins labeled with nitroxides, with a very high R (*ca.*, >1000) are quite rare. In contrast, the solid-state probe fusinite can be used in a relatively high spin concentration without toxic effects (Vahidi *et al.*, 1994). The fusinite line shape is not affected by particle concentration (with the possible exception of a susceptibility effect on the EPR spectrum at very high concentrations), and therefore a better R is achievable in EPR experiments.

From Equation (38) it is clear that the coefficient $(c_{11})^{1/2}$ is another factor affecting accuracy of line width - oxygen concentration measurements. This coefficient depends on the line shape as well as the spectral window, ΔB. Smirnova *et al.* (1995) carried out two model calculations to demonstrate these effects. In the first calculation, various degrees of inhomogeneous (Gaussian) broadening (measured as $\Delta B^G{}_{p\text{-}p}$) were added to a synthetic Lorentzian first-derivative spectrum. The peak-to-peak width of the Lorentzian component, $\Delta B^L{}_{p\text{-}p}$, was kept the same; there were 1024 data points. The spectral window was $\Delta B=35.5\cdot\Delta B^L{}_{p\text{-}p}$. All spectra were normalized by the peak-to-peak amplitude. The coefficients $(c_{11})^{1/2}$ were calculated for parameter values corresponding to $\chi^2(\vec{a})=0$ (*i.e.*, exactly at the minimum and in the absence of noise) and plotted as a function of relative inhomogeneous broadening $\Delta B^G{}_{p\text{-}p}/\Delta B^L{}_{p\text{-}p}$ (Figure 19).

The values of $(c_{11})^{1/2}$ calculated for the same parameter vector $\vec{a}$ in presence and absence of a random noise are about the same if the noise has a zero mean. Indeed, from analysis of α_{kl} given by Press *et al.* (1986) and using χ^2 as defined by Equation (16), one derives:

$$\alpha_{kl} = \sum_{j=1}^{N}\left[\frac{\partial^2 I(B_j,\vec{a})}{\partial\alpha_k\alpha_l} - \left(E(B_j) - I(B_j,\vec{a})\right)\frac{\partial^2 I(B_j,\vec{a})}{\partial\alpha_k\alpha_l}\right] \tag{39}$$

The first term in this sum is noise-independent. For a successful model, the second term is modulated by random noise and tends to be cancelled (*ibid*). Therefore, the coefficients α_{kl} and the matrix $[C]$ will be negligibly altered in presence of the random noise. It is important to note that our treatment, in contrast to that by Press *et al.* (1986), assumes the standard deviation of the noise σ to be the same for all data points and does not include it in χ^2. This results in noise-independent coefficients c_{kl}.

Figure 19 shows that after a small lag period, $(c_{11})^{1/2}$ is proportionally increasing with increasing of the relative magnitude of inhomogeneous broadening to the spectrum measured as $\Delta B^G{}_{p\text{-}p}/\Delta B^L{}_{p\text{-}p}$.This means that for the same signal-to-noise ratio and the oxygen sensitivity coefficient sensitivity coefficient β, the experimental error for oxygen broadening is increasing with the relative increase of inhomogeneous contribution $\Delta B^G{}_{p\text{-}p}/\Delta B^L{}_{p\text{-}p}$. Therefore, the EPR oxygen probes without or with a small inhomogeneous broadening would provide a better accuracy for the same β and R.

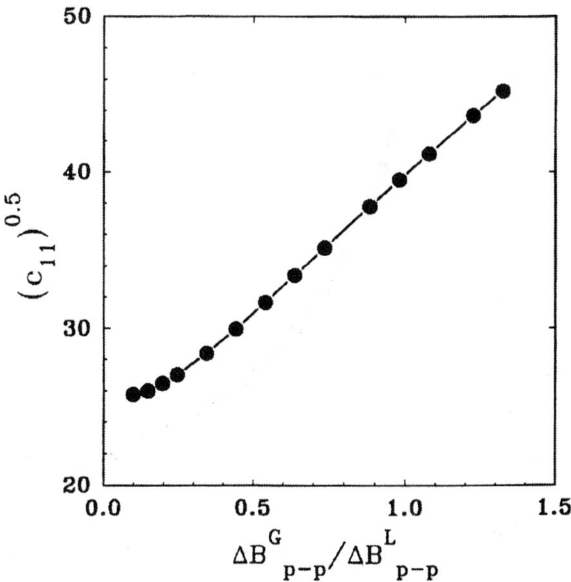

Figure 19. The coefficient $(c_{11})^{1/2}$ as a function of the relative inhomogeneous broadening $\Delta B^G_{p-p}/\Delta B^L_{p-p}$. calculated for amplitude-normalized model spectra. Details are given in text. Reproduced with permission from Smirnova *et al.*, 1995.

The second model calculation was carried out to demonstrate the effect of the width of the spectral window on the accuracy of the Lorentzian line width ΔB^L_{p-p} extracted from fitting a Voigt line shape. Model first-derivative EPR spectra (1024 data points) were constructed as convolutions of Lorentzian and Gaussian functions with equal peak-to-peak line widths ($\Delta B^G_{p-p}=\Delta B^L_{p-p}$) and simulated on spectral windows of a variable interval ΔB. All spectra were amplitude-normalized. Coefficients $(c_{11})^{1/2}$ were calculated for parameter values corresponding to $\chi^2(\vec{a})=0$ and plotted as a function of the width of the relative spectral interval $\Delta B/\Delta B^G_{p-p}$ (Figure 20). The plot shows that the coefficient $(c_{11})^{1/2}$ decreases (and therefore accuracy improves) as the width of the relative spectral window $\Delta B/\Delta B^G_{p-p}$ increases. A significant decrease in $(c_{11})^{1/2}$ is observed at the beginning of the curve with a more moderate decrease as the spectral window increases further. Although

Figure 20 demonstrates that larger spectral windows can improve the accuracy in line width EPR oximetry experiments, in practice, the width of spectral intervals for data acquisition is always limited for technical and other reasons (*e.g.*, baseline problem, the model being accurate only on a limited interval, *etc.*). Additionally, it has been assumed that the spectrum remains properly digitised over all spectral windows used in the simulations.

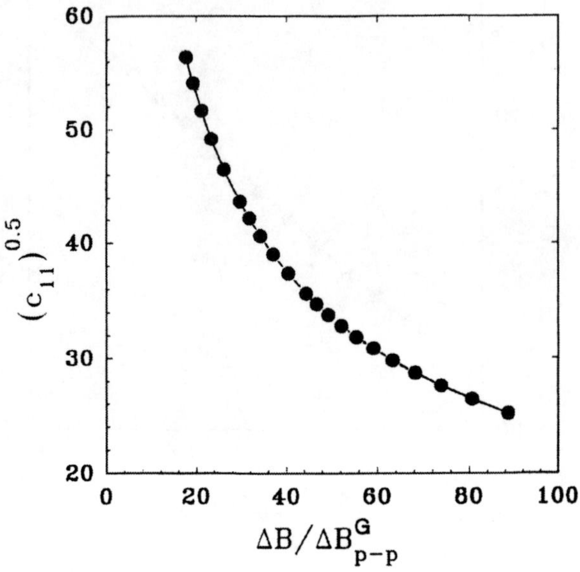

Figure 20. The coefficient $(c_{11})^{1/2}$ as a function of the width of the relative spectral interval $\Delta B/\Delta B^{G}_{p\text{-}p}$ for amplitude-normalized model EPR spectra of a Voigt shape. Details are given in text. Reproduced with permission from Smirnova *et al.*, 1995.

Although the results we have summarized here were obtained for EPR oximetry experiments, they are still valid for other experiments and can be briefly summarized as follows:

1) The accuracy of line broadening measurements using convolution-based fitting is proportional to signal-to-noise for properly digitized spectra and the broadening function.

2) The accuracy decreases if the width of the "inhomogeneous" envelope increases.

3) It is important to record the wings of the spectra for better accuracy.

Of course, all these observations are valid only when the convolution equation properly describes the experimentally observed broadening of EPR spectra.

Finally, but not lastly, the use of Equation (19) for fitting uniform broadening effects automatically accounts not only for all inhomogeneous broadening of a spectroscopic origin but also for broadening caused by sub-optimal instrumental parameters such as time constant and modulation amplitude. Because both time constant and modulation result in an additional uniform broadening across the EPR spectrum, they are described by a convolution integral. Thus, as long as both *I(B)* and *P(B)* spectra (see Equation (19)) are recorded with the same instrumental parameters, modulation and time constant broadening are automatically accounted for. It is easy to see that when fitting with Equation (19), one does not need to know what those instrumental settings are and the extracted broadening is undistorted by, for example, "over-modulation" (see also Smirnova *et al.*, 1995).

3.3.5 Measurements of Spin Label Concentrations: Membrane Partitioning and Nitroxide Bioreduction Experiments

There are at least two types of EPR spin label experiments that are based on measurements of the spin label concentration rather than the line width. One type is an EPR partitioning experiment when a nitroxide spin label is redistributed between several phases. By measurements of partitioning coefficients by EPR, important conclusions can be drawn about the phase state and, *e.g.*, phase diagrams of membrane systems (*e.g.*, McConnell, 1976). Another type of experiment involves measurements of nitroxide bioreduction kinetics that can provide insight into cellular redox processes (*e.g.*, for a review see Kocherginsky and Swartz, 1995).

Typically, intensities of the EPR signals are determined from peak height/line width measurements or after digital double integration of the CW spectra. Both methods can be inaccurate in case of inhomogeneously broadened lines. Even for EPR spectra with a good signal-to-noise ratio and absence of dispersion contribution, the digital integration might be inaccurate because of the broad wings of the Lorentzian function. For example, digital integration of a Lorentzian function over a spectral window of 5 line widths gives the EPR intensity that is only ≈89% of the value calculated over the spectral window of 50 line widths; such wide spectral windows are not always practical. The double-integrated intensity of EPR signals could be also estimated from a product of line-height and line-width-squared. However, this estimate is only valid for the same ratio of $\Delta B^G_{p\text{-}p}$ *to* $\Delta B^L_{p\text{-}p}$ (for Voigt line shape) and cannot be used if this ratio is varied during the experiment (*e.g.*, $\Delta B^L_{p\text{-}p}$ is varied with concentration or temperature).

If the EPR spectra from a spin probe partitioned in different domains all fall into the fast motional limit, then the convolution-based least squares algorithm can be used to accurately measure not only line widths and magnetic parameters but also intensities of the overlapping spectra. One example of this is a study of partitioning of a nitroxide Tempo (2,2,6,6-tetramethyl-1-piperidinyloxy) in large unilamellar liposomes (LUV) prepared from a phospholipid DPPC (1,2-dipalmitoyl-sn-glycero-3-phosphatidylcholine) carried out at 95 GHz (Smirnov *et al.*, 1995, Smirnov and Smirnova, 2002). At this frequency, the spectra from Tempo arising from aqueous and lipid phases are clearly resolved (see also Figure 21) and this guarantees the uniqueness of the fit (*ibid.*). Although the resolution for aqueous and lipid components is superb at 95 GHz, the spectra also contain some dispersive components, which make double integration even more inaccurate.

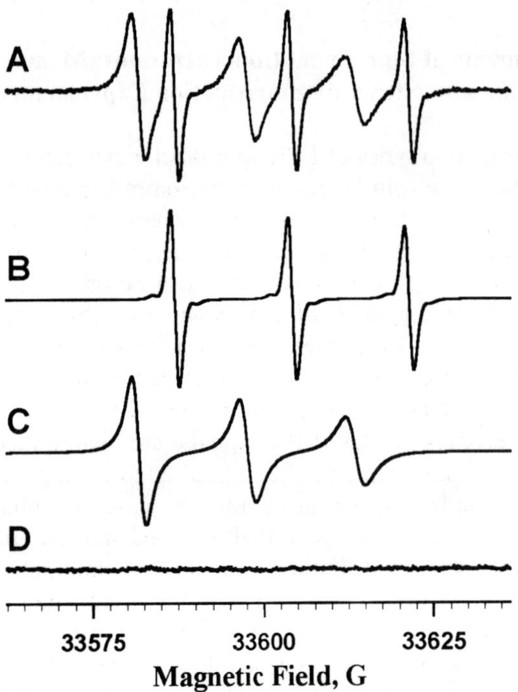

Figure 21. (A) Experimental 95 GHz EPR spectrum of multibilayer vesicles made from 1,2-dipalmitoyl-sn-glycero-3-phosphatidylcholine (DPPC) and labeled with a small nitroxide, Tempo (2,2,6,6,-tetramethylpiperedine-1-oxyl). Liposomes prepared in a phosphate-buffered saline (0.15 M NaCl, 5 mM phosphate buffer, pH 7.0) with a final concentration of DPPC in aqueous media of 200 mg/ml; concentration of Tempo was 620 μM.(B). Reproduced with permission from Smirnov and Smirnova, 2001.

Figure 21 shows how the aqueous and the hydrocarbon phase spectra can be effectively separated from the experimental 95 GHz spectra through the fitting methods we described in this chapter. The residual of the fit, which is the difference between the experimental spectrum and the simulations, shows no systematic deviation between the experiment and the fit (Figure 21D). The standard deviation of the residual is consistent with the spectrometer noise. From these fits the double integrals of each of the two spectra are calculated directly from the fitting parameters and the errors are derived using standard covariance matrix method. Thus, the partitioning coefficients obtained this way are free from uncertainties caused by a standard double integration and also corrected for dispersive components.

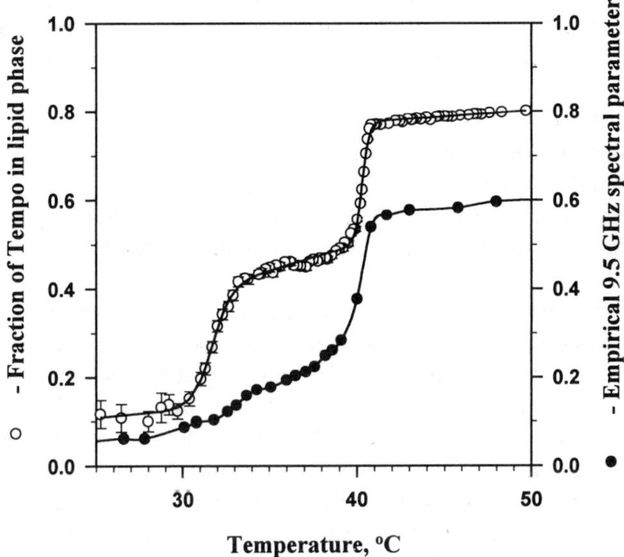

Figure 22. Temperature dependence of fraction of Tempo molecules partitioned in the lipid phase of the bilayer (f_w, open circles) as determined from least squares convolution fitting of high resolution W-band EPR spectra is compared with an empirical X-band spectral parameter f_X (filled circles). Solid line is the least-squares fit of the f_w data to a phase transition model. Error bars were estimated as described in the Section 3.3.4. Reproduced with permission from Smirnov and Smirnova, 2001.

Accurate partitioning parameters collected over a range of temperatures provide a much clearer picture of phase transitions in the phospholipid bilayers than empirical parameters derived from the X-band spectra. This is illustrated in Figure 22. The fraction of Tempo molecules in the lipid phase of the phospholipid membrane (f_w, Figure 22, open circles) was determined from least-squares simulations of the high resolution W-band (95 GHz) EPR data. Note that here f_w is defined as a ratio of Tempo molecules partitioned in the lipid phase to the total number of Tempo molecules in the sample. This definition is different than one used by Smirnov and co-workers in their W-band EPR study (1995). The fraction f_w is compared with an empirical parameter f_x previously employed in low-resolution X-band partitioning studies. The parameter f_x is typically defined as a ratio of intensities of a partially split high-field nitrogen hyperfine component of the Tempo spectrum observed at X-band (for definition see also Smirnov *et al.*, 1995).

The f_W data clearly show two transitions. The first transition at $T \approx 40.3$ °C corresponds to a ripple-structure ($P_{\beta'}$) - fluid bilayer-structure (L_α) phase transition and the temperature we measured is in agreement with the results of previous EPR studies (*e.g.*, Shimshick and McConnell). The second transition at $T \approx 31.7$ °C corresponds to the gel-structure ($L_{\beta'}$) - ripple-structure ($P_{\beta'}$) phase transition (or pre-transition). The observed phase transition temperatures are close to those reported from calorimetric data (Chen and Sturtevant, 1981). In a contrast, the f_x plot reveals only the ripple - fluid-bilayer transition while the second transition is hardly noticeable. It is also clear that the empirical parameter f_x is not proportional to the actual fraction of Tempo molecules partitioned in the lipid phase (f_w, Figure 22). Thus, least-squares simulations is the most accurate way to compare double-integrated intensities of EPR spectra components corresponding to different species.

Even when only one type of paramagnetic species is present in the spectrum, least-squares simulations still offer a better way to follow kinetics of changes in line width and double-integrated intensities over the measurements of a few spectral parameters from the spectra. One example of the experiments which would benefit from the use of fitting method, are nitroxide bioreduction studies in which spin probe concentration in a cell suspension (for review of nitroxide bioreduction see, *e.g.*, Kocherginsky and Swartz, 1995) or in an animal tissue (*e.g.*, Swartz and Halpern, 1998) is monitored as a function of time.

Redox reaction with participation of nitroxides (nitroxides may undergo biological reduction to hydroxylamines and oxidation to oxoammoniums, neither of which contribute to the EPR signal or line width of the remaining nitroxide (Swartz *et al.*, 1986)) can provide important insight into cellular redox processes. A number of studies have shown that nitroxides are reduced by ubiquinone by the electron transport system (Quintanilha and Packer, 1977; Chapman *et al.*, 1985; Chen *et al.*, 1989), but other redox reactions may be equally important.

Since both nitroxide and oxygen concentration affect the EPR line width, these parameters are difficult to separate experimentally. Generally, this perceived problem has been avoided by making rapid measurements (so that the concentration of the nitroxide does not change), by using nitroxides that do not enter the cell and thus are not reduced or oxidized rapidly, or the problem has been ignored. Typically, nitroxide reduction experiments have been carried out in the following way. A cellular suspension containing a spin probe at a concentration of about 10-100 μM was drawn into a glass or a gas-permeable plastic capillary. The capillary was fixed inside EPR resonator and maintained at a desired temperature. The magnetic field of the EPR spectrometer was set to the peak of one (typically the central) of the three nitrogen hyperfine lines of the nitroxide EPR spectrum and the height of this line was monitored as a function of time. Also it was often assumed that the amount of nitroxide is proportional to the signal height and the line width changes are neglected. Typically, complex kinetics (neither zero nor first order) were observed in such experiments, so the rates were determined using the first few minutes of the decay curve (which typically extends over half-an-hour).

Experimental difficulties with this method include: (i) the use of high concentrations of cells (10^8/ml is characteristic) and subsequent depletion of nutrients from and acidification of the cell culture media, (ii) settling of the cells along the bottom of the capillary tube even when the cavity is tilted by 90° so that the sample is horizontal and, most importantly, (iii) complicated kinetics of nitroxide intensity observed in such experiments was difficult to relate to biological function.

Some of the problems stated above can be resolved by measuring both line width and amplitude in the same experiment as described by Morse and Smirnov (1995). This method is based on least-squares fitting we described in this chapter as applied to a sequential array of data collected in course of nitroxide bioreduction experiments. The experimental protocol also uses many fewer cells, so the cell crowding is no longer a factor and local depletion of nutrients and nitroxides does not occur. Briefly, a cellular suspension at ca.5×10^6 cells/ml concentration is placed in quartz capillary (0.8 or 1.0 mm I.D.), which is then sealed and subsequently placed inside the cavity of an EPR spectrometer. EPR spectra are collected sequentially and stored as separate files. Initial simulation parameters for the first spectrum in a sequence are entered by an operator and subsequent spectra are fitted in automatic sequential mode: best fit parameters for the last spectrum are used as a first fitting approximation for the next spectrum in the sequence.

The spectra of a nitroxide spin label under conditions of bioreduction experiments can be simulated explicitly by accounting for all proton hyperfine splittings. However, some common nitroxides, like Tempol (2,2,6,6-tetramethylpiperidine-N-oxyl-4-ol) have a complicated proton hyperfine pattern which becomes only partially resolved upon depletion of oxygen in the capillary. This makes a simpler Gaussian envelope approximation impossible to apply. However, as shown by Morse and Smirnov (1995), approximation given by Equation (19) can be used instead. It is important to note here that Equation (19) remains strictly valid only when the spectra are described by a broadening effect that is uniform across the spectra. In a nitroxide bioreduction experiment, the EPR line width is affected by changes in both oxygen and spin label concentrations. While the first effect is precisely described by Equation (19), the second phenomenon, which arises from spin exchange between identical paramagnetic species, is generally not (Molin *et al.*, 1980; Robinson *et al.*, 1999). Nonetheless, at moderate spin label concentrations typically employed in nitroxide bioreduction experiments (≤1 mM), the nitroxide-nitroxide spin exchange effects are well approximated by an additional Lorentzian broadening. For example, Halpern and co-workers (1993) have shown that for aqueous solution of 4-hydro-3-carbamoyl-tetraperdeutero-methyl -3-pyrrolinyl-1-oxy (in this partially deuterated nitroxide only the ring hydrogen was retained) the Lorentzian line width was proportional to the nitroxide concentration (167±3 mG/mM). The only other effect observed was that the hyperfine splitting on the retained hydrogen decreased with the rate of 26.5±0.2 mG/mM. This means that for small concentration changes (*ca.*0.1-0.2 mM), the effect on proton hyperfine constants is typically of the order of a few mG. Thus, the changes in the envelope function *f(B)* in Equation (19) due to spin exchange on proton hyperfine constants can be neglected and the spectra can be accurately simulated using a convolution of an experimental spectrum measured at the lowest oxygen and nitroxide concentrations with the Lorentzian broadening function (Morse and Smirnov, 1995).

Figures .23-24 show typical results of Tempol bioreduction experiments for two types of cells - Baby Hamster Kidney, BHK, and adenovirus transformed hamster cells, 983.2 cells (*ibid.*). All spectral parameters were extracted by fitting experimental spectra to Equation (19). Initially, Lorentzian line width for both types of cells decreases linearly with time. This is in full accord with the Michaelis-Menten kinetics model for the respiration. The lines have sharp breaks: at 45 min for the BHK (Figure 23, squares) and at 94 min for the 983.2 cells (Figure 23, circles). These breaks correspond to the moments when all oxygen was used in the sample; the decrease in line widths after this time arises solely from a decrease in nitroxide concentration. Figure 24 shows kinetics of the nitroxide double-integrated intensity for both types of cells. The kinetics has an apparent zero order with respect to Tempol and the rate is independent of oxygen concentration. This is in a sharp contrast to the results of previous studies which have shown that generally the nitroxide reduction rates vary with oxygen concentrations ((*e.g.*, see Kocherginsky and Swartz, 1995 and references therein). Although those studies were carried out with different types of cells, the data of Morse and Smirnov (1995) suggest that the prior results may have been a result of overcrowding of the cells in the sample tubes. Another reason for inaccuracies is that changes in the line width are typically ignored unless accurate least-squares fitting of spectral shapes is employed.

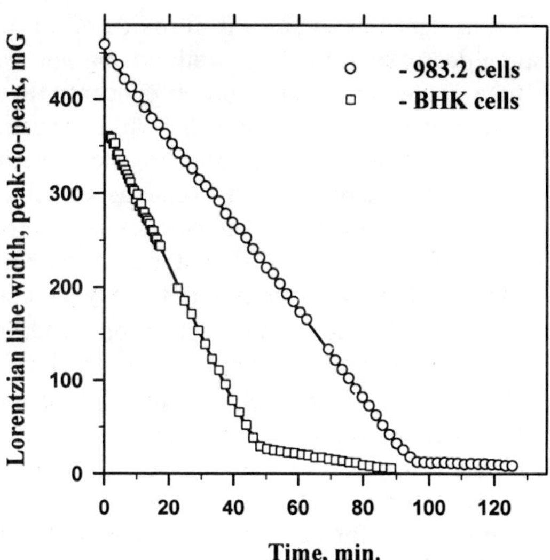

Figure 23. Kinetics of Lorentzian line width of Tempol EPR signal for samples of Baby Hamster Kidney (BHK, open circles) and adenovirus-transformed hamster (983.2; open circles) cells. The line width was measured by least-squares convolution-based simulations of the central (m_I=0) nitrogen hyperfine component as described in the text. Estimated errors are less than the size of the symbols on the plot. Reproduced with permission from Morse and Smirnov, 1995.

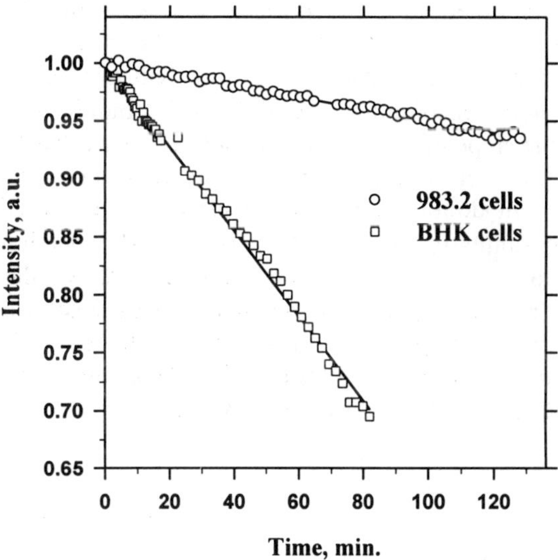

Figure 24. Kinetics of double-integrated intensity of Tempol EPR signal for the set of data shown in Figure 23: BHK cells - open squares, 983.2 cells - open circles. Estimated errors are less than the size of the symbols on the plot. Reproduced with permission from Morse and Smirnov, 1995.

From Figure 23 it is clear that in an EPR nitroxide respiration experiment there is always some decrease in the line width arising not from decrease in oxygen but due to a reduction in spin probe concentration. Because the nitroxide concentration can be monitored in the same experiment, it is easy to separate line width data corresponding to the kinetics of oxygen broadening from nitroxide-nitroxide spin exchange effects. For example, from Figure 24 the nitroxide bioreduction kinetics is a zero order. So the slopes in Figure 23 before the break should be corrected by the slopes observed after the break. Using this correction, the oxygen consumption rate for BHK cells was $27.3 \cdot 10^{-16}$ moles O_2/min/cell (Morse and Smirnov, 1995). This agrees well with that measured using an oxygen electrode in another experiment ($21.5 \cdot 10^{-16}$ moles O_2/min/cell). Although line width correction for spin exchange is small, it might be necessary for accurate determination of the Michaelis constant which is defined by the data at the very bottom of the respiration curve. Thus, accurate measurements of spin label concentration are very useful for correcting the line width oximetry results for nitroxide-nitroxide spin exchange effects.

Another example of the utility of the convolution-based least-squares fitting *vs.*peak-height measurements is shown in Figure 25. In this experiments, bioreduction of 100 μM of deuterated nitroxide ^{15}NPDT (2,2,6,6-tetramethylpiperidine-N-oxyl-4-one) by BHK cells ($4.2 \cdot 10^5$ cells/ml) was monitored over several hours (Morse and Smirnov, 1995). Figure 25A shows kinetics of the Lorentzian line width obtained from automatic least-squares fitting of 450 sequential EPR spectra. A sharp break in the slope seen at 3.2 hours corresponds to exhaustion of oxygen in the sample tube. Oxygen consumption rate of $23.8 \cdot 10^{-16}$ moles O_2/min/cell was calculated from the slope of line width data before the break after correcting for the slope arising from reduction in nitroxide concentration. Figure 25B shows the change in double-integrated intensity of the ^{15}NPDT signal as obtained from the simulation; the initial rate of ^{15}NPDT reduction is $2.8 \cdot 10^{-16}$ moles ^{15}NPDT/min/cell. For comparison, Figure 25C shows the line height (or peak-to-peak amplitude) data for the same set of spectra as in Figure 25A-B. Note that the slope is positive and curvilinear until oxygen is exhausted; after this point, the slope is negative. Thus, measurements of the peak height alone can yield some erroneous results.

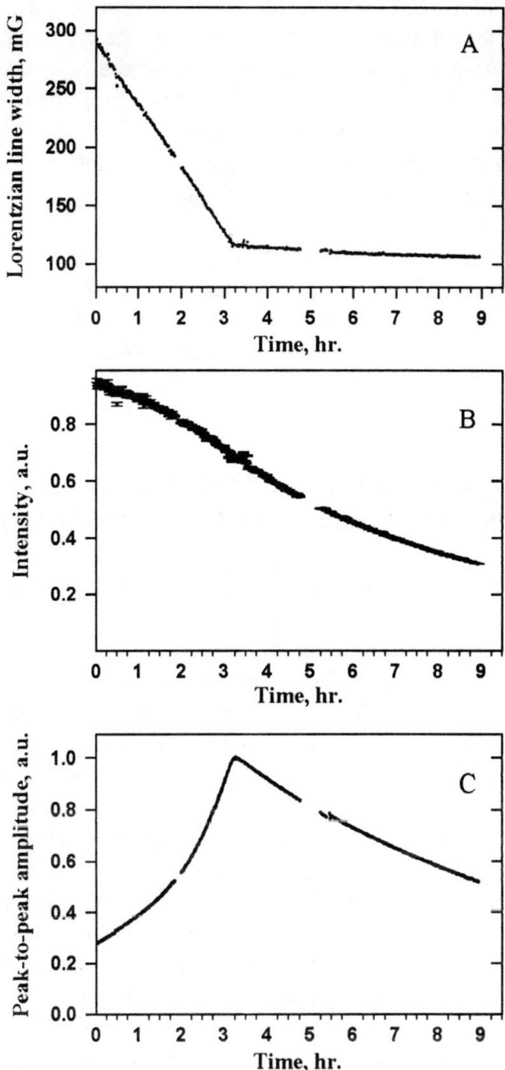

Figure 25. Kinetics of Lorentzian line width (A), spectral intensity (B) and amplitude (C) derived from automatic least-squares fitting of a set of sequential 450 EPR spectra measured in a course of a nitroxide bioreduction experiment with BHK cells. Initial concentration of perdeuterated nitroxide ^{15}NPDT (2,2,6,6-tetramethylpiperidine-N-oxyl-4-one) was 100 μM. Estimated error bars are shown if exceed the size of the symbols on the plot. Reproduced with permission from Morse and Smirnov, 1995.

The improvements in accuracy of spectral parameters obtained by least-squares is very beneficial because it allows one to use cellular suspensions at lower concentrations. The cells in the experiments of Morse and Smirnov

(1995) survive and even thrive under these experimental conditions. Cells were generally found attached to the surface of the capillary and some even spread out and took on a shape characteristic of their growth phase. Thus, it was assumed that the information on nitroxide bioreduction was obtained from cells in approximately their normal state. Recently Rapoport and co-workers (1999) described an application of this method to study effect of ultrasound on kinetics of bioreduction of a spin-labeled antibiotic gentamicin by gram-negative bacteria.

We shall mention here that automatic convolution-based fitting is fully applicable to other EPR kinetic problems to improve the accuracy of signal intensity measurements. In the past we have routinely used this method to monitor kinetics of spin-adduct formation. Other important applications include but not limited to formation of radical formation in rapid mixing experiments and kinetics of spin-labeled protein folding/denaturing.

3.4 Dipolar Broadening and Distance Measurements in Proteins, Membranes, and Disordered Matrices

The use of spin labeling EPR for distance determination in proteins and membranes is an area of contemporary interest because these measurements can be carried out with disordered systems (no protein crystals are required) and in many cases at physiologically relevant conditions. The method is based on detecting spin-spin interactions (exchange and/or dipolar) arising from pairs of interacting spins such as two nitroxides (*e.g.*, Rabenstein and Shin, 1995; Mchaourab *et al.*, 1996; Hustedt *et al.*, 1997) or a nitroxide and a paramagnetic metal ion (Raitsimring *et al.*, 1992; Voss *et al.*, 1995). Spin labels (both nitroxides or metal ions) are either introduced into a protein by site-directed mutagenesis methods (this includes engineering of a metal ion site as described by Voss *et al.*, 1995) or native binding sites can be also used. The latter includes labeling, *e.g.*, of native cysteine residues with thiol-specific nitroxide reagents or the use of naturally occurring paramagnetic metal ion centers as a reference points for distance measurements to the sites labeled with nitroxides. Recently methods of distance measurements based on nitroxide spin-spin interactions and applications of these methods to proteins were reviewed by Hustedt and Beth (1999) and also in this Series (Eaton *et al.*, 2001).

Generally, dipolar spin-spin interaction results in anisotropic effects on the EPR spectra. However, under certain conditions, and particularly in the motional narrowing regime, the dipole-dipole interaction will give rise to line width broadening that could be considered as uniform across the EPR spectrum. Particularly, if the correlation time, τ_c, of a stochastic process which modulates the dipole-dipole interaction is much less than reciprocal of the frequency shift produced by the dipolar field (*i.e.*, fast motion limit condition), then the dipolar interaction will produce a relaxation effect. For like spins S with the Larmor frequency ω_S, the dipolar relaxation is given by:

$$\frac{1}{T_{2dd}} = \gamma^4\hbar^2 S(S+1)\left[\frac{3}{8}J^{(0)}(0) + \frac{15}{4}J^{(1)}(\omega_S) + \frac{3}{8}J^{(2)}(2\omega_S)\right] \quad (40)$$

where $J^{(k)}(\omega)$ are the spectral densities of the random functions $F^{(k)}$ (Abragam, 1961). Equation (40) was also derived under assumption that the Redfield conditions are satisfied. The particular form of the spectral densities $J^{(k)}(\omega)$ depends on the model of the random process. For example, for like spins that are localized on a protein, the dipolar interaction primarily is modulated by the rotation of the interspin vector and:

$$\frac{1}{T_{2dd}} = \gamma^4\hbar^2 S(S+1)\frac{\tau_c}{r^6}\left[3 + \frac{5}{1+\omega_S^2\tau_c^2} + \frac{2}{1+4\omega_S^2\tau_c^2}\right] \quad (41)$$

where r is the interspin distance. In this case τ_c is the rotational correlation time of the protein molecule. This particular form of Equation (40) was derived assuming that the motion of the molecule can be modelled by an isotropic rotation of a rigid sphere (*ibid.*).

Mchaourab and coworkers (1997) developed a method based on Equation (41) to measure distances in spin label $S=½$ pairs attached to a protein. Using a series of double mutants of a small globular protein, T4 lysozyme, these workers showed that the distances calculated from the Equation (41) agree well with the crystal structure data. In this model study, all spin label EPR spectra were either in the fast motional limit or approaching it. As was pointed by Mchaourab *et al.* (1997), detailed modelling of the spectra would require a rigorous solution of the relaxation problem. But from the T4 data it appears that the dipolar relaxation can be treated independently so the dipolar effects can be approximated as uniform broadening, ΔB_d, across the spectrum. Thus the dipolar effect can be modelled by Equation (19) in which $P(B)$ is the sum of normalized EPR spectra from single-labeled protein mutants. It is worthwhile to mention here that if the local motion of spin labels occurs on the time scale much faster than the overall protein rotation, then the local motion will result in a reduction of the dipolar coupling depending on the angle between the fastest rotation axis and the interspin vector. Thus, in general, the equation (41) should somewhat overestimate the distance r between the labels. However, because r is proportional to $(\Delta B_d)^{-1/6}$ these corrections are typically relatively small with the exception of a few special cases, which are beyond the scope of this review.

Recently, the convolution-based approach to evaluate the strength of dipolar interaction was utilized by Sun and coworkers (1999) to probe proximity between the periplasmic loops in the lactose permease. One conclusion of their work was that the fitting algorithm based on Equation

(19) offers more sensitivity in the analysis of dipolar interaction than the use of normalized peak-height parameters.

Random rotation of the interspin vector is only one of several possible stochastic processes, which could modulate the dipole-dipole interaction and lead to dynamic dipolar broadening. Two other common sources of stochastic modulations are translational diffusion and fluctuating dipolar field due to a short spin-lattice relaxation time. Smirnova *et al.* (1998) gave an example of the latter by reporting a spin-labeling EPR study of interactions of .two Gd^{3+} complexes with multilamellar DMPC (1,2-dimyristoyl-*sn*-glycero-3-phosphocholine) bilayers. One of the complexes, Gd-DOTAP (1,4,7,10-tetraazacyclododecane-N-(n-pentyl)-N',N'',N'''-tri-acetic acid), was sufficiently lipophilic to partition in the phospholipid phase of the bilayer. This was confirmed by an experimental observation of broadening of EPR spectra from a series of doxyl-labeled stearic acids upon addition of Gd-DOTAP to the membrane samples (*ibid.*).

Basically, two magnetic interactions, Heisenberg spin exchange and dipole-dipole coupling, could be responsible for such a broadening. However, the spin exchange contribution to a spin label spectrum is likely to be small for an *S*-state ion complex and particularly for Gd^{3+}, whose paramagnetic inner *4f* orbitals are well shielded by the outer *5s* and *5p* electrons (Hyde and Sarna, 1978). For example, Powell *et al.* (1996) demonstrated that at high magnetic fields, concentration broadening of EPR spectra from similar Gd^{3+} complexes is well described by the collisional dynamic dipolar mechanism.

Dipolar interaction between doxyl stearic acid labels and a lipophilic Gd^{3+} complex partitioned in the bilayer could be modulated by several stochastic processes. Let us assume that dynamic properties of the spin labels and the lipophilic Gd-DOTAP complex are similar to those of the lipids. Rotational correlation time of these molecules is of the order of a few ns depending on temperature. Correlation time for lateral diffusion, τ_{tr}, can be estimated as $\tau_{tr}=d^2/D\approx64$ ms assuming that the lateral diffusion coefficient D is similar to that of the phospholipids ($\approx$10-14 m^2s^{-1} for the membrane in the gel phase; Cevc and Marsh, 1987) and the distance of the nearest approach d=8 Δ (according to data of Powell *et al.* (1996) for Gd-Gd distances for similar complexes). However, the electronic relaxation time T_{1e} for Gd^{3+} complexes at ca.0.34 T magnetic field is much shorter (0.5-0.2 ns). Thus, in this case the electronic spin-lattice relaxation of the Gd^{3+} ion is the leading mechanism of dipolar relaxation for the membrane spin labels.

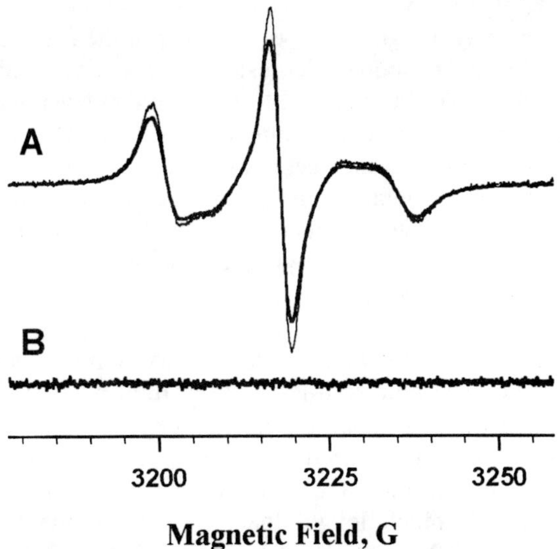

Figure 26. Experimental X-band spectra from a cholestane spin label in nitrogen-equilibrated DMPC aqueous dispersion at 30.8 °C (A) and in presence of 10 mM Gd-DOTAP; pH=9.5 (B). The spectrum (B) is fitted to an additional Lorentzian-broadening model of as described in the text. The best-fit Lorentzian broadening is $\delta(\Delta B^{L}_{p\text{-}p})$=0.095±0.005 G.(C) is the residual, the difference between the experimental and the simulated spectrum. Reproduced with permission from Smirnova *et al.*, 1998.

The dipolar relaxation mechanism we consider here will produce a broadening effect on the spin label spectrum that depends on the interspin distance as r^{-6}. The same mechanism was described by Leigh (1970) for spin label - metal ion interactions. However, in the Leigh model the nitroxide spectrum was immobilized because of the cryogenic temperatures required to observe the effect (at ambient temperatures the electronic relaxation time of many paramagnetic metal ions is too short). In the rigid limit the nitroxide spectrum is anisotropic and the Leigh model predicts broadening that is not uniform across spin label spectra and thus cannot be described by the convolution equation (19). The Gd^{3+} ion has unusually long electronic relaxation time even in solution at room temperatures. At these temperatures, EPR spectra of doxyl stearic acid and many other membrane probes are approaching the fast motional limit. In addition, for membrane spin labels and Gd^{3+} complexes that are somewhat randomly distributed through the membrane, there is no unique orientation of the interspin vector with respect to the nitroxide axes. This would also lead to further averaging of the dipolar relaxation effect. Thus, the dipolar interactions of the membrane spin probes with lipophilic Gd^{3+} complexes could be modelled as a uniform broadening across the spin label spectra as given by Equation (19). Indeed, data presented by Smirnova and coworkers (1998) confirmed that the convolution model is valid for a series of doxyl-labeled stearic acids and even for the nitroxides that display a high degree of ordering in the membrane phase. As an example, Figure 26 shows the results of spectral simulations under Equation (19) for a DMPC membrane in a fluid bilayer phase (30.8 °C) labeled with a cholestane (CSL) spin label at 1 mol%. The Lorentzian broadening model describes the spectrum quite well; the residual of the fit (difference between simulated and experimental spectrum) shows no significant deviations between experiment and fit. Using a series of lipid bilayers labelled with different doxyl stearic acids, Smirnova *et al.* (1998) monitored the dipolar broadening caused by lipophilic Gd^{3+} complexes across the bilayer.

Another property of the dynamic dipolar relaxation mechanism, which is determined by the electronic spin-lattice relaxation of the Gd^{3+} ion, is that the broadening of the spin label spectra is independent of viscosity and is proportional to the Gd^{3+} concentration. Indeed, in the phospholipid bilayers the effect of dipolar relaxation due to collisions is minimal because typically the collision time is much longer than the rate of the Gd^{3+} electronic relaxation (≈64 ms vs.X-band relaxation time of 0.5-0.2 ns respectively). Only in water at room temperature does the time of molecular collisions become comparable to that of the Gd^{3+} electronic relaxation. Thus, for more viscous fluids the dynamic dipolar relaxation becomes independent upon the viscosity. This effect was previously observed by Hyde and Sarna (1978) for $GdCl_3$. For Gd^{3+} – spin label pairs the dynamic dipolar relaxation is proportional to r^{-6}. If the spins are randomly distributed, then the relaxation of a particular spin label is given as an average over the volume. This will lead to $<r>^{-3}$ dependence on the average interspin distance $<r>$. Thus, the dipolar broadening in this regime is proportional to the Gd^{3+} concentration.

In some cases the convolution model also works well to model effects of static dipolar interaction on the magnetic resonance spectra. In fact, in his pioneer work Pake (1948) used a convolution of a dipolar profile obtained by averaging of the static dipolar interaction over all the interspin vector orientations with a Gaussian line to model the experimental spectra. Clearly, this method is strictly valid only when the experimental spectrum in the absence of interaction is isotropic. For immobilized nitroxides, this is not the case. However, even for immobilized nitroxide pairs the convolution model can serve as a useful approximation as was shown by Rabenstein and Shin (1995) and Steinhoff *et al.* (1991, 1997). Rabenstein and Shin (*ibid.*) suggested that an EPR spectrum from a nitroxide pair in a typical protein experiment could be approximated as a convolution of the spectrum of two non-interacting labels with an average Pake dipolar pattern. This averaging could be caused by a distribution of inerspin distances and mutual spin orientations that might be typical for spin labels attached to a protein. Instead of fitting the convolution equation in a manner we described here, Rabenstein and Shin (1995) applied Fourier deconvolution to estimate an average dipolar broadening function, from which an average dipolar splitting and distance are then calculated (*ibid.*). Correction for impurities of single-labeled proteins was also included in their method. A similar method was independently developed by Steinhoff *et al.* (1991, 1997). It also approximates the EPR spectrum of a nitroxide dimmer as a convolution integral, Equation (19), of an EPR spectrum from two non-interacting labels and the dipolar broadening function. The latter is constructed digitally assuming a Gaussian distribution for the interspin distances. Then the experimental spectrum that is broadened by dipolar interaction is fitted digitally to the convolution equation (19). Both distance parameters - the interspin distance and the breadth of distance distribution - are adjusted during the least-squares fitting procedure.In principle, this method uses all the experimental data points and reduces the experimental data to just a few parameters while the Fourier deconvolution method requires filtering of high frequency noise (*e.g.*, with a square filter as decribed by Rabenstein and Shin (1995)). Thus, any sharp features of the dipolar broadening function cannot be reconstructed by Fourier deconvolution because of limited bandwidth after the filtering. This bandwidth limitation also confines the application of Fourier deconvolution method to EPR spectra with a large (at least a few Gauss) dipolar broadening (however, for short interspin distances ≤10 Δ which lead to a large dipolar broadening, contributions of the spin exchange should be also considered). On the other hand, through examples from EPR oximetry we have demonstrated that the convolution fitting approach works well even for relatively small broadening effects caused by large spin separations in the pairs. Thus, the use of convolution-based fitting might be advantageous to somewhat extend the distance range of this CW EPR method.

More recently, Altenbach *et al.* (2001) carried out empirical evaluation of the approach initially developed by Rabenstein and Shin (1995) for the rigid lattice to estimate the inter-residue distances in spin labelling proteins at physiological temperatures where the correlation time of the interspin vector is insufficiently slow to average out the dipolar interaction. They have described a user-interactive convolution-deconvolution procedure to model the EPR spectra of double-labeled protein side-chains. Similar to Rabenstein and Shin (1995), the Fourier deconvolution was carried out first, but then the recovered broadening function was interpreted as a normalized sum of Pake patterns. This interpretation was carried out with participation of an user and the results were verified by direct convolution with the experimental spectrum (Altenbach *et al.*, 2001).

It is worthwhile to note here that although the modeling of the static dipolar interaction in the nitroxide pairs by a convolution of the EPR spectrum from non-interacting spins and a weighted/averaged Pake's dipolar profile gave good results in the model studies of small peptides (Rabenstein and Shin, 1995), liquid protein solutions (Altenbach *et al.*, 2001) and frozen nitroxide solutions (Steinhoff, 1991), it is unlikely to be applicable for the cases when the spin labels adopt unique mutual orientations and/or exhibit narrow distributions of distances and orientations (Hustedt and Beth, 1999). In the latter case the methods of Hustedt *et al.* (1997) should be applied instead. It seems to us that one reason why the convolution methods for modeling the static dipolar interaction are so successful is that the nitroxide rigid limit EPR spectra at X-band (9.5 GHz) are dominated by the central feature, which is a superposition of A_{xx}, A_{yy}, g_{xx}, g_{yy}, and g_{zz} components. In these spectra only the A_{zz} components are clearly resolved, but the peak-to-peak amplitude of those is smaller than that of the central feature. Thus, the results of the convolution-based fitting or Fourier deconvolution will be dominated by the central component, which is approximately "isotropic" because it contains the contributions from all the nitroxide orientations with respect to the magnetic field. This is in a sharp contrast to the nitroxide EPR spectra at high magnetic fields, which clearly resolve all the *g*-components. For these spectra the dipolar broadening typically is not uniform and thus cannot be modeled by the convolution equation (Smirnov, unpublished results).

Clearly, if one seeks only the distance information from the EPR spectra of spin pairs, then the use of species with isotropic or close to isotropic narrow-line spectra simplifies the analysis. This was recently demonstrated by Smirnov and Sen (2001) on example of EPR line shape studies of Gd^{3+}-doped glasses. These are silicate glasses, which, when doped with rare-earth ions, are considered as a promising active media for new miniature optical devices such as optical amplifiers, microchip lasers, and lossless splitters. To obtain the information on the average distances between the rare earth ions, the silicate glasses were doped with 0.058 to up to 4.64 wt% Gd_2O_3. While conventional X-band EPR spectra of Gd^{3+} in such glasses are rather broad spreading from 0 to *ca.* 5000 G and exhibit a complicated fine structure pattern, the spectral shapes are greatly simplified by increasing the resonant field/frequency of EPR experiment to 95 GHz (W-band). At this frequency only a single peak corresponding to $+1/2 \rightarrow -1/2$ is observed. The narrow line width of such a peak for diluted ion systems (*ca.*30 G) facilitates the observation of concentration-induced broadening effects. Smirnov and Sen (2001) have shown that such broadening effects can be modelled with a convolution equation using a Lorentzian broadening function. Also, it was found that concentration dependence of the Lorentzian broadening agrees well with the results of statistical theory of Grant and Strandnerg (1964) that is based on the Anderson-Weiss approach. It has been also shown that for a homogeneous distribution of Gd^{3+} the average distances can be determined in the range, which spreads to at least 40Å. For high Gd concentrations Smirnov and Sen (2001) observed some deviations between the broadening observed at 4.64% doping level and that predicted from a uniform ion distribution. This observation has been ascribed to be an onset of clustering of Gd^{3+} ions

4. CONCLUSIONS

In summary, we have shown that the convolution-based fitting is very useful to model spectra obtained in a variety of EPR experiments ranging from studies of rotational dynamics to EPR oximetry and distance measurements in proteins. The convolution algorithm we described here is computationally efficient because of the use of the Fast Fourier Transform algorithm. Calculation of the partial derivative matrix required for the Levenberg - Marquardt minimization is carried out efficiently and accurately using properties of the Fourier transform. The approach presented here is quite general and can easily accommodate various broadening models and/or envelope functions. For example, the broadening function could have a dispersive component to account for contamination of experimental absorption spectrum by dispersion one, or constructed from a weighted Pake's dipolar pattern if the broadening is caused by dipolar interaction.

The convolution algorithm also allows one to use an experimental spectrum as an envelope, so two spectra can be "compared" and parameters of broadening be accurately measured even when the broadening is just a few percent of the overall spectral width. This feature of the algorithm allowed us to extend line-width-based EPR oximetry experiments to nitroxide probes exhibiting intermediate and restricted anisotropic motion and, thus, to complement the T_1-oximetry data available from continuous-wave saturation or pulsed and saturation-recovery EPR. The accuracy of the technique (which is proportional to spectral signal-to-noise ratio) appears to be sufficient for many practical applications, such as measurements of oxygen permeability profile through phospholipid bilayers, determination of membrane location of spin-labeled molecules, and effects of various biochemical compounds on membrane structure and oxygen permeability.

The use of experimentally measured spectra as envelopes for convolution-based fitting is very attractive because this model does not require calculating the whole spectrum in order to extract the broadening parameters. This feature reduces the problem to one line width parameter fitting, which is more accurate and far less time consuming than the multiparameter optimization required for simulation of spin-label spectra according motional model. This reduction in fitting parameters could be very useful for simulation analysis of EPR spectra that have a distribution of parameters or several species with different broadening. Finally, application of convolution-based fitting to protein distance measurements in both static and dynamic modes should be further explored to evaluate the applicability of this model.

5. ACKNOWLEDGMENTS

The authors wish to thank Dr.R.L.Belford (University of Illinois) for many fruitful discussions and invaluable editorial comments at the earlier stages of this work. I also appreciate discussions with my colleagues from the Illinois EPR Research Center - Drs.R.B.Clarkson, M.J.Nilges, S.W.Norby and very stimulating collaborations with Drs.P.D.Morse (II) (Illinois State University), L.H.Sutcliffe (Norwich, UK), D.G., Gillies (University of Surrey, UK), N.Rapoport (University of Utah), and many others. This work was supported by North Carolina State University, North Carolina Biotechnology Center (grant number 2001-ARG-0033 to A.S.), the DOE Contract DE-FG02-02ER15354 (to A.S.), and the NSF (MCB-0196326 to T.S.). Resources for the initial stages of this work were provided by the Illinois EPR Research Center, an NIH biotechnology research resource (RR01811).

6. REFERENCES

Abragam, A., 1961, *The Principles of Nuclear Magnetism*, Chapter VIII, Oxford, London, pp.289-304.

Altenbach, C., Oh, K.-J., Trabanino, R.J., Hideg, K., and Hubbell, W.L., 2001, Estimation of Inter-Residue Distances in Spin Labeled Proteins at Physiological Temperatures: Experimental Strategies and Practical Limitations.*Biochem.***40**: 15471.

Bales, B.L., 1989, Inhomogeneously broadened spin-label spectra, in *Spin Labeling, Theory and Applications* (L.J.Berliner and J.Reuben, eds.), *Biol.Magn.Reson.* **8**, Plenum, New York, pp.77-130.

Bales, B.L., Wajnberg, E., and Nascimento, O.R., 1996, Temperature-Dependent Hyperfine Coupling Constant of the Dianion Radical of Fremy's Salt, a Convenient Internal Thermometer for EPR Spectroscopy*J.Magn.Reson.A.*, **118**: 227.

Ball, W.E., Kuester, J.L., and Mize, J.H.(eds.), 1973, *Optimization Techniques with FORTRAN*, (McGraw-Hill, New York) 240 p.

Boyer, S., and Clarkson, R.B., 1994, Electron paramagentic resonance studies of an active carbon - The influence of preparation procedure on the oxygen response of the linewidth.*Colloids and Surfaces A*, **82**:217.

Britton, M.M., Fawthrop, D.G., Gillies, D.G., Sutcliffe, L.H., Wu, X., and Smirnov, A.I., 1997, Matched spin probes for the study of the overall motion of model lubricants, *Magn.Reson.Chem.***35**: 493.

Budil, D.E., Earle, K.A., and Freed, J.H., 1993, Full determination of the rotational diffusion tensor by electron paramagnetic resonance at 250 GHz *J.Phys.Chem.***97**: 1294.

Budil, D.E., Lee, S., Saxena, S., and Freed, J.H., 1996, Nonlinear-least-squares analysis of slow-motion EPR spectra in one and two dimensions using a modified Levenberg-Marquardt algorithm *J.Magn.Reson.A* **120**: 155.

Cevc, G., and Marsh, D., 1987, *Phospholipid Bilayers. Physical Principles and Models*, Wiley, New York., p.375.

Chapman, D.A., Killian, G.J., Gelerinter, E and Jarrett, M.T., 1985, Reduction of spin-label TEMPONE by ubiquinol in the electron transport chain of intact rabbit spermatozoa, *Biol.Reprod.***32**: 884.

Chen, K., Glockner, J.F., Morse II, P.D., and Swartz, H.M., 1989, Effects of Oxygen on the metabolism of nitroxide spin labels in cells, *Biochem.***28**: 2496.

Clarkson R.B., Ceroke, P.J., Norby, S.-W., Odintsov, B., 1999, Stable particulate paramagnetic materials as oxygen sensors in EPR oximetry, in *In Vivo EPR. Theory and Applications* (L.J.Berliner ed.), *Biol.Magn.Reson.* **20**, Plenum, New York, *in press.*

Dean, J.A., 1985, *Lange's Handbook of Chemistry*, 13th Edition, McGraw-Hill Book Company, New York, p.10-5.

Duling, D.R., 1994, Simulation of multiple isotropic spin-trap EPR spectra, *J.Magn.Reson.B* **104**: 105.

Dunham, W.R., Fee, J.A., Harding, L.J., Grande, H.J., 1980, Application of fast Fourier transforms to EPR spectra of free radicals in solution, *J.Magn.Reson.***40**: 351.

Earle, K.A., Budil, D.E., and Fred, J.H., 1993, 250-GHz EPR of nitroxides in the slow-motional regime - Models of rotational diffusion.*J.Phys.Chem.***97**: 13289.

Eaton, S.S., Eaton, G.R., and Berliner, L.J.'Distance Measurements in Biological Systems by EPR', Kluwer Academic, 2001, *Biol.Magn.Reson.*, vol.19, 450 p.

Evans, J.C., Morgan, P.H., and Renaud, R.H., 1978, Simulation of electron spin resonance spectra by fast Fourier transform. A novel method of calculating spectra to include substitution, superhyperfine coupling, instrument time constant and modulation broadening in fluid and polycrystalline media, *Anal.Chim.Acta* **103**: 175.

Freed, J.H., 1964, Anisotropic rotational diffusion and electron spin resonance linewidths, *J.Chem.Phys.* **41**: 2077.

Grant, W.J.C., and Strandnerg, 1964, *Phys Rev.***135**: A715.

Grinberg, O.Ya., Smirnov, A.I., and Swartz, H.M., 2001, High Resolution Multi-Site EPR Oximetry: The Use of Convolution-Based Fitting Method.*J.Magn.Reson.***152**: 247.

Halpern, H.J., Peric, M., Yu, C., and Bales, B.L., 1993, Rapid quantitation of parameters from inhomogeneously broadened EPR spectra *J.Magn.Reson.A* **103**: 13.

Hustedt, E.J.and Beth, A.H., 1999, Nitroxide spin-spin interactions: applications to protein structure and dynamics, *Annu.Rev.Biophys.Biomol.Struct.*, **28**: 129.

Hustedt, E.J., Smirnov, A.I., Laub, C.F., Cobb, C.E., Beth, A.H., 1997, Molecular distances from dipolar coupled spin-labels - the global analysis of multifrequency continuous wave electron paramagnetic resonance data, *Biophys.J.*, **76**: 1861.

Hyde, J.S.and Sarna, T., 1978, Magnetic interactions between nitroxide free radicals and lanthanides or Cu2+ ions, *J.Chem.Phys.*, **68**: 4439.

Hyde, J.S., and Subczynski, W.K., 1984, Simulations of ESR spectra of the oxygen-sensitive spin-label probe CTPO, *J.Magn.Res.*, **56**: 125.

Hyde, J.S., and Subczynski, W.K., 1989, Spin label oximetry, in *Spin Labeling, Theory and Applications* (L.J.Berliner and J.Reuben, eds.), *Biol.Magn.Reson.* **8**, Plenum, New York, pp.399-425.

Hyde, J.S., Swartz, H.M., and Antholine, W.E., 1979, The Spin-probe-spin-label method, in *Spin Labeling II, Theory and Applications* (L.J.Berliner, ed.), Academic Press, New York, pp.71-114.

Jones, M.T., 1963, Electron spin exchange in aqueous solutions of $K_2(SO_3)_2NO$, *J.Chem.Phys.* **38**:2892.

Kocherginsky, N., and Swartz, H.M., 1995, *Nitroxide Spin Label Reactions in Biology and Chemistry*, CRC Press, Boca Raton, FL, 270 p.

Kuppusamy, P., Chzhan, M., Vij, K., Shteynbuk, M., Lefer, E., Giannella, D.J., and Zweier, J. L., 1994, 3-Dimensional spectral spatial EPR imaging of free radicals in the heart - a technique for imaging tissue metabolism and oxygenation, *Proc.Natl.Acad.Sci.U.S.A.* **91**: 3388.

Kusumi, A., Subczynski, W.K., and Hyde, J.S., 1982, Oxygen transport parameter in membranes as deduced by saturation recovery measurements of spin-lattice relaxation times of spin labels, *Proc.Natl.Acad.Sci.USA.***79**:1854.

Lai, C.-H., Hopwood, L.E., yde J.S.and Lukiewicz, S., 1982, ESR studies of O_2 uptake by chinese hamster ovary cells during the cell cycle, *Proc.Natl.Acad.Sci.U.S.A.*, **79**:1166.

Leigh, J.S., 1970, ESR Rigid-lattice line shape in a system of two interacting spins, *J.Chem.Phys.***52**:2608.

Liu, K.J., Gast, P., Moussavi, M., Norby, S.W., Wu, M., and Swartz, H.M., 1993, Lithium phthalocyanine: a new probe for EPR oximetry in viable biological systems *Proc.Natl.Acad.Sci.U.S.A.***90**:5438.

McConnell, H.M., 1976, Molecular Motion in biological membranes, in *Spin Labeling, Theory and Applications* (L.J.Berliner, ed.), Academic Press, New York, pp.525-561.

Mchaourab, H.S., Oh, K.J., Fang, C.J., and Hubbell, W.L, 1997, Conformation of T4 lysozyme.in solution - hinge-bending motion and the substrate-induced conformational transition studied by site-directed spin labeling, *Biochem.* **36**:307.

Molin, Yu.N., Salikhov, K.M., Zamaraev, K.I., 1980, *Spin Exchange*, (Springer, New York), 242 p.

Morse, P.D., and Smirnov, A.I., 1995, Simultaneous ESR measurements of the kinetics of oxygen consumption and spinlabel reduction by mammalian cells *Magn.Reson.Chem.*, **33**: S46.

Nilges, M.J., Walczak, T., and Swartz, H.M., 1989, 1 GHz in vivo ESR spectrometer operating with a surface probe, *Phys.Med.***2-4**:195.

Nilges, M.J., Smirnov, A.I., Clarkson, R.B., and Belford, R.L., 1999, Electron Paramagnetic resonance spectrometer with a low noise amplifier, *Appl.Magn.Reson.*, **16**:167.

Ottaviani, M.F., 1987, Analysis of resolved EPR spectra of neutral nitroxide radicals in ethanol and pyridine: The dynamic behavior in fast motional conditions, *J.Phys.Chem.***91**:779.

Pake, G.E., 1948, Nuclear Resonance absorption in hydrated crystals: fine structure of the proton line, *J.Chem.Phys.***16**:327.

Peric, M., Halpern H.J., 1994, Fitting of the derivative Voigt ESR line under conditions of modulation broadening, *J.Magn.Reson.A*, **109**:198.

Popp, C.A., and Hyde, J.S., 1981, Effects of oxygen on EPR spectra of nitroxide spin-label probes of model membranes, *J.Magn.Reson.*, **43**:249.

Portis, A.M., 1953, Electronic structure of *F* centers: saturation of the electron spin resonance, *Phys.Rev.***91**: 1071.

Powell, D.H., Ni Dhubhghaill, O.M., Pubanz, D., Helm, L., Lebedev, Ya.S., Schlaepfer, W., Merbach, A.E., 1996, Structural and dynamic parameters obtained from O-17 NMR, EPR, and NMRD studies of monomeric and dimeric Gd^{3+} complexes of interest in magnetic resonance imaging - an integrated and theoretically self consistent approach, *J.Am.Chem.Soc.***118**:9333.

Press, W.H., Teukolsky, S.A., Vetterling, W.T., and Flannery, B.P., 1986, *Numerical Recipes in Fortran*, Second Edition, (Cambridge Univ.Press, Cambridge), pp.531-536.

Rabenstein, M.D., Shin, Y.-K., 1995, Determination of the distance between two spin labels attached to a macromolecule, *Proc.Natl.Acad.Sci.USA* **92**:8239.

Raitsimring, A., Peisach, J., Lee, H.C., and Chen X, 1992, Measurements of distance distribution between spin labels in spin-labeled hemoglobin using an electron spin echo method, *J.Phys.Chem.***96**: 3526.

Rapoport, N.Ya., Smirnov, A.I., Pitt, W.G., and Timoshin, A.A, 1999, Bioreduction of spin-labeled gentamicin and tempone by gram-negative bacteria: kinetics and effect of ultrasound *Arch.Biochem.Bioph.*, **362**:233.

Robinson, B.H., Mailer, C., and Reese, A.W., 1999, Linewidth analysis of spin labels in liquids - I.Theory and data analysis, *J.Magn.Reson.*, **138**:199.

Robinson, B.H., Mailer, C., and Reese, A.W., 1999, Linewidth analysis of spin labels in liquids - II.Experimental*J.Magn.Reson.*, **138**:210.

Quintanilha, A.T., and Packer, L., 1977, Surface localization of sites of reduction of nitroxide spin-labeled molecules in mitochondria, *Proc.Natl.Acad.Sci.U.S.A.*, **74**: 570.

Schreier-Muccillo, S., Marsh, D., and Smith, I.C.P., 1976, Monitoring the permeability profile of lipid membranes with spin probes, *Arch.Biochem.Biophys.***172**: 1.

Silsbee, R.H., 1966, Fourier transform analysis of hyperfine structure in EPR, *J.Chem.Phys.* **45**: 1710.

Smirnov, A.I., and Belford, R.L., 1995, Rapid quantitation from inhomogeneously broadened EPR spectra by a fast convolution algorithm, *J.Magn.Reson.A*, **98**:65.

Smirnov, A.I., Clarkson, R.B., and Belford R.L., 1996, EPR linewidth (T_2) method to measure oxygen permeability of phospholipid bilayers and its use to study the effects of low ethanol concentrations *J.Magn.Reson.B.,* **111**:149.

Smirnov A.I., Norby S.W., Clarkson R.B., Walczak T., and Swartz H.M., 1993, Simultaneous multi-site EPR spectroscopy in vivo, *Magn.Res.Med.***30**:213.

Smirnov A.I., Norby S.W., Weyhenmeyer J.A., Clarkson R.B., 1994, The Effect of temperature on the respiration of cultured neural cells as studied by a novel EPR technique *Bioch.Bioph.Acta* **1200**:205.

Smirnov, A.I., Smirnova, T.I., and Morse II, P.D., 1995, Very High frequency electron paramagnetic resonance of 2,2,6,6-tetramethyl-1-piperidinyloxy in 1,2-dipalmitoyl-*sn*-glycero-3-phosphatidylcholine liposomes: partitioning and molecular dynamics, *Biophys.J.***68**:2350.

Smirnov, A.I.and Sen, S., 2001, High Field EPR of Gd^{3+}-doped Glasses: Line Shapes and Average Ion Distances in Silicates, *J.Chem.Phys.***115**: 7650.

Smirnov, A.I.and Smirnova T.I., 2001, Resolving Domains of Interdigitated Phospholipid Membranes with 95 GHz Spin Labeling EPR.*Appl.Magn.Reson.***21**: 453.

Smirnova T.I., Smirnov A.I., Clarkson R.B., and Belford R.L., 1995, Accuracy of oxygen measurements in T_2 (linewidth) EPR oximetry, *Magn.Reson.Med.***33**:801.

Smirnova, T.I., Smirnov, A.I., Clarkson, R.B., and Belford, R.L., 1995, W-band (95 GHz) EPR spectroscopy of nitroxide radicals with complex proton hyperfine structures: fast motion *J.Phys.Chem.***99**:9008.

Smirnova, T.I., Smirnov, A.I., Belford, R.L., and Clarkson, R.B., 1997, Interaction of MRI gadolinium contrast agents with phospholipid bilayers as studied by 95 GHz EPR, *Acta Chemica Scandinavica*, **51**:562.

Smirnova, T.I., Smirnov, A.I., Clarkson, R.B., Belford, R.L., Kotake, Y., and Janzen, E.G., 1997, High-frequency (95 GHz) EPR spectroscopy to characterize spin adducts, *J.Phys.Chem.B*, **101**:3877.

Smirnova, T.I., Smirnov, A.I., Clarkson, R.B., Belford, R.L., 1998, Lipid-MRI contrast agent interactions: a spin-labeling and a multifrequency EPR study, *J.Am.Chem.Soc.*, **120**:5060.

Steinhoff, H.-J., Dombrowsky, O., Karim, C., and Schneiderhahn, C., 1991, Two dimensional diffusion of small molecules on protein surfaces: an EPR study of the restricted translational diffusion of protein-bound spin labels, *Eur.Biophys.J.***20**:293.

Steinhoff, H.-J., Radzwill, N., Thevis, W., Lenz, V., Brandenburg, D., Antson, A., Dodson, G., and Wollmer, A., 1997, Determination of interspin distances between spin labels attached to insulin: comparison of electron pramagnetic resonance data with the X-ray structure, *Biophys.J.***73**: 3287.

Subczynski, W.K., and Hyde, J.S.,1984, Concentration of oxygen in lipid bilayers using a spin-label method, *Biophys.J.***45**: 743.

Subczynski, W.K., Hyde, J.S., and Kusumi, A., 1989, Oxygen permeability of phosphatidylcholine-cholesterol membranes, *Proc.Natl.Acad.Sci. U.S.A.***86**:4474.

Sun, J., Voss, J., Hubbell, W.L., and Kaback, H.R., 1999, Proximity between periplasmic loops in the lactose permease of Escherichia coli as determined by site-directed spin labeling, *Biochem.*, **38**:3100.

Swartz, H.M., and Clarkson, R.B., 1998, The measurements of oxygen in vivo using EPR techniques, *Phys.Med.Biol.*, **43**: 1957.

Swartz, H.M., and Halpern, H., 1998, EPR Studies of Living Animals and Related Model Systems (*In* Vivo EPR) in *Spin Labeling. The Next Millennium* (L.J.Berliner ed.), *Biol.Magn.Reson.* **14**, Plenum, New York, pp.367-408.

Swartz, H.M., Liu K.J., Goda, F., and Walczak, T., 1994, India ink - a potential clinically applicable EPR oximetry probe, *Magn.Reson.Med.***31**:229.

Swartz, H.M., Boyer, S., Brown, D., Chang, K., Gast, P., Glockner, J.F., Hu, H., Liu, K.J., Moussavi, M., Nilges, M.J., Norby, S.W., Smirnov, A.I., Walczak, T., Wu, M., Clarkson, R.B., 1992, The use of EPR for the measurement of the concentration of oxygen in vivo in tissues under physiologically pertinent conditions and concentrations, in *Oxygen transport to tissues XIV* (W.Erdmann, D.F.Bruley, eds.), Plenum, New York, pp.221-228.

Swartz, H.M., Sentjurc, M, and Morse II., P.D., 1986, Cellular metabolism of water-soluble nitroxides: effect on rate of reduction of cell/nitroxide ratio, oxygen concentrations, and permeability of nitroxides, *Biochim.Biophys.Acta*, **888**:82.

Vahidi, N., Clarkson, R.B., Liu,K.J., Norby, S.W., and Swartz, H.M., 1994, In vivo and in vitro EPR oximetry with fusinite: a new coal-derived, particulate EPR probe, *Magn.Reson.Med.***31**:139.

Voss, J., Hubbell, W.L., Kaback, H.R., 1995, Distance determination in proteins using designed metal ion binding sites and site-directed spin labeling: application to lactose permease of Esherichia coli, *Proc.Natl.Acad.Sci. USA,* **92**:12300.

Wertheim, G.K., Butler, M.A., West, K.W., and Buchanan,D.N.E., 1974, Determination of the Gaussian and Lorentzian content of experimental line shapes, *Rev.Sci.Instrum.***45**: 1369.

Windrem, D.A., and Plachy, W.Z., 1980, The diffusion-solubility of oxygen in lipid bilayers, *Biochim.Biophys.Acta*, **600**: 655.

Chapter 7

1D and 2D Electron Spin Resonance Imaging (ESRI) of Transport and Degradation Processes in Polymers

Mikhail V. Motyakin and Shulamith Schlick
Department of Chemistry, University of Detroit Mercy, Detroit, Michigan 49219-0900, USA

"Feeling stones while crossing a river"
Chinese proverb

1. INTRODUCTION

Paul Lauterbur was the first to propose, in his notebook entry of 2 September 1971, the use of magnetic field gradients in order to "tell exactly where an NMR signal came from"; his first proton density map appeared soon afterwards (Lauterbur, 1973). He named the technique *zeutmatography*, from the Greek word for "joining together", meaning to join the magnetic field gradient and the corresponding radiofrequency in a nuclear magnetic resonance (NMR) experiment. This connection allowed the encoding of spatial information in the NMR spectra. The use of magnetic field gradients to separate the resonant frequencies corresponding to the various spatial elements led to the development of NMR imaging (NMRI) or, in current language, magnetic resonance imaging (MRI). Against a back-drop of general skepticism, NMRI has blossomed into an essential, and routine, diagnostic procedure in medicine that provides an image of previously hidden anatomic parts. Applications of NMRI to Materials Science and other important disciplines, although not as dramatic as the medical applications, are extensive, ubiquitous, steadily developing, and well documented (Blumler *et al.*, 1998). The wonderful story on the discovery of NMR imaging has been told recently (Kevles, 1997).

The imaging method based on magnetic field gradients can also be applied to imaging of unpaired electron spins via electron spin resonance

(ESR) spectroscopy. The advantages of ESR imaging (ESRI) are the specificity for the detection of paramagnetic species, and the high sensitivity. Papers describing the feasibility of ESRI started to appear in the late 1970s and continue to this day. The early instrumentation, software, and applications of ESRI have been described in a 1991 monograph (Eaton *et al.*, 1991). The challenges in the application of the Lauterbur method to ESRI are numerous: First, higher gradients are needed compared to NMRI, usually 100-1000 times larger. Second, the ESR spectra are often complex and contain hyperfine splittings that add complexity to ESRI experiments. Third, and perhaps the most important challenge of all, most systems do not contain stable paramagnetic species on which imaging is based; ESR imaging is usually performed on radicals produced by irradiation, paramagnetic transition metal ions, or stable nitroxide radicals as dopants.

As clearly seen in the 1991 monograph, the early efforts laid the foundation for the hardware necessary for ESRI and developed the software necessary for image reconstruction in 1D (spatial) and 2D (spectral-spatial and spatial-spatial). These studies also investigated the feasibility of ESRI experiments in a variety of "phantom" samples, and discussed and estimated the spatial resolution. The resolution in most ESRI experiments is about the same as in the NMRI experiments, of the order of 50-100 μm, but can vary widely, depending on the ESR line shapes and line widths.

Biological applications of ESRI have taken advantage of the developing low-frequency ESR instrumentation in order to overcome the dielectric losses of aqueous solutions (Halpern and Bowman, 1991; Halpern *et al.*, 1994; Halpern *et al.*, 1999). Among the recent biological advances we note: 3D spatial and 4D spectral-spatial ESRI at 1.2 GHz of rat hearts perfused with nitroxide free radicals, with submillimeter resolution (Zweier and Kuppusamy, 1998); and 3D spatial ESRI at 0.7 GHz of nitroxides in brains and livers of rats and mice, with a resolution of ≈1 mm (Oikawa *et al.*, 1996).

The information on the spatial distribution of paramagnetic molecules in a sample deduced from ESRI experiments has been used successfully for measurement of the translational diffusion. Diffusion coefficients of paramagnetic diffusants can be deduced from an analysis of the time dependence of the concentration profiles along a selected axis of the sample. The determination of diffusion coefficients for spin probes in liquid crystals and model membranes, and the effect of polymer and probe polydispersity, have been described in a series of papers by Freed and coworkers (Moscicki *et al.*, 1991; Freed, 1994; Xu *et al.*, 1996). These papers represent an effort to move beyond phantoms, and to extract quantitative information from ESRI experiments.

In our laboratory efforts to develop ESRI methods (hardware and software) have focused on applications to important problems in Materials Science, with emphasis on polymeric systems. We started by developing 2D spectral-spatial ESRI, in order to measure the translational diffusion coefficient of the probes, D, as well as the spatial dependence of the rotational correlation time, τ_c, in one experiment. The line shape and the intensity variation along the sample depth were determined in 2D ESRI (spectral-spatial) from a complete set of projections, using a convoluted back-projection algorithm (Kweon, 1993). The imaging experiments were based on nitroxides (Schlick *et al.*, 1995; Gao *et al.*, 1996; Malka and Schlick, 1997) and paramagnetic Mo^V complexes (Gao and Schlick, 1996; Kruczala *et al.*, 1996) as the paramagnetic diffusants. In some experiments deuteriated spin probes were used as tracers, in order to reduce the line widths and thus improve the spatial and spectral resolution. The 2D ESRI experiments required, however, a higher probe concentration than that used in ESR spectroscopy; the broadening of the lines by spin-spin interactions was a complication when trying to simulate and deduce the motional mechanism. For these reasons the 2D spectral-spatial experiments were used for the determination of the diffusion coefficients of guests in polymeric systems. The goal of the 2D experiments was to assess the effects of polymer concentration (in solutions), network content (in swollen cross-linked polymers), temperature, and solvent on the transport properties of the tracers, and to compare the diffusion coefficients with results obtained by other methods. The initial goal of determining both D and τ_c in one experiment was not achieved. However, we were able to obtain D values, to experience first hand the problems associated with ESRI in one and two dimensions, to prepare the necessary spin probes, and to develop the appropriate software for image reconstruction. Because the time required for data acquisition was rather long, typically $\approx$20 min for a complete set of projections, only relatively slow diffusion processes were measured by 2D ESRI. Our earlier work has been summarized in a recent chapter (Schlick *et al.*, 1998a).

Developments in the imaging software and in the experimental techniques in our laboratory made possible improvements in the accuracy of the determination of D values (Marek *et al.*, 2002).

Two recent applications of ESRI will be described in this Chapter:

(a) The diffusion coefficients, D, at 300 K of nitroxides probes that differ in their hydrophobicity were determined by 1D ESRI, in the various phases (micellar, hexagonal, lamellar, and reverse micellar) of a self-assembling polymeric surfactant. In these experiments the ESR spectra measured in the presence of magnetic field gradients ("1D images") were simulated by convoluting the ESR spectrum measured in the absence of the gradient with the concentration profile calculated for a given D and time t during the diffusion process.

(b) The time evolution of concentration profiles and line shapes of radicals in polymers exposed to UV irradiation or thermally degraded were determined by 1D and 2D spectral-spatial ESRI.

These two applications were selected, and are presented here, because they represent an approach to solving interesting scientific problems that take advantage of the unique strengths of ESR imaging.

This chapter consists of five major parts. The Introduction is Section 1. Section 2 describes the ESR spectra in the presence of magnetic field gradients, the hardware necessary for ESRI, and the data needed and collected in 1D and 2D spectral-spatial ESRI. In Section 3 we present the 1D ESRI experiments performed for the measurement of D in a self-assembled polymeric surfactant; this section includes sample preparation, data acquisition, and data processing leading to the D values in the various phases of the surfactant. In Section 4 we focus on the application of ESRI for the elucidation of the degradation and stabilization of a copolymer, poly(acrylonitrile-*co*-butadiene-*co*-styrene) (ABS). Section 5 presents our conclusions, both on the present practice of ESRI in our laboratory, and on the potential for future applications of the technique.

2. ESR IMAGING: HARDWARE, DATA ACQUISITION AND SOFTWARE

2.1 ESR Spectra in the Presence of Field Gradients

ESR spectroscopy can be transformed into an imaging method for samples containing unpaired electron spins, if the spectra are measured in the presence of magnetic field gradients. In an ESR imaging experiment the microwave power is absorbed by the unpaired electrons located at point x when the resonance condition, equation 1, is fulfilled.

$$\nu = (g\beta_e / h)(H_{res} + xG_x) \tag{1}$$

Here ν is the microwave frequency, g is the spectroscopic g-factor for the electron, β_e is the Bohr magneton, h is the Planck constant, and G_x is the linear magnetic field gradient (in Gauss/cm) at x. As in NMR imaging, the field gradients produce a correspondence between the location x and the resonant magnetic field H_{res}. If the sample consists of two point samples, for example, the distance between the samples along the gradient direction can be deduced if the field gradient is known (Berliner and Fujii, 1985).

The spatial resolution, Δx, is an important parameter in imaging, and can de defined in various ways, as discussed recently (von Kienlin and Pohmann, 1998); obviously, the resolution depends on the line width and line shape. One way to define the spatial resolution is to specify the accuracy of the sample location. Usually, however, Δx is expressed as the ratio of the line width to the field gradient, $\Delta H/G_x$; this definition implies that two signals separated by one line width due to the field gradient can be resolved.

2.2 Hardware for ESR Imaging

The ESR imaging system in our lab consists of the Bruker 200D ESR spectrometer equipped with two Lewis Coils (George Associates, Berkeley, U.S.A, type 503D), and two regulated DC power supplies (Kikusui Electronic Corp., Japan, model PAE 35-30). The two sets of coils, each consisting of a figure-eight coil, are fixed on the poles of the spectrometer magnet. Depending on the electrical wiring of the two sets, the coils can supply a maximum linear field gradient of ≈320 G/cm in the direction parallel to the external magnetic field (z axis), or ≈250 G/cm in the vertical direction (along the long axis, x, of the microwave cavity), with a control current of 20 Amperes applied to each power supply. The coils are positioned so that the zero point of the gradient field coincides with the center of the microwave cavity.

To calibrate and check the linearity of the field gradient, we have prepared a sample consisting of two specks of the radical 2,2 diphenyl-1-picryl hydrazyl (DPPH) at a distance of 10 mm. In the absence of the gradient a single line is detected. Two lines, however, are obtained in the presence of gradients, and the separation increases with increasing gradient. The excellent straight line obtained by plotting the separation in Gauss as a function of the current indicated the linearity of the gradient. This calibration is repeated regularly.

2.3 Intensity Profiling from 1D ESRI

In the general case, the sample contains a distribution of paramagnetic centers along a given direction, for example x. The ESR spectrum in the presence of the magnetic field gradient is a superposition of signals from paramagnetic centers located at different positions. In 1D ESRI the intensity profile is obtained from two spectra: $F_0(H)$ measured in the absence of magnetic field gradient, and $F(H)$ measured in the presence of the gradient.

Mathematically, $F(H)$ is a convolution of $F_0(H)$ with the distribution function of the paramagnetic centers (Ewert and Thiessenhusen, 1991; Yakimchenko *et al.*, 1995), equation 2,

$$F(H) = \int_{-\infty}^{\infty} F_0(H - H^*)C(H^*)dH^* \tag{2}$$

where $H^*=H_0-xG$, $H_0=h\nu/g\beta_e$, and $C(H^*)$ is the intensity distribution (profile) of the paramagnetic centers along the gradient direction.

Equation 2 can be written using convolution symbol, ⊗:

$$F(H) = C(H) \otimes F_0(H) \tag{3}$$

The determination of the distribution function becomes a deconvolution problem, which has been solved in a number of ways (Jansson, 1984). The most commonly used method is based on the Fourier Transform (FT). Taking the FT of both sides of equation 3 we obtain equation 4,

$$F(q) = C(q) \cdot F_0(q) \tag{4}$$

where q is the variable in the Fourier space. Dividing both sides of equation 4 by $F_0(q)$, we obtain $C(q)$, which can be presented in real space using the inverse FT:

$$C(H) = \int_{-\infty}^{\infty} e^{2\pi iqH} \left\{ \frac{F(q)}{F_0(q)} \right\} dq \tag{5}$$

The general problem with deconvolution is the division by zero or by values near zero. The simplest solution is the limitation of the division by some window function, $Y(q)$. In this case, the basic equation for the deconvolution based on FT can be presented as shown in equation 6.

$$C(H) = \int_{-\infty}^{\infty} e^{2\pi iqH} \left\{ \frac{F(q)}{F_0(q)} Y(q) \right\} dq \tag{6}$$

The most widely used windows are Wiener (Moscicki *et al.*, 1991), Hamming (Lucarini *et al.*, 1996a and 1996b; Lucarini and Pedulli, 1997) and rectangular (Ewert and Thiessenhusen, 1991) functions. The window function acts as a smoothing factor and leads to some loss of high frequency information in the distribution function.

Various optimization methods are viable alternatives to the deconvolution method. The process starts by assuming an initial distribution, which can be described by some parameters. In diffusion studies, for example, the initial distribution is calculated as a function of the diffusion coefficient, diffusion time, and other parameters that define the sample configuration (Pilar *et al.*, 1999). Alternatively, the distribution function, *C(H)*, obtained from equation 6 can be used as a starting function (Lucarini and Pedulli, 1996a). Optimization methods use the convolution of this initial distribution function with the experimental spectrum in the absence of magnetic field gradient in order to calculate the spectrum in the presence of the gradient. The deviation between the calculated and the experimental spectra is then minimized by an optimization procedure.

Another approach for the determination of the distribution function from certain prior knowledge is to follow the principles of the maximum probability method (Jansson, 1984). A useful formulation is to maximize the so-called entropy functional, equation 7 (Smirnov *et al.*, 1991; Smirnov *et al.*, 1999). The attractive advantage of the maximum entropy method is the smoothness of the calculated distribution function.

$$\int C(H)\ln\{C(H)\}dH \rightarrow \max \tag{7}$$

Each method for the determination of spatial information has its advantages and disadvantages, and can be chosen in different experimental situations. We will illustrate the use of these various methods for a specific case in Section 4.

It is essential to note that the convolution expressed in equation 2 is correct only if the ESR line shape *has no spatial dependence.* As will be seen below, this requirement has dictated the conditions for data acquisition in the 1D ESRI study of degradation processes described in Section 4.

2.4 Line Shape Profiling from 2D Spectral-Spatial ESRI

This imaging method was essential for determining the spatial variation of line shapes of radicals formed during polymer degradation, *vide infra.* Data acquisition and image reconstruction have been recently described in detail (Schlick *et al.* 1998); therefore only a general outline is presented here.

Each image was reconstructed from a complete set of projections collected as a function of the magnetic field gradient, using a convoluted back-projection algorithm (Kweon, 1993). The number of points for each projection (1024) was kept constant. The maximum experimentally accessible projection angle, α_{max}, depends on the maximum gradient G_{max} according to $\tan\alpha_{max}=(L/\Delta H)\ G_{max}$, where L is the sample length, and ΔH is the spectral width. The maximum sweep width is $SW_{max}=\sqrt{2}\Delta H/\cos\alpha_{max}$. For a width $\Delta H\approx 65\ G$ (which is typical for the slow-motional spectral component of a nitroxide radical presnet in irradiated polymers), a sample length of 4 mm, and a maximum field gradient of 250 G/cm along the vertical axis, $\alpha_{max}= 57^{\circ}$ and $SW_{max} = 169\ G$. A complete set of data for one image consisted of 64 projections, taken for gradients corresponding to equally spaced increments of α in the range -90° to +90°; of these projections, typically 41 or 43 were experimentally accessible projections and the rest were projections at missing angles (for α in the intervals 60° to 90°, and -60° to -90^0). The projections at the missing angles were assumed to be the same as those at the maximum experimentally accessible angle α_{max}. Recently, a second stage was added, and the projections at the missing angles were finalized by the projection slice algorithm (PSA) with 2 iterations (Motyakin and Schlick, 2002a and 2002b).

The first-derivative ESR spectra taken in the presence of gradients were numerically integrated and multiplied by the square of the sweep width in order to obtain a constant integrated intensity, as required by the image reconstruction algorithm; the 1024 points for each spectrum were compressed by averaging to 256 points. The reconstruction algorithm produced a spectral-spatial image on of a 256x256 grid. For display, the number of points for each coordinate can be chosen (64, 128 or 256 points); in most cases we displayed the 2D images with 16 or 128 points for the spatial coordinate, and 128 or 256 for the spectral coordinate.

3. DIFFUSION IN SELF-ASSEMBLED POLYMERIC SURFACTANTS BY 1D ESRI

3.1 Polymeric Surfactants

The triblock copolymers poly(ethylene oxide)-*b*-poly(propylene oxide)-*b*-poly(ethylene oxide), $EO_mPO_nEO_m$, (commercial name Pluronics) have been extensively studied in the last few years, because of their interesting aggregate structures in aqueous solutions, and their numerous applications in drug release systems, detergents, cosmetics, treatment of burns, and water purification (Alexandridis, 1997; Chu and Zhou, 1996; Alexandridis and Lindman, Eds., 1997; Batrakova *et al*, 1999). The self-assembling is due to the different hydrophobicity of the two blocks, PEO and PPO, in water as solvent. Detailed phase diagrams of several Pluronics have become available (Alexandridis *et al.*, 1996; Svensson *et al.*, 1998), and progress has been achieved in the general understanding of the process of self-assembly (Zhou and Chu, 1998).

We have initiated a study of the Pluronics using ESR spectroscopy of nitroxide spin probes as a source of structural and dynamical data; ESR imaging for measurements of the translational diffusion coefficients, *D*, of the probes; and rheological measurements for an evaluation of the macroscopic properties (Zhou and Schlick, 2000; Malka and Schlick, 1997; Caragheorgheopol *et al.*, 1997; Caragheorgheopol and Schlick, 1998). The ESR measurements were based on nitroxide spin probes varying in size, charge, hydrophilicity, and position of the nitroxide group relative to the probe head group. Most of our studies have focused on $EO_{13}PO_{30}EO_{13}$ (Pluronic L64) and $EO_6PO_{34}EO_6$ (Pluronic L62), whose phase diagrams in aqueous solutions have been deduced by ^2H NMR, polarizing microscopy, and ocular inspection (Alexandridis *et al.*, 1996): In the vicinity of 300 K, the main phases detected in L64 with increasing polymer content are L_1 (micellar), H (hexagonal), L_α (lamellar) and L_2 (reverse micellar).

The nitroxide spin probe ESR method offers a rich source of information on the dynamics, polarity, and self-assembly in these polymeric surfactants. Important parameters are the isotropic nitrogen hyperfine splitting, a_N, which is polarity sensitive, the line shapes (which reflect the dynamics), and the correlation times τ_c deduced directly from the spectra or from simulations. By this approach we have determined the local hydration and the hydration gradient in the nonpolar (PO) and polar (EO) domains; followed changes in the hydration of the EO blocks at various distances from the hydrophobic core in L62 and L64 aggregates on a scale of $\leq$20 Å; and explained the mechanism of phase transitions.

The specific sites selected by the nitroxide probes in the self assembled system are also reflected in the rate of transport of the probes: in the diffusion coefficients, *D*. The importance of diffusion data for solvent molecules in self-assembled surfactants in binary systems (water as solvent), or in ternary systems (water and "oil") has been demonstrated (Lindman *et al.*, 1989; Lindman *et al.*, 1995). Pulsed-field gradient spin-echo nuclear magnetic resonance (PFGSE NMR) has emerged as the method of choice because of its ability to evaluate the diffusion coefficients of both solvents in a ternary mixture containing the surfactant. As shown below, we have demonstrated that by ESRI it is possible to measure diffusion coefficients, D, of guests present in low concentrations, and located in specific sites (Degtyarev and Schlick, 1999; Malka and Schlick, 1997).

3.2 Measurement of Diffusion Coefficients by 1D ESRI

The spatial distribution of paramagnetic centers, $C(x)$, is encoded in the ESR spectra via the magnetic field gradient G. The time evolution of the distribution function is a result of the translational diffusion of paramagnetic centers. From the distribution function, $C(x,t)$, of paramagnetic centers at time t in the diffusion process, D_x can be determined. As the concentration of the paramagnetic centers is low, the translational diffusion is expected to obey Fick's Law, equation 8 (Crank, 1993):

$$\frac{\partial C(x,t)}{\partial t} = D_x \frac{\partial^2 C(x,t)}{\partial x^2} \tag{8}$$

The solution of the diffusion equation depends on the initial distribution of paramagnetic centers. In the diffusion experiments presented here the solution was obtained by assuming an initial thickness h for the diffusant and an infinitely long layer of the polymer solution along x. The boundary conditions at $t=0$ are: $C=C_0$ for $x\leq h$, and $C=0$ for $x>h$. The solution is given by equation 9, and the error function $erf(x)$ by equation 10.

$$C(x,t) = \frac{1}{2} C_0 \left\{ erf \frac{h+x}{2\sqrt{D_x t}} + erf \frac{h-x}{2\sqrt{D_x t}} \right\} \tag{9}$$

$$erf(x) \equiv \frac{2}{\sqrt{\pi}} \int_0^x \exp(-\xi^2) d\xi \tag{10}$$

The distribution function $C(x,t)$ at time t in the diffusion process was calculated from equation 9 with D_x and h as parameters, convoluted according to equation 2 with the ESR line shape $F_0(H)$, and compared with the experimental line shape, $F(H)$; the best visual fit was used to extract the diffusion coefficient, D_x. No correction for the sensitivity profile of the ESR cavity was needed, because the initial diffusant layer was in the center of the microwave cavity, and only the initial stages of the diffusion process were used to deduce the diffusion coefficients.

3.3 Experimental Details

The diffusion coefficients, *D,* were measured by 1D ESRI at 300 K in the various phases of L64 for a hydrophilic probe, 4-(N,N,N-trimethyl) ammonium-2,2,6,6-tetramethyl-piperidine-1-oxyl iodide, (CAT1), which resides in the water-rich domains, and for the hydrophobic probe 5DSE, the methyl ester of doxylstearic acid where 5 indicates the carbon atom to which the doxyl group is attached. Data for these two probes were compared with results obtained by 2D spectral-spatial ESRI for the spin probe perdeuterio-2,2',6,6'-tetramethyl-piperidone-N-oxide, PDTEMPONE or PDT. The probes are shown below. The range of *D* values that were measured in this study was $1.0\times10^{-5}cm^2s^{-1}$ - $2.0x10^{-8}cm^2s^{-1}$.

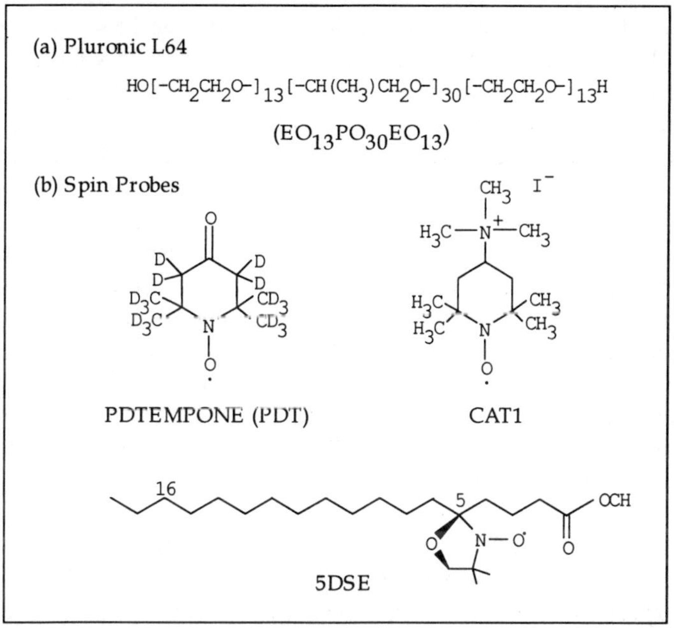

(a) Pluronic L64

$HO[-CH_2CH_2O-]_{13}[-CH(CH_3)CH_2O-]_{30}[-CH_2CH_2O-]_{13}H$

$(EO_{13}PO_{30}EO_{13})$

(b) Spin Probes

PDTEMPONE (PDT)

CAT1

5DSE

The spin probes CAT1 and 5DSE were dissolved in methanol to a concentration of $\approx 2x10^{-2}$ mol/L and then divided into several vials. After evaporation of the solvent in air, a corresponding amount of L64 solution was added to yield a final spin probe concentration of $\approx 2x10^{-3}$ mol/L. The solution containing the spin probe was transferred with a syringe to the bottom of capillary tubes (1.5 mm o.d., $\approx$1 mm i.d.) to a height of <1mm;

then the polymer solution was added very carefully so as not to disturb the probe layer, to a height of ≈20-30 mm in order to apply the infinite model of diffusion, equation 9. The capillary was sealed with teflon tape that also held a thread for sample positioning, and the sample was inserted into an ESR quartz sample tube (4 mm o.d.) and placed in the ESR cavity with the boundary between the two layers in the center of the cavity, within ±1 mm.

3.4 1D ESR Images

X-band ESR spectra of the probes at 300 K were measured in L64 aqueous solutions as a function of the polymer content. All spectra were in the motional narrowing regime and consisted of three narrow signals. The ESR spectra of CAT1 exhibited well-resolved proton hyperfine splittings in all solutions, except that containing 90 % w/w polymer. The spectral resolution increased in the presence of the polymer; this behavior is expected when the spin-rotation relaxation mechanism, which is dominant in the aqueous solutions of the probe, is progressively suppressed at higher viscosities (Caragheorgheopol and Schlick, 1998). The ESR line widths of 5DSE in L64 solutions were larger than for CAT1, and only minor changes were detected for polymer concentrations in the range 20-100 % w/w.

The a_N value is considered an indicator of the local polarity, and therefore of probe location (Malka and Schlick, 1997; Caragheorgheopol and Schlick, 1998). For CAT1 the a_N value is similar to that in pure water, 16.7 G, in solutions containing less than 40% w/w L64, and decreases to 15.9 G in solutions containing 90 % L64. This variation can be rationalized by assuming that the probe is located in the water trapped between the polymer chains in the hydrated EO regions of the aggregates. The spin probe 5DSE is not soluble in neat water and is solubilized only because of its intercalation in the polymer aggregates (Szajdzinska-Pietek *et al.*, 1994; Szajdzinska-Pietek and Schlick, 1996); the a_N values for 5DSE vary between 14.0 and 13.4 G, indicating a hydrophobic site (Griffith and Jost, 1976). Our analysis has indicated that PDT is located at the EO/water interface; in L64 the a_N values of PDT at 300 K vary between 16.0 G in neat water to 14.7 in neat L64. The three probes CAT1, 5DSE, and PDT reflected therefore different regions of the self-assembled system.

ESR spectra in the absence and presence of magnetic field gradients are presented for CAT1 and 5DSE in Figure 1A and 1B, respectively. For CAT1 the two lower 1D ESRI spectra in Figure 1A were recorded with a magnetic field gradient of 68 G/cm, after 6 and 14 min of diffusion, respectively. For 5DSE the gradient was 45 G/cm, and the diffusion times were 54 and 452 min, respectively.

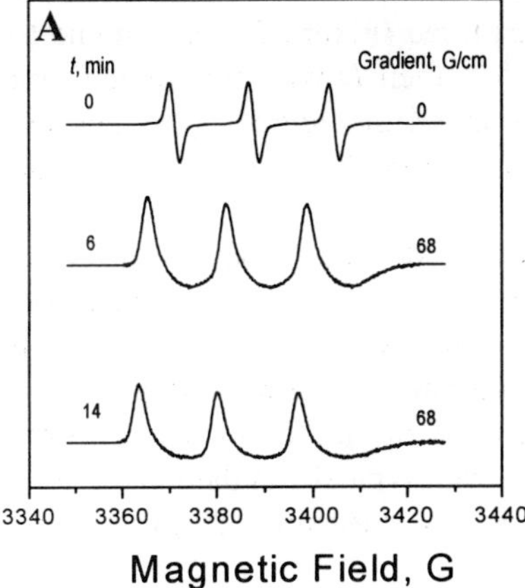

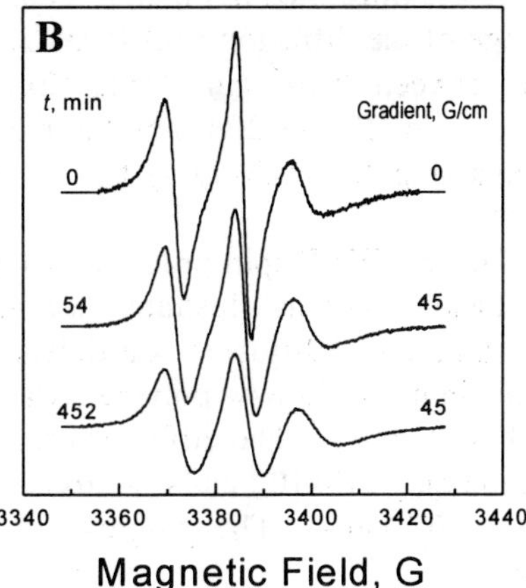

Figure 1. (A) X-Band ESR spectra of CAT1 at 300 K in aqueous L64 (polymer concentration 20 % w/w) in the absence of gradient (top), and for a gradient of 68 G/cm after 6 and 14 min of diffusion (middle and bottom). (B) X-Band ESR spectra of 5DSE at 300 K in aqueous L64 (polymer concentration 80% w/w) in the absence of gradient (top), and for a gradient of 45 G/cm after 54 and 452 min of diffusion (middle and bottom).

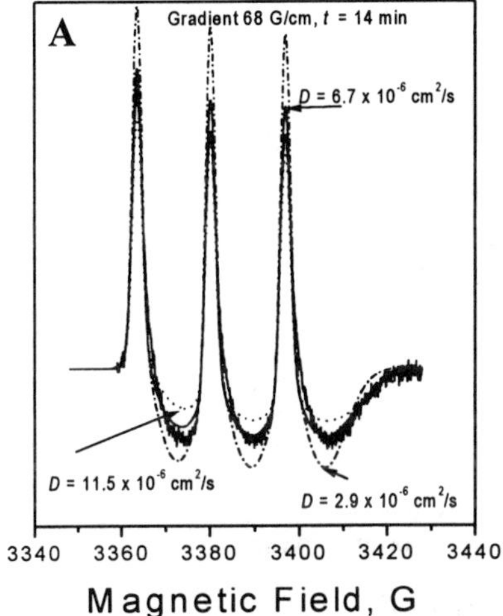

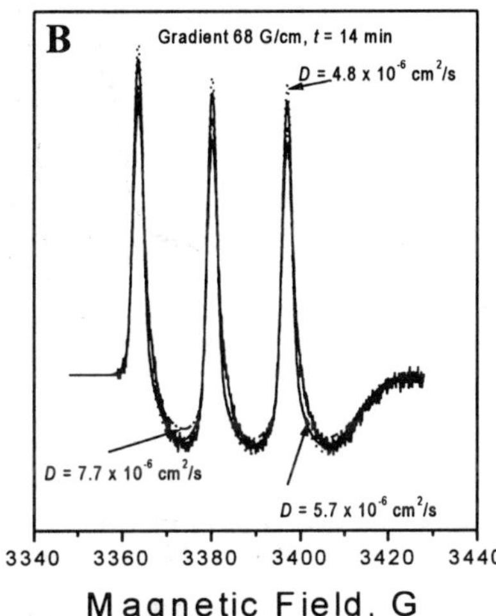

Figure 2. Simulation of the lower ESR spectrum shown in Figure 1A for CAT1 in aqueous L64 (polymer concentration 20 % w/w) by assuming the indicated values of the diffusion coefficient, *D*: (**A**), to establish the range of the *D* values. (**B**), for *D* values in the vicinity of the best fit.

The simulation of the ESR line shapes in the presence of gradient for CAT1 is illustrated in Figures 2A and 2B. For a diffusion time of 14 min, the range of the diffusion coefficient D was explored (Figure 2A) and found to be between $3x10^{-6}$ cm^2s^{-1} and $12x10^{-6}$ cm^2s^{-1}; then D was determined in a finer grid (Figure 2B). For the results presented in Figure 2 (the solution containing 20 % w/w L64), the D value chosen from the best fit was $6.3x10^{-6}$ cm^2s^{-1}.

For the 5DSE spin probe it was important to deduce D values for longer diffusion times; as illustrated in Figure 3, where we present the range of the D values. The diffusion coefficient for the data given in Figure 3 (solution containing 80 % w/w polymer) was then determined in a narrow grid to be $0.13x10^{-6}$ cm^2s^{-1}. The diffusion coefficients of 5DSE in the L64 solutions are in the range $10x10^{-8}$ cm^2s^{-1} - $20x10^{-8}$ cm^2s^{-1}, with typical standard deviations of $3x10^{-8}$ cm^2s^{-1}. The progress of diffusion was measured during several hours.

The variation of the diffusion coefficients for CAT1, 5DSE, and PDT are plotted as a function of the polymer concentration in Figure 4.

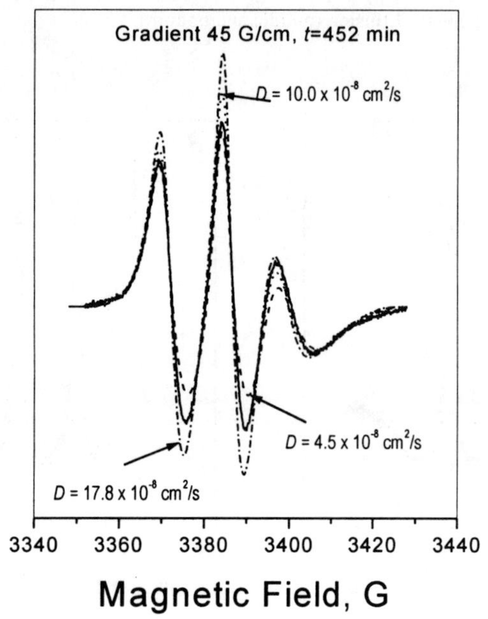

Figure 3. Simulation of the lower ESR spectrum shown in Figure 2B for 5DSE in aqueous L64 (polymer concentration 80% w/w) for the indicated D values, after a diffusion time of 452 min.

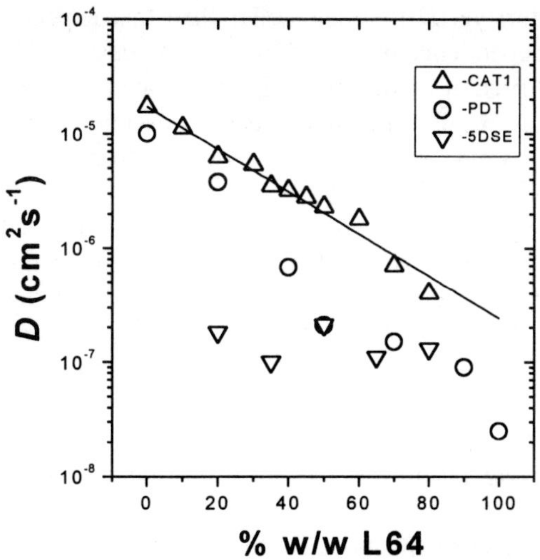

Figure 4. Variation of the diffusion coefficient, *D*, with L64 concentration for the spin probes CAT1, PDT, and 5DSE. T=300 K.

3.5 Effect of Probe Location on Transport Properties

The spin probes CAT1 and 5DSE offer a contrasting transport behavior in aqueous solutions of L64. Although the molecular masses *M* are similar, 340 for CAT1 (213 for the cation) and 414 for 5DSE, the corresponding *D* values at each polymer content are very different. The ratio D_{CAT1}/D_{5DSE} is 35 in the micellar phase (polymer contents 20 and 35 % w/w), 11 in the hexagonal phase (polymer content 50 % w/w), and 3.1 in the mixed L_2+L_α phase (polymer content 80 % w/w). Moreover, the D_{5DSE} values are similar to the diffusion coefficients of the polymer chains, D_{L64}, measured by PFGSE NMR (Zhang *et al.*, 1995). For instance in the L64 solution containing 20 % w/w polymer (L_1 phase), D_{5DSE}=0.18x10^{-6} cm^{2}s^{-1}, and $D_{L64}\approx$0.25x10^{-6}cm^{2}s^{-1}; in the polymer solution containing 35 % w/w L64, D_{5DSE}=0.10x10^{-6} cm^{2}s^{-1}, and $D_{L64}\approx$0.05x10^{-6}cm^{2}s^{-1}. The similar ranges of D_{5DSE} and D_{L64}, together with the very low a_N values for 5DSE, clearly indicated that the probe is located in the polymer aggregates and diffuses slowly, together with the polymer chains.

The D_{CAT1} values show a marked decrease with increasing polymer content, and are in the range 17.3x10^{-6} cm^{2}s^{-1} (in neat water) to 0.4x10^{-6} cm^{2}s^{-1} (solution containing 80 % w/w polymer). In neat water D_{CAT1} is close to the self-diffusion of water molecules, 25×10^{-6} cm^2s^{-1} at 300 K (Yasunaga and Ando, 1993). The molecular weight of the probe cation is more than 10 times that of water; yet D is much larger than that expected for either an $M^{1/3}$ or an $M^{1/2}$ dependence predicted by some models (de Gennes, 1979). It seems that the diffusion of the probe is accelerated by that of the solvent (water) (Schlick *et al.*, 1998b); this effect is possible because the probe is located in the water regions, as deduced from the ESR spectra (Caragheorgheopol and Schlick, 1998).

The dependence of D_{CAT1} on polymer content follows the expression $D=D_0\exp(-aw_2)$, where D_0 is the diffusion coefficient of the probe in neat water, a is a constant, and w_2 is the weight fraction of the polymer. The straight line in Figure 5 shows the excellent agreement between this expression and the experimental results for CAT1. This behavior is consistent with the expression suggested by Phillies for diffusion of probes (Phillies, 1989): $D=D_0\exp(-\alpha c^{\nu})$, where c is the polymer concentration, and α and ν are constants. Data for CAT1 therefore suggest a behavior typical of a solvent.

From these considerations it is clear that the major factor that controls the transport rate of the guests is their location. ESRI is the method of choice for the determination of the diffusion coefficients for guests present in low concentrations and located in various regions of a self-assembled systems.

3.6 Obstruction Factor

Because CAT1 is located in the solvent domains, it is logical to assume that the self-assembled polymer aggregates decrease the rate of diffusion of the probe, compared to the value in the neat solvent. In this context it is interesting to apply the concept of "obstruction", which is expressed as the "obstruction factor" or "effective diffusion coefficient", $D_{eff}=D/D_0$. The dependence of D/D_0 has been calculated as a function of φ_2, the volume fraction of the obstructing particles, for particles of different geometries: spheres, long prolates, and oblates with different axial ratios. For non-interacting particles, the limiting value of D/D_0 is 2/3 in large oblates (Jonsson *et al.*, 1986). For the diffusion of D_2O in isotropic solutions of the nonionic surfactants $C_{12}EO_4$ and $C_{12}EO_3$, the concept of obstruction was considered as a way to distinguish between the various shapes of the aggregates (Nilsson and Lindman, 1984): D_{eff} values close to unity were taken as an indication of spherical aggregates, and D_{eff} in the range 0.6-0.7 as evidence for the presence of large oblate particles. For the diffusion of water in the presence of poly(methylmethacrylate) (PMMA) latex spheres, the D_{eff} values as a function of the obstructing volume fraction, φ_2, in the range 0 to 0.5 is slightly lower than, but close to, the calculated values.

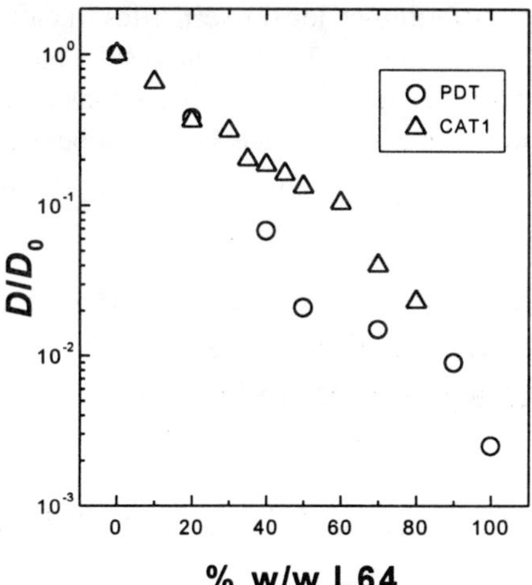

Figure 5. Variation of the obstruction factor D/D_0 at 300 K for CAT1 and PDT as a function of the L64 concentration. D_0 is the diffusion coefficient of each probe in neat water.

The obstruction factors D/D_0 for CAT1 and PDT are plotted in Figure 5 as a function of L64 content. (This type of data is not available for 5DSE, because the probe is not soluble in neat water.) For both probes the ratio D/D_0 decreased significantly even at low polymer concentrations, and was 0.4 in the isotropic micellar phase (L_1, 20 % w/w polymer). In the hexagonal and lamellar phases the D/D_0 factors are different for the two probes, and much lower, in the range 0.1-0.01; the lowest value, 0.01, was measured for PDT in the isotropic reverse micellar phase, L_2. A reasonable explanation for these very low D/D_0 values is the interaction between the probe and the solvent and/or polymer. The presence of water inside the aggregates, and the location of the probes in water that interacts with the aggregates ("bound water"), can lead to lower rates of transport. These results emphasize the complexity of the systems consisting of EO and PO blocks, and the different behavior compared to the systems that contain strictly hydrophobic obstructing particles such as PMMA, or to self-assembled systems such as surfactants of the type C_mEO_n, where the hydrophobic and hydrophilic parts of the molecule are more clearly defined than in the Pluronics.

The results presented in this section suggest that the transport properties of guests in complex fluids are intimately related to the *local* sites selected by the guests, and to the interactions on the molecular level in the system. These considerations are important when designing drug release systems, and also when considering the specific domain of a biological system where the drugs are selectively located. An interesting case was described in a recent study of the partitioning of local anesthetics in cationic surfactants as model membranes (Matsuki *et al.*, 1998). The results have indicated that the probe site (in water or in the micellar phases) depends on the hydrophobicity of the drug, and a good correlation was established between the partitioning into the hydrophobic environment and the anesthetic potency of the drugs.

4. SPECTRAL PROFILING OF DEGRADATION AND STABILIZATION PROCESSES IN POLYMERS BY 1D AND 2D ESRI

4.1 Degradation and Stabilization of Polymers

Polymeric materials exposed to heat, mechanical stress, and ionizing or UV irradiation undergo degradation in the presence of oxygen due to the formation of reactive intermediates such as free radicals R·, RO· and ROO·, and hydroperoxides ROOH (O'Donnell, 1989). The degradation process can be accelerated by chromophores, free radicals, and metallic residues from the polymerization reactions. Often the deleterious effects are not immediately detected, but develop over longer periods. The gradual changes in the polymer properties observed in many polymers and the ultimately grave results are due to trapped radicals that react slowly, to peroxy radicals that decompose in time with formation of reactive radicals and gas molecules, and to trapped gases that lead to local stresses and to cracking. While the time scale of these changes may vary, the final results are dramatic: degradation of the structure and collapse of the mechanical properties. The accelerated rate of ozone depletion in the stratosphere due to environmental factors is expected to raise the level of UVB radiation, ≈290-330 nm, (McKenzie *et al.*, 1999), thus adding severity to the problem of degradation and urgency to the need for solutions.

Hindered amine stabilizers (HAS) rank among the most important recent developments for light and heat stabilization of polymeric materials (Pospisil, 1995). Nitroxides and hydroxylamine ethers are the major products of reactions involving HAS. The HAS-derived nitroxides are thermally stable, but may react with free radicals (as "scavengers") to yield diamagnetic species; the hydroxylamine ethers can regenerate the original nitroxide, thus resulting in an efficient protective effect. The pivotal role of nitroxides (added as such, or derived from HAS) has been documented recently in low-molecular mass model systems (Brede *et al.*, 1998). In polymers the situation is more complicated, and the equilibrium concentration of the nitroxides varies markedly with the nature of the polymer matrix, thermal history, humidity, and type of HAS (Gerlock and Bauer, 1984; Gerlock *et al.*, 1986). At the present time, the complex protective mechanism offered by HAS is known only in broad terms (Pospisil, 1995). The presence of the nitroxide radicals is an advantage in ESR because of the radical stability and the well-established methods for interpretation and simulation of their spectra. These advantages are valid also for ESRI studies; the additional challenge in the ESRI measurements is to translate spectroscopic and imaging data into the chemistry, kinetics, and mechanism of degradation.

We present the application of 1D and 2D ESRI in HAS-containing ABS polymers exposed to UV radiation in the presence of oxygen. The goal is to deduce the intensity and line shape profiles of HAS-derived nitroxides: to achieve *spectral profiling*. The behavior of the HAS-derived nitroxides was compared in UV-irradiated and in thermally-treated samples.

(a) Repeat Units in ABS Polymers

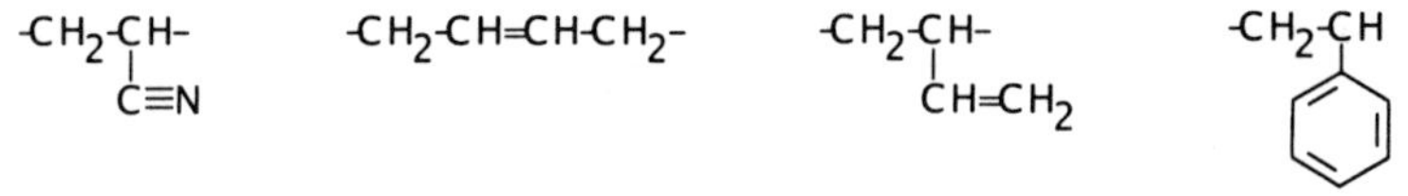

$-CH_2-CH(C{\equiv}N)-$ $-CH_2-CH{=}CH-CH_2-$ $-CH_2-CH(CH{=}CH_2)-$ $-CH_2-CH(C_6H_5)$

(b) Hindered Amine Stabilizer (HAS): Tinuvin 770

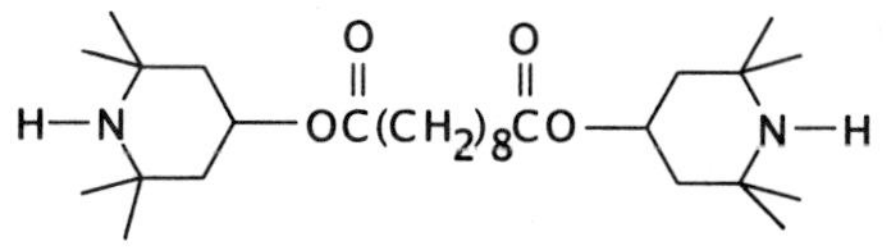

$H-N \quad -OC(=O)(CH_2)_8C(=O)O- \quad N-H$

4.2 Accelerated Polymer Degradation

Accelerated degradation experiments were performed by exposure to UV irradiation of ABS (Magnum 342 EZ, from Dow Chemical Company) doped with 2 % w/w (bis(2,2,6,6-tetramethyl-4-piperidinyl) sebacate), the HAS known as Tinuvin 770 (Ciba Specialty Chemicals Corporation). The polymer and the HAS were shaped into 10cmx10cmx0.4cm plaques in a injection molding machine at 210°C. The plaques were exposed on one side to UV radiation in an Atlas Ci35 Weather-ometer® that allowed simultaneous control of the temperature and humidity. The spectrum of the Xe irradiation source in the chamber, from 295 nm into the IR region, is similar to the solar spectrum, and the irradiation intensity, 0.45 W/m^2, is comparable with the average summer noontime solar intensity in Miami, FL. The plaques were irradiated continuously. For the ESR imaging experiments we cut from the plaques cylindrical samples ≈4 mm long and 4 mm in diameter, with the cylinder axis along the direction of the UV radiation. The samples were placed in the ESR resonator with the symmetry axis along the field gradient and the irradiated part on top (Motyakin *et al.*, 1999 and 2000; Kruczala *et al.*, 2000).

4.3 Comparison of Methods for Simulation of 1D ESR Images

As mentioned in Section 2, we have simulated the 1D ESRI images using various methods, and deduced the corresponding intensity profiles. A comparison is shown in Figure 6. The ESR spectrum in the inset was obtained after UV irradiation of ABS in the weathering chamber for 403 h.

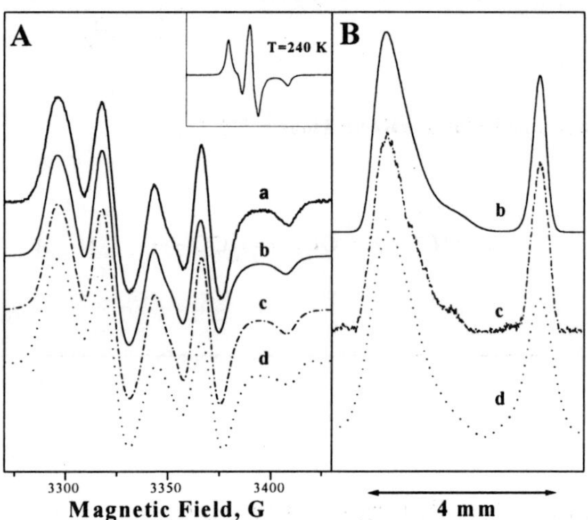

Figure 6. Comparison of methods for image deconvolution. (**A**) The experimental ESR spectrum in the presence of a gradient of 152 G/cm is **a**. The calculated images are as follows: **b** by maximum entropy method, **c** by the optimization method, and **d** by FT method. (**B**) **b**, **c**, and **d** are the corresponding intensity profiles. The inset is the experimental spectrum in the absence of magnetic field gradient at 240 K obtained by UV irradiation of ABS in the weathering chamber for 403 h.

As seen in the **d** images and profiles, the main disadvantage of Fourier Transform method is the loss of high frequency information in the distribution function due to the use of the (Hamming) window function. For this reason the intensity profile is broadened and the edges are not well reproduced. The two main features in the profiles are significantly sharper when the optimization of the fit is performed by the Monte Carlo procedure, as seen in the calculated spectrum and profile **c** in Figure 6, but the noise level is higher. The profile obtained by the maximum entropy method, **b**, is very smooth and the edges are sharp.

The main goal of optimization methods is to find the best fit between the experimental and the calculated spectra. If this criterion is followed, the optimization method is the best (Lucarini *et al.*, 1996a). In terms of profiles, however, the best result is obtained by the maximum entropy method.

From this comparison it appears that the optimization and maximum entropy have advantages, and the specific choice depends on the intensity and line shapes.

It is important to note that the simulation of 1D ESR images is in reasonable agreement with experimental spectra measured at 300 K in irradiated ABS, where the line shapes of the nitroxides have a strong spatial variation. The simulations were considerably improved, however, when the 1D images were measured at 240 K, a condition that eliminated the line shape variation. We caution therefore that in all studies by 1D ESRI it is essential to check first for spatial line shape variations, by 2D ESRI or by sample sectioning, for instance; and to perform 1D measurements only after the spatial variation of the line shapes has been avoided.

4.4 ESR Spectra of HAS-Derived Nitroxides in ABS

In Figure 7 we present X-band ESR spectra at 300 K of ABS containing Tinuvin 770 for the indicated UV irradiation times in the weathering chamber. All spectra, except that corresponding to the longest irradiation time, consist of a superposition of two components, from nitroxide radicals differing in their mobility: a "fast" component (F) responsible for the 3-line spectrum with a total width of 32.2 G (upward arrow in second from the top spectrum), and a "slow" component (S) with a spectral width of 64.2 G (downward arrows). This result indicates the presence of the HAS-derived nitroxide radicals in two different environments. It is reasonable to assume that the fast and slow components reflect nitroxides located respectively in low-T_g domains dominated by polybutadiene sequences ($T_g \approx 200$ K), and in high-T_g domains dominated by polystyrene ($T_g \approx 370$ K) or polyacrylonitrile sequences ($T_g \approx 360$ K). It is clear from the spectra shown in Figure 7A that the relative intensity of the "fast" component decreases with increasing irradiation time, and is negligible for an irradiation time of 2325 h.

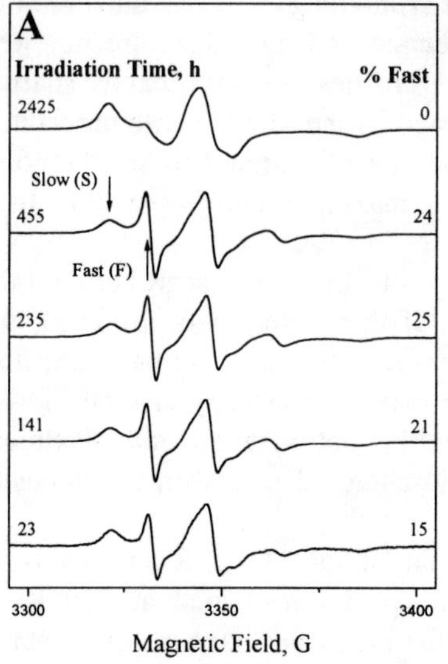

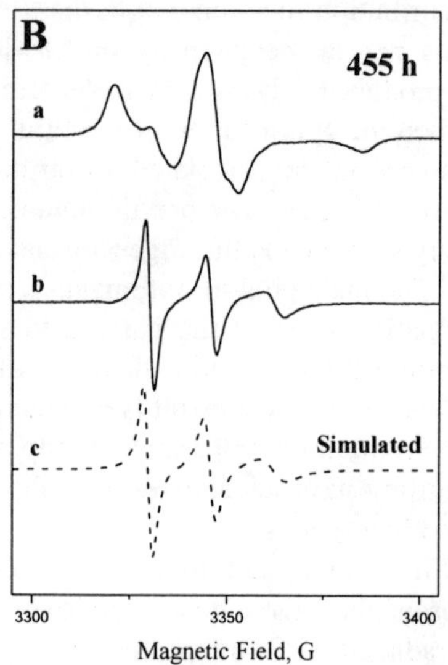

Figure 7. (**A**) X-Band ESR spectra at 300 K of ABS containing HAS for the indicated irradiation times. Upward and downward arrows point respectively to the extreme low-field features of the "fast" and "slow" spectral components. The percentage of the "fast" component, %F, in each case is also indicated (see text). (**B**) Spectrum **a**: X-band ESR at 300 K of the irradiated ABS layer (thickness 0.8 mm) after UV (Xe source) irradiation during 455 h. Spectrum **b**: ESR spectrum of the "fast" component obtained by deconvolution of the composite spectrum shown for irradiation time 455 h. Spectrum **c**: Calculated F spectrum (dotted line), based on the following magnetic parameters: g_{xx}=2.0088, g_{yy}=2.0061, g_{zz}=2.0027, A_{xx}=6.3 G, A_{yy}=5.8 G, A_{zz}=33.6 G, $R_{\parallel}=6.3\times10^{8}$ rad/s, $R_{\perp}=2.5\times10^{7}$ rad/s, and $\theta=90^{0}$; the line widths were $\Delta H_{\parallel}$=0.84 G, $\Delta H_{\perp}$=1.24 G, and the line shape were a mixture of Gaussian and Lorentzian. $R_{\parallel}$ and $R_{\perp}$ are the parallel and perpendicular components, respectively, of the rotational diffusion tensor; θ is the tilt angle, between the symmetry axis of the rotational diffusion tensor and the *z* axis of the nitroxide.

In Figure 7B we present the ESR spectrum at 300 K (spectrum **a**) of the directly irradiated layer of thickness 0.8 mm, for ABS containing HAS (irradiation time 455 h); the spectrum consists of the S component only. A similar result was obtained for the directly irradiated layers in the other samples. By subtracting the S component from the composite spectra given in Figure 7A, we deduced the line shape of the F component, as shown in Figure 7B (spectrum **b**). After performing a large number of such "spectral titrations", we became convinced that the line shapes of the F and S components do not vary with irradiation time. The line widths of the three signals for the F component (spectrum **b** in Figure 7B) increase from low to high magnetic field. This line shape is typical of a dynamical mechanism that involves an anisotropic rotation about the N-O bond of the nitroxide (Griffith and Jost, 1976), but can also be explained by a distribution of *g*- and ^{14}N hyperfine tensor components (Schlick *et al.*, 1989). We propose that the former motional mechanism is reasonable for a large molecule such as Tinuvin 770. In Figure 7B we also present a simulated F spectrum (dotted line, spectrum **c**), calculated with the magnetic parameters given in the caption and using the nonlinear-least-square analysis method of simulation (Budil *et al.*, 1996); the calculated spectrum is in excellent agreement with the experimental spectrum **b**. In the simulation we used an extreme separation $2A_{zz}$ value that is the same, within experimental error, as that measured at 120 K (68.0±0.4 G); and an ^{14}N isotropic hyperfine splitting, a_N, identical to that measured at 400 K (30.5±0.1 G). Simulation of the F and S components (spectra **a** and **b** in Figure 7B) indicated that the corresponding rotational correlation times, τ_c, are $\approx 4 \times 10^{-9}$ s/rad and $\approx 5 \times 10^{-8}$ s/rad, respectively. We note that only the simulation of the F component is presented.

Based on the separate spectra for the S and F components shown in Figure 7B, we reproduced all the experimental spectra shown in Figure 7A and, by double integration, deduced the ratio of the two components in the composite spectra. The relative intensity of the F component, %F, as a function of irradiation time increases to a maximum, decreases, and becomes negligible at the longest irradiation time. We propose that the decrease of the relative intensity of F with irradiation time is due to the consumption of the HAS-derived nitroxide radicals located in the butadiene-rich domains of the polymer; this scenario is plausible, as the butadiene component is expected to be more vulnerable to degradation, compared to the other repeat units in ABS (Jouan and Gardette, 1992; Carter and McCallum, 1994). The same logic can explain the absence of the F component near the irradiated side of the sample, where the high UV intensity produces more damage to the polymer.

The temperature variation of the ESR spectra in the range 120-420 K indicates that the transition temperature for the "slow" component (T_{50G}) is 410±3 K. The fast component appears as a 3-line spectrum at ≈280 K; the splitting changes from 34.3 G at 280 K to 30.5 G at 400 K. The transition temperature for the F component occurs at a higher temperature compared to T_g of pure polybutadiene; this result can be explained by the fact that the probe site is not pure polybutadiene, but a mixture that contains the other components also (Daniels *et al.*, 1990). Such situations were described in our previous papers (Pilar *et al.*, 1993; Muller *et al.*, 1994; Abetz *et al.*, 1995).

4.5 Intensity Profiling from 1D ESRI

The concentration profiles were obtained by optimization (Lucarini *et al.*, 1996a; Marek, 2002), or by the maximum entropy (Smirnov *et al.*, 1991 and 1999) methods.

The presence of two spectral components in the (bulk) irradiated sample and the absence of the "fast" component near the irradiated side suggested a spatial variation of the line shapes, which presents a problem in 1D ESRI experiments, when applying the convolution described in equation 2. The problem was solved when we analyzed the temperature dependence of the composite spectra, such as those shown in Figure 7A: In all composite spectra the F component was detected only at and above 260 K. Below 260 K both spectral components, F and S, are in the slow motional regime and have identical line shapes. In order to deduce the concentration profiles, all ESR spectra and the corresponding ESR images were therefore measured at 240 K, and the spatial variation of the line shape was thus avoided. To the best of our knowledge this is the first time that the spatial line shape variation was detected and dictated the temperature used for image collection in 1D ESRI experiments.

In Figure 8A we present the ESR spectrum of ABS containing Tinuvin 770 and the corresponding ESR image obtained in the presence of a field gradient of 101 G/cm for an irradiation time of 455 h, both measured at 240 K. In Figure 8B we present the concentration profiles obtained by deconvolution for HAS-derived nitroxides in ABS at the indicated UV irradiation times.

The larger nitroxide concentration near the outer planes of the sample and the gradual increase of the nitroxide concentration at the non-irradiated side clearly indicated the combined effects of oxygen and UV radiation, and showed the regions where the chemistry takes place. We note that the same results were obtained when the back of the samples was covered with aluminum foil. The higher nitroxide concentration at the back of the sample (not directly irradiated) at irradiation times above 455 h can be explained by a smaller loss of nitroxide radicals due to formation of >NO·*, which transforms into diamagnetic species such as >NOH (Keana *et al.*, 1971).

The results presented in Figures 7 and 8 led to several important conclusions. *First*, we demonstrated that both the concentration and the ESR line shapes of nitroxides in UV-irradiated ABS polymers vary along the irradiation direction. *Second*, the concentration of the HAS-derived nitroxide is low in the interior of the sample but high in a region of ≈1 mm on the irradiated sides. *Third*, the polybutadiene-rich domains in ABS polymers are most vulnerable to UV irradiation; the "fast" nitroxide component near these domains is consumed and decreases in intensity with irradiation time, and its concentration is lower where the UV irradiation is more intense. The protective mechanism of HAS appears to be specific to the system, and sensitive to the morphology and thermal transitions (for instance the glass temperature T_g) of the polymer. Additional experiments on the kinetic processes responsible for the observed concentration profiles, and on the degradation mechanism of HAS-containing ABS polymers are carried out in our laboratories.

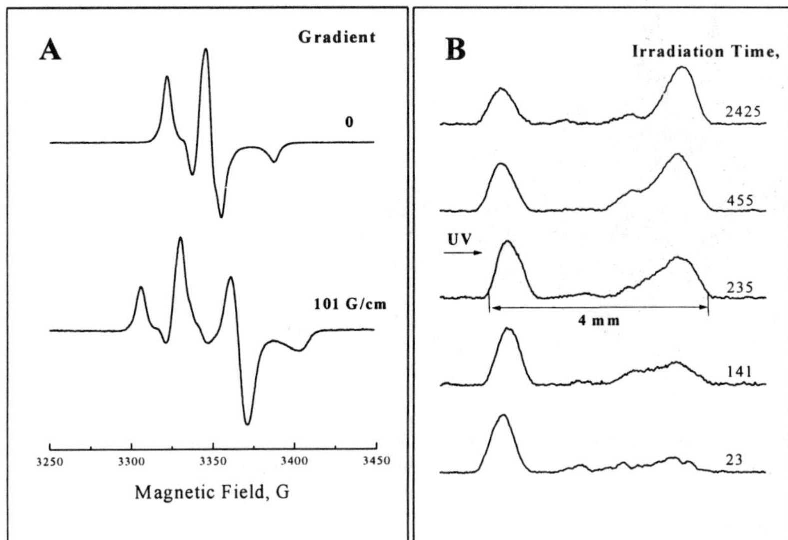

Figure 8. (**A**) X-band ESR spectrum at 240 K of ABS after UV irradiation in the weathering chamber for 455 h, and the corresponding ESR image measured at 240 K with a magnetic field gradient of 101 G/cm. (**B**) Intensity profiles deduced by the optimization method of ABS containing HAS for the indicated UV irradiation time in the weathering chamber. An arrow indicates the irradiated side of the sample.

4.6 Line Shape Profiling from 2D Spectral-Spatial ESRI

In order to shed light on the spatial variation of the two spectral components, we performed 2D ESRI on the UV-irradiated polymer, and compared the results with those of the thermally degraded ABS polymer (Kruczala *et al*, 2000). In Figures 9 and 10 we present 2D spectral-spatial perspective and contour images of nitroxide radicals in ABS UV-irradiated for 70 h and 934 h, respectively, in the weathering chamber at 65^{0}C. The ESR intensity is presented in absorption mode. To the right of Figures 9 and 10 we also present spectral slices (in the derivative mode) at the indicated depth of the sample.

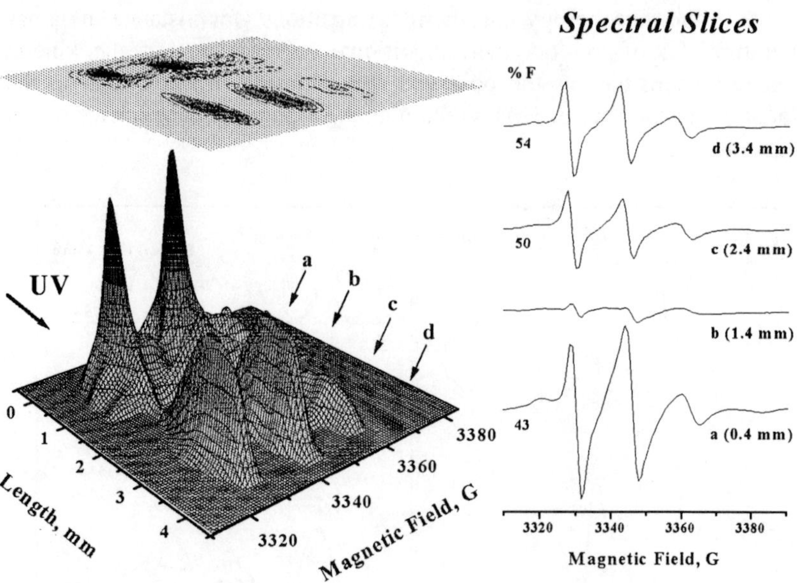

Figure 9. 2D spectral-spatial contour (top) and perspective (bottom) plots of HAS-derived nitroxides in ABS after 70 h UV irradiation in the weathering chamber, presented in absorption. The spectral slices **a**, **b**, **c** and **d** for the indicated depths are presented in the derivative mode.

The contour plots showed very clearly the distribution of the signal intensity, and the negligible signal intensity in the sample interior, as also seen in the concentration profiles deduced from 1D ESRI. The spectral slices indicated not only the line shape variation but also the relative intensity of each signal as a function of depth. For the short irradiation time, the directly irradiated part of the sample shows a composite spectrum with %F≈43%; on the non-irradiated side the relative intensity of the F component is larger, ≈50%, at 2.4 mm from the irradiated edge, and ≈54% at the non-irradiated edge. After 934 h of irradiation, the irradiated side contained no fast component, and %F decreased to ≈14% and ≈18%, respectively, near and at the non-irradiated side. We note that section "**b**" in Figure 10 represents a very weak signal and the line shapes are distorted; for this reason %F was not calculated for this section.

The major conclusion from the 2D images presented in Figures 9 and 10 is that the nitroxide in the butadiene-rich domains is consumed rapidly on the irradiated side, and its intensity is negligible after 934 h of irradiation.

On the non-irradiated side the degradation is much slower, but even on this side a decrease of %F was detected, from 50% to 14% at 2.4 mm from the irradiated side, and from 54% to and 18% at the irradiated side, when the irradiation time increased from 70 h to 934 h.

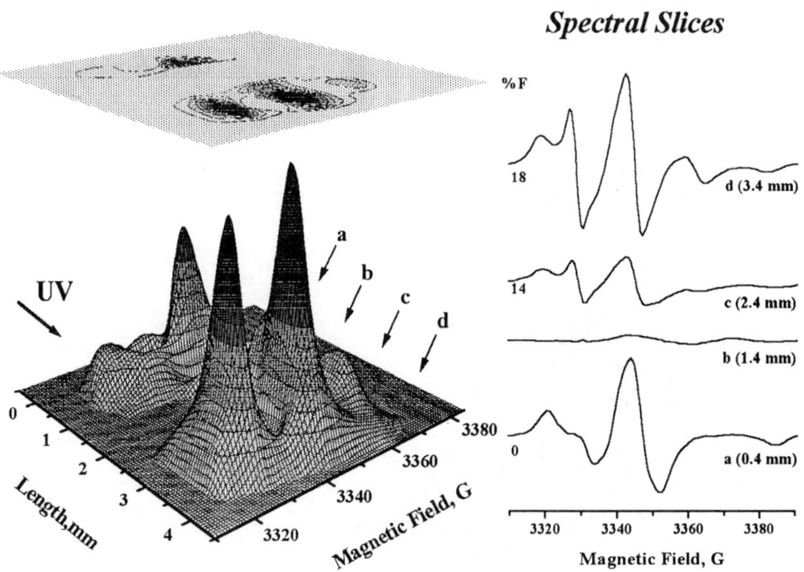

Figure 10. 2D spectral-spatial contour (top) and perspective (bottom) plots of HAS-derived nitroxides in ABS after 934 h UV irradiation in the weathering chamber, presented in absorption. The spectral slices a, b, c and d for the indicated depths are presented in the derivative mode.

In Figure 11 we present the 2D spectral-spatial perspective and contour images, and the spectral slices, of nitroxides in ABS thermally degraded at 60^{0}C during 796 h. In contrast to the UV-irradiated samples, the nitroxide distribution is essentially homogeneous along the sample depth; moreover, no spatial variation of the line shapes was detected, and on the average %F=27±3%. This F content deduced by 2D ESRI is in agreement with results of sectioning the sample, and weighing and determining the %F in each slice; the average F content in the cut sample is 26%. An almost homogeneous intensity profile was detected recently for thermally-produced radicals in polyimide resins (Ahn *et al.*, 1997).

Some of the features of thermal degradation are visually clearer in the first derivative mode presentation of the 2D images, Figure 12, where the F and S components are indicated by arrows. The intensity of the nitroxide signal is significantly lower in the thermally degraded polymer compared to the UV-irradiated samples, by a factor of ≈15. The rate of oxygen transport is commensurate with the rate of thermal degradation, and the process in not diffusion-limited as in the case of the UV-irradiation. The 2D spectral-spatial ESR images reflect the degradation mechanism and the difference between UV- and thermal degradation of the ABS polymer.

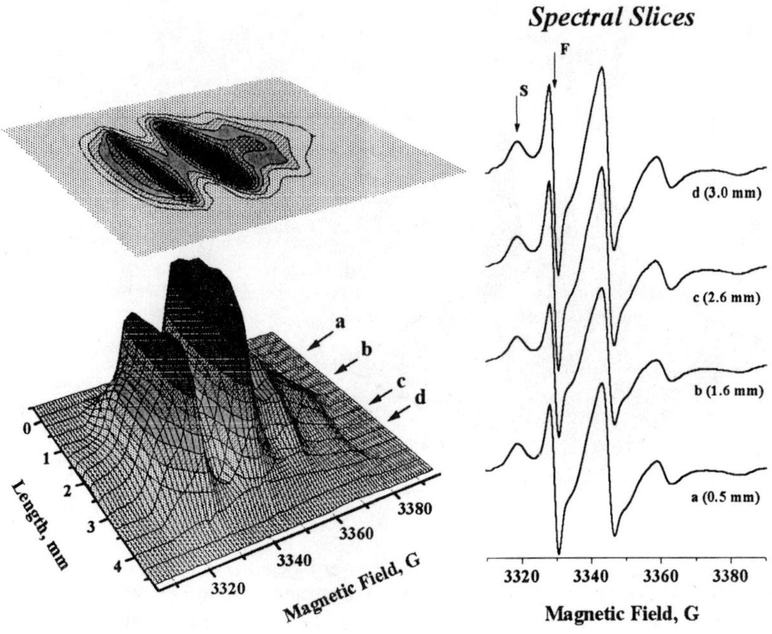

Figure 11. 2D spectral-spatial contour (top) and perspective (bottom) plots HAS-derived nitroxides in ABS after 796 h thermal degradation at 60^0 C, presented in absorption. The spectral slices **a**, **b**, **c** and **d** for the indicated depths are presented in the derivative mode; these slices were obtained by averaging signals from sections within ± 0.2 mm of the position indicated by the arrow in the perspective plot.

Methods for measuring the spatial distribution of polymer properties due to degradation have been developed by other groups: Density profiling measures the change in density, which is expected to increase in aged samples, along the irradiation depth (Gillen *et al.*, 1986); and modulus profiling measures (via the tensile compliance) the tensile modulus, which decreases during degradation (Gillen *et al.*, 1987; Gillen *et al.*, 1996). The spatial variation of these properties is an excellent indicator of degradation. The *nondestructive spectral profiling* made possible by ESRI, which was demonstrated in this study, is expected to be exceptionally sensitive to early events in the degradation process, and to be therefore of predictive value, and a dependable indicator of things to come.

Additional, novel aspects of ESRI as applied to polymer degradation have emerged from our most recent studies: 1. The spatial distribution of nitroxide radicals and of the corresponding line shapes depend on the wavelength of the UV radiation to which the polymer was exposed (Motyakin and Schlick, 2001). 2. Spatial heterogeneity of the radical distribution was detected during thermal degradation of ABS at higher temperatures, 80-120^{0}C. (Motyakin and Schlick, 2002a and 2002b). 3. Finally, simulation of the 1D image by analytical functions has allowed the detection of "skin effects" in the radical profiles, which hint at the importance of the molding process for sample preparation on the sample morphology and spatial effects (Motyakin and Schlick, 2002b).

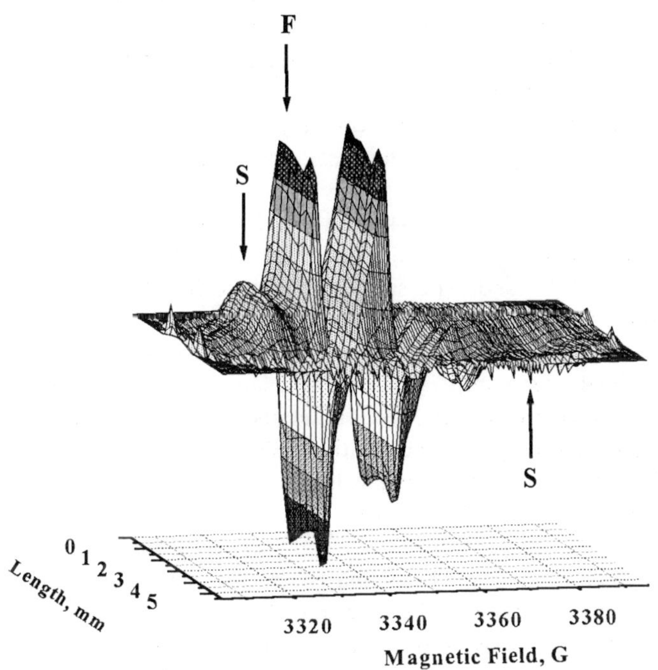

Figure 12. 2D spectral-spatial ESRI perspective plot of HAS-derived nitroxides in ABS after 309 h thermal degradation at 60^{0}C, presented in the first derivative. Note the homogeneous distribution of the radicals. Signals for the S and F components are indicated by arrows.

5. CONCLUDING REMARKS

Important recent contributions together with the work presented in this chapter have demonstrated that ESRI has moved beyond phantoms, and has become a method for obtaining specific quantitative information, some of which cannot be obtained by other methods. Progress has been made in the deconvolution software, in the software for image reconstruction in 2D and 3D ESRI, and in the type and intensity of the gradient fields.

But challenges remain. In the diffusion measurements, while in principle only two spectra are needed for each D value, and these can be collected in <10 min, sample preparation is not easy. In our laboratory numerous samples, typically 10, had to be prepared in order to extract one value of the diffusion coefficient. These problems are due to the difficulty to layer the probe solution and the host solution or gel without mixing; and the formation of a meniscus when preparing samples for diffusion in thin capillaries (~1 mm i.d.). Such problems are more serious in the case of aqueous systems. Efforts are currently in progress in our laboratory to correct for capillarity effects by modification of the optimization software for deconvolution of 1D images (Marek, 2002). A more radical solution is to perform the diffusion measurements at L-band. Recent ESRI experiments at low frequency, as well as the examination of the frequency dependence of the sensitivity (Eaton *et al.*, 1998), signal to an increased optimism for diffusion imaging at low ESR frequency.

The type of work described in Section 4, application of ESRI to the problem of polymer degradation, is of great potential and is well suited for study by ESRI at X-band. The sample is easy to handle, has no dielectric losses, and the radicals studied have adequate stability at and above ambient temperature. Moreover, this type of study is expected to be of interest not only to the ESR community, but also to a large number of scientists and engineers involved in prediction of materials lifetimes and challenged by serious problems of polymer deterioration (Clough *et al.*, 1996).

In his introduction to the recent volume on magnetic resonance imaging Lauterbur wrote that "NMR, with or without imaging, is a powerful but fragile technique" (Lauterbur, 1998). This "fragile" technique has, however, become a major tool in medicine chemistry and other disciplines. Practitioners of ESRI can only be encouraged by this development.

The advantage of ESR is its higher sensitivity, compared to NMR. A major problem in ESRI is the need for paramagnetic species. It is the authors' opinion that two types of problems are favored as candidates for study by ESRI: (a) Problems that are difficult, or impossible, to solve by other methods. An example is the diffusion of specifically located probes that are present in very low concentrations; such an approach was illustrated in Section 3. (b) Problems in which the paramagnetic species are part of the system. Examples are catalytic systems that contain paramagnetic transition metal ions, or radicals produced by irradiation. The HAS-derived nitroxides in ABS polymers, as described in Section 4, perform a double role: they probe the morphology of the system, in terms of glass transition characteristics and dynamics; and also reflect on the degradation process. The remaining important challenge is to relate the information extracted from the ESRI experiments to the complex degradation kinetics and mechanism.

6. ACKNOWLEDGEMENTS

ESR imaging studies in our laboratory were generously supported by the Polymers Program of the National Science Foundation; additional support came from a University Research Grant from Ford Motor Company, and the American Association of University Women (AAUW). We are grateful to John L. Gerlock (Ford Research Laboratory) for sharing with us his considerable insight in polymer degradation processes, and M. Lucarini (University of Bologna, Italy), Alex Smirnov (University of Illinois at Urbana-Champaign) and Grzegorz Mazur (Jagiellonian University, Cracow) for their help and advice with the simulation software. Additional thanks are due to Alex Smirnov for making available to us the software package from Scientific Software Services (Bloomington, IL).

7. REFERENCES

Abetz, V., Muller, G., Stadler, R., and Schlick, S., 1995, The glass transition of mixtures of polystyrene with alkyl-terminated oligostyrene. Experimental evidence for microphase separation in a polymer blend, *Macromol. Chem. Phys.* **196**: 3845-3855.

Ahn, M.K., Eaton, S.S., Eaton, G.R., and Meador, M.A.B., 1997, Electron paramagnetic resonance imaging of the spatial distribution of free radicals in PMR-15 polyimide resins, *Macromolecues* **30**: 8318-8321.

Alexandridis, P., 1997, Poly(ethylene oxide)/poly(propylene oxide) block copolymer surfactants, *Curr. Opin. Colloid Interface Sci.* **2**: 478-489.

Alexandridis, P., and Lindman, B., (eds.), 1997, *Amphiphilic block copolymers: Self-assembly and applications*, Elsevier, Amsterdam.

Alexandridis, P., Zhou, D., and Khan, A., 1996, Lyotropic liquid crystallinity in amphiphilic block copolymers: temperature effects on phase behavior and structure of poly(ethylene oxide)-b-poly(propylene oxide)-b-poly(ethylene oxide) copolymers of different compositions, *Langmuir* **12**: 2690-2700.

Batrakova, E., Lee, S., Li, S., Venne, A., Alakhov, V., and Kabanov, A., 1999, Fundamental relationships between the composition of Pluronic block copolymers and their hypersensitization effect in MDR cancer cells, *Pharm. Research* **16**: 1373-1379.

Berliner, L. J., and Fujii, H., 1985, Magnetic resonance imaging of biological specimens by electron paramagnetic resonance of nitroxide spin labels, *Science* **227**: 517-519.

Blümler, P., Blümich, B., Botto, R., and Fukushima E., (eds.), 1998, *Spatially resolved magnetic resonance: methods, materials, medicine, biology, rheology, ecology, hardware*, Wiley-VCH, Weinheim.

Brede, O., Beckert, D., Windolph, C., and Gottinger, H.A., 1998, One electron oxidation of sterically hindered amines to nitroxyl radicals: intermediate amine radical cations, aminyl, α-amino-alkyl, and aminylperoxyl radicals, *J. Phys. Chem. A* **102:** 1457-1464. This paper provides a detailed summary of the current literature on the protective mechanism of HAS.

Budil, D.E., Lee, S., Saxena, S., and Freed, J.H., 1996, Nonlinear-least-squares analysis of slow-motion EPR spectra in one and two dimensions using a modified Levenberg-Marquardt algorithm, *J. Magn. Reson. A* **120**: 155-189.

Caragheorgheopol, A., Pilar, J., and Schlick, S., 1997, Hydration and dynamics in reverse micelles of the triblock copolymer $EO_{13}PO_{30}EO_{13}$ in water/*o*-xylene mixtures: a spin probe study, *Macromolecule* **30**: 2923-2933.

Caragheorgheopol, A., and Schlick, S., 1998, Hydration in the various phases of the triblock copolymers $EO_{13}PO_{30}EO_{13}$ (Pluronic L64) and $EO_6PO_{34}EO_6$ (Pluronic L62), based on electron spin resonance spectra of cationic spin probes, *Macromolecule* **31**: 7736-7745.

Carter, R.O.III, and McCallum, J.B., 1994, Photoacoustic infrared spectroscopy of an ABS as an automotive interior material, *Polym. Degrad. Stab.* **45**: 1-10.

Chu, B., and Zhou, Z., 1996, Ch. 3, Physical chemistry of polyoxyalkylene block copolymer surfactants, in *Nonionic surfactants* (Nace, V.P., ed.), pp. 67-143, Marcel Dekker, New York.

Clough, R.G., Billingham, N.C., and Gillen, K.T., (eds.), 1996, *Polymer durability: degradation, stabilization and lifetime prediction*, Adv. Chem. Series 249, American Chemical Society, Washington, D.C.

Crank, J., 1993, *The mathematics of diffusion*, Clarendon Press, Oxford, U.K.

Daniels, E.S., Dimonie, V.L., El-Aasser, M.S., and Vanderhoff, J.W., 1990, Preparation of ABS latexes using hydroperoxide redox initiators, *J. Appl. Polym. Sci.* **41**: 2463-2477.

Degtyarev, E.N., and Schlick, S., 1999, Diffusion coefficients of small molecules in various phases of Pluronic L64 measured by one-dimensional electron spin resonance imaging (1D ESRI), *Langmuir* **15**: 5040-5047.

Eaton, G.R., Eaton, S.S., and Ohno, K., (eds.), 1991, *EPR Imaging and in vivo EPR*, CRC Press, Boca Raton, FL.

Eaton, G.R., Eaton, S.S., and Rinard, G.A., 1998, Ch. 4, Frequency dependence of EPR sensitivity, in *Spatially resolved magnetic resonance: methods, materials, medicine, biology, rheology, ecology, hardware* (Blümler, P., Blümich, B., Botto, R., and Fukushima, E., eds.), pp. 65-74, Wiley-VCH, Weinheim.

Ewert, U., and Thiessenhusen, K.-U., 1991, Ch. 11, Deconvolution for the stationary-gradient method, in *EPR imaging and in vivo EPR* (Eaton, G.R., Eaton, S.S, and Ohno, K., eds.), pp. 119-126, CRC Press, Boca Raton, FL.

Freed, J. H., 1994, Field gradient ESR and molecular diffusion in model membranes, *Annu. Rev. Biophys. Biomol. Struct.* **23**: 1-25.

Gao, Z., Pilar, J., and Schlick, S., 1996, ESR Imaging of tracer translational diffusion in polystyrene solutions and swollen networks, *J. Phys. Chem.* **100**: 8430-8435.

Gao, Z., and Schlick, S., 1996, Binding and transport of Mo^V in perfluorinated ionomers swollen by ethanol using ESR and ESR imaging, *J. Chem. Soc. Faraday Trans.* **92**: 4239-4254.

de Gennes, P. G., 1979, *Scaling concepts in polymer physics*, Cornell University Press, Ithaca, NY.

Gerlock, J.L., and Bauer, D.R., 1984, ESR measurements of free radical photoinitiation rates by nitroxide termination, *J. Polym. Sci. Polym. Letters Ed.* **22**: 447-455.

Gerlock, J.L., Bauer, D.R., and Briggs, L.M., 1986, Photo-stabilisation and photo-degradation in organic coatings containing a hindered amine light stabiliser, *Polym. Degrad. Stab.* **14**: 53-71.

Gillen, K.T., Clough, R.L., and Dhooge, N.J., 1986, Density profiling of polymers, *Polymer* **27**: 225-232.

Gillen, K.T., Clough, R.L., and Quintana, C.A., 1987, Modulus profiling of polymers, *Polym. Degrad. Stab.* 17: 31-47.

Gillen, K.T., Clough, R.L., and Wise, J., 1996, Ch. 34, Prediction of elastomer lifetimes from accelerated thermal-aging experiments, in *Polymer durability: degradation, stabilization and lifetime prediction* (Clough, R.G., Billingham, N.C., and Gillen K.T., eds.), pp. 557-575, Adv. Chem. Series 249, American Chemical Society, Washington, D.C.

Griffith, O.H., and Jost, P.C., 1976, Ch.12, Lipid spin labels in boplogical membranes, in *Spin labeling: theory and applications* (Berliner, L.J., ed.), pp. 453-523, Academic Press, New York. The lowest a_N value for a doxyl stearic acid methyl ester in the table on p 501 is 14 G, in mineral oil as solvent. We have measured at 300 K an a_N value of 13.9 G for 10DSE in the lamellar phase of L64 (see Zhou, L., 1997, *M. Sc. Thesis*, University of Detroit Mercy).

Halpern, H.J., and Bowman, M.K., 1991, Ch. 6, Low-frequency EPR spectrometers: MHz range, in *EPR imaging and in vivo EPR* (Eaton, G.R., Eaton, S.S, and Ohno, K., eds.), pp. 45-63, CRC Press, Boca Raton, FL.

Halpern, H.J., Yu, C., Peric, M., Barth, E., Grdina, D.J., and Teicher, B., 1994, Oxymetry deep in tissues with low-frequency electron paramagnetic resonance, *Proc. Natl. Acad. Sci. USA*, **91**. 13047-13051.

Halpern, H.J., Chandramouli, G.V.R., Barth, E.D., Williams, B.B., and Galtsev, V.E., 1999, Approaches to problems in high resolution in vivo spectral spatial imaging with radiofrequency EPRI, *Curr. Top. Biophys.* **23**: 5-10.

Jansson, P.A., 1984, *Deconvolution with applications in spectroscopy*, Academic Press, Orlando, USA.

Jouan, X., and Gardette, J.L., 1992, Photo-oxidation of ABS: Part 2 – Origin of the photodiscoloration on irradiation at long wavelengths, *Polym. Degrad. Stab.* **36**: 91-96.

Jonsson, B., Wennerström, H., Nilsson, P.G., and Linse, P., 1986, Self-diffusion of small molecules in colloidal systems, *Colloid & Polymer Sci.* **264**: 77-88.

Keana, J.F.W., Dinerstein, R.J., and Baitis, F., Photolytic studies on 4-hydroxy-2,2,6,6-tetramethylpiperidine-1-oxyl, a stable nitroxide free radical, *J. Org. Chem.* **36**: 209-211.

Kevles B.H., 1997, Ch. 8, Naked to the bone: Medical imaging in the twentieth century, pp. 173-200, Rutgers University Press: New Brunswick, NJ.

Kruczala, K., Gao, Z., and Schlick, S., 1996, 2D spectral-spatial ESR imaging of diffusion based on Mo(V), *J. Phys. Chem.* **100**: 11427-11431.

Kruczala, K., Motyakin, M.V., and Schlick, S., 2000, 1D and 2D spectral-spatial electron spin resonance imaging (ESRI) of degradation and stabilization processes in poly(acrylonitrile-butadiene-styrene) (ABS), *J. Phys. Chem. B*, 104: 3387-3392.

Kruczala, K., Varghese, B., Bokria, J.G., and Schlick, S., 2003, Thermal aging of heterophasic propylene- ethylene copolymers: Morphological aspects based on ESR, FTIR, and DSC, *Macromolecules*, **36**: 1899-1908.

Kruczala, K., Bokria, J.G., and Schlick, S., 2003, Thermal aging of heterophasic propylene-ethylene copolymers: Spatial and temporal aspects or degradation based on ESR, ESR Imaging, and FTIR, *Macromolecules*, **36**: 1909-1919.

Kweon, S.-C., 1993, Development of hardware and software for image reconstruction in an ESR imaging system, *M. Sc. Thesis*, University of Detroit Mercy.

Lauterbur, P., 1973, Image formation by induced local interactions: Examples employing nuclear magnetic resonance, *Nature* **242**: 190-191.

Lauterbur, P., 1998, Foreword, in *Spatially resolved magnetic resonance: methods, materials, medicine, biology, rheology, ecology, hardware* (Blümler, P., Blümich, B., Botto, R. and Fukushima, E., eds.), Wiley-VCH, Weinheim.

Lindman, B., Shinoda, K., Olson, U., Anderson, D., Karlström, G., and Wennerström, H., 1989, On the demonstration of bicontinuous structures in microemulsions, *Colloids and Surfaces* **38**: 205-224.

Lindman, B., Olson, U., and Söderman, O., 1995, Ch. 8, in *Dynamics of solutions and fluid mixtures by NMR* (Delpuech, J.J., ed.), pp. 345-395, Wiley, New York.

Lucarini, M., Pedulli, G.F., Borzatta, V., and Lelli, N., 1996a, X-Band EPR imaging of polymers irradiated with UV light, *Res. Chem. Intermed.* **22**: 581-591.

Lucarini, M., Pedulli, G.F., Borzatta, V., and Lelli, N., 1996b, The determination of nitroxide radical distributions in polymers by EPR imaging, *Polym. Degrad. Stab.* **53**: 9-17.

Lucarini, M., and Pedulli, G.F., 1997, Use of EPR to image nitroxyl radicals in HALS stabilized polymers, *Angew. Makromol. Chem.* **252**: 179-193.

Lucarini, M., Pedulli, G.F., Motyakin, M.V., and Schlick, S., 2003, Electron spin resonance imaging of polymer degradation and stabilization, *Progress Polym. Sci.*. **28**: 331-340.

Malka, K., Schlick, S., 1997, Hydration, dynamics, and transport across the phase diagram of aqueous $EO_{13}PO_{30}EO_{13}$ (Pluronic L64) by spin probe ESR and ESR imaging, *Macromolecules* **30**: 456-465.

Marek, A., Labsky, J., Konak, C., Pilar, J., and Schlick, S., 2002, Translational diffusion of paramagnetic tracers in HEMA gels and in concentrated solutions of polyHEMA by 1D electron spin resonance imaging, *Macromolecules* **35**:5517-5528.

Matsuki, H., Kaneshina, S., Kamaya, H., and Ueda, I., 1998, Partitioning of charged local anesthetics into model membranes formed by cationic surfactant: effect of hydrophobicity of local anesthetic molecules, *J. Phys. Chem. B* **102**: 3295-3304.

McKenzie, R., Connor, B., and Bodeker, G., 1999, Increased summertime UV radiation in New Zealand in response to ozone loss, *Science* **285**: 1709-1711.

Moscicki, J. K., Shin, Y. K., and Freed, J. H., 1991, Ch. 19, The method of dynamic imaging by EPR, in *EPR Imaging and in vivo EPR* (Eaton, G.R., Eaton, S. S., and Ohno, K., eds.), pp. 189-219, CRC Press, Inc., Boca Raton.

Motyakin, M.V., Gerlock, J.L., and Schlick, S., 2000, Electron spin resonance imaging (ESRI) of degradation and stabilization processes in polymers: Poly(acrylonitrile-butadiene-styrene) (ABS) polymers, in *Ageing Studies and Lifetime Extension of Materials* (Mallinson, L.G., ed.), pp.353-358, Kluwer/Plenum Publishing Corporation, New York.

Motyakin, M.V., Gerlock, J.L., and Schlick, S., 2001, Electron spin resonance imaging (ESRI) of degradation and stabilization processes: Behavior of a hindered amine stabilizer in UV-exposed poly(acrylonitrile-butadiene-styrene) (ABS) polymers, *Macromolecules* **32**: 5463-5467.

Motyakin, M.V., and Schlick, S., 2002a, Thermal degradation at 393 K of poly(acrylonitrile-butadiene-styrene) containing a hindered amine stabilizer: A study by 1D and 2D electron spin resonance imaging (ESRI) and ATR-FTIR, *Polym. Degrad. Stab.* **76**: 25-36.

Motyakin, M.V., and Schlick, S., 2002b, Electron spin resonance imaging (ESRI) and ATR-FTIR study of poly(acrylonitrile-butadiene-styrene) (ABS) containing a hindered amine stabilizer and thermally treated at 353 K, *Macromolecules* **35**: 3984-3992.

Muller, G., Stadler, R., and Schlick, S., 1994, An ESR study of the local structure and dynamics in poly(styrene-co-maleic anhydride)/poly(methyl vinyl ether) blends and semi-IPNs, *Macromolecules* **27**: 1555-1561.

Nilsson, P.-G., and Lindman, B., 1984, Nuclear magnetic resonance self-diffusion and proton relaxation studies of nonionic surfactant solutions. Aggregate shape in isotropic solution above the clouding temperature, *J. Phys. Chem.* **88**: 4764-4769.

O'Donnell, J.H., 1989, Ch. 1, Radiation chemistry of polymers, in *The effects of radiation on high-technology polymers* (Reichmanis, E., and O'Donnell, J.H., eds.), pp. 1-13, American Chemical Society, Washington, D.C.

Oikawa, K., Ogata, T., Togashi, H., Yokoyama, H., Ohya-Nishiguchi, H., and Kamada, H., 1996, A 3D and 4D ESR imaging system for small animals, *Appl. Radiat. Isot.* **47**: 1605-1609.

Phillies, G. D. J., 1989, The hydrodynamic scaling model for polymer self-diffusion, *J. Phys. Chem.* **93**: 5029-5039.

Pilar, J., Sikora, A., Labsky, J., and Schlick, S., 1993, Phase structure of poly(styrene-co-acrylic acid)/poly(ethylene oxide) blends by spin label EPR and DSC, *Macromolecules* **26**: 137-143.

Pilar, J., Labsky, J., Marek, A., Konak, C., Schlick, S., 1999, Translational diffusion of paramagnetic tracers in poly(1-vinylpyrrolidone) (PVP) hydrogel and concentrated aqueous solutions by 1D electron spin resonance imaging, *Macromolecules* **32**: 8230-8233.

Pospisil, J., 1995, Aromatic and heterocyclic amines in polymer stabilization, *Adv. Polym. Sci.* **124**: 87-189.

Schlick, S., Harvey, R.D., Alonso-Amigo, M.G., and Klempner, D., 1989, Study of phase separation in IPNs using nitroxide spin labels, *Macromolecules* **22**: 822-830.

Schlick, S., Pilar, J., Kweon, S.-C., Vacik, J., Gao, Z., and Labsky, J., 1995, Measurements of diffusion processes in HEMA-DEGMA hydrogels by ESR imaging, *Macromolecules* **28**: 5780-5788.

Schlick, S., Eagle, P., Kruczala, K., and Pilar, J., 1998a, Ch. 17, Electron spin resonance imaging (ESRI) of transport processes in polymeric systems, in *Spatially resolved magnetic resonance: methods, materials, medicine, biology, rheology, ecology, hardware* (Blümler, P., Blümich, B., Botto, R., and Fukushima, E., eds.), pp. 221-234, Wiley-VCH, Weinheim.

Schlick, S., Gao, Z., Matsukawa, S., Ando, I., Fead, E., and Rossi, G., 1998b, Swelling and transport in polyisoprene networks exposed to benzene-cyclohexane mixtures: a case study in multicomponent diffusion, *Macromolecules* **31**: 8124-8133. An interesting result of this study was that the diffusion coefficient of cyclohexane at 300 K increases from 1.43×10^{-5} cm^2s^{-1} (neat solvent) to 2.03×10^{-5} cm^2s^{-1} in a 1/1 v/v mixture containing benzene, but that of benzene, 2.32×10^{-5} cm^2s^{-1}, remains unchanged.

Schlick, S., Kruczala, K., and Bokria, J.G., 2003, ESR imaging of degradation in heterophase polymers: The case of propylene-ethylene copolymers (HPEC), *Polym. Mat. Sci. Eng. (Proc. ACS Div. PMSE)*, in press.

Smirnov, A.I., Yakimchenko, O.E., Golovina, H.A., Bekova, S.K., and Lebedev, Y.S., 1991, EPR imaging with natural spin probes, *J. Magn. Reson.* **91**: 386-391.

Smirnov, A.I., Belford, R.L., and Morse, R., 1999, Magnetic resonance imaging in a hands-on student experiment using an EPR spectrometer, *Concepts Magn. Resonance* **11**: 277-290.

Szajdzinska-Pietek, E., Schlick, S., and Plonka, A., 1994, Self-assembling of perfluorinated polymeric surfactants in water. ESR spectra of nitroxide spin probes in Nafion solutions, *Langmuir* **10**: 1101-1109.

Szajdzinska-Pietek, E., and Schlick, S., 1996, Ch. 7, ESR spectroscopy of perfluorinated ionomers as swollen membranes and solution, in *Ionomers: characterization, theory, and applications* (Schlick, S., ed.), pp. 135-163, CRC Press: Boca Raton, FL.

von Kienlin, M., and Pohmann, R., 1998, Ch. 1, Spatial resolution in spectroscopic imaging, in *Spatially resolved magnetic resonance: methods, materials, medicine, biology, rheology, ecology, hardware* (Blümler, P., Blümich, B., Botto, R., and Fukushima, E., eds.), pp. 3-20, Wiley-VCH, Weinheim.

Xu, D., Hall, E., Ober, C.K., Moscicki, J.K., and Freed, J.H., 1996, Translational diffusion in polydisperse polymer samples studied by dynamic imaging of diffusion ESR, *J. Phys. Chem.* **100**: 15856-15866.

Yakimchenko, O.E., Degtyarev, E.N., Parmon, V.N., and Lebedev, Ya. S., 1995, Diffusion in porous catalyst grains as studied by EPR imaging, *J. Phys. Chem.* **99**: 2038-2041.

Yasunaga, H., and Ando, I., 1993, Dynamic behavior of water in hydro-swollen crosslinked polymer gel as studied by PGSE ^{1}H NMR and pulsed ^{1}H NMR, *Polymer gels and networks* **1**: 83-92.

Zhang, K., Lindman, B., and Coppola, L., 1995, Melting of block copolymer self-assemblies induced by a hydrophilic surfactant, *Langmuir* **11**: 538-542. This study includes the diffusion coefficients of the surfactant (L64) in the isotropic phases only (L_1 and L_2). The *D* values we quoted in this Chapter were read from a plot of log*D* vs L64 content (Figure 4 in above paper) and are therefore only approximate values.

Zhou, S., and Chu, B., 1998, Water-induced micellar structure change in Pluronic P103/water/o-xylene ternary system, J. *Polym. Sci. Part B: Polym. Phys*. **36**: 889-900.

Zhou, L., and Schlick, S., 2000, Electron spin resonance (ESR) spectra of amphiphilic spin probes in the triblock copolymer $EO_{13}PO_{30}EO_{13}$ (Pluronic L64): hydration, dynamics, and order in the polymer aggregates, *Polymer* **41**: 4679-4689.

Zweier, J.L. and Kuppusamy, P., 1998, Ch. 34, EPR Imaging of the rat heart, in *Spatially resolved magnetic resonance: methods, materials, medicine, biology, rheology, ecology, hardware* (Blümler, P., Blümich, B., Botto, R., and Fukushima, E., eds.), pp. 373-388, Wiley-VCH, Weinheim.

Chapter 8

Peptide Aggregation and Conformation Properties as Studied by Pulsed Electron-Electron Double Resonance

Pulsed ELDOR in spin labeled peptides.

Yuri D.Tsvetkov
Institute of Chemical Kinetics and Combustion Russian Academy of Sciences, Siberian Branch, Novosibirsk, 630090 Russia E-mail: tsvetkov@ns.kinetics.nsc.ru

Abbreviations used: Aib, α-aminoisobutyric acid; ESE, electron spin echo; PELDOR, pulsed electron-electron double resonance; TEMPONE, 2,2,6,6-tetramethyl-1-piperidinyloxy-4-one; TFE, 2,2,2-trifluoroethanol.

Abstract: Pulsed ELDOR is described as a technique for structural studies of spin labeled peptides in frozen solutions and for determining crystallographic distances on the order of 10 Å between spins by measuring magnetic dipole interactions. Results from studies of double labeled trichogin GA IV analogues demonstrate conformational variations with solvent polarity, and aggregates of four peptides are detected in weakly polar (frozen) solvents. An aggregate model of four 3_{10}-helical trichogin molecules has been proposed.

1. INTRODUCTION

The structural properties of peptides provide very important information for understanding the biophysical behavior of complex peptide-membrane systems. Spin labeling of the peptide chain and analysis of molecular dynamics via electron spin resonance (ESR) spectroscopic methods have traditionally been used to provide useful information with regard to structural properties of proteins, both in solution and within biomembranes. Prior to 1990, electron-electron double resonance (ELDOR) and saturation transfer ESR methods were used with spin label probes to study the supramolecular interactions of proteins (*cf.* Berliner, 1976, 1979, 1989; Dalton, 1985). But since 1990, advances in time domain ESR have opened up new vistas for the study of proteins via spin probes. These include two-dimensional Fourier Transform ESR, which may be used to examine motional dynamics, and electron spin echo (ESE, also known as pulsed) ELDOR, which allow one to measure crystallographic distances between spins via magnetic dipolar interactions. This chapter addresses some of the experimental and theoretical problems associated with the application of pulsed ELDOR to the study of protein structure, as specifically applied to aggregation phenomena and resultant activity of certain peptides (*cf.* Arsher *et al.*, 1991).

One such family of membrane active peptides are the peptaibols, which exhibit antibiotic activity via the formation of ion channels and resultant disruption of membrane integrity. They are typically oligomers of less than 20 amino acid residues, and a high proportion of these are non-standard. Besides the occurrence of non-standard residues within the polypeptide chain, the N- and C-terminal amino acids of the polypeptides are also modified by an alkyl group and hydroxyl group, respectively. Trichogin GA IV is a member a peptaibols subfamily (*cf.* Chugh & Wallace, 2001) that isolated from the mold *Trichoderma longbrachiatum*. This 10-residue polypeptide contains α-aminoisobutyric acid (Aib) as the non-standard amino acid residue. It is terminated on the C-end by a 1,2-amino alcohol (leucinol) and on the N-end by an *n*-octanoyl group (Auvin-Guette, *et al.*, 1992).

Trichogin GA IV displays a number of unique characteristics. Surprisingly, despite of its short main-chain length, it exhibits membrane-modifying properties on liposomes that are comparable to those shown by the longer-chain peptaibols (Auvin-Guette, *et al.*, 1992). Conformational studies in organic solvents (Auvin-Guette, *et al.*, 1992; Toniolo & Benedetti, 1991) suggested that trichogin GA IV has a right-handed, mainly α-helical structure and that the peptide helix has an amphophilic character. Recently these results were confirmed using a spin labeling technique (Miick *et al.*, 1991; Friori *et al.*, 1993). So far, however, only conventional continuous wave ESR (cw-ESR) technique has been applied to the study of spin labeled peptaibols.

Following a precedent set by Hanson *et al.* (1992), this chapter reviews current results obtained via pulsed electron-electron double resonance (PELDOR), as applied to the problem of conformation and aggregation of spin-labeled trichogin peptides. This work (*cf.* Milov *et al.*, 1998; 1999; 2000a–d; 2001; Maryasov *et al.*, 1998; Maryasov & Tsvetkov, 2000) started in Novosibirsk in 1998 in collaboration with Professor C. Toniolo of the University of Padova and Dr. J. Raap of Leiden University.

2. THEORETICAL BASIS OF PELDOR SPECTROSCOPY

2.1 Magnetic Dipolar Interactions in Inhomogeneously Broadened Spectra

ESR is widely used for the investigations of structure and dynamics of paramagnetic centers that are associated with biologically important compounds and/or structures. These paramagnetic centers may be inherent to the structure, such as a metal ion or free radical intermediate associated with an enzyme, or an extrinsic spin probe whose location is designed (*cf.* Eaton & Eaton, 1989; Hubbell *et al.*, 2000). For the purposes of determining (supra)molecular dimensions in biological structures, crystallographic distances between centers may be measured indirectly via magnetic dipole-dipole interactions. Unfortunately, magnetic dipolar interactions are determined from cw-ESR spectral linewidths (*cf.* Bales, 1989), and therefore the limits to resolving line width will, in turn, limit the range of crystallographic distances that may be determined. The problem is that the magnitude of dipole-dipole (d-d) interactions, measured as a broadening of the ESR line width, is proportional to $1/R^3$, where R denotes the distance between unpaired electrons, and this proportionality factor becomes immeasurably small for the distances greater then 10 Å. For example, d-d broadening is approximately 0.05 mT at $R = 15$ Å, and the effect is therefore lost among other inhomogeneous sources of ESR line broadening (anisotropic g-factor, hyperfine interactions, *etc.*). As an alternative, the pulsed ESR technique, particularly methods based on Electron Spin Echo (ESE), is a superior method of measuring dipolar interactions (Salikov *et al.*, 1976, Salikov & Tsvetkov, 1979; Tsvetkov, 1985).

The single frequency, double pulse ESE method makes it possible to get information about weak d-d interactions by studying the dependence of the ESE signal amplitude on the time between microwave pulses, denoted as τ. This method was successfully applied in 1970s and 1980s for the determination of distances between paramagnetic centers and spatial distribution of radicals in irradiated materials and the other systems (Salikov & Tsvetkov, 1979; Tsvetkov, 1985), but the application of the ESE technique to a particular system is subject to certain constraints. The first of these constraints is spectrometer deadtime, which limits the minimum measurable distance between paramagnetic centers (or the maximum concentration thereof). Secondly, if a given paramagnetic center is surrounded by magnetic nuclei, then the anisotropic hyperfine interaction will modulate the ESE signal decay envelope (Dikanov & Tsvetkov, 1992) and be difficult to distinguish modulations from the d-d interaction. These difficulties may be overcome by introducing a pulse at a second frequency into the pulse scheme (Milov *et al.*, 1981).

An inhomogenously broadened electron magnetic resonance spectrum (Figure 1b) represents an envelope of overlapping spin packets, each of which corresponding to a resonant field/frequency combination or *g*-value. In a given experiment in which the spectrometer is operated at a single field/frequency combination (corresponding to some *g*-value within the spectral envelope), a spin packet is selected whose spins, collectively denoted as A, will nutate under the action of the microwave pulses the one field/frequency combination (designated ν_a). Neglecting spectral diffusion, all other unexcited spins, denoted as B preserve their orientation (Figure 1a). In a conventional two-pulse ESE experiment with the spectrometer

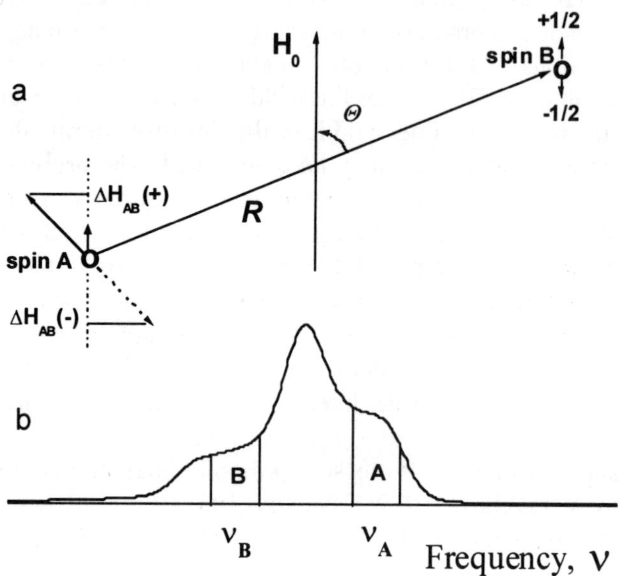

Figure 1. a) Dipole-dipole interaction between A and B spins. Spins B produces a local field at spin A position $\Delta H_{AB}(\pm)$ which depends upon B spin projection ±1/2 on external field H_0; b) Inhomogeneously broadened ESR spectrum, represented in frequency domain. Spins A and B excited by the pulses at ν_A and ν_B, respectively.

operating at frequency ν_A, the A spins are stimulated and contribute to the echo amplitude. If, however, a second spin packet B is stimulated by a pumping pulse at frequency ν_B (Figure 1b), then the echo amplitude will be affected by the degree to which the spins of packets A and B interact (*i.e.* via magnetic dipolar interactions). In practice, the ν_B pulse is temporally positioned between the two pulses producing the ESE signal at frequency ν_A (Figure 2a). The time interval between the first pulse at ν_A and the pumping pulse at ν_b is

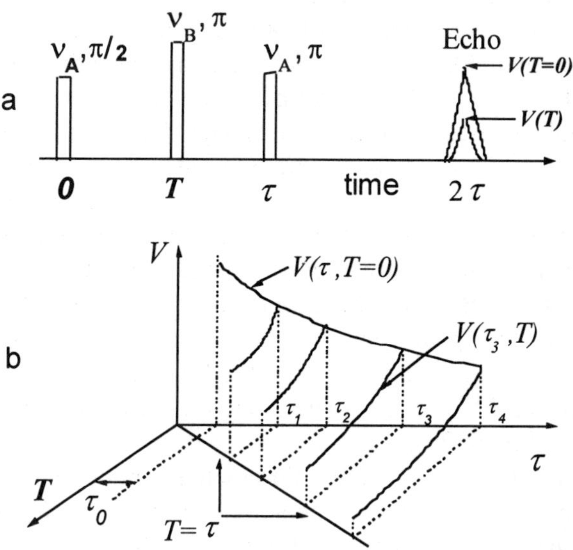

Figure 2. a) Sequence of $\pi/2$ and π mw pulses at ν_A separated by time interval τ produce ESE signal at time 2τ, $V(T=0)$. Then pumping π pulse at ν_B applied at time T, the amplitude of ESE signal changes up to $V(T)$ value; b) ESE signal decay when time interval τ changes $V(\tau,T=0)$, τ_0 – dead time in ESE two pulse method; τ_1, τ_2... τ_i- starting fixed time intervals in PELDOR experiments, $V(\tau_i,T)$ - an example of PELDOR signal decay.

denoted by T. The pulse acting on spins B causes a change in the *z*-projections of spins B and, as a result, a change in the local magnetic fields at the sites of the spins A (Figure 1a). The magnitude of this change is determined by the d-d interaction of spins A and B. A change in the local fields induced by the pumping pulse, causes additional dephasing of spins A which affects the magnitude of the ESE signal compared to that in the absence of the pumping pulse. It is assumed that in the system under investigation, the spin-lattice relaxation time (T_1) is much longer than phase relaxation time of the spins (T_f). A similar approach to measuring d-d interactions via NMR spectroscopy has also been described (Emshwiller *et al.*, 1960).

In cw-ESR spectroscopy the method of exciting the spin system at two frequencies is known as ELDOR spectroscopy. We chose to call the pulsed version of ELDOR as PELDOR, but one can find several complex pulsed ESR techniques for dipolar investigations (Borbat & Freed, 1999; Kurshev *et al.*, 1989; Saxona & Freed, 1996; Jeschke *et al.*, 1998). One such technique goes by the acronym DEER in ESE (Double Electron-Electron Resonance in ESE). These various pulse schemes have some advantages and limitations when compared to PELDOR, and many are reviewed by Schweiger & Jeschke (2001), but in our opinion PELDOR is the experimentally simplest to apply and interpret. PELDOR applications in free radical research are summarized by Tsvetkov (1989) and Milov *et al.* (1998).

In the PELDOR method one measures the signal amplitude variation with T, denoted here as $V(T)$, for a fixed time interval τ between the pulses that produce the spin echo signal (Figure 2a). Maintaining a fixed time τ avoids the masking influence of dynamic and relaxation processes on the ESE signal from spins A and allows one to explicitly determine the d-d interaction with spins B from the T-dependence of the ESE amplitude (Figure 2b), PELDOR signal decay. It is necessary to emphasize that the minimum T value (or maximum $V(T)$ amplitude) in this case is not limited by the dead time of ESE spectrometer, but only by the duration of the microwave pulses (Figure 2b).

In order to analyze the PELDOR spin-labeled peptide data presented in Section 3, we now present some theoretical results for $V(T)$ decay of different model spatial organizations of the spin system.

2.2 Mathematical Formulation of the Dipole-Dipole Interaction

2.2.1 Case 1: Spatially oriented spin pairs

For a single pair of spins A and B, fixed in space and coupled by the d-d interaction (Figure 1a) the PELDOR signal decay law has the form (Milov *et al.*, 1998):

$$V(T) = V(0)[1 - p_b(1 - \cos(DT))] \tag{1}$$

In this case p_b is the fraction of spin B population that flips under the action of the pumping pulse, or the degree of ESR spectrum excitation at ν_B. D is the d-d splitting (in rad/s) of the resonance spin A due to interaction with spin B,

$$D = \frac{\gamma^2 \hbar}{R^3}(1 - 3\cos^2(\Theta)) + J \tag{2}$$

where γ is the gyromagnetic ratio for the electron; ħ is Planck's constant; R is the distance between the paramagnetic centers; and Θ is the angle between the direction of the external magnetic field H_0 and the vector which connects the paramagnetic centers (Figure 1a). In order to make Equation (2) generally applicable, the exchange interaction J is included. As written, Equations (1) and (2) reveal that this method allows detection of rather weak d-d interactions, since by definition

$$|D| \approx \frac{1}{\tau} \ll |2\pi(\nu_A - \nu_B)|. \tag{3}$$

It is worth mentioning that the lower limit of D is set by the phase relaxation rate $1/T_f$. Equation (1) contains the information necessary for analysis of the simplest systems, but this must be averaged correctly for the different disordered systems.

2.2.2 Case 2: Uniformly distributed paramagnetic centers

Let us consider the case when paramagnetic centers are distributed uniformly rather than in pairs over some range of distances. The PELDOR signal decay in this case should be calculated by using the Markoff method (Abragam, 1961) taking into account the influence of all spins B in the sample. If the dynamics of spins B is considered to be independent of each other, then instead of Equation (1) we get for this case

$$V(T) = V(0)\left\langle \prod_j (1 - p_b[1 - \cos(D_j T)]) \right\rangle_{\Theta R \nu} \tag{4}$$

where the subscript j numbers paramagnetic centers of type B. The angular brackets denote the averaging over their spatial (angle and distance) distribution and the shape of the ESR absorption line. In this general case we assume that the paramagnetic centers are localized isotropically in the volume. Calculations show that for this case PELDOR decay in the a sample with a paramagnetic center concentration C, distributed uniformly over a specified volume:

$$V(T) = V(0)\exp\left(-\frac{8\pi^2}{9\sqrt{3}}\gamma^2\hbar p_b CT\right) = V(0)\exp[-2p_b \Delta\omega_{1/2}T] \tag{5}$$

where $\Delta\omega_{1/2}$ is the d-d broadening of the ESR line. This expression completely concurs with the general expression for ESE signal decay (Salikov *et al.*, 1976; Salikov & Tsvetkov, 1979).

2.2.3 Case 3: Polyoriented spin pairs

Equation 1, which expresses the PELDOR amplitude decay $V(T)$ in the case of spatially oriented pairs with fixed distance, can be used to obtain the decay function in the case of fixed distance but polyoriented systems. If the mutual orientations of the fixed radical pairs are arbitrary the following result is obtained after averaging of Equation (1) (Milov *et al.*, 1981, 1998).

$$V(T) = V(0)[1 - p_b(1 - F(T))] \tag{6}$$

$$F(T) = \left[\frac{\pi(c^2 + s^2)}{6\omega_D T}\right]^{1/2} \cos\left[(\omega_D + J)T - \arctan\left(\frac{s}{c}\right)\right] \tag{7}$$

where c and s are the Fresnel integrals

$$c = \int_0^v \cos\left(\frac{\pi}{2}x^2\right)dx, \quad s = \int_0^v \sin\left(\frac{\pi}{2}x^2\right)dx, \tag{8}$$

and

$$\nu = \left(\frac{6\omega_D T}{\pi}\right)^{1/2} \tag{9}$$

$$\omega_D = \frac{\gamma^2\hbar}{R^3} \tag{10}$$

Equation 7 shows that the d-d and exchange interactions will lead to modulation of the PELDOR echo amplitude. The form of Equation 7 also shows that this amplitude modulation will be damped with increasing T at a rate that depends on the dipolar interaction ω_D (Figure 3) and the dispersion of distances separating the interacting spin pairs. In the latter case, the deviation of distance between paramagnetic centers from a fixed value R is illustrated in Figure 3; the effect can be significant at even a small deviation $\delta R/R = 0.11$, and modulation will disappear at $\delta R/R = 0.25$.

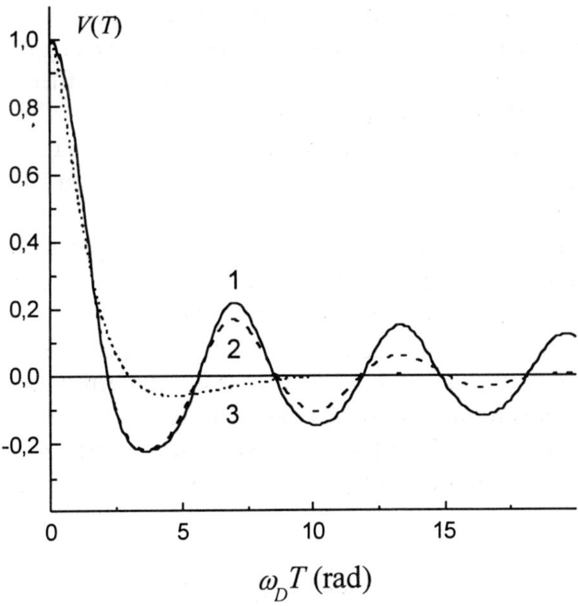

Figure 3. Simulated modulation the PELDOR echo decay function $V(T)$ for polyoriented fixed distance spin pairs, using Equation 6 parameterized with different spreads in distances: (1) $\delta R/R$=0; (2) $\delta R/R$=0.1; (3) $\delta R/R$=0.25

2.2.4 Case 4: PELDOR in spin groups

In complex spin systems such as spin pairs, spin clusters or aggregates, two types of d-d interactions exist. They are intercluster (interpairs, interclusters, *etc.*) and intracluster (between spins in pairs, clusters, *etc.*) dipolar spin couplings. The corresponding PELDOR decays will be denoted as $V(T)_{inter}$ and $V(T)_{intra}$. If it is assumed that intercluster and intracluster d-d couplings are independent, then the total PELDOR decay can be represented as the product (Milov *et al.*, 1998; Salikov *et al.*, 1979).

$$V(T) = V(T)_{intra} V(T)_{inter} \quad (11)$$

This, in principle, enables one to resolve the influence of intercluster and intracluster d-d couplings on the relaxation in complex systems. From the preceding analyses, Equations (1) and (6) correspond to $V(T)_{intra}$ and Equation (5) corresponds to $V(T)_{inter}$ in the case of d-d relaxation of diluted frozen solutions of radical pairs or biradicals. Sometimes the term $V(T)_{\text{inter}}$ may have a complex structure such as $V(T)_{\text{inter}} = \prod_i V_i$ if the «inter» system can be divided into in several independent subsystems.

In many cases spatial groups of paramagnetic centers (*i.e.* clusters or aggregates) may be formed in which the distance between particles is much smaller then between the groups, for example, due to different clustering processes or upon radiolysis or photolysis of solid. The results of theoretical calculations of the PELDOR signal decay for this situation make it possible to estimate the geometrical parameters of the groups and the number of spins, N, in them.

For simplicity, it is assumed that all paramagnetic centers under investigation are in identical groups and that each group contains N paramagnetic molecules. In order to neglect the probability that more than one spin B flips in the group caused by a pumping microwave pulse at ν_B, a small the value of $(p_b N)$ is taken. In this case the correlation between the positions of spins B in the groups can be neglected. As a result, the relations 1 and 2 of Milov *et al.* (2000b) are reduced, and the PELDOR signal decay due to dipole-dipole coupling of paramagnets within (intra) the groups will have the form:

$$V(T)_{\text{int}\,ra} = V(0)[1 - p_b \langle 1 - \cos(DT) \rangle_{\Theta R}]^{N-1} \tag{12}$$

where $V(T)_{\text{intra}}$ is the PELDOR signal decay function, and $<.......>_{\Theta R}$ indicates averaging over all possible values of angles and distances between spins inside the group.

The averaging in Equation (12) due to a random orientation of spin groups leads to a fast decay of the echo signal at time $T \leq T^*$ corresponding to the value of the mean d-d coupling of spins and effective distance R_{eff}

$$T^* = \frac{R^3_{\text{eff}}}{\gamma^2 \hbar} \tag{13}$$

According to Equation (12), the function $V(T)_{\text{intra}}$ tends to its limit V_p at $T \geq T^*$ (Figure 4a). The V_p value can be obtained from Equation (12) at $\langle \cos(D\,T) \rangle =$ 0:

$$V_p = (1 - p_b)^{N-1} \cong 1 - (N-1)p_b \tag{14}$$

Equation (14) makes it possible to estimate the number of spin labels in the group, N. To this end it is necessary to know the V_p and p_b values (Figure 4b).

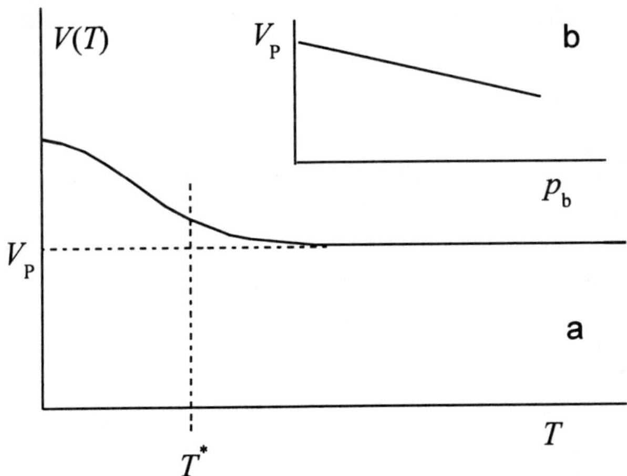

Figure 4. a) PELDOR decay function $V(T)$ in the case of isolated group with N spins in the group and mean distance between spins, R_{eff}; b) Determination of number of spins in group, N, from the experiments with different degree of ESR spectrum excitation p_b by pumping pulse.

When there is a distribution over the number of spins in a group, the method described above gives the effective number of spins (Milov *et al.*, 1984). The PELDOR method was used to extract information about the statistics of spin number in groups (Ponmarev *et al.*, 1988, 1990). The solution to this problem is based on the fact that in the presence of a distribution of the number of spins in groups, the PELDOR signal will involve only spins from groups in which the second pulse at ν_A excites only one spin. The groups, in which two or more spins are excited, will fail to contribute to echo signal due to complete dephasing of the entire spin ensemble at moment 2τ. The dependence of the number of spins on p_a (the fraction of spins forming the echo signal at ν_A) will give information on the spin number-distribution function.

2.3 Determination of the parameter p_b

When analyzing the PELDOR decay kinetics it is necessary to keep in mind the physical meaning and the methods for determination of the parameter p_b. The value of p_b depends upon the turning angle of the B spins' magnetization due to the pumping pulse action. This turning angle, θ, depends on the difference between the frequencies of pumping pulse and spins under action. When the line width of the spins is small in comparison with the pumping pulse magnitude, all the B spins rotate by the same angle under pumping pulse action. In this case p_b takes a simple form:

$$p_b = \sin^2(\theta/2), \tag{15}$$

where $\theta = \gamma H_1 \tau_p$. The terms H_1 and τ_p correspond to the amplitude and duration of the pumping pulse (experimental parameters), respectively. For broad ESR lines it is necessary to average Equation (15) on the line shape function. According to Mims (1965), the p_b value corresponding to an ESR spectrum with the line shape function $I(\nu)$ (*i.e.* see Figure 1) could the represented (in the case of rectangular pulses) as

$$p_b = \int_{-\infty}^{+\infty} \frac{\omega_1^2 I(\nu)}{\omega_1^2 + 4\pi^2(\nu_B - \nu)^2} \sin^2\left(\frac{\tau_p}{2}\sqrt{\omega_1^2 + 4\pi^2(\nu_B - \nu)^2}\right) d\nu \tag{16}$$

where $\omega_1 = \gamma H_1$, and ν_B is the frequency of the pumping pulse.

This means that it is possible to calculate p_b corresponding to a given experimental condition if the pulse shape parameters are accurately known. In practice, however, it is easier to experimentally determine p_b from the $V(T)$ decay (see Equation 5) in the case of uniformly distributed paramagnetic centers with the same $I(\nu)$. In this latter case an exponential decay occurs, and if the concentration of paramagnetic centers is known, the rate of decay will give the p_b-value. For example, a sample of uniformly distributed nitroxyl spin labels may be used as a standard from which p_b is measured. It is also possible to obtain p_b using Equation (14) in the case when N is fixed (synthetic radical pairs for example).

2.4 Characteristic distances from PELDOR experiments

It is possible to estimate the range of distances R that are accessible for measurements using the PELDOR oscillation decay (Figure 2). The smallest possible value of d-d coupling, ω_D, is determined by the largest time interval between microwave pulses at ν_A, τ_{max}, at which it is still possible to detect the two pulse ESE signal, denoted $V(0)$ (*cf.* Figure 2b). This time corresponds to the phase relaxation time T_f and depends upon the paramagnetic center concentration, temperature and other properties of the spin-system. Estimates can be made from the modulation amplitude of the electron spin echo by so-called matrix protons, in which case the limiting process is spin diffusion in the nuclear system (Salikov *et al.*, 1976; Salikov & Tsvetkov, 1979). The phase relaxation time $T_f \sim \tau_{max}$ for an organic matrix will therefore be less then ~5 μs. In order to observe single oscillation period in the PELDOR decay function $V(T)$ it is necessary to have $\omega_D \tau_{max} = 2\pi$, then the maximum distance R_{max}, as determined from Equation (10) is

$$R_{max} = \left[\frac{\gamma^2 \hbar}{2\pi} \tau_{max} \right]^{1/3} . \tag{17}$$

This estimation gives $R_{max} \sim 65$ Å at $\tau_{max} = 5$ μs.

Assuming the detection bandwidth of spectrometer is wide enough, the minimum value of R, or maximum ω_D value is determined by the amplitude of the microwave field, H_1, at ν_B in PELDOR. This field should be enough to excite both lines of the Pake dipolar doublet separated at ω_D (Maryasov & Tsvetko, 2000). For this case $\omega_D = 2\gamma H_1$ and as $2\gamma H_1 \tau_p = \pi$, where τ_p is the duration of pumping pulse, and we have

$$R_{min} = \left[\frac{\gamma^2 \hbar}{2\pi} \tau_p \right]^{1/3} . \tag{18}$$

At $\tau_p = 30$ ns this gives $R_{min} = 10$ Å, and therefore these estimates show that PELDOR oscillation decay provides the possibility to make measurements of distances in the range from 10 Å to 65 Å.

Another important theoretical scenario occurs when the distance between the paramagnetic centers is determined according to some distribution function $n(R)$. It appears possible to estimate the parameters of this function by analysis of PELDOR decay up to distances of about 100 Å (Milov *et al.*, 1981).

When performing PELDOR experiments at ν_A and ν_B frequencies one should have in mind the constraints imposed by the ESR line width and pulse duration. Pulses at ν_A and ν_B should be located in the frequency domain within the ESR line width $\Delta\nu$ and their duration τ_p should be not too small. The excitation profiles across the entire ESR spectrum should not overlap spectra. This requires: $(\nu_A - \nu_B) < \Delta\nu$ and $(\nu_A - \nu_B)\tau_p >> 1$ and prevents measurements by PELDOR for narrow ESR lines and narrow microwave pulses in the time domain. In our studies of nitroxyl radicals $\Delta\nu \cong 300$ MHz, so that for $(\nu_A - \nu_B) \cong 100$ MHz and $\tau_p \leq 40$ ns satisfy the requisite conditions. For narrow ESR lines, other pulse methods may be more convenient to use, for example the «2+1» method (Kurshev *et al.*, 1989).

3. APPLICATION OF PELDOR TO SPIN PROBE INTERACTIONS

The following sections present cw-ESR and PELDOR results from our investigations of single and double labeled trichogin type peptides. These results pertain to glassy state peptide solutions frozen at 77 K. Experimental measurements of d-d interactions performed on liquid phase samples will appear in the future publications.

3.1 Experimental Details

The PELDOR experiments are performed using a conventional ESE spectrometer equipped with a provision for producing microwave pulses at a second frequency ν_B (Milov *et al.*, 1981). The microwave power of both sources is applied to a bimodal reflection resonator, and the PELDOR signal (echo) arising at frequency ν_A is recorded. The detection design also allows one to determine the shape, duration and amplitude of the additional microwave pulse at frequency ν_B.

A bimodal cavity resonator with two TE_{102} modes oriented at right angles (Huisjen & Hyde, 1974) is used to simultaneously drive the sample at the two frequencies. In our experience, these resonators are very experimentally convenient because they allow one to use either a Dewar flask or a gas flow system to cool the sample as in the usual ESE experiments. A schematic diagram of the resonator and Dewar flask is shown on Figure 5. In our case, the isolation between resonators at ν_A and at ν_B (at frequencies $\nu_A = 9.4$ GHz and $\nu_A - \nu_B = 100$ MHz) was about 20 dB with a quartz Dewar flask introduced into the resonator. The quality factors of the resonator were $Q_a = 220$ and $Q_b = 150$. Durations of the first and the second pulse were 40 and 70 ns, respectively. Duration of the pumping pulse was about 40 ns. In experiments we used standard 5 mm o.d. sample tubes 5 mm of 5-6 mm length.

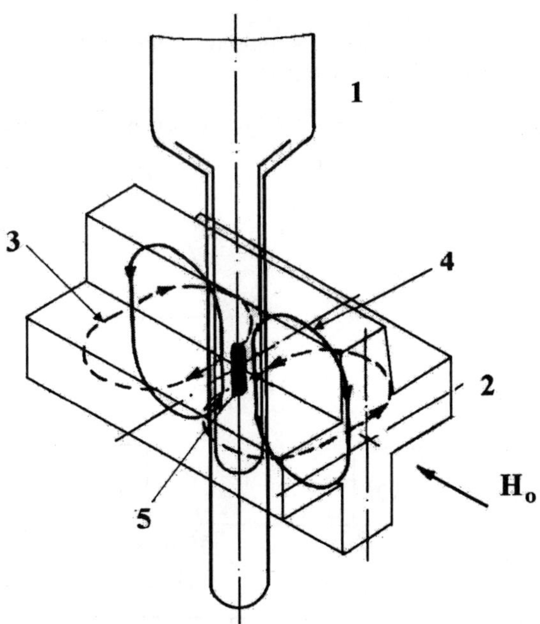

Figure 5. The position of a Dewar flask 1 in the TE_{102} modes of bimodal resonator 2 for PELDOR experiments. A schematic drawing of H_1 fields orientation 3 and 4 in the external magnetic field H_0 and specimen position 5 are shown. (Reproduced from Milov *et al.*, 1998).

The spin labeled analogues of trichogin GA IV were used to examine their chain conformation and aggregation. In each of these the spin label TOAC replaced one or more of the Aib groups in peptide chain at positions 1, 4, 8, and the C terminal leucinol (1,2 amino alcohol) is replaced by Leu-OMe group (Table 1). It is also possible to change n-Octanoyl N-group to Fmoc. We shall denote the specific trichogin analogues modified by TOAC and FTOAC by numbers according to Table 1 (*cf.* Hanson *et al.*, 1996). In those experiments requiring spin dilution, we used nonlabeled trichogin peptide, Tric-OMe. The stable nitroxyl radical 2,2,6,6-tetramethyl-1-piperidinyloxy-4-one (TEMPONE) was used as a reference standard nitroxide radical.

Table 1. Spin-labeled and N-terminal substituted trichogin GA IV (Tric-OMe), structures and its notations used in this paper.

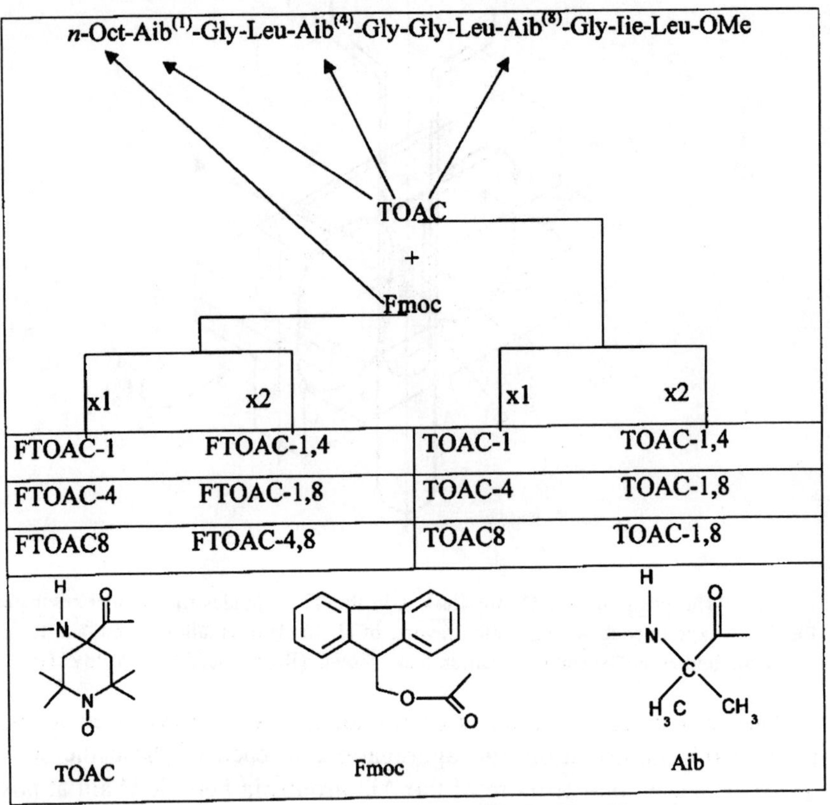

n-Oct-Aib(1)-Gly-Leu-Aib(4)-Gly-Gly-Leu-Aib(8)-Gly-Iie-Leu-OMe

TOAC + Fmoc

x1	x2	x1	x2
FTOAC-1	FTOAC-1,4	TOAC-1	TOAC-1,4
FTOAC-4	FTOAC-1,8	TOAC-4	TOAC-1,8
FTOAC8	FTOAC-4,8	TOAC8	TOAC-1,8

3.2 Conformational properties of trichogin peptides in polar glassy solutions

3.2.1 Intermolecular d-d interactions in mono and double- labeled peptides

The cw-ESR spectra of spin-labeled peptides FTOAC-1, FTOAC-8, FTOAC-1,8 are compared in Figure 6 with the spectrum of the nitroxyl radical TEMPONE. A 7:3 (v/v) mixture of $CHCl_3$/DMSO was used as the glass-forming matrix at 77 K, and all spectra are typical of nitroxide radicals

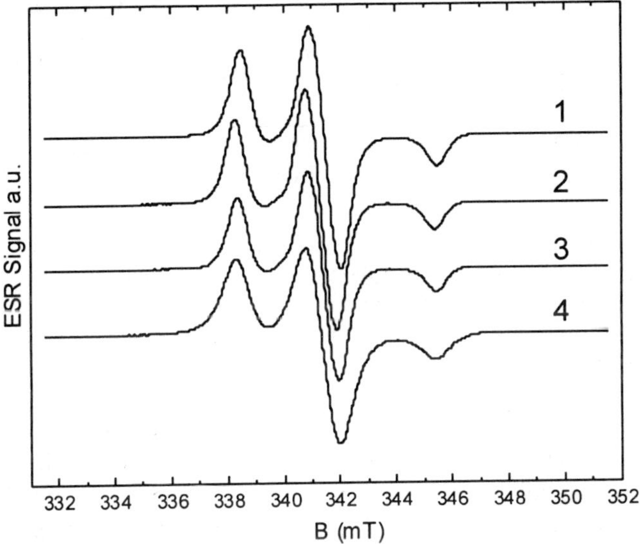

Figure 6. CW-ESR spectra of spin labeled peptides and TEMPONE in $CHCl_3$/DMSO (7:3) at 77 K: (1) TEMPONE; (2) FTOAC8; (3) FTOAC 1; (4) FTOAC-1,8. Spin label concentrations were ~10^{-2} M. (Reproduced from Milov *et al.*, 1999).

in frozen glassy solutions. The spectrum of double labeled peptide FTOAC-1,8 is broadened by approximately 0.3 mT more than the line width of the mono-labeled peptides. This broadening of the cw-ESR spectrum, however, is too small to be analyzed correctly.

That the PELDOR technique is much more sensitive than cw-ESR to weak dipolar and exchange spin couplings is illustrated in Figure 7. These data clearly show that the PELDOR echo amplitude decays differently among the samples that are otherwise indistinguishable via cw-ESR. The observed exponential decay is typical for a uniform random distribution of spin labels in the sample volume and conforms to Equation (5).

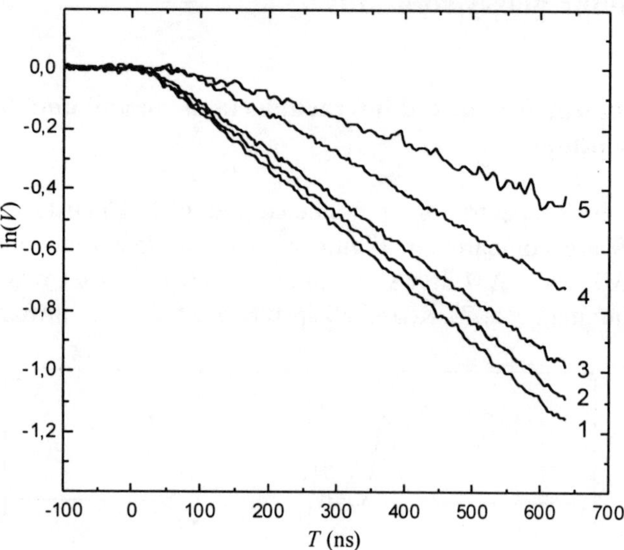

Figure 7. PELDOR signal decay due to intermolecular d-d interaction at 77 K. Concentrations of paramagnetic compounds in $CHCl_3$/DMSO (7:3) are as follows: (1) 7.2×10^{18} cm^{-3} FTOAC-1; (2) 6.6×10^{18} cm^{-3} TEMPONE; (3) 6×10^{18} cm^{-3} FTOAC8. Curves 4, 5 are differences of curves 1, 2 and 3, 2, respectively from Figure 8 (see text). (Reproduced from Milov *et al*., 1999).

From slopes of the ln $V(T)$ decay profiles in Figure 7 (curves 1-3) the value of d-d broadening is about 0.02 mT. This d-d broadening detected among mono-labeled peptides is one order of magnitude smaller than the broadening measured by cw-ESR lines between the biradical FTOAC-1,8 and the mono-radicals in the same concentration range (ca. 10^{-2} M of unpaired electrons). One may conclude that the 0.3 mT broadening observed in the cw-ESR spectrum of the biradical must be attributed to the coupling of spins inside the peptide molecule.

3.2.2 Intramolecular d-d interactions in double-labeled peptides

Returning to the PELDOR echo decay of the double-labeled peptide FTOAC-1,8 (Figure 8), we see that by varying the biradical concentration the decay envelope may be resolved into at least three contributing phenomena. The first of these corresponds to a fast decay during the first 50 ns, followed by a period of weak damping oscillations, and finally by a slow decay at T>50 ns. The last clearly depends on the concentration of the peptide and should be attributed to intermolecular coupling of spin labels, *i.e.* the dipolar spin coupling between labels belonging to different biradical molecules. The first two characteristics must originate from the coupling inside the biradical peptide.

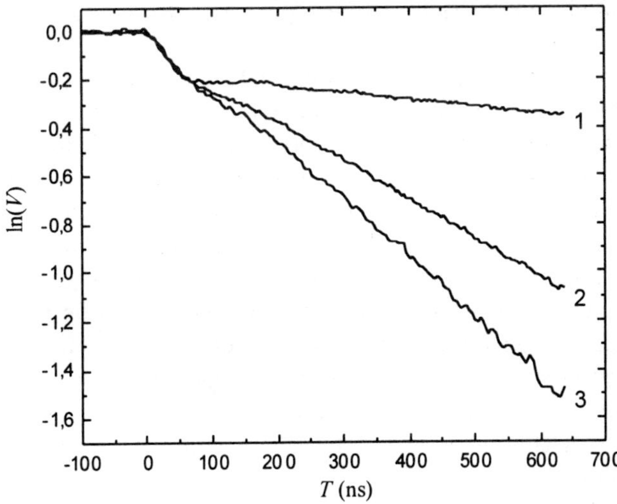

Figure 8. PELDOR signal decay at 77 K of double-labeled FTOAC-1,8 in $CHCl_3$/DMSO (7:3) at different mean concentrations of spins in a sample. (1) 1.3×10^{18} cm^{-3}; (2) 7.6×10^{18} cm^{-3}; (3) 1.2×10^{19} cm^{-3}. (Reproduced from Milov *et al.* 1999).

The contributions of both intramolecular and intermolecular couplings of spin labels to the PELDOR signal decay can be separated using Equations (5) and (11) as follows. Let $V_1(T)$ and $V_2(T)$ be PELDOR signals corresponding to concentrations C_1 and C_2 of biradicals, respectively. Taking the logarithm of Equation (11) and using Equation (5) in its exponential form $V = \exp[-Cf(T)]$, we can obtain system of two equations.

$$\begin{cases} \ln(V_1(T)) = \ln(V(T)_{intra}) - C_1 f(T) \\ \ln(V_2(T)) = \ln(V(T)_{intra}) - C_2 f(T) \end{cases} \tag{19}$$

It follows that subtraction of the appropriate experimental $\ln V$ *vs.* T plots makes it possible to exclude the intramolecular contribution due to the relaxation. Only the intermolecular coupling of biradicals will remain by this procedure. Curves 4 and 5 in Figure 7, for example, are obtained by subtracting curves shown in Figure 8 from each other (respectively curve 2 minus curve 1, and curve 3 minus curve 2). Of note is the fact that curves 4 and 5 in Figure 7 show small but significant differences compared with the linear plots obtained for single labeled compounds (TEMPONE and peptides

FTOAC-8, FTOAC-1; see Figure 7, curves 1-3). At short times (T<300 ns) the linearity in the $\ln V$ *vs*. T plots (see curves 4 and 5 in Figure 7) is distorted significantly. Relaxation caused by intermolecular d-d interaction has little effect in this time range as compared with case of randomly distributed molecules. This means that strong intermolecular d-d interaction (at short intermolecular distances) in this system occurs more rarely. This effect indicates that close approach of some spin labels for the double spin-labeled molecules is hampered.

The nitroxide fragments of the single labeled peptides (and TEMPONE) are accessible to each other, and their distribution in space is random, but addition of one more spin label per peptide molecule leads to the loss of such accessibility. The simplest explanation is that TOAC labels become charged in the peptide chain, and the resultant coulomb interaction prevents the nearest approach of biradicals. The Onsager radius for biradicals should be in this case four times more than that for monoradicals. Estimations show that electrostatic interactions could be responsible for this phenomenon. A similar effect was observed for the spin-labeled polyvinylpyridine polymer (Milov & Tsvetkov, 1997), and electrostatic repulsion between the positively charged polymer molecules was considered as a possible reason. This phenomena was studied in detail for charged nitroxide radicals (Milov & Tsvetkov, 2000).

The main difficulty in this case is that the TOAC residue possesses a large electrical dipole moment (the spin label TEMPO has the value of it 3.14 D; Rozantsev, 1970), but seems not to be charged easily. The interaction energy of electrical dipoles strongly depends on the orientations of dipole moments of both labels and of the vector connecting them. Attraction as well as repulsion is both possible. But the correlation of mutual orientations of attracting dipoles is possible in this case and the spatial distribution of the double labeled peptides might be not random. Further experiments are needed to prove the electrostatic nature of the observed phenomenon, for example, by using media with different polarity.

3.2.3 Peptide secondary structure from distance determination

The relaxation curves for the biradical peptide FTOAC-1.8 in $CHCl_3$/DMSO, shown in Figure 8, can be used to extract the effect of intramolecular d-d interaction on the PELDOR signal. As follows from Equation (19), the intramolecular contribution to the PELDOR echo amplitude is

$$\ln(V(T)_{intra}) = \ln(V_1(T)) - \frac{C_1}{C_1 - C_2}\left[\ln(V_1(T)) - \ln(V_2(T))\right] \qquad (20)$$

The factor in square brackets is proportional to the contribution caused by interparticle interactions. Figure 9 shows the dependence of the PELDOR amplitude on T, after subtracting the intermolecular contribution from the experimental curve 1 (V_1) in Figure 8. This is the mean value obtained for two estimations of intermolecular contribution when respectively curve 2 and curve 3 in Figure 8 were used as V_2. The concentration dependent factor in Equation (20) was estimated each time from experimental data. The dependence of $V(T)_{intra}$ versus T obtained this way is given by dots in Figure 9 and shows an oscillating decay.

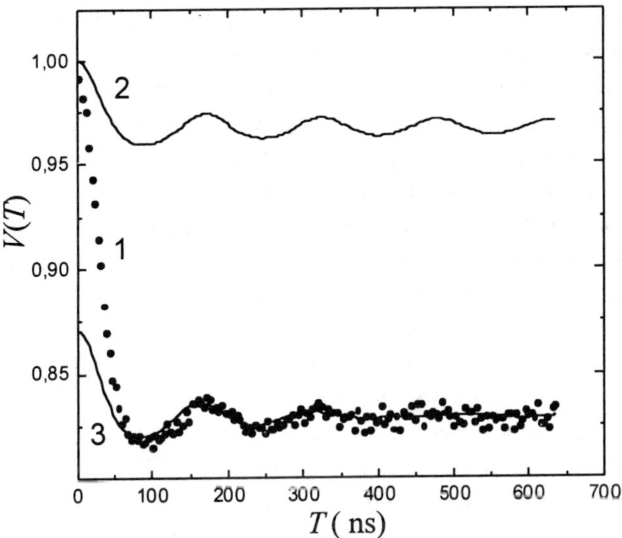

Figure 9. PELDOR signal decay of double-labeled FTOAC-1,8 caused by intramolecular d-d interaction between spin labels. Curve 1 (dots) is obtained from experimental data given in Figure 8 using Equation 20 (see text for details). Curve 2 shows calculation results for peptide with fixed distance between unpaired electrons 19.7 Å. Curve 3 presents PELDOR signal for uniform distribution over distance in the range 18.7-20.7 Å. The relation between the experimental signal amplitude at T=0 and the fitting curve indicates a 25% fraction of such peptides. Unfitted initial PELDOR signal decay at short time (T<70 ns) is caused by the rest of the molecules. (Reproduced from Milov *et al.*, 1999).

The PELDOR amplitude modulation indicates that the peptides assume a conformation in which the spin labels are far removed with only a small variation in distance (*cf.* Equations 6-10). The period of the PELDOR modulation corresponds to a frequency of 6.53 ± 0.2 MHz, which, according to Equation (10), corresponds to a distance between unpaired electrons of R = 19.7 Å. This analysis has neglected exchange for the following reasons: in order to influence the PELDOR signal at a detectable level, the absolute value of J must be more than 2×10^5 s^{-1}. (JT_{max}~1). Direct exchange interaction caused by overlapping wavefunctions of nitroxyl's p-electrons becomes less than this value at distances greater than 8-10 Å (Friori *et al.*, 1993). Indirect exchange interaction decrease exponentially with the number of single bonds between unpaired electrons, ca. one order of magnitude per one bond (Parmon *et al.*, 1980), and therefore, because the radical fragments are interconnected by a long chain (more than 20 single bonds), the exchange interaction must be very small.

The modulation amplitude decreases with time T so fast that it seems reasonable to conclude that the width of the distance distribution is not negligible. This is demonstrated by the simulated decay envelope (curve 2 in Figure 9), which was modelled by using a single conformation without any spread in the spin-to-spin distance. In this simulation we set p_b as to match the simulated and experimental modulation amplitudes, and yet it is clear that the signal intensity, modulation amplitude or modulation decay are still quite distinct between the experimental data and the calculated curve. This discrepancy between model and experiment was removed by introducing a narrow uniform distribution for the interspin distance in the range between 18.7 and 20.7 Å (curve 3 in Figure 9). This fitting could be interpreted as evidence for multiple conformations of the TOAC spin label, differently oriented in the space (Milov *et al.*, 1999).

The fraction of the biradicals FTOAC-1,8, that produce PELDOR modulation effects is roughly estimated to be about 25% from the relation between the p_b value used for this simulation and its estimation from experimental data ($p_b = 1 - V(T \to \infty) \approx 0.17$). The remainder of the molecules (about 75%) have longer distances between the radical fragments with broad distribution over this parameter and a fluidly decaying PELDOR signal (this part of the signal is not fitted in Figure 9).

In order to follow the conformational (or R-distance) changes of peptide due to substitution of N-terminal group together with the changes of glassy solution properties we studied double labeled peptides TOAC-1,8 and FTOAC-1,8 in several polar solvents: methanol, ethanol, TFE and $CHCl_3$/DMSO mixtures of different ratio. All measurements were performed on 2×10^{-3}M solutions cooled at 77 K (Milov *et al.*, 2000a).

After the $V(T)_{inter}$ subtraction from the general PELDOR decay by the same procedure as described above, the $V(T)_{intra}$ part for these measurements is shown in Figure 10.

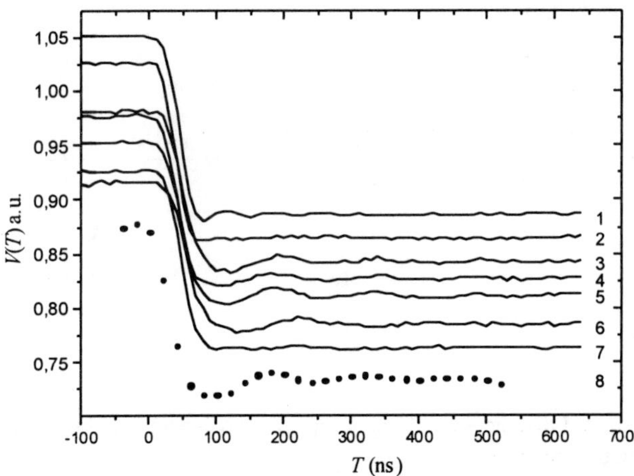

Figure 10. PELDOR signal decay due to intramolecular d-d interactions in double-labeled peptides:(1) TOAC-1,8 in TFE; (2) FTOAC-1,8 in TFE; (3) TOAC-1,8 in $CHCl_3$/DMSO(1:1); (4) FTOAC-1,8 in $CHCl_3$/DMSO(7:3); (5) TOAC-1,8 in $CHCl_3$/DMSO(7:3); (6) TOAC-1,8 in ethanol; (7) FTOAC-1,8 in ethanol; (8) TOAC-1,8 in methanol. (Reproduced from Milov *et al.*, 2000a).

In these data only the TOAC-1,8 samples exhibit modulation of the PELDOR signal decay, and only when the peptide is suspended in the $CHCl_3$/DMSO solution. This modulation pattern does not change when we changed the ratio $CHCl_3$/DMSO from 7:3 to 1:1 by volume in the glassy matrix. In other solvents, the PELDOR signal from FTOAC-1,8 is not modulated in the T range 100-700 ns. As before, the oscillation period $V(T)_{intra}$ obtained from the $CHCl_3$/DMSO samples yield a distance R between spin labels, and from the initial modulation amplitude the relative fraction of peptides in the sample that correspond to this distance. All these data are collected in Table 2.

Table 2. PELDOR results of distance determination (in Å) and corresponding fraction of double labeled trichogins (in %) adopted fixed distance conformation in different glassy matrix at 77 K

Peptide	Glassy Matrix			
	$CHCl_3$/DMSO	CF_3CH_2OH	CH_3OH	CH_3CH_2OH
FTOAC-1,8	19.7 25%			
TOAC-1,8	19.7 30%	15.3 15%	19.7 36%	21.8 31%

Table 3. Calculated distances (in Å) between spin labels in TOAC-1,8 type trichogin in different helix conformations

Reference	Conformation			
	α	3_{10}	2_7	2_5
7	11.0	14.0	22.0	28.0
38	12.04	13.98		
39	10.06	14.08		

The distance R obtained from the PELDOR decay should correspond to the distance between labels at the particular peptide chain conformation. In order to calculate this parameter for different molecular conformations we use the known structure parameters of these peptides obtained by X-ray analysis (Monaco *et al.*, 1999). When the backbone torsion angles of peptide FTOAC-1,8 were adjusted to generally accepted values for secondary structures like α-, 3_{10}-, 2_7- and 2_5-helix, the distances between the labels could be calculated. The calculated values (Milov *et al.*, 1999; Monaco *et al.*, 1999; Anderson *et al.*, 1999) are collected in Table 3. It is evident from the calculated and measured R values that TOAC-1,8 and FTOAC-1,8 in all frozen solutions, except in TFE, adopt a ribbon like conformation 2_7-helix type. In the case of the TFE matrix TOAC-1,8 adopts a 3_{10} conformation, while FTOAC-1,8 has no fixed distance conformers with the exception of $CHCl_3$/DMSO solutions.

This observation seems to be rather unusual in view of the fact that Aib- and TOAC-residues are generally accepted to be strong promotors of α- and 3_{10}-helical conformations in peptaibol molecules (Karle & Balaram, 1990; Toniolo & Benedetti, 1991; Toniolo *et al.*, 1998). Only the fully extended structure has been reported for a homo-tripeptide based on a C_α-amino acid (Aubry *et al.*, 1994). The crystallography data (Monaco *et al.*, 1999) shows that the distance between spin-labels in TOAC-1,8 is $R = 11.4$ Å and from the half-field cw-ESR signal intensity analysis R was found (Anderson *et al.*, 1999) for this peptide as 12.0 Å.

PELDOR data for TOAC-1,8 and FTOAC-1,8 evidently shows that the conformation type strongly depends upon the properties of the solvents and the structure of the peptide. It is not clear at the moment if one should attribute these effects to conformational properties of the particular peptide at low temperatures, or to the structure and polarity of these solvents that form a glassy matrix. It is evident, however, that the comparison between the data obtained by different methods will be possible only for the same experimental conditions (temperature, matrix, *etc.*)

3.3 Self-assembly of trichogin peptides in weakly polar glassy solutions

In this section we examine self-assembly of spin labeled trichogin GA IV analogues by using PELDOR method. It is generally assumed that the membrane modifying properties of peptaibols are due to formation of amphiphilic helix bundles with polar groups pointing to the inside of the bundle and hydrophobic groups projecting towards the hydrophobic membrane. But in spite of a variety of investigations, experimental data on the formation of peptide clusters in the phospholipid bilayer or even in membrane mimicking hydrophobic solvents are still lacking. As it was shown in the preceding section, trichogin does not aggregate in polar glass-forming solvents. In the following sections data are presented in which PELDOR was used to investigate the effects of less polar solvents on the self-assembling properties of this particular antibiotic peptide.

3.3.1 Aggregation of peptides: detection and investigation

As the first step of our investigation of spin labels d-d couplings we have measurements for frozen solutions of FTOAC-4 and TOAC-1 in chloroform-toluene (7:3). This solvent mixture is appropriate because it contains components of low polarity, the peptides are readily dissolved, and it forms a transparent glass upon freezing to 77 K. An additional solvent system, chloroform-toluene-ethanol (3.5:1.5:5), was also examined in order to reveal the effect of a polar solvent on the intermolecular interaction between peptides. We also present PELDOR data on the double labeled peptide FTOAC-1,8 in frozen methanol solution, which will be used as the reference for the determination of experimental parameters in the analysis of the PELDOR results for peptide FTOAC-4 and TOAC-1.

cw-ESR spectra of the spin labeled peptides FTOAC-4, TOAC-1, FTOAC-1,8 are shown in Figure 11. The microcrystalline powder spectra (curves 1 and 2) are simple singlets, whose widths are caused by strong d-d and exchange interactions between spin labels at short distances within the crystals. The spectra of frozen peptide solutions (curves 3-7) are typical for nitroxide radicals under these conditions. The spectral shapes indicate that freezing of solutions does not cause peptides to segregate into a separate phase.

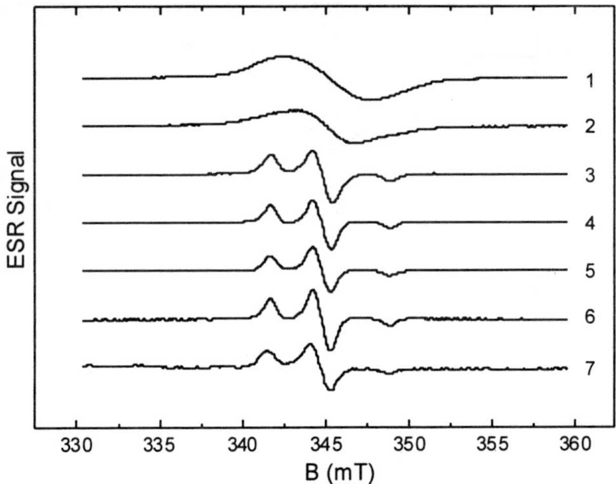

Figure 11. CW-ESR spectra of peptides FTOAC-4, TOAC-1, FTOAC-1,8 in the solid phase at 77 K: (1 & 2) dry powder of peptides FTOAC-4 and TOAC-1; (3 & 4) peptides FTOAC-4 and TOAC-1 in a chloroform-toluene mixture; (5 & 6) peptides FTOAC-4 and TOAC-1 in a chloroform-toluene-ethanol mixture; (7) double-labeled peptide FTOAC-1,8 in methanol. (Reproduced from Milov *et al.*, 2000b).

Curves 1 and 2 in Figure 12 illustrate the PELDOR signal decays, $V(T)$, for two single labeled peptides FTOAC-4 and TOAC-1 in glassy chloroform-toluene. A comparison between these results and phase relaxation data for polar solvents (Figure 7) reveals two unusual features. The first peculiarity is a fast decrease in the amplitude of $V(T)$ at the initial time region up to $T \approx 150$ ns. A fast decrease of the $V(T)$ in this time region indicates the existence of compact groups of spin labels in the system under study instead of a uniform spatial distribution of spins (Figure 4). A strong d-d coupling of spin labels within a group leads to a fast dephasing at short times. This type of dependence was repeatedly observed in glassy solutions of biradicals (Milov *et al.*, 1998; Ponomarev *et al.*, 1998) and double-labeled peptides (Milov *et al.*, 1999). The latter represent a simple case of groups consisting of two spins that is exemplified by curve 5 in Figure 12 obtained for a frozen solution of double labeled peptide FTOAC-1,8 in methanol. By comparing curves 1, 2 and 5 in Figure 12 one sees that, within the initial time region, the depth of the fast PELDOR signal decay of peptides FTOAC-4 and TOAC-1 is much greater than that of peptide FTOAC-1,8 whose structure contains only two spin labels. According to Equation (14), a substantially greater depth for peptides FTOAC-4 and TOAC-1, relative to the corresponding value for the biradical, indicates that these compounds form aggregates of more than two peptide molecules.

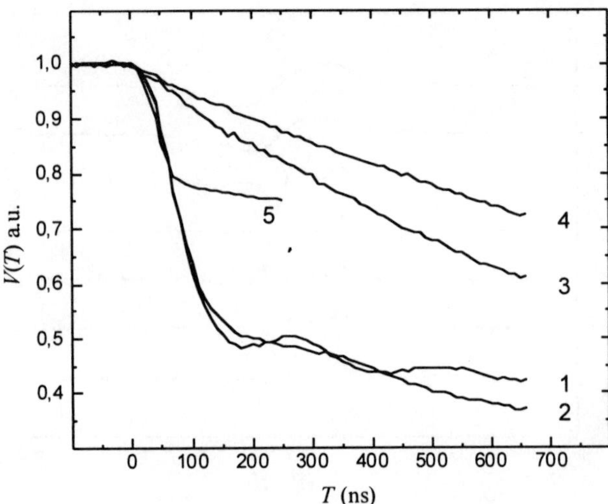

Figure 12. PELDOR signal decay for glassy peptide solutions at 77 K: (1 & 2) peptides FTOAC-4 and TOAC-1 in a chloroform-toluene mixture, respectively; (3 & 4) peptides FTOAC-4 and TOAC-1 in a chloroform-toluene-ethanol mixture, respectively; (5) double-labelled peptide FTOAC-1,8 in methanol. (Reproduced from Milov *et al.*, 2000b).

The second remarkable feature of Figure 12 is the presence of a slower decrease in the time region at $T>150$ ns, which is accompanied by signal modulation. The slow decrease of $V(T)$ is related to the d-d coupling of spin labels belonging to different aggregates. Since this coupling is almost independent of the interaction between spin labels within the aggregate, the total decay of the PELDOR signal can be considered as the product of two time dependencies: $V = V_{\text{inter}} V_{\text{intra}}$ (see Equation 11). The $V(T)$ modulation that is observed in curves 1 and 2 of Figure 12 are related to the interaction of labels within aggregates (V_{intra}) and are of the same origin as the PELDOR modulation obtained from biradicals and double labeled peptides which have been measured in frozen polar glassy solutions. These modulations indicate that the aggregates of peptides FTOAC-4 and TOAC-1 have fragments with a fixed structure in which the distances between spin labels have a minor spread. These distances can be estimated from the period of oscillations.

Adding ethanol to the otherwise low polarity mixture of chloroform-toluene markedly changes the behavior of the PELDOR signal decay, as shown in curves 3 and 4 of Figure 12. As compared with curves 1 and 2, the signal decay now has no fast decay component at short T and no oscillations. Curves 3 and 4 in Figure 12 can be described by a simple exponential decay, typical for a random distribution of spin labels in the bulk (part 2). Transitions from dependencies 1 and 2 in Figure 12 to 3 and 4 after addition of ethanol to the chloroform-toluene mixture strongly indicate dissociation of peptide aggregates into their monomeric constituents.

Figure 13 shows the cw-ESR spectra of peptides FTOAC-4 and TOAC-1 in the same solvent mixtures at room temperature. Spectra 1 and 2 are broader than spectra 3 and 4, which allows one to conclude that peptide aggregates do also exist in liquid solutions at room temperature. The broadening effect can be caused by additional spin relaxation in the aggregates due to d-d or exchange interactions of spin labels. Similar cw-ESR spectra were observed for some double-labeled peptides in other solutions at room temperature (Friori *et al.*, 1993). An additional contribution to the linewidth from the anisotropy of the hyperfine interaction and g tensors is also possible in the case of a slow rotational mobility of aggregates. A detailed analysis of these effects is currently in progress.

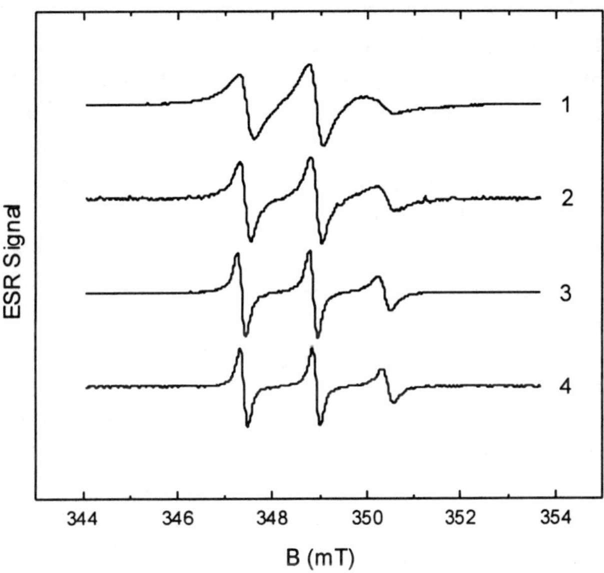

Figure 13. Solution CW-ESR spectra of peptides FTOAC-4 and TOAC-1 solution at room temperature: (1 & 2) peptides FTOAC-4 and TOAC-1 in a chloroform-toluene mixture; (3 & 4) peptides FTOAC-4 and TOAC-1 in a chloroform-toluene-ethanol mixture. (Reproduced from Milov *et al.*, 2000b).

3.3.2 Quantitative Estimates of Peptide Aggregation

The number of peptide molecules in the aggregate can be estimated by applying the theoretical models of PELDOR decay (Section 2) to curves 1 and 2 in Figure 12). To do so we shall use the formulaic model corresponding to the scenario in which the spins occur in groups, for which it is necessary to obtain the limiting value of PELDOR decay function V_p (see Equation 14).

It is difficult to make a reliable determination of V_p directly from the experimental decay $V(T)$ (Figure 12), because $V(T)$ is the product of $V(T)_{intra}$ and $V(T)_{inter}$. The $V(T)_{intra}$ function, after the fast decay at short T, can have oscillations around its V_p value if a fraction of the spins in an aggregate has a fixed distance. At the same time, $V(T)_{inter}$ smoothly depends on T in a similar way as appears in curves 3 and 4 (Figure 12). This means, unfortunately, that the increasing influence of the $V(T)_{inter}$ part of the decay and the incomplete attenuation of the oscillation do not permit us to get the precise value of V_p from the experimental curves 1 and 2 at long T values, as predicted by calculations (Figure 4). However, at short T values the contribution to the total $V(T)$ decay by $V(T)_{inter}$ is relatively small and we can assume that $V(T) \sim V(T)_{intra}$. This means that the depth of the initial fast decay of the experimental $V(T)$ function at short T corresponds to the V_p value with an uncertainty determined by the oscillation amplitude. Therefore the V_p value was taken from the mean $V(T)$ at $T = 150$ ns for curves 1 and 2 (Figure 12). The fast decay is by then over and the slow decay of $V(T)_{inter}$ is not significant, due to the low interaction between spin labels in different aggregates. This estimate gives the mean value of $V_p = 0.51 \pm 0.01$ for both peptides FTOAC-4 and TOAC-1.

In order to determine the p_b value for peptides FTOAC-4 and TOAC-1, we used the $V(T)$ function of the double labeled peptide FTOAC-1,8 (curve 5 in Figure 12). This approach is possible since the p_b value is only a function of the ESR line shape and pumping microwave pulse parameters according Equation (16). Therefore, the value of p_b is the same for all peptides under investigation because the ESR spectra and the experimental PELDOR parameters are the same for all peptides. We chose the double labeled peptide FTOAC-1,8 in methanol as in this case we have no modulation in the $V(T)$ decay function. This property gives us the opportunity to determine p_b from the depth of the initial decay at short T. For double labeled peptide FTOAC-1,8 $N = 2$ and the V_p value obtained is 0.8 ± 0.01. After substituting $N = 2$ and $V_p = 0.8$ into Equation (14), we obtain $p_b = 0.2$, which can be used in the estimation of N for peptides FTOAC-4 and TOAC-1.

By substitution of $V_p = 0.51 \pm 0.01$ and $p_b = 0.2 \pm 0.01$ into Equation (14), we obtain $N = 4 \pm 0.3$ for both peptides. Thus, the quantitative estimate shows that peptides FTOAC-4 and TOAC-1 form aggregates consisting of four molecules within a chloroform-toluene mixture. Note that the assumption we have made above about the absence of a spread in N needs additional verification. If N varies for different aggregates, only a mean effective value can be determined using the method that has been developed (Milov *et al.*, 1984; Ponomarev *et al.*, 1988).

It is possible to roughly estimate a mean value for distances between the labels within the aggregate from the time T^* of the fast decay function $V(T)$ (see Equation 13). This time corresponds to the mean value of d-d interaction. For T^* in the range 100–150 ns, this will give $R_{eff} \cong 30–36$ Å and gave us an upper limit for the aggregate dimension.

More correct distance values between spin labels in the aggregates can be calculated from the frequency of oscillations of the PELDOR signal. The experimental values of oscillation frequencies are 2.51×10^7 rad/s for peptide FTOAC-4 and 1.85×10^7 rad/s for peptide TOAC-1. Using Equation (10) for the oscillation frequency, we obtain a distance of 23.5 Å for peptide FTOAC-4 and 26.0 Å for peptide TOAC-1. The estimated error in the R depends on the error in the measurement of oscillation frequencies and does not exceed 1.5%.

Although the position of the spin label in the primary structures of peptides FTOAC-4 and TOAC-1 is different, the difference between the observed distances of the respective peptides is not large. As mentioned above, oscillations indicate the existence of rigid structural fragments in the aggregates without rather small any spread in distances between spin labels. Not all the possible distances between spin labels are manifested in our experiments. We are likely to observe oscillations due to pairs of spin labels with relatively small distances, because these pairs are located in rigid fragments of the aggregate. Therefore, it should be particularly emphasized that the values of distances obtained from oscillations can only characterize the size of the rigid part of the aggregate structure.

3.4 Aggregation effect as general properties of peptides in weakly polar glassy solutions

As described in previous section for peptides FTOAC-4 and TOAC-1, the oscillation frequency and amplitude of PELDOR signal decay depend upon the position of spin labels in the peptide structure and the difference in the structure of terminal peptide groups which could affect the structure of aggregates. These preliminary observations require additional investigation of PELDOR relaxation effects in spin labeled peptides aggregates: changing the structure of peptides, the position of label and the properties of solvents (polarity, ability to form complexes, *etc.*). To start this program we have studied the d-d relaxation for the group of trichogin peptides TOAC and FTOAC types (Table 1), both singly labeled in positions 1, 4 and 8 and doubly-labeled 1,8. As before, the latter peptide FTOAC-1,8, was used to measure experimental parameters based on the study of the intramolecular d-d interaction of its two spin labels. A solution of the nitroxyl radical TEMPONE was used to examine the random distribution of spins in solid glass.

As a solvent for peptides and TEMPONE, we used the chloroform-toluene mixture in a 1:1 ratio by volume. In addition, to elucidate the influence of solvent properties, the solutions of peptide TOAC-4 were also studied in mixtures of chloroform-decalin in a 1:1 ratio, dichloroethane-toluene in a 1:2 ratio and tetrachloromethane-toluene in a 1:2 ratio by volume. To avoid the aggregation, we have studied solutions of FTOAC-1,8 in polar solvent ethanol. In addition to glass composition differences, we also recorded a wider time T domain, which allowed a more detailed study of peculiarities in the PELDOR relaxation decay $V(T)$. For comparison under the same conditions, the data are given for TOAC-1 and FTOAC-4 peptides studied in previous part.

In Figure 12 the $V(T)$ decay corresponding to TOAC-1 and FTOAC-4 in the non-polar glass mixture featured a rapid amplitude decrease at short times ($T < 150$ ns), which indicates the formation of aggregates from mono spin-labeled peptides. The same rapid decay of ELDOR signal amplitude was found for all mono-labeled peptides in various solutions studied in this part. A strong d-d interaction of spin labels in the aggregates leads to fast decay of the PELDOR signal during the initial time period followed by oscillations caused by peptides in the aggregates with a mutual regular arrangement of spin labels inside the aggregates.

Figure 14 shows the effect of spin label aggregation on the PELDOR signal decay. The $\ln(V/V_{inter})$ curves for all spin labeled peptides in the chloroform-toluene mixture was obtained by subtracting the V_{intra} part from the PELDOR decay in the manner described by Milov *et al.* (2000d). The data obtained for TOAC-4 in other solvents are given in Figure 15. The modulation frequencies and the corresponding distances calculated using Equation (10) are presented in Table 4 along with the V_p values measured from the curves shown in Figures 14 and 15. The mean quantities of labels in aggregate, N are calculated using Equation (14). This quantity was determined for all peptides by using $p_b = 0.17$ calculated from Equation (14) for FTOAC-1,8 containing two spin labels and served as the N-reference specimen.

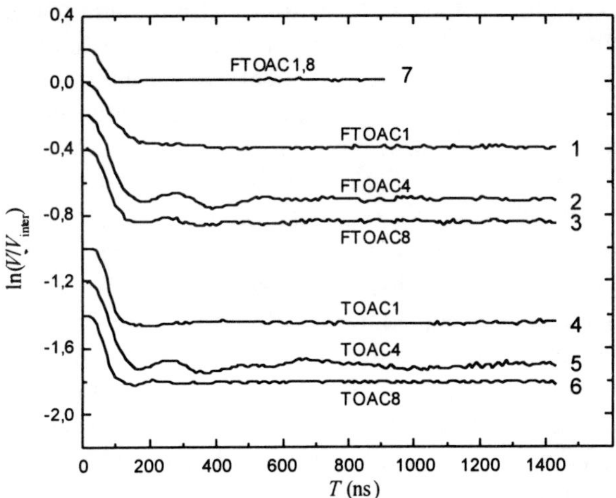

Figure 14. PELDOR signal decay due to d-d interaction of spin labels in aggregates of mono labeled peptides in a glassy frozen (77 K) mixture of chloroform-toluene (1:1): 1- FTOAC-1; 2- FTOAC-4; 3 – FTOAC8; 4 – TOAC-1; 5 – TOAC-4; 6 – TOAC8; 7 – FTOAC-1,8 in ethanol. (Reproduced from Milov *et al.*, 2000d).

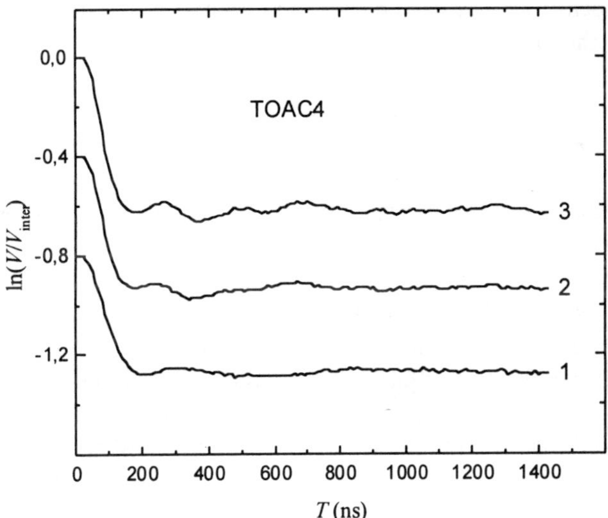

Figure 15. PELDOR signal decay due to d-d interaction of spin labels in aggregates for glassy solutions of single labeled peptide TOAC-4 in the solvents: (1) chloroform/cis-trans decalin 1:1; (2) tetrachloromethane/toluene 1:2; (3) dichloroethane / toluene 1:2. (Reproduced from Milov *et al.*, 2000d).

The $\ln(V/V_{inter})$ plots versus T shown for single labeled peptides in Figure 14 are different for the structure of terminal group and the position of spin label in peptide. Curves 1-3 refer to peptides containing Fmoc as a N-terminal group, curves 4-6 refer to the peptides with the N-terminal n-octanoyl group. As follows from the Figure 14, the structure of the terminal group has a weak effect on $V(T)$ whereas the oscillation amplitude and frequency strongly depend on the position of spin label in the peptide structure. Note that the oscillation amplitude and frequency measured for TOAC-1 and FTOAC-4 are close to those given earlier (Figure 12). The minor difference can be assigned to the different compositions of solvents.

The occurrence of two modulation frequencies in the case of TOAC-4 peptide is of interest (Table 4). This can indicate either the existence of two fixed distances inside the aggregate or the possible existence of the two different types of aggregates with different distances between spin labels. Unfortunately, the data available do not allow us to choose between these possibilities.

Figures 14 and 15 for TOAC-4, TOAC-1 and FTOAC-4 indicate that the aggregate properties depend on the solvent nature and composition. It appears from the figures that all parameters, such as frequency, amplitude and modulation damping rate, change when one solvent of the binary solvent mixture is changed. This is best observed when comparing the $V(T)$ decay for TOAC-4 solutions in the chloroform-toluene (Figure 14, curve 5) and chloroform-decalin (Figure 15, curve 1) mixtures. After substituting toluene by decalin, the modulation frequencies decrease and the modulation damping rate increases, which can be assigned to some 'loosening' of aggregates and disorder in their structure. It is noteworthy that despite some variation in modulation amplitude and damping that are attributable to peptide structure and solvent properties, the distances between spin labels (Table 4) in aggregates are grouped around the values of 23, 26 and 33 Å.

It is worthwhile to emphasize that the use of experimental V_p values for determining the number of labels in aggregates from Equation (14) is sure to give the correct number of labels in the aggregate providing that all aggregates in solution contain an identical number of labels. In the case of a distribution in the number of labels within the population of aggregates or in the presence of some fraction of monomeric peptide molecules, the V_p value depends not only on the parameter of the pumping pulse p_b but also on the similar parameter p_a of pulses forming echo signal (Milov *et al.*, 1998; Ponomarev *et al.*, 1990). The effective number of N found from the experimental V_p value will be in this case lower than the true mean number of labels in aggregates. In this case, the parameters of the distribution in the number of labels in aggregates can be estimated by the approach given by Milov *et al.* (1998) and Ponomarev *et al.* (1988). This approach will need additional measurements with changes of the p_b and p_a parameters.

Table 4. Experimental frequencies of oscillations, distances between spin labels *R*, V_p and *N* values for different spin labeled peptides.

Peptide	Solvent Mixtures	Frequency, MHz	R(Å)	$V_p \pm 2\%$	N ± 4%
FTOAC-1,8	Ethanol			0.17	2
FTOAC-1	Chloroform:Toluene (1:1)			0.67	3.14
FTOAC-4	Chloroform:Toluene (1:1)	3.85±0.15	23.7±0.3	0.6	3.75
FTOAC-8	Chloroform:Toluene (1:1)	5.0±0.25	21.8±0.36	0.64	3.4
TOAC-1	Chloroform:Toluene (1:1)	2.5±0.3	27.4±1.1	0.63	3.5
TOAC-4	Chloroform:Toluene (1:1)	4.3±0.2 1.5±0.2	22.9±0.35 32.5±1.4	0.6	3.75
TOAC-4	Chloroform:cis,trans Decalin (1:1)	3.3±0.2 1.2±0.2	25±0.5 35.0±1.9	0.62	3.56
TOAC-4	CCl_4:Toluene (1:1)	4.3±0.2 1.5±0.2	22.9±0.35 32.5±1.4	0.58	3.92
TOAC-4	$ClCH_2CH_2Cl$:Toluene (1:1)	4.3±0.2 1.5±0.2	22.9±0.35 32.5±1.4	0.54	4.3
TOAC-8	Chloroform:Toluene (1:1)	4.5±0.2	22.6±0.33	0.66	3.23
FTOAC-4	Chloroform:Toluene (7:3)	4.0±0.16	23.5±0.3	0.51	4.0
TOAC-1	Chloroform:Toluene (7:3)	2.9±0.1	26.0±0.4	0.51	4.0

The values of the effective number of spin labels *N* in aggregates (Table 4) vary from 3.1 to 4.3 depending on the peptide structure and solvent composition. As mentioned above, this distribution in the number of labels can be related to the distribution in the number of labels in aggregates or the presence of the uncontrolled number of unbounded peptide molecules in solution. A change in these parameters with peptide structure and solvent composition can lead to the corresponding changes in the *N* value determined from Equation (14). In addition, a certain error in the measurement of *N* can be caused by differences between the p_b values of the studied peptides and p_b for double-labeled peptide FTOAC-1,8 chosen as a reference with $N = 2$. These differences can arise from the fact that the p_b value is the convolution of the absorption line shape of ESR spectrum with the pumping pulse parameters (Equation 16) and the deviations of the determined value of *N* from its true value can be due to the slight differences in ESR spectra. Therefore we have at least two reasons for a spread in N values.

3.5 Conformational structure of trichogin inside the aggregates

It has shown in the preceding sections how PELDOR may be used to analyze the structure of a supramolecular cluster or aggregates of spin-labeled peptides in frozen glassy solutions by extracting information about the intermolecular distances between spin labels from d-d couplings. Based on the experimental data obtained we showed that the number of peptide molecules in aggregate of trichogin GA IV is close to four.

The data on magnetic d-d relaxation obtained by PELDOR in frozen glassy solutions of single-labeled peptides indicate that the aggregates include peptide chains with a fixed structure whereas the intermolecular distances between spin labels show minor spreading. This observation allows one to assume a fairly ordered spatial structure of the peptide within the tetrameric cluster. It is therefore of interest to experimentally determine the secondary structure adopted by the peptide within the aggregates.

One approach to secondary structure analysis within the aggregate is to examine the intramolecular magnetic dipole coupling of two spin labels incorporated at well-defined positions in the peptide chain. In this way one can get information about the distance between labels of an individual peptide molecule, which should differ among the peptide conformational states (Table 3), and therefore comparing experimental and calculated distances can provide the evidence of a particular peptide conformational state. Towards this end we have studied the intramolecular d-d couplings of aggregated TOAC-1,8 (Table 1; *cf.* Milov *et al.*, 2001).

Intramolecular d-d couplings from TOAC-1,8 are resolved from intermolecular d-d couplings by spin dilution using the unlabeled peptide Tric-OMe. This mixture was then dissolved in the weakly polar chloroform-toluene solvent (7:3) mixture, under which conditions aggregates are formed. With adequate spin dilution, each peptide aggregate contributing to the PELDOR data will contain at most one double-labeled peptide per aggregate, since the fraction of aggregates with two and more double-labeled peptides will be negligible, and therefore the resultant PELDOR data will correspond to intramolecular d-d interactions coming from a single represtenative peptide within an aggregate. For comparison purposes, we shall use PELDOR data obtained from TOAC-1,8 in a polar (non-aggregate forming) solution of TFE.

Figure 16 shows the cw-ESR spectra of TOAC-1,8 peptide. The spectrum in the dry powder form at 77 K (curve 1) is a single line whose width between the extreme is 3.2 mT. By contrast, the ESR spectrum in frozen glassy TFE ($\cong 10^{-2}$ M, curve 2) features an anisotropic triplet that is typical of TOAC labeled peptide in a diluted glassy solution. The difference between spectra 1 and 2 is probably due to the strong exchange coupling of spin labels due to their high concentration in the powder.

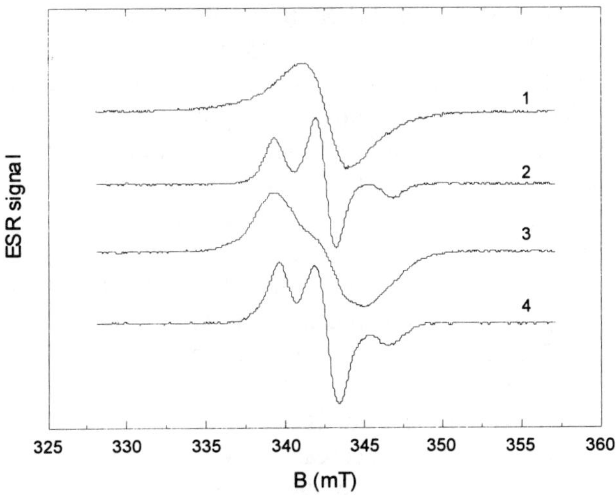

Figure 16. CW-ESR spectra of peptide TOAC-1,8 at 77 K: (1) peptide TOAC-1,8 in the dried powder; (2) solution of peptide TOAC-1,8 in TFE; (3) solution of peptide TOAC-1,8 in a chloroform/toluene mixture (7.3); 4 - mixture of peptides TOAC-1,8 and Tric-OMe (ratio 1:10) in a frozen chloroform-toluene (7:3). (Reproduced from Milov *et al.*, 2001).

Of interest is the considerable difference between the spectra of peptide TOAC-1,8 in polar TFE and in weakly polar chloroform-toluene solution (Figure 16, curves 2 and 3). In the case of the chloroform-toluene mixture a substantial broadening of the components and a general change in the spectral shape are seen. Since the difference in concentration of peptide TOAC-1,8 in the two solutions is negligible, this change might be explained by formation of aggregates, as has been found previously for single spin-labeled trichogin peptides in nonpolar solvents. Note that in this case the effect of aggregation leads to a more considerable change in the ESR spectral shape than it has been revealed for the aggregates consisting of single spin-labeled peptides (Figure 11). Thus, the cw-ESR data for the double spin-labeled peptide TOAC-1,8 for the first time definitely confirm the previously observed phenomenon of aggregation of single spin-labeled peptides in apolar solvents.

Curve 4 in Figure 16 refers to a mixture of double labeled peptide TOAC-1,8 and unlabeled peptide Tric-OMe in a frozen chloroform-toluene mixture. The ratio of double labeled to unlabeled peptides is 1:10. It turns out that a decrease in the fraction of double labeled peptide molecules in aggregates, owing to the addition of unlabeled peptide, causes a substantial narrowing and higher resolution of the ESR spectrum. These changes are related to a decrease in the intermolecular d-d coupling of spin labels with a decreasing fraction of spin-labeled peptide molecules inside the aggregates. Note that the ESR spectral shape becomes close to the ESR spectrum of peptide TOAC-1,8 in TFE. This is an obvious result from the point of view of magnetic d-d interactions in the tetramer, as the dilution from eight to two spins per cluster is the same as that taking place in the dissociation of the aggregate to the monomeric peptide in a polar solvent.

Our conclusions regarding the aggregation of peptide TOAC-1,8 in chloroform-toluene are confirmed by the PELDOR data. Figure 17 shows the PELDOR signal amplitude decay $\ln V(T)$ for a mixture of peptides TOAC-1,8 and Tric-OMe in a chloroform-toluene (1:3) glass at 77 K. Curves 1-3 were obtained for different total concentrations of the peptide while keeping the ratio of peptide TOAC-1,8 to Tric-OMe constant. The concentration-independent fast decrease of $V(T)$ at $T < 100$ ns, followed by weak signal oscillations, refers to the coupling of spin labels inside the peptide cluster. At $T > 100$ ns, the decrease of the amplitude is dependent on peptide TOAC-1,8 concentration and can be referred to the interaggregate spin coupling signal.

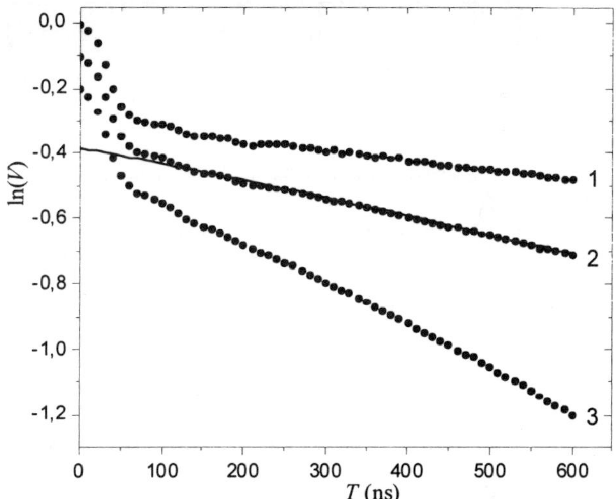

Figure 17. Dependence of ln(V) on T for glassy solutions of mixtures of peptides TOAC-1,8 and Tric-OMe (ratio 1:3) in chloroform-toluene at 77 K. The concentrations of peptide TOAC 1,8 for curves 1, 2 and 3 are 1×10^{-3} M, 2.5×10^{-3} M and 5×10^{-3} M, respectively. (Reproduced from Milov *et al.*, 2001).

This time dependent behavior of $V(T)$ makes it possible to resolve intra- and inter-aggregate d-d interactions by using Equation (16). To this end, after the initial fast signal decay, the experimental dependence ln(V) was considered to decay with time only due to the spin coupling between aggregates and is represented by the second-order polynomial (Milov *et al.*, 1999; 2000c) with regard to T in the form of a smooth non-oscillating curve $\ln V_{inter}$. As an example, this dependence is shown in the form of a smooth curve for experimental curve 2. As in the previous part, by subtracting the $\ln V_{inter}$ dependence from the experimental curve 2 we found the V/V_{inter} ratio close to V_{intra} and sufficient to estimate the oscillation frequencies as well as the V_p values.

Curves 1-5 in Figure 18 illustrate the dependence of V/V_{inter} on T at 77 K, for different ratios of double labeled and unlabeled peptides in chloroform-toluene (curves 1-4) as well as for peptide TOAC-1,8 in TFE (curve 5). In general, the data show the characteristic features of spin aggregation, namely, a fast decrease in the PELDOR signal at T<100ns followed by passage beyond the limiting V_p value, accompanied in some cases by rapidly attenuating modulation.

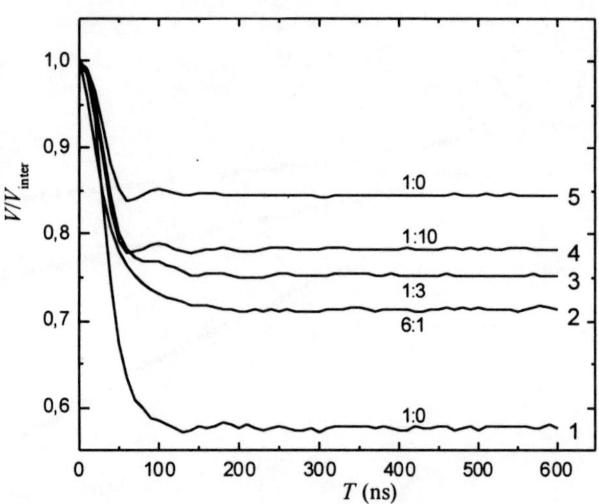

Figure 18. Dependence of V/V_{inter} on T for the frozen glassy solutions of the mixtures of peptides TOAC-1,8 and Tric-OMe at different ratios of peptides: (1) TOAC-1,8 in chloroform-toluene; (2) TOAC-1,8/Tric-OMe (6:1) in chloroform-toluene; (3) TOAC-1,8/Tric-OMe (1:3) in chloroform-toluene; (4) TOAC-1,8/Tric-OMe (1:10) chloroform-toluene; (5) TOAC-1,8 in TFE. (Reproduced from Milov *et al.*, 2001).

By comparing curves 1 and 5 in Figure 18, it is seen that the V_p value for peptide TOAC-1,8 in the aggregated form (curve 1) is smaller than that for the same peptide in the monomeric form (curve 5). The V_p value observed for curve 5 corresponds to two spin labels per monomeric peptide. According to our data presented in the preceding sections, the number of molecules in the aggregate is about four. Therefore, the number of spins per aggregate for curve 1 should be about eight. According to Equation (14), when p_b is constant V_p will decrease with increasing number of spin labels, N, in the aggregate. This corresponds qualitatively in the change of the experimental V_p value (Figure 18, curves 1-4). It is necessary to stress that it is rather difficult to calculate quantitatively N values from curves 1-4, Figure 18. This analysis is complicated by the p_b dependence upon the labeled/unlabeled peptide ratio which corresponds to the dependence of ESR spectra shape upon this value (Figure 16).

Figure 18 shows that dilution of double labeled peptide TOAC-1,8 by unlabeled peptide Tric-OMe leads to both an increase in the V_p value and the appearance of oscillations in the dependence of V/V_{inter} on time T. The modulation period determined for curve 4 recorded for the peptides TOAC 1,8/Tric-OMe mixture (at a 1:10 ratio) in chloroform-toluene is 75 ± 5 ns. This value is, within experimental error, equal to the modulation period derived from curve 5 for peptide TOAC-1,8 in TFE. This coincidence looks accidental and shows only that the conformation of TOAC-1,8 in TFE and in the aggregate are closely related.

An increase in V_p and changes in the ESR spectra by dilution of double labeled peptide TOAC-1,8 by unlabeled peptide Tric-OMe correspond to a decrease in the mean number of spin labels in the aggregates. Thus, at a fairly high degree of dilution by unlabeled molecules, the fraction of aggregates containing more than one double spin-labeled peptide molecule becomes small. Then, the main contribution to the dependence of V/V_{inter} on time T will be made by the intramolecular coupling of labels in the isolated double spin labeled molecules that are hidden in the aggregates. This makes it possible to estimate the intramolecular distance between two spin labels.

The modulations of V/V_{inter} on time T in Figure 18 indicate that some fraction of peptide TOAC-1,8 molecules, in aggregates as well as in TFE, has a fixed distance structure in which the intramolecular distances between spin labels are fixed with a minor spread. In this case the mean distances R between spin labels in this fraction can be estimated from the experimental modulation frequencies using the Equation (10). In particular, using this equation and the observed modulation period Δ value 75 ± 5 ns, we get the intramolecular distance between spin labels of peptide TOAC-1,8 in the aggregate, *i.e.* $R = 15.7$ Å. According to Equation (10), the error for R, $dR =$ 2.3% is calculated from the experimental error of the oscillation period ($d\Delta/\Delta = 0.07$) using the relation $dR/R = d\Delta/3\Delta$.

The estimation of distance between spin labels for double labeled peptide TOAC-1,8 in TFE (Figure 18, curve 5) gives the value 15.4 Å. Within experimental error, this value is equal to the between labels distance of 15.3 Å obtained earlier for double labeled peptide in TFE (Table 2). The analysis performed in previous part (Table 3) shows that this distance most closely corresponds to a 3_{10}-helical structure.

Two remarks have to be made at this stage. First, the peptide conformation is, considering the flexible nature of peptides, most likely the average of different segmental conformations. Second, the conformation is in agreement with the secondary structure that has been proposed earlier based on a set of intermolecular distances between spins of single labeled peptides in the aggregate measured with the PELDOR technique. Thus, the data obtained confirm the possibility of the 3_{10}-helical conformation for at least a part of the peptide molecules in the aggregate. The coincidence of modulation frequencies and the close values of modulation amplitudes for peptide TOAC-1,8 in the aggregated state and glassy TFE indicate a similar spatial structure of peptide TOAC-1,8 under these two different conditions.

As mentioned in the previous section, when studying the frozen TFE solution of peptide TOAC-1,8, the observed modulation amplitude is much smaller than that expected from theoretical calculations for a pair of spins at a fixed distance (Figure 3). A similar behavior is shown by peptide molecules included in aggregates (Figure 18, curve 4). This is indicative for the fact that the aggregated labeled peptides do not all have the 3_{10}-helical structure. Indeed, some fraction of them has a significant spread of intramolecular distances between spin labels and therefore gives only a fast decay in $V(T)_{intra}$ without any modulation around V_p. The fast decay of V_{intra} at short T should be attributed to all types of spin labeled peptides.

The relationship between the depth of the fast decay and the modulation amplitude makes it possible to estimate the fraction of aggregated peptides with an 3_{10}-helical conformation much more accurately than earlier in previous part (Table 2). To determine this fraction, we use the equations for the modulation amplitude in the case of a fixed pair of spins randomly oriented in the magnetic field, Equations 6-10. In the absence of any exchange interaction between these spins (J=0), these two equations are simplified, and for $T > 1/\omega_D$ the V_{intra} behavior will have the form

$$V_{intra} \cong 1 - p_b\left[1 - \left(\tfrac{\pi}{12}\omega_D T\right)^{1/2} \cos\left(\omega_D T - \tfrac{\pi}{4}\right)\right] \tag{21}$$

where $\omega_D = 2\pi/\Delta = \gamma^2 \hbar / R^3$ and $p_b = 1 - V_p$. According to Equation (21), when the spread of frequency, $\delta\omega_D$, is not important, the attenuation of the V_{intra} oscillation approaches $1/T^{1/2}$. The analysis of the experimental oscillation attenuation (Figure 18, curve 4) shows that it is faster than $1/T^{1/2}$ and this means that we have to consider some spread in oscillation frequencies and a corresponding spread of distances δR around the mean value R = 15.7 Å obtained from the experimental oscillation period.

In order to estimate the frequency spread let us assume for simplicity that this spread is small in comparison with the oscillation frequency, and the frequency distribution function is a Lorentzian line centered at the frequency $\omega_0 = \gamma^2 \hbar / R_0^3$ with the half-width $\delta\omega_{1/2}$ at the half-height ($\delta\omega_{1/2} \ll \omega_0$). In this case, the mean value of the decay function, $\langle V_{intra} \rangle$, could be obtained by integration of Equation (21) as.

$$\langle V_{intra} \rangle = \int_{\omega_D} \frac{V_{intra}\,\delta\omega_{1/2}}{\pi(\delta\omega_{1/2}^2 + (\omega_D - \omega_0)^2)} d\omega_D = 1 - p_b[1 - A_{(T)} \cos(\omega_0 T - \pi/4)] \tag{22}$$

$$A_{(T)} \cong (\pi/12\omega_0 T)^{1/2} \exp(-\delta\omega_{1/2} T) \tag{23}$$

where ω_0 is the mean value of the oscillations frequency and $A_{(T)}$ is the oscillation amplitude. Equations (22) and (23) makes it possible to estimate $\delta\omega_{1/2}$ from the experimental $A_{(T)}$ decay and therefore to find the corresponding distance spread $\delta R_{1/2} = \pm\delta\omega_{1/2}/3$

The ratio of the modulation amplitudes $A_{(T)}$ at the peak points of the $\cos(\omega_0 T - \pi/4)$ function which are separated by the oscillation period $\Delta = 2\pi/\omega_0$ will be

$$\frac{A_{(T+\Delta)}}{A_{(T)}} = \left(\frac{T}{T+\Delta}\right)^{1/2} \exp(-\delta\omega_{1/2}\Delta) \tag{24}$$

This gives for the distance spread the following relations

$$\delta R_{1/2} / R_0 = \pm\left(1/6\pi\right) \ln z \tag{25}$$

$$Z = A_{(T+\Delta)} \frac{(T+\Delta)^{1/2}}{A_T T^{1/2}} \tag{26}$$

where Z is the attenuation of V_{intra} modulation amplitude due only to the distance spread between spin labels.

From the experimental V_{intra} measurements (Figure 18 curve 4) it is evident that the Z value is practically the same at any oscillations period chosen. In our particular case $Z = 0.83 \pm 0.02$ and according to Equations (25) and (26) this gives the corresponding distance spread between spin labels $\delta R_{1/2}/R_0 = \pm 0.01$. This value reflects qualitatively the geometric rigidity of the corresponding peptide conformations that are responsible for the observed oscillations in V_{intra}. It is worthwhile to mention that the distance spread measured by this way is less than the experimental error of the mean distance measurements from the experimental modulation frequency.

The Equation (22) now could be used in order to estimate the fraction of aggregated peptides with a 3_{10}-helical conformation. Using it, the general dependence of V_{intra} on T for $T > 1/\omega_0$ may be written in the form

$$\langle V_{\text{intra}} \rangle \cong X(1 - p_b[1 - A_{(T)} \cos(\omega_0 T - \tfrac{\pi}{4})]) + (1 - X)(1 - p_b) \tag{27}$$

where X is the fraction of peptides with a 3_{10}-helical structure. The first term in Equation (27) is related to 3_{10}-helical peptides, while the second is related to peptides having other types of conformation. After the fast decay of $\langle V_{\text{intra}} \rangle$, when $T > 1/\omega_0$, the peak values of Equation (10) are reached by $\cos(\omega_0 T - \pi/4) \cong 1$. In this case Equation (27) makes it possible to estimate the value of X:

$$X = \frac{\langle V_{\text{int}\,ra} \rangle - V_p}{p_b A_{(T)}} = \left(\frac{12\omega_0 T}{\pi} \right)^{\frac{1}{2}} \exp(\delta\omega_{\frac{1}{2}} T) \frac{\langle V_{\text{int}\,ra} \rangle - V_p}{p_b} \tag{28}$$

By substitution of the experimental values of $\omega_0 = 8.37 \times 10^7$ rad/s, V_{intra} and T (at the first peak of curve 4 in Figure 19), exp ($\delta\omega_{1/2}T$) = 1.28 and $p_b = (1-V_p) = 0.22$ into Equation (28), we obtain the fraction of aggregated peptides with 3_{10}-helical conformation, *i.e.* $X = 0.19$ (± 0.03). This value is close to the estimated fraction of peptides with 3_{10}-helical structures for TOAC-1,8 in TFE solution, as it has been reported in previously part.

From our data there is no direct evidence to assign a conformation to the remaining fraction (≈80%) of aggregated peptides. In comparison with the cw-ESR spectra of mono-labeled peptides (Figure 11), the cw-ESR spectrum of the spin diluted double-labeled peptide in the aggregate (Figure 16, curve 4), indicates an additional broadening of the lines due to dipole coupling between labels inside the peptide. This means that the molecules of the main fraction of our aggregated spin system exist with a comparatively large spread of distances between the spin labels. This spread even could include some fraction of distances which may correspond to α-helical or mixed α/3_{10}-helical structures. However, more experiments are needed in order to get additional information on the secondary structure of these aggregated peptides.

In other types of frozen solutions, TOAC-1,8 may adopt different conformations depending upon the nature of solvent. Indeed, the intramolecular distance between spin labels may vary from 15.3 Å (for TFE) to 21.8 Å for ethanol (Table 2). A somewhat different situation was observed in MeOH/EtOH glass at 77 K. By analyzing the cw-ESR spectra at half-field ($g \cong 4.0$), it was reported that trichogin GA IV can exist in a mixed α/3_{10}-helical conformation in equilibrium with unfolded conformers (Anderson *et al.*, 1999). The same type of mixed helical conformation was found in the crystal state by X-ray diffraction analysis for trichogin GA IV (Toniolo *et al.*, 1994) and the TOAC-4,8 analog (Crisma *et al.*, 1997). It may be concluded that short peptides as trichogin exhibit conformational flexibility depending upon temperature and the nature and organization of surrounding molecules into the matrix (glass, crystalline, *etc.*).

Molecular model of aggregated trichogin peptides. The values of distances found for aggregates of two different trichogin analogues, one labeled at the first and the other at the fourth position of the peptide chain, establish a specific set of constraints that may be used to build a molecular model (Milov *et al.*, 2000). X-ray diffraction analysis of TOAC-4,8 trichogin revealed two independent molecules in the $P2_1$ asymmetric unit (Monaco *et al.*, 1999; Crisma *et al.*, 1997). The N-terminal region of each molecule folds in a 3_{10}-helical conformation, while the central and C-terminal regions are mainly α-helical. From cw-ESR studies of three different double labeled trichogin analogues it was concluded that the overall secondary structure of these lipopeptaibol analogues in solution remains essentially unchanged (Monaco *et al.*, 1999).

An aggregate model was constructed from four 3_{10}-helices by adjusting the helical axes in pairs with the polar sides pointing to the center of the tetrameric peptide cluster (Figure 19).

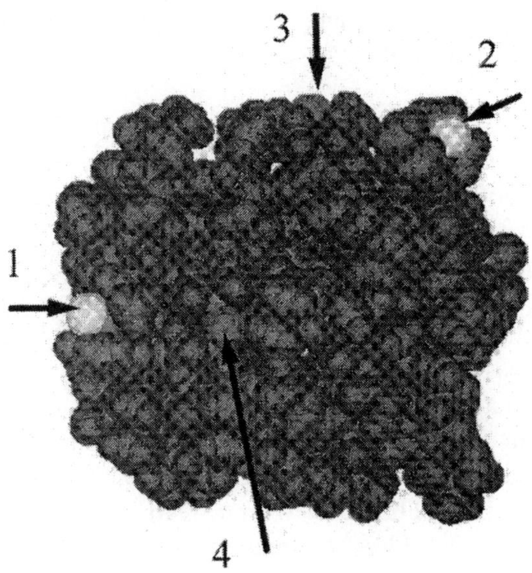

Figure 19. A molecular model for the aggregate containing four TOAC-1,4 molecules with the polar sides pointing to the center. The red and blue marked parts contains two head-to-tail oriented peptide molecules, respectively. The marked by arrows atoms corresponds to the spin label oxygen inside the peptide chains. Positions 1, 2 and 3, 4 correspond to 1 and 4 TOAC positions in peptide chain, respectively.

After steps of energy minimization a model was produced with the following average inter-residue distances: TOAC-1...TOAC-1: 26 Å and TOAC-4... TOAC-4: 22 Å. These values are consistent with the distances obtained from the PELDOR experiments: 26.0 Å (peptide TOAC-1) and 23.5 Å (peptide FTOAC-4) and in agreement with 3_{10} conformation state. The exterior of the aggregate appears to be highly hydrophobic. The interior of the peptide cluster leaves room for several solvent molecules. At each end, the polar cavity is closed by four different hydrophobic groups, *i.e.* the two side chains of Leu-11 and the two n-octanoyl groups (or the two Fmoc groups in peptide FTOAC-4). The question on how solvent molecules interact to stabilize these aggregates has to be examined in further detail.

4. CONCLUSION

The results given in this paper demonstrate the potential and select applications of the pulsed electron-electron double resonance in electron spin echo (PELDOR) method when used with spin labels for studying the structure of macromolecules in a solid phase. As compared with cw-ESR, the PELDOR technique extends the range of measured distances between spin labels up to several tens of angstrom. This provides information on both the structure of studied molecules and the presence and structure of their aggregates.

The use of PELDOR technique for studying the spin-labeled analogues of trichogin GA IV in frozen solutions shows that depending on peptide structure and the polar properties of a solvent, 15-35% of peptide molecules can have of 2_7 or 3_{10} helical conformations with a minor spread in distances between spin labels. In frozen weakly polar solvents, such as the mixtures of chloroform-toluene, decalin-toluene, toluene-CCl_4, aggregates of the spin labeled trichogen analogues were found. These aggregates could be dispersed by increasing the polarity of the solvent mixture. It was found that the aggregate structure mainly depends on solvent composition and, to a lesser degree, on the type of terminal groups. Estimating the distances between spin labels in aggregates from the PELDOR data shows that the distances between the labels group, with a small spread, are situated around the values of 23, 26 and 33 Å. Estimation of the number of spin labels in aggregates, based on the behavior of PELDOR signal at the long time interval, gives the lower boundary of the number of peptide molecules in aggregates to within 3.1–4.3. The cw-ESR results indicate the possible existence of aggregates in the same solutions, but in the liquid phase at room temperature.

The PELDOR method combined with the cw-ESR technique was used to study the double TOAC spin labeled trichogin GA IV aggregates diluted by the unlabeled peptide. The magnetic dipole-dipole interaction of spin labels for these aggregates was experimentally investigated in glassy chloroform-toluene mixture as a function of the content of unlabeled peptide. At high degree of dilution with unlabeled peptide oscillations of the PELDOR signal were observed due to intramolecular coupling of the spin labels. The intramolecular distance between spin labels (inside the peptide molecule) was determined from the oscillation frequency to be 15.7Å with a small spread in distances. This distance is close to that calculated for the peptide in the 3_{10} conformation. The fraction of 3_{10}-helices was estimated from the oscillation amplitude to be about 19% of the total number of spin-labeled peptides in solution. The remainder of the peptide molecules exhibit a fairly great spread in distances between spin labels.

A molecular model of the aggregate with four 3_{10}-helical trichogin molecules has been proposed which is consistent with some experimentally obtained distance data.

5. ACKNOWLEDGMENTS

I thank Dr. A. D. Milov for help in preparing this chapter. Participants in the research described in this chapter include Drs. A.D. Milov, Jan Raap, A. Maryasov, and R. Samoilova. Prof. C. Toniolo and his collaborators Dr. F. Formaggio and Dr. M. Crisma (Biopolymer Research Centre CNR, Department of Organic Chemistry, University of Padova, Italy) provided the spin labeled peptides. I am also grateful to Prof. G. Millhauser and Dr. M. Bowman for helpful discussions of these results and fruitful comments.

Permission to reprint figures from our publications was granted by Elsevier Science, Springer Verlag and American Chemical Society.

This work was supported by The Netherlands Organization for Scientific Research (NWO) project 047.009.018, US Civilian Research and Development Foundation for the Independent States of the Former Soviet Union (CRDF), Grant RCI-2056 and by the Russian Basic Research Foundation, grants 00-03-40124, 99-03-33149, 00-15-97321.

6. REFERENCES

Abragam, A., 1961, *Principles of Nuclear Magnetism*, Clarendon Press: Oxford. Chapter 4.

Anderson, D., Hanson, P., McNulty, J., Millhauser, G., Monaco, V., Formaggio, F., Crisma, M., and Toniolo, C., 1999, *J. Am. Chem. Soc.*, **121**: 6919.

Arsher, S. J., Ellena, S.F.,and Cafiso, D.S., 1991, *Biophys*. **60**: 389.

Aubry, A., Del Duca, V., Pantano, M., Formaggio, F., Crisma, M., and Kamphuis, 1994, *Lett. Pept. Sci.*, **1**: 157.

Auvin-Guette, C., Rebuffat, S., Prigent, Y., and Bodo, B., 1992, *J. Am. Chem. Soc.*, **114**: 2170.

Bales, B.J., 1989, In *Biological Magnetic Resonance, Vol. 8: Spin Labeling Theory & Practice* (L.J. Berliner & J. Reuben, eds.), Plenum, New York. Ch. 2.

Berliner, L.J., ed., 1976, *Spin Labeling Theory & Applications*, Academic Press, New York.

Berliner, L.J., ed., 1979, *Spin Labeling II*, Academic Press, New York.

Borbat P., and Freed J., 1999, *Chem. Phys. Lett.*, **313**: 145.

Chugh, J.K., and B.A. Wallace, 2001, *Biochem. Soc. Trans.*, **29**: 565.

Crisma, M., Monaco, V., Formaggio, F., Toniolo, C., George, C., and Flippen-Anderson, J. L., 1997, *Lett. Pept. Sci.*, **4**: 213.

Dalton, L.R., ed., 1985, *EPR and Advanced EPR Studies of Biological Systems*, CRC Press, Boca Raton.

Dikanov, S. A., and Tsvetkov, Yu. D., 1992, *Electron Spin Echo Envelope Modulation (ESEEM) Spectroscopy*, CRC Press: Boca Raton.

Eaton, G.R., and S.S. Eaton, 1989, In *Biological Magnetic Resonance, Vol. 8: Spin Labeling Theory & Practice* (L.J. Berliner & J. Reuben, eds.), Plenum, New York. Ch. 7.

Emshwiller, M., Hahn, E.I., and Kaplan, D., 1960, *Phys. Rev.*, **118**: 414.

Friori, W. R., Miick, S. M., and Millhauser, G. L., 1993, *Biochemistry*, **32**: 11957.

Hanson, P., Millhauser, G., Formaggio, F., Crisma, M., and Toniolo, C., 1992, *J. Am. Chem. Soc.*, **118**: 2170.

Hanson, P., Martinez, G., Millhauzer, G., Formaggio, F., Crisma, N., Toniolo, C., and Vita, S., 1996, *J Am. Chem. Soc.*, **118**: 271.

Hubbell, W.L., D.S. Cafiso, and C. Altenbach, 2000, *Nature Struct Biol.*, **7**: 735.

Huisjen, M., and Hyde, J., 1974, *Rev. Sci. Instr.*, **45**: 669.

Jeschke, G., Painnier, M., Godt, A., and Spiess, H., 199$, *Chem. Phys. Lett.* **331**: 243.

Karle, I., and Balaram, P., 1990, *Biochemistry*, **29**: 6747.

Kurshev, V. V., Raitsimring, A. M., and Tsvetkov, Yu. D., 1989, *J. Magn. Reson.*, **81**: 441.

Maryasov, A. G., Tsvetkov, Yu. D., and Raap, J., 1998, *Appl. Magn. Reson.*, **14**: 101.

Maryasov, A. G., and Tsvetkov Yu. D., 2000, *Appl. Magn. Reson.*, **18**: 583.

Miick, S.M., Martinetz, G.V., Fiori, W. R., Todd, A. P., and Millhauser, G. L., 1991, *Nature*, **359**: 653.

Milov, A. D., Salikhov, K. M., and Schirov, M. D., 1981, *Fiz. Tverd. Tela (Leningrad)*, **23**: 975.

Milov, A. D., Ponomarev, A. D., and Tsvetkov, Yu. D., 1984, *Chem. Phys. Lett.*, 67.

Milov, A. D., and Tsvetkov, Yu. D., 1997, *Appl. Magn. Reson.*, **12**:_495.

Milov, A. D., Maryasov, A. G., and Tsvetkov, Yu. D., 1998, *Appl. Magn. Reson.*, **15**: 107.

Milov, A. D., Maryasov, A. G., Tsvetkov, Yu. D., and Raap, J., 1999, *Chem. Phys. Lett.* **303**: 135.

Milov, A. D., Maryasov, A.G., Samoilova, R. I., Tsvetkov, Yu. D., Raap, J., Monaco V., Formaggio F., Crisma, M., and Toniolo, C., 2000a, *Dokl. Akad. Nauk.* **370**: 265.

Milov, A. D., Tsvetkov, Yu. D., Formaggio, F., Crisma, M., Toniolo, C., and Raap, J., 2000b, *J. Am Chem. Soc.*, **122**: 3843.

Milov, A. D., and Tsvetkov, Yu. D., 2000c, *Appl. Magn. Reson.*, **18**: 217.

Milov, A. D., Tsvetkov, Yu. D., and Raap J., 2000d, *Appl. Magn. Reson.*, **19**: 215.

Milov, A. D., Tsvetkov, Yu. D., Formaggio, F., Crisma, M., Toniolo, C., and Raap, J., 2001, *J Am. Chem. Soc.*, **123**: 3784.

Mims, W., 1965, *Rev. Sci. Instr.*, **10**: 78.

Monaco, V., Formaggio, F., Crisma, M., Toniolo, C., Hanson, P., Millhauser, G., Gegrge, C., Deschamps, J., and Flippen-Anderson, J., 1999, *Bioorg. Med.Chem.*, **7**: 119.

Parmon, V. N., Kokorin, A. I., and Zhidomirov, G. M., 1980, *Stable Biradicals*, Nauka: Moscow (in Russian).

Ponomarev, A. B., Milov, A. D., and Tsvetkov, Yu. D., 1988, *Khim. Fiz*, **7**: 1673.

Ponomarev, A. B., Milov, A. D., and Tsvetkov, Yu. D., 1990, *Khim. Fiz,*, **9**: 498.

Rozantsev, E. G., 1970, *Free Iminoxyl Radicals*. Khimia: Moscow (in Russian).

Salikhov, K. M., Semenov, A. G., and Tsvetkov, Yu. D., 1976, *Electron Spin Echo and Its Applications*; Nauka: Novosibirsk (in Russian).

Salikhov, K. M., and Tsvetkov, Yu. D., 1979, In *Time Domain Electron Spin Resonance*, Kevan, L., Schwartz R., Eds.; Wiley: New York. p232.

Saxona, S. and Freed J., 1996, *Chem. Phys. Lett.*, **251**: 102.

Schweiger, A., and Jeschke, G., 2001, *Principles of Pulsed Electron Paramagnetic Resonance*, Oxford University Press, Oxford.

Toniolo, C., and Benedetti, E., 1991, *Trends Biochem. Sci.* **16**: 350.

Toniolo, C., and Benedetti, E., 1991, *Macromolecules*, **24**: 4004.

Toniolo, C., Peggion, C., Crisma, M., Formaggio, F., Shui, X., and Eggleston, D. S., 1994, *Nature Struct. Biol.*, **1**: 908.
Toniolo, C., Crisma, M., and Formaggio, F., 1998, *Biopolymers* (*Pept. Sci*), **47**: 153.
Tsvetkov, Yu. D., 1985, *Usp. Khim.* (Russ.) **52**: 184.
Tsvetkov, Yu. D., 1989, In: *Pulsed ESR: A New Field of Applications*, Keijzers C., Reijerse E., Schmidt J. Eds.; Amsterdam: North Holland. p.206.

Index